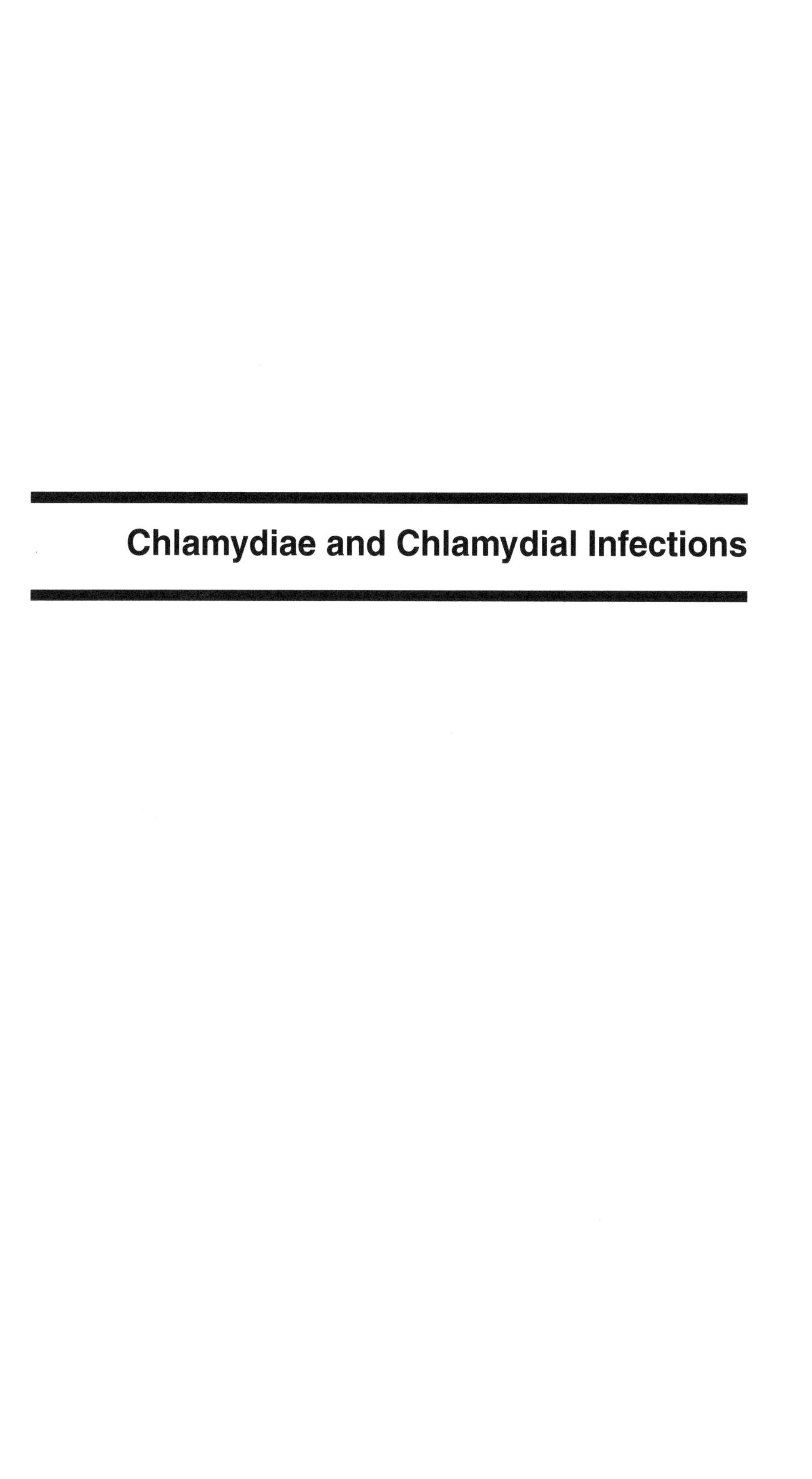

Chlamydiae and Chlamydial Infections

RIVER PUBLISHERS SERIES IN RESEARCH AND BUSINESS CHRONICLES: BIOTECHNOLOGY AND MEDICINE

Indexing: All books published in this series are submitted to the Web of Science Book Citation Index (BkCI), to CrossRef and to Google Scholar.

Combining a deep and focused exploration of areas of basic and applied science with their fundamental business issues, the series highlights societal benefits, technical and business hurdles, and economic potentials of emerging and new technologies. In combination, the volumes relevant to a particular focus topic cluster analyses of key aspects of each of the elements of the corresponding value chain.

Aiming primarily at providing detailed snapshots of critical issues in biotechnology and medicine that are reaching a tipping point in financial investment or industrial deployment, the scope of the series encompasses various specialty areas including pharmaceutical sciences and healthcare, industrial biotechnology, and biomaterials. Areas of primary interest comprise immunology, virology, microbiology, molecular biology, stem cells, hematopoiesis, oncology, regenerative medicine, biologics, polymer science, formulation and drug delivery, renewable chemicals, manufacturing, and biorefineries.

Each volume presents comprehensive review and opinion articles covering all fundamental aspect of the focus topic. The editors/authors of each volume are experts in their respective fields and publications are peer-reviewed.

For a list of other books in this series, visit www.riverpublishers.com

Chlamydiae and Chlamydial Infections

Svetoslav P. Martinov

National Diagnostic and Research Veterinary Medical Institute – Sofia
Bulgaria

LONDON AND NEW YORK

Published 2018 by River Publishers
River Publishers
Alsbjergvej 10, 9260 Gistrup, Denmark
www.riverpublishers.com

Distributed exclusively by Routledge
4 Park Square, Milton Park, Abingdon, Oxon OX14 4RN
605 Third Avenue, New York, NY 10017, USA

Chlamydiae and Chlamydial Infections / by Svetoslav P. Martinov.

Routledge is an imprint of the Taylor & Francis Group, an informa business

ISBN 978-87-93609-51-8 (print)

Contents

Preface

Chlamydiae are a group of obligate intracellular microorganisms with a homogeneous group-specific antigenic structure and a unique mode of development. The infections caused by them are characterized by an unprecedented wide spread throughout the world, a very broad range of hosts among domestic and wild-animal species and humans, and a variety of clinical manifestations. Some members of the Chlamydiaceae family have been shown to be zoonotic pathogens, and for others, there is a presumption of zoonotic potential – an issue that is the focus of the scientific community in current and future research. Certain avian chlamydial strains are characterized by a very high virulence that has long placed them among agents with biological weapon characteristics and bioterrorist threat. The veterinary and economic importance of the problem is determined by the existence and consequences of an impressive number of diseases in farm animals, including abortions and related conditions, respiratory, ocular, arthritis, mastitis, reproductive disorders, infertility, and latent and persistent infections. Of particular importance are the health and social aspects of chlamydiae among human populations associated with such prevalent diseases as sexually transmitted chlamydial infections, leading to a significant rate of infertility, trachoma with a number of cases of blindness, lymphogranuloma venereum, and the above-mentioned zoonotic infections.

The aim of this book is the review of certain aspects of chlamydial agents and Chlamydia-induced diseases relating to the biological, morphological, and antigenic properties of organisms, genes and genomic structure, interactions with cells, clinical forms and manifestations, pathology, diagnosis, epidemiological peculiarities in animals and humans, immunity, and vaccines. Historical notes and taxonomy are also covered.

The unique nature of the chlamydial agents is analyzed from the standpoint of modern microbiology, cell biology and molecular biology. The complex and controversial issue for the development of chlamydia is discussed by presenting different viewpoints and hypotheses.

The epizootiological and epidemiological particularities, links, and relationships are detailed and emphasized. An epizootiological characteristic of active foci of chlamydial infection in domestic ruminants is presented, containing multiple components. Detailed analyzes are presented on epidemics of ornithosis in humans associated with outbreaks of avian chlamydosis in waterfowl. The host barriers may be looser for some chlamydial species. The important fact of expanding the range of certain chlamydial species among new hosts is noted with all the resulting epidemiological consequences. Attention is drawn to some vague or poorly explored issues of epidemiological nature in different species and clinical conditions.

The accepted and proven methods for the isolation, cultivation, identification and differentiation of *Chlamydia spp.* have been presented with specific results with experimental and field clinical and pathological materials. The diagnostic efficiency of some methods for direct detection of the pathogen as well as the use of complex etiological diagnostics involving several methods has been shown. Adequate attention is paid to modern DNA-detection methods.

Various clinical diseases and symptoms are described in farm mammals and birds, some ornamental birds, guinea pigs, koalas, reptiles, and amphibians.

Finally, the book discusses the vaccines and immunoprophylaxis of Chlamydia-induced diseases in animals and humans. This is one of the main issues that however remains unresolved for most diseases. Relatively good results are obtained from vaccinations against chlamydial abortion in sheep and feline chlamydiosis. There remains a question of detecting suitable novel chlamydial antigens as promising potential vaccine candidates and the overall experimentation and testing of the resulting products.

A large number of references have been selected to provide original information and further sources of information. For obvious reasons, it would not be possible in one edition to quote all the publications on the issues under discussion.

I have intended the book for the specialized audience of veterinary and medical researchers, professionals, microbiologists, virologists, epidemiologists, molecular biologists, and students of veterinary and human medicine; however, due to the global implications of chlamydial diaseases, this work would also be very informative to a broader range of readers.

Professor Svetoslav P. Martinov,
DVM, Ph.D., D.Sci.

List of Figures

List of Tables

List of Abbreviations

ABs	Aberant bodies
ADCC	Antibody – dependant cellular cytotoxicity
AMI	Antibody mediated protection
BGM	Buffalo green monkey cells
BHK- 21	Baby hamster kidney cell line
Bp	Base pairs
CB	Condense body
CC	Cell culture
CCI	Cytoplasmatic chlamydial inclusion
CDC	Centers for Disease Control
CDS	Coding sequences
CE	Chicken embryo
CEF	Chicken embryo fibroblasts
CF	Complement fixing
CFT	Complement fixation test
CRP	Cystein rich protein
COMC	Chlamydia trachomatis outer membrane complex
CPAF	Chlamydial Protease-like activity factor
CPE	Cytopathic effect
Ct	Chlamydia trachomatis
CT	Computer tomography
Cs	Chlamydia suis
DEM	Direct electron microscopy
DFA	Direct fluorescent antibody test
DNA	Deoxyribonucleic acid
DPI	Days post infection
EAE	Enzootic abortion of ewes
EB	Elementary body
EC-ReA	Enterocolitis reactive arthritis
EIA	Enzyme immunoassay
ELISA	Enzyme-linked immunosorbent assay

EM	Electron microscopy
ERIC - PCR	Enterobacterial repetitive intergenic consensus sequence-based PCR
FATs	Fluorescent Antibody Tests
FcRn	Fc receptor
FHCS	Fitz-Hugh-Curtis syndrome
GMT	Geometric mean titer
GP	Guinea pig
GPIC	Guinea pig inclusion conjunctivitis
HeP-2	Human epithelial cells
IgA, IgG, IgM	Immunoglobulin classes
IB	Initial body
i.c.	Intracerebral inoculation
IF	Immunofluorescence
IFN	Interferon
IFU	Inclusion forming units
IIF	Indirect immunofluorescence
IMIF	Indirect microimmunofluorescence
i.n.	Intranasal inoculation
i.p.	Intraperitoneal inoculation
i.v.	Intravenous inoculation
IUDR	5-iodo-deoxyuridine
IVF	In-vitro fertilization
kD	Kilodalton
LGV	Lymphogranuloma venereum
LM	Light microscopy
L929	Mouse fibroblasts cell line
LPS	Lipopolysaccharide
LT	Lymphocyte transformation
MAb	Monoclonal antibody
McCoy A	Human synovial cells
McCoy B	Mouse fibroblasts cell line
MCI	Membrane of chlamydial inclusion
MIFT	Microtiter indirect immunofluorescence test
MOMP	Major outer membrane protein
m.w.	Molecular weight
NAATs	Nucleic acid amplification tests
NGU	Nongonococcal urethritis
Nm	Nanometer

NWM	New born white mice
OEA	Ovine enzootic abortion
OIE	World organization for Animal Health
OD	Optical density
Omp A	Outer membrane protein A gene
PAGE	Polyacrilamid gel electrophoresis
PBMC	Peripheral blood mononuclear cells
PCR	Polymerase chain reaction
PGU	Postgonococcal urethritis
PID	Pelvic inflammatory disease
pIgR	Polymeric immunoglobulin receptor
PLT	Psittacosis-Lymphogranuloma venereum-Trachoma group
POMP	Polymorphic outer membrane protein
PHYLYP	Phylogeny inference package
PR-ReA	Post-respiratory reactive arthritis
PZ	Plasticity zone
RAPD-PCR	Randomly amplified polymorphic DNA analysis
RB	Reticulate body
REP-PCR	Repetitive element PCR; Repetitive extragenic palindromic sequence polymorphism – PCR
RS	Reiter's syndrome
RNA	Ribonucleic acid
SBE	Sporadic bovine encephalomyelitis
SD	Sodium dodecyl sulphate
SEM	Scanning electron microscopy
STI	Sexually transmitted infections
Tc^R	Tetracycline resistant
TCR	T cell receptor
TEM	Transmission electron microscopy
Tg	Transgenic
TTS	Type III secretion
UG-ReA	Urogenital reactive arthritis
UKS	Uncultured koala Chlamydiales
UPGMA	Unweighted pair group method
Vero cells	African green monkey kidney cells
YS	Yolk sac

Introduction

In the world literature, more than 60 mammalian chlamydial diseases, chlamydial infections in 465 species of domestic and wild birds and 30 chlamydia-induced diseases in humans are known. This is an unprecedented case of widespread infectious diseases. The uniqueness of chlamydia pathology consists mainly in the fact that the agents of the individual diseases are so close in their biological properties that they are represented by only the single genus *Chlamydia*, which includes all currently recognized species.

Chlamydia can be qualified as a biological paradox. One of the paradoxical qualities is their uniform group-specific antigenic structure. Behind this antigenic unity lies their wide range of pathogenicity. Due to the fact that some avian chlamydial strains have long been part of the arsenal of biological weapons, the strategic importance of these microorganisms should not be overlooked. Avian chlamydiosis is listed by the OIE as a notifiable disease.

Although chlamydiae and chlamydial infections were discovered a long time ago, they are still under-researched and relatively little known to broad circles of microbiologists, virologists, and epidemiologists. For example, there are many uncertainties about the factors of pathogenicity and virulence, and scientific research and hypothetical proposals for new candidate genes playing a role in these important biological properties of chlamydia agents continue. The issue of the mode of development of the chlamydiae is controversial and debatable. After 1930, when the concept of the so-called cycle of development of these micro-organisms based on binary division and sequential conversion was postulated, other hypotheses appeared trying to explain the mode of propagation of chlamydia – through a development cycle, or a kind of morphogenesis. Regardless of diametrically opposed ideas about the mode of development of Chlamydia organisms, there exists a unidirectional view of the uniqueness of the mechanisms associated with it and the reproduction of these infectious agents. Nowadays, identifying the genes involved in the various phases of the chlamydial developmental cycle is considered to be the important condition for a more complete understanding of the problem. The issues of interaction between the chlamydiae and host cell are essential. Recently, they have gained a new focus since it has become clear that the membrane of the chlamydial inclusion (MCI) in the cell, which until recently did not attract much attention, undoubtedly plays a key role in the relationship of the pathogen with the cells [161]. In this connection, the discovery of gene-encoding proteins present in the MCI of *Chlamydia psittaci* and *Chlamydia trachomatis* is an important development [162–164]. The isolation and cultivation of chlamydia is an important scientific and practical problem that a large number of studies have been devoted to. Particular attention is paid to the peculiarities of cultivation in cell cultures related to the difficulty of penetrating the chlamydial agents into the cell. The sensitivity of the biological models, the duration of the laboratory procedures, and the safety of the personnel are essential. There was a need to adopt new approaches to the classic methodology for culturing Chlamydia, a proper treatment of the inocula to ensure high multiplicity of the infection, and the introduction of methods for the early indication and identification of isolated strains. The antigens, antigen analysis, and antigenic differentiation of chlamydia are the target of many studies. A key point in connection with these and other problems is the production of sufficient amounts of highly purified and highly concentrated chlamydial suspensions and the construction of optimal schemes for obtaining specific sera against them. The principles of classical diagnosis of chlamydia have become routine, but diagnosis is not sufficiently updated and fast. Serious difficulties are due to the lack of sensitive and species-specific commercial tests. There has been a need to develop new serological tests. Laboratory practice requires a wider penetration of methods for direct detection of the causative agent. While electron microscopic (EM) diagnostic is widely used in viruses, in chlamydia it is in the diagnostic arsenal only in individual laboratories. Importance and high efficiency have both earlier EM indication and identification of chlamydial isolates in chicken embryos and cell cultures, and the direct electron microscopy of chlamydia in clinical and pathological material from animals and humans. Significant advances in research and diagnosis of chlamydial infections have been achieved with the development and deployment of DNA detection systems – different variants of PCR-technique, restriction enzyme analysis, DNA microarray technology, and others. Studies on the genomic structure, genes and molecular regulation of the development of Chlamydiae create major new opportunities to decode a number of properties and mechanisms of chlamydial agents and induced by these diseases. Sequencing of the Chlamydiae genome has provided new means to analyze the biology of these organisms from a molecular and structural point of view

[168–170, 182]. From the standpoint of pathogenesis, important results are the identification of new multi-gene family of chain membrane proteins and the set of components for secretion system of type III-factors that play a role in the virulence of chlamydia. The research seeking to decipher the function of genes in the chlamydial developmental cycle continues. These examples and a number of other scientific facts not mentioned here testify to the usefulness and prospects of molecular and biological studies of chlamydia and chlamydial infections.

The issue of developing vaccines against chlamydial diseases in animals and humans has long been the focus of the scientific community. This, however, proved to be a difficult and complex issue, as in a number of studies, unsatisfactory or controversial results were obtained. In fact, there are commercial vaccines for two diseases – chlamydial abortion in sheep and chlamydiosis (conjunctivitis and pneumonitis) in cats. These vaccines have relative efficacy because they reduce disease severity and duration but do not prevent shedding of the agent. At this stage, the issue of obtaining vaccines against human infections caused by *C. trachomatis* remains unresolved, despite progress in the detection of novel chlamydial antigens, which in tests of both mouse and guinea pigs models are emerging as promising potential vaccine candidates. There is currently no vaccine to protect against *Chlamydia pneumoniae*.

From the epidemiological point of view, the important fact about the worldwide dissemination of chlamydioses, with all the influences and consequences of a veterinary, medical, social, economic, and environmental nature, is undoubtedly highlighted. Nonetheless, the epidemiological features and significance of many of them remain vague and insufficiently studied.

Small and large ruminants, pigs, and birds are very susceptible to infection with Chlamydiaceae, resulting in a large number of foci of chlamydial infections. The research of these diseases in domestic animals is placed on a broad experimental basis. Despite significant achievements, many aspects of epizootiology, pathogenesis, and immunity in these diseases are still unclear. In addition to clinically manifested diseases, chlamydia in farm animals and pets can cause clinically inapparent chlamydial infections that cause chronic inflammatory reactions and dysfunctions on different organ levels. These conditions are probably more economically important than rare outbreaks of severe chlamydial disease [633]. It should be added that the latent form of chlamydial infection is not static. It is important to identify the factors that can be the potential activators of latent infection and its transformation into a clinical disease. The pathogenetic importance of the simultaneous presence of more than one chlamydial species in an animal's organism remains unclear, as well as the role of chlamydial agents in mixed infections such as co-pathogens and by-standers. Some new findings indicate that the host barriers may probably be looser for many Chlamydiaceae members and infections occur in unexpected animal hosts. Chlamydial abortion in ewes (EAE, OEA) is recognized as a major veterinary problem in many countries. The existence of other chlamydial clinical diseases in ruminants has been proven beyond doubt – pneumonia, keratoconjunctivitis, polyarthritis, mastitis, and intestinal encephalomyelitis in cattle and buffaloes. It is imperative to fully disclose the nosogeography of these infections in a given country through complex etiological studies to detect new foci of chlamydial infections and to track the dynamics of the old. However, greater and more targeted attention is advised on other clinical forms than the well-known EAE. There is insufficient knowledge of pathogenesis and protective immune mechanisms in pigs at this stage. Progress in this regard may create preconditions for searching for potential antigens with a view to developing vaccines.

Drug therapy for acute chlamydial diseases is based on the use of tetracyclines, macrolides and quinolones. However, it is problematic to treat persistent chlamydial infections that do not respond to these preparations. One insufficiently researched question is the real susceptibility of dogs to chlamydia infections and the actual prevalence of these infections among canine populations. There are similar uncertainties about the supposed but unproved etiological involvement of *Chlamydia felis* in reproductive disorders in cats. Of interest is the chronic chlamydial infection in this animal species with two hypothetical view of its occurrence – as a consequence of the presence of persistent chlamydiae or as a result of repeated re-infections.

Studies on the chlamydial infections in wild birds and mammals shed light on the natural-focal nature of Chlamydia–induced diseases. It should be emphasized, however, that these studies are still of an episodic nature. It can be said that the natural foci in chlamydial infections are not studied in detail, as has been done with a number of viral infections and the Q fever. At present we know little about emerging chlamydial infections in wild animals.

The issue of the zoonotic importance of chlamydia of bird and mammalian origins is of utmost significance. The role of the avian strains *C. psittaci* in infectious human pathology has long been established. There are known epidemics of psittacosis in 1929–1930 in Europe with numerous deaths. Nowadays the role of decorative and other categories of wild birds as sources of chlamydial infection for humans has not diminished, especially in sporadic and family cases. At risk of infection are the owners of exotic birds, the staff of pet shops and zoos, and veterinary specialists. At the same time, epidemic outbreaks of ornithosis in humans have been recorded in connection with the development of industrial poultry farming. The pattern of the disease has changed and acquired the character of occupational disease for the people working in the poultry and poultry-processing industry as well as for the veterinary staff serving these activities.

The issue of the role of mammalian chlamydia as a disease agent in humans has long been the focus of veterinary and medical experts. In most cases, there was a suspected etiological connection between clinically ill sheep, goats, and cattle (abortions and related conditions, respiratory diseases, sporadic encephalomyelitis), and diseases in humans who were in contact with such animals. Reports of specific serological results, agent isolation, and clinical descriptions have been published. In several cases, some well-documented, intra-laboratory infections with chlamydia of sheep or bovine origin are described. Notwithstanding these earlier data, some contemporary authors believe that not enough is currently known about the zoonotic relevance of bovine chlamydiosis. This does not apply to the widely recognized etiological role of *Chlamydia abortus* from aborted sheep and goats in abortion in women. There is uncertainty about the possible transmission of *Chlamydia suis* from diseased pigs to humans. It should be emphasized, however, that the zoonotic potential of this pathogen is still insufficiently studied and underestimated. The third chlamydial species (except *C. psittaci* and *C. abortus*) which is considered to be zoonotic is *Chlamydia felis*. Here, however, there is no lack of opinion that the zoonotic significance of feline chlamydiosis is insufficiently convincing and controversial, as well as the perception that the suspicions of *C. felis*'s zoonotic role arise from the fact that a rare publication on the subject is a direct consequence of under-diagnosing of human cases of *C. felis* infections.

A large proportion of chlamydial diseases in humans is associated with *C. trachomatis*. The interest in them over the last decades has increased enormously. This is largely due to the fact that chlamydia attacks various organs and systems – the urogenital tract, the respiratory tract, including newborns, eyes, joints, and so on. From here comes the great interest in chlamydia in various branches of medicine. In chronological order, we will first point to trachoma – the world's leading cause of acquired blindness. Typical of the trachoma is that it is a big problem in areas and countries with low income and poor infrastructure and sanitation. Lymphogranuloma venereum (LGV) is a generalized sexually transmitted disease. It is most common in countries with tropical and subtropical climates. Of particular importance are sexually transmitted chlamydial infections, manifested in a number of clinical forms and syndromes. These infections are extremely common around the world, with an estimated more than 130 million people infected each year. *C. trachomatis* sexually transmitted infection (STI) is the leading cause of infertility in women in the United States.

Significant health problems are human infections caused by *Chlamydia pneumoniae*, which are spread through respiratory droplets and cause pharyngitis, bronchitis, atypical pneumonia, and coronary artery disease. Originally considered a human pathogen, it is now well known that *C. pneumoniae* is also pathogenic to other hosts – horses, koalas, reptiles, and amphibians. This is another fact that confirms the breadth and dynamism of scientific knowledge in the field of chlamydia and chlamydial infections.

When discussing some of the questions in the book, we include our own observations, results, and interpretations, so I hope the reader can benefit from our experience.

I believe that the book "Chlamydiae and Chlamydial Infections" will expand and enrich the knowledge on the subject of veterinary and medical researchers, professionals, microbiologists, virologists, epidemiologists, molecular biologists and students of veterinary and human medicine. This work will provide current information on important problems with veterinary, health, social, zoonotic, and economic significance and for a wider range of readers from society as a whole.

1

Historical Notes

The first information about eye diseases in humans caused by *Chlamydia trachomatis* have been found in papyrus of ancient Egypt of 10th century BC. In 1st century BC, the Sicilian doctor Diascarides introduced the term "trachoma," which means "rugged." Halberstädter Prowazek [1] made a step towards revealing the true cause of the disease in 1907, which detected inclusions in the cytoplasm of conjunctival cells in the experimentally infected orangutans. These researchers indicate the similarity of the inclusions in experimental infection of animals with those from conjunctival swabs from children with non-gonococcal ophthalmia.

In the late 19th century appears the suspicion of a link between a disease of exotic birds – mainly parrots – and pneumonia in humans who had been in contact with such birds. The name of the disease "psittacosis", introduced by Morange [2] in 1895, derives from the Latin word "psittacus" – parrot. A number of reports in this period and in the early 20th century described psittacosis in humans manifested as an epidemic and the role of parrots as the source of infection is established beyond doubt.

In 1929–1930 there was an outbreak of psittacosis in the form of pandemics sweeping no less than 12 countries in America and Europe. As a first restrictive measure, a ban was declared on imports of parrots from tropical countries. Intense research efforts on the etiology of disease led Levinthal [3] to the discovery of spherical basophil cells in the tissues of the affected parrots.

Almost simultaneously, Coles [4] and Lillie [5] observed similar formations in the reticuloendothelial cells of infected people and birds. First Bedson et al. [6] proved the etiological link between these cells and psittacosis in 1930, and later established polymorphism and filterability of stimuli [7, 8].

The first cultivation of the agent of psittacosis in the yolk sac (YS) of chicken embryo (CE) is the work of Yanamura and Meyer in 1941 [9]. In 1938, Haagen and Maurer [10] set the beginning of research on chlamydiosis of fulmars and other non-psittacine birds called "ornithosis" and Gönner, 1941 [11] first found chlamydial infection in mammals (mice).

Five years earlier (1936), Greig [12] described abortion in sheep and name it "enzootic" (EAE). However, the author suggests that abortions are due to nutritional deficiency. The infectious nature of the disease and the involvement of chlamydia were established in 1950 by Stamp [13]. These pioneering studies signaled the start of extensive studies on the prevalence of infections with chlamydia in a number of avian and mammalian species having important zoonotic and economic significance.

The cytological period of studying the agent of trachoma in man (*C. trachomatis*) was completed after the first cultivation of the agent in YS of CE in 1957 by Tang et al. [14]. Two years later (1959), Jones et al. [15] reported the first isolation of *C. trachomatis* from the urogenital tract of humans, which initiated a broad research on sexually transmitted chlamydial infections having global distribution and important medical and social significance. Gordon and Quan (1965) [16] introduced the cultivation of chlamydiae in cell cultures (CCs). The difficulties in the cultivation of chlamydia led to the introduction of the serological methods.

The following scientific searches revealed that part of the human chlamydial strains having similar characteristics to those of *Chlamydia psittaci*, constitute a special chlamydial species initially called TWAR – Grayston et al. (1986) [17] and later *Chlamydia pneumoniae*. The perception that *C. pneumoniae* is pathogenic only for humans has been revised and three variants (biovars) are now known to affect people, koalas and horses respectively [138].

The cited publications and other studies not mentioned here sparked great international interest. These were followed by research in many countries [18–29].

In Bulgaria, the first studies on chlamydia are associated with the names of Kuyumdjiev et al. [30, 31]. From 1956 to 1959, the authors found antibodies against ornithosis antigen in the blood sera of pigeons, turkeys, ducks, cattle, sheep, and humans. Vachev et al. [32] reported psittacosis in two parrots. Mincheva and Genov [33] found ornithosis in chickens and Genov and Savov [34] – psittacosis in parrots and ornithosis in pigeons and canaries in the zoo – Sofia and in the 30- to 40-day-old chicks in six poultry farms. The same authors reported positive serological results against the antigen of *C. psittaci* in 26 workers from a bird slaughterhouse and in a veterinary officer at the zoo. Nikolov (1963–1965) [35, 36] established serological ornithosis in ducks and turkeys and perform epidemiological studies among workers in the poultry and poultry processing plants. The same author reports on the use of indirect complement fixation test (CFT) in serological diagnostics of ornithosis. Ognianov [37], Ognianov et al. [38], and Semerdziev et al. [39, 40] carried out studies in the 1960s and 1970s on some chlamydial disease in domestic ruminants with the names "viral abortion in sheep" and "neorickettsia." A major share of these studies was on abortions in sheep, which represented the first in Bulgarian serological, virological, and epidemiological studies in this disease. The reports of acute "neorickettsia" (chlamydiosis) in buffalo calves clinically manifested by encephalomyelitis [38], the serological evidence of chlamydia in some eye diseases in humans, and the described case of chlamydial infection in acne-like dermatitis in a woman deserve special attention [41]. In other early studies, Guenov [42] and Natschefff et al. [43] reported infection with the "virus" of ornithosis in pigs with inflammation of the pericardium.

Over an extended period (1977–2013), Martinov carried out large-scale complex studies on Chlamydia agents and Chlamydia-induced diseases in animals and man [44–54]. The first detection and characterization of sexually transmitted infections with *C. trachomatis* in Bulgaria in the late 1970s and the early 1980s was also made by Martinov et al. [55–60].

2

Nature of Chlamydial Organisms and Taxonomy

The question of the nature of chlamydial agents is a controversial scientific problem in the past and present. Their essence was defined broadly – from being considered genuine viruses to being considered as bacteria. In all hypotheses about the nature of chlamydiae, however, it is invariably stated that their main characteristic is the unique mechanism of their intracellular development. Knowledge about the biological properties of Chlamydiae and especially their interaction with the cell-host underlie the notion that these pathogens differ from the bacteria and the rickettsiae [21, 61–67].

The name of chlamydial organisms has been the subject of many discussions and attempts to change. Initially and until the mid-1960s, a number of authors consider them viruses. The other names used are Miyagawanella (Brumpt, 1938), Chlamydia (Jones, Rake, Stearns), Bedsonia (Meyer, 1953), Pararickettsia (Nicolau, 1955), Rakeia (Levaditi et al., 1964) Neorickettsia (Giroud, 1968), Favrella, Rickettsiaformis, Prowazekia, Group PLT (Psittacosis–Lymphogranuloma venereum–Trachoma). The last name has been widely promoted by Giroud [68] and other French researchers.

The disturbing disparity in the names of the chlamydiae and their classification were overcome after the adoption of the proposal made by Page [69, 70] for the establishment of a new order Chlamydiales with family Chlamydiaceae and single genus *Chlamydia*, uniting all members of the group PLT. The obvious differences in the biological properties of isolates in the genus *Chlamydia* led to the differentiation of two species of chlamydiae: *Chlamydia psittaci* and *Chlamydia trachomatis*. A study by Gordon and Quan [71] served as a basis for this distinction on a number of isolates considering the morphology of the inclusions of the pathogen in the infected cell and the presence or absence of glycogen in these inclusions. Agents forming dense compact inclusions are placed in group A (*C. trachomatis*), and those forming diffuse inclusions which are not stained with the glycogen by treatment with iodine, are assigned to group B (*C. psittaci*). To these two criteria Lin and Moulder [72] added the indicator inhibition using sodium sulfadiazine: *C. trachomatis* is inhibited by the compound, and *C. psittaci* is not affected. In the establishment of the genus *Chlamydia,* the similarities in its representatives in morphologic and antigenic regard, the chemical composition, metabolic properties, method of reproduction, and sensitivity to antibiotics are taken into account, without forgetting important serological differences and the ability to cause a variety of clinical conditions. According to the decision of the legal committee of the International Association of microbiological companies of January 1, 1980, the name Chlamydia becomes mandatory, irrespective of its semantic inconsistency and the terminological muddle arisen between Chlamydiales and Chlamydobacteriales (tunicates bacteria) and between chlamydiae and chlamydospores in mushrooms.

Up until the end of the 1990s, the group of chlamydiae comprised four species: *Chlamydia psittaci*, *Chlamydia trachomatis*, *Chlamydia pneumoniae*, and *Chlamydia pecorum* [73].

In the late 1980s, the first published data on DNA – DNA hybridization of chlamydial strains revealed major differences in *C. psittaci*. These studies found that most isolates *C. psittaci* from different host animals have <70% hybridization similarity between them [74]. By those criteria, the species *C. psittaci* included 12 serovars, while 3 biovars and 12 serovars were established for *C. trachomatis*. DNA – DNA hybridization revealed low sequence homology (30% between the genomes of the trachoma and mouse biovars of *C. trachomatis* [74]). Obviously the nucleic acid-based characterization technique reveals that *C. psittaci* and *C. trachomatis* are rather heterogeneous species.

The mentioned facts did not satisfactorily solve the problem of classification of *C. psittaci* [75]. The reason for this lies in the extremely wide range of hosts, the presence of numerous clinical conditions, and striking differences in pathogenic potential which possess individual strains of this chlamydial species. In addition to the above, there are no commercial reagents for differentiating *C. psittaci* from *C. pecorum*. There are no such reagents to distinguish strains of *C. psittaci* [75].

Everett et al. [76] investigated the genetic relationships, ecological and evolutionary relationships among all known chlamydial species. For this purpose, they collected large amounts of data for the ribosomal genes and also for the genes expressing the major outer membrane protein (MOMP) (>100,000 bp). Analyses of 16S and 23S rRNA showed that family Chlamydiaceae contains two genealogical lines and nine species-specific groups.

The two genealogical lines, based on the established taxonomic criteria were announced for two new genera and the nine species-groups for nine different chlamydial species. Later Bush and Everett [77] received additional sequence data on the genes of GroEL chaperonin, small cysteine-rich lipoprotein, 60 kDa cysteine-rich protein (ompB) and KDO-transferase. The authors considered the above data to confirm the newly proposed taxonomy. According to Everett et al. [76], the received data, representing the basis for reclassification of chlamydia organisms, are in agreement with three groups of research of other authors: to the data on not genetically typing; studies of MOMP; and studies of the 16S rRNA. An important point in relation to the reclassification of chlamydiae based on the analysis of ribosomal and coding genes is that the proposed grouping in two genera and nine species have been supported by analyses of the antigenic properties phenotype, endonuclease restriction, virulence, and clinical disease induced by the respective chlamydial species.

According to the proposal of Everett et al. [76], there are two genera in the family Chlamydiaceae: *Chlamydia* and *Chlamydophila*. Genus *Chlamydia* includes three species: *C. muridarum*, *C. suis*, and *C. trachomatis*. Genus *Chlamydophila* brings together six species: *C. abortus*, *C. caviae*, *C. felis*, *C. pecorum*, *C. pneumoniae*, and *C. psittaci*.

In an open letter, Schachter et al. – a total of 31 scientists in the field of chlamydiae research [78] – reject the proposal of Everett et al. [76] for a new classification of chlamydial organisms. The authors believe that radical changes in the taxonomy of chlamydia are not needed for the following reasons: (1) Although the 16S rRNA is useful for studies on the evolution of microorganisms, this marker is not always the best choice for the formation of a species, especially since Everett's proposal is based on minor differences in this respect. Much more useful would be the genus-specific genes. (2) The determination of a new genus *Chlamydia* (*Chlamydophila*) is unnecessary. This act ignores the unique and very conservative biological characteristic of all chlamydial agents incorporated into one single genus. (3) There are no specific biological differences between the chlamydial pathogens and there are no new biological markers for their generic or species differentiation. (4) The name change chlamydia, which has been affirmed in the public domain for decades, will create chaos not only among the public but also among medical professionals and scientists in the field of chlamydia instead of bringing stability to the microbiological nomenclature and classification [78].

The quoted opinion of a large group of scientists was followed by a prolonged discussion in the scientific community. Overall, the opinions supporting the position of Schachter et al. prevail in most cases [78]. According to Stephens et al. [79] the previous taxonomic separation of chlamydia based on ribosomal sequences is neither consistent with the natural history of the organism revealed by genome comparisons, nor widely used by the chlamydia research community 8 years after its introduction; thus it is proposed to reunite chlamydiae into a single genus *Chlamydia*. In publications from 2010, Greub [80, 81] reported that all members of the Subcommittee on the Taxonomy of Chlamydiae consider that a single genus *Chlamydia* should be used for species of the genera *Chlamydia* and *Chlamydophila*. This position is supported by growing evidence. As an example, the percentage of 16S rRNA gene sequence similarity between two members of the genus *Chlamydophila*, *Chlamydophila felis*, and *Chlamydophila pneumoniae*, is 95.1%, whereas this similarity is 94.9% between *Chlamydophila felis* and *Chlamydia trachomatis* which were classified in different genera by Everett et al. [76]. Kuo et al. [82] and the OIE [83] indicated that the taxonomy of family Chlamydiaceae is currently under consideration. Until recently, nine different genotypes of *C. psittaci* based on the ompA gene coding for the MOMP were distinguished. Seven of these “classical serovars” are thought to predominantly occur in a particular order or class of Aves and two in non-avian hosts, i.e., genotype A in psittacine birds, B in pigeons, C in ducks and geese, D in turkeys, E in pigeons, ducks, and others, E/B in ducks, turkeys, and pigeons, F in parakeets, WC in cattle, and M56 in rodents [83]. A significant part of the bird genotypes have also been identified sporadically in isolates from cases of zoonotic transmission to humans, particularly A, B, and E/B [224–226].

In recent publications from 2014 and 2015, Sachse et al. [84, 85] presented evidence for the existence of new family members of Chlamydiaceae – *Chlamydia avium* sp. nov. (pigeons and psittacine birds) and *Chlamydia gallinacea* sp. nov. and proposed their inclusion into the single genus *Chlamydia*. Moreover, the authors made a proposal of a single genus *Chlamydia*, to include all currently recognized species [85]. To these new data, should be added *Candidatus Chlamydia ibidis*, from the digestive tracts of feral ibises in France [1098] and *Candidatus Chlamydia sanzinia* from a captive snake [1099].

Apparently, there is a growing view that family Chlamydiaceae is composed of a single genus, *Chlamydia*, that thus far comprises 11 species: *C. abortus* (sheep, goats, cattle), *C. avium* (pigeons and psittacine birds),

C. caviae (guinea-pigs), *C. felis* (cats), *C. gallinacea* (poultry), *C. muridarum* (mouse, hamster), *C. pecorum* (sheep, cattle), *C. pneumoniae* (humans and others), *C. psittaci* (birds and others), *C. suis* (swine), and *C. trachomatis* (humans).

Modern molecular genetic techniques provide opportunities to continue the detection and differentiation of various genotypes within a chlamydia species. For example, the avian strains of *C. psittaci* are currently divided into at least 15 outer membrane protein A gene (*omp A*)-based genotypes, each one tending to be associated with certain bird species [227]. The use of an expanded gene set encompassing phylogenetic markers capable of differentiating taxa at the species, genus and family level would be very useful when changing or refining the taxonomy of Chlamydiaceae family [1100].

3

Morphology, Mode of Development, and Interaction with the Cells

3.1 Morphology of Chlamydiae in Light Microscopy

The first morphological studies began with the discovery of the causative agent of trachoma. Halberstädter and Prowazek [28] studied scraped material from the conjunctiva of patients, stained them with Giemsa and located cytoplasmic inclusions in the epithelial cells, which were later named after the discoverers. In these bodies are seen two types of particles – tiny, with dimensions of 0.25 nm, colored violet–red, and larger, colored in blue. The authors called the small particles infectious "elementary bodies (EBs)." Later similar cytoplasmic inclusions were found in scrapings from another type of conjunctivitis (non-trachoma) in adults and infants (blennorrhea) – Lindner et al. [86], Schmeichler [87], Stargardt [88], as well as in cervicitis in mothers of newborns and in not gonococcal urethritis in men [89, 99]. Subsequent numerous research on light microscopic level confirmed the dimorphism of Chlamydia organisms. In Chlamydiae, the EBs and the larger forms, the so-called "initial bodies" (IBs) have different staining properties, allowing them to be differentiated in light microscopy (CM). In staining using the Giemsa, Macchiavello, Gimenez, and May-Grundwald methods, EBs acquire red or purplish-red color. The staining properties of the IBs are basophil. In Giemsa staining and Macchiavello, they have blue or blue–dove color.

Light microscopic methods are still used at present – both for direct observation of appropriate clinical and pathological materials of animals and humans, as well as for morphological indication of Chlamydiae in CEs and CCs [74–76, 90, 91]. Terskih [21] and Carlson [91] achieve early detection of inclusions of *C. trachomatis* in CC by acridine-orange. This method determines the presence of RNA and DNA in cytoplasmatic inclusions depending on the degree of maturity of the IBs and EBs contained in them. Shatkin and Mavrov [92] reported a method for determining the average number of cytoplasmatic inclusions in cytological preparation. Using this method they displayed a typical one-step growth of *C. trachomatis* in the monolayer of CC L-929. According to these data the accumulation of inclusions begins between the 12th and 24th hour and reaches its maximum at the 60th hour. Cytological diagnostic criteria have been proposed by some authors, but have been the subject of conflicting opinions and are not generally accepted. Discussed have been the issues of correlation between the cytological methods and isolation of the pathogen, the timing of detection and counting of the inclusions, the small dimensions of chlamydial EBs as a serious obstacle to LM examination, the false-positive results and so on. Another methodological difficulty is the possibility that some vacuoles in preparations might be mistakenly considered chlamydial inclusions.

3.2 Electron Microscopic Morphology

The morphological studies by LM revealed the dimorphism of Chlamydiae and the polymorphism of EM surveys. The EBs are monomorphic units, but they are only part of the chlamydial population. The electron microscopy of Chlamydiae goes back to the 1950s (Crocker and Williams, 1955) [93]. Transmission (TEM) and scanning electron microscopes (SEM) have been used. In TEM, the methods of ultrathin sections and negative contrasts are used and in SEM, the methods of freeze-etching, drying, and fracturing. By means of all these methods, it has been found that chlamydial bodies are quite heterogeneous in their morphological features. EM – immunogold labeling is also used, which is an unequaled technique for the identification, localization, and distribution of proteins, antigens, and other macromolecules of interest, at an ultrastructural level [94].

The EBs are spherical particles with a diameter of 200–300 nm. They have a complete genome and are therefore infectious. They comprise DNA and RNA. Their main features are observed in ultrathin sections of infected CE and CC – a three-layer membrane and interior contents consisting of nucleoid and chlamydia plasma. These bodies are osmiophilic and have high electron density. The triple layer membrane is thick, about 10 nm. It consists of two osmiophilic layers – external and internal, separated by an average osmiophobic electron-transparent layer. According to Murray et al. [95, 96], EBs have a rigid outer membrane that is highly densified with cross disulfide bonds. This membrane determines the stability of EBs against threatening environmental conditions when chlamydiae are outside the eukaryotic host cells, for it contains peptidoglycan in a small amount, which contributes to the strength of the membrane. The nucleoid of the EBs is compact, osmiophilic, and electron-dense. As genetic material comprises a tightly packed DNA and RNA, the nucleoid is eccentrically positioned. There are also a number of deviations from its typical morphology. Some authors [97] noted that virulent strains have a more compact, clearly shaped nucleoid. There is evidence that the nucleoid is close fitting on chlamydial histone protein that is a major DNA-binding protein regularly encountered in higher plants and animal cells, but it is unusual for microorganisms [98].

The second component of the internal contents of EBs is the chlamydial plasma consisting of small (~5 nm) 70S ribosomes responsible for the protein synthesis. The chlamydial plasma is highly osmiophilic and has average electron density.

According to some authors, in ultrathin sections, the morphology of the EBs resembles the morphology of Gram-negative bacteria. In negatively stained preparations, however, EBs have a very specific ultrastructure, which has nothing to do with the morphology of the bacteria [23, 99, 100]. In such preparations of chlamydial suspensions, EBs have a characteristic appearance of the fallen rubber ball or torn envelopes. In negatively stained suspensions of the *C. psittaci* strain Cal 10, Matsumoto [101] observed the "envelopes" of EBs which had dark centrally located rosettes with circular symmetry, made of nine subunits.

3.3 Initial Bodies, Condense Bodies, and Reticulate Bodies

Besides EBs, other genetically immature forms are characteristic of the chlamydial population, designated by the group term "IBs". In descending degree of genetically-infectious maturity, IBs are as follows: (a) condense (CBs) which have an electron-dense center similar to the in nucleoid of EB and net-like-plasma; (b) reticulate (RBs), which have only net-like-plasma. They have significantly larger sizes than EBs – from 400 to 800 nm and even up to 1 micron. They have an oval shape, but are very diverse. Their membrane is thinner than that of the EBs: its outer part is 7 nm, and the plasmatic, 5 nm. Russian authors [97] observed a slight microcapsule on the outer membrane of IB having polysaccharide nature. IBs mainly contain RNA and less DNA and so, because of their genetic inferiority, they probably are non-infectious. In the plasma of CBs and RBs numerous small ribosomes are found. An important morphological feature of IB is that they are less osmiophilic and have a lower electron density. Also characteristic is the greater rigidity of the membranes of the IBs than that of the EBs, so that the forms of the IBs are flowing. According to Murray et al. [95, 96] the RB is the non-infectious, intracellular, metabolically active replicating form of the chlamydia and possesses a fragile membrane that lacks the disulfide bonds listed in EB. Ward [102] observed RBs of *C. trachomatis* (LGV 404), on the outer membranes of which are discovered growths and rosettes similar to those found in EBs, but with greater optical density. These appendages extend from the surface of chlamydial cell to the membrane of the formed chlamydial inclusions. Similar growths in the form of groups (clusters) were seen in RBs of *C. psittaci* Cal 10 [101].

Electronic microscopic methods give better opportunities for the observation and study of the morphology of chlamydial inclusions in various biological models. For example, Callegos-Avila et al. [139] demonstrate the usefulness of TEM in the analysis of semen samples in infertile men. The authors observe the intracellular persistence of *C. trachomatis* in leucocytes in the form of inclusions. Some of the chlamydial bodies have been found degraded in phagolysosomes. The phagocytized spermatozoa present several kinds of damage, like vacuoles within the chromatin and acrosomes detached and with protuberances. Another form of damage is the observed pattern of a loose fibrilar–microgranular chromatin network [139].

3.4 Aberrant Forms, Chlamydial Membranes, and Miniature Particles

Along with the typical EBs and IBs in the infected CE and CC, a variety of atypical or aberrant forms of chlamydia bodies are detected, as well as large amounts of osmiophilic membranous structures. The latter are arranged separately or are associated with the aberrant and degenerative EBs and IBs [99, 100, 103–106].

There is evidence of the link between the use of certain antibiotics (penicillin, ampicillin) in *in vitro* systems and the formation of aberrant chlamydial bodies which are abnormally sized and shaped. RB-like structures are most frequently seen. Several studies suggest the association of the aberrant bodies (Abs) with the persistence of the chlamydiae [104, 106, 107]. According Santalucia and Farinaty [106], the infection persists because (1) the aberrant RBs of *C. trachomatis* can be transferred from one cell to another during division, (2) the aberrant forms are capable of persisting for 2–3 years,

(3) the Abs are relatively inert regarding the biochemical processes and less sensitive to antimicrobials.

Popov and Martinov [99, 100, 108, 109], and Martinov and Popov [22, 23] in the studies of YS of CEs, CCs, ovine placentas and fetal parenchymal organs infected with *C. psittaci*, *C. trachomatis*, and *C. abortus* regularly observed aberrant EBs, which, by their form, are as follows: EBs without membrane; EBs with double membrane; EBs with myelin-like membrane; homogeneously dense EBs; degenerative EBs; giant EBs. The atypical or aberrant IBs are the following: giant IBs; vacuolized IBs; budding IBs; IBs with double membranes; IBs with several membranes; IBs without membrane; IBs with myelin-like membrane; amorphous IBs; metamorphic IBs; degenerative IBs.

Pospischil et al. [110] reported aberrant forms in the gastrointestinal tract of pigs spontaneously or experimentally infected with *Chlamydia suis*. These findings, as well as the above-cited data testify that aberrant bodies occur *in vivo* in pigs and sheep. The question of which factors favor the formation of abnormal or aberrant forms of the Chlamydiae *in vivo* remains unresolved.

Tanami and Yamada [111] first described the miniature forms of bodies of *C. psittaci*. Later, Phillips et al. [112] observed "miniature reticular bodies" along with typical EBs and IBs in the chlamydial inclusions.

In ultrathin sections from sediments of the lightest factions of suspensions of *C. psittaci* and *C. trachomatis* propagated in the yolk sacs of CEs and pelleted at 60,000 g, we found miniature chlamydial particles. Their dimensions are from 50 to 200 nm. Some are empty spherical membranes without inner content, while others have an internal content. It is interesting to note that fractions containing miniature chlamydial particles caused the formation of specific antibodies in guinea pigs. Miniature particles were also found in ultrathin sections from the yolk sacs, where they were observed like budding on IBs, or were located freely between the cells [23, 99, 100]. Similar findings are missing or rare in the descriptions of chlamydial ultrastructure in the majority of publications.

3.5 Fine Structure of Chlamydia Organisms

Morphological features such as specialized surface formations, specialized structures in the inner content and the macromolecular construction of structures relate to the concept of fine structure. The elements of fine structure are revealed using TEM and SEM. Scanning electron microscopy is prioritized. The construction of the macromolecules such as DNA is studied using a biomolecular electron microscope. Regardless of methodological provision, the data on the fine structure of Chlamydia organisms are still small.

- ***Specialized surface structures.*** Two types of structures of a model *C. psittaci* are described by SEM – rod-shaped formations of the plasmatic membrane and hemispherical structures with a diameter of about 30 nm [113, 114]. Nichols et al. [115] give another view on the surface structures based on studies of ultrathin sections by means of a TEM. They describe two types of surface structures of *C. trachomatis* in a cell line from mouse fibroblasts: (a) hemispherical formations with a diameter of 40/30 nm, as the hemispheres are detected on the plasmatic membrane of certain EBs; (b) rod-shaped growths coming out of small indentations on the plasmatic membrane of IBs with a length of 70–90 nm. They pierce the cell membrane and can be observed on the surface. Both types of formations are found separately in EBs and IBs. They depend on the rigidity of the plasma membrane and are associated with the degree of maturity of chlamydial bodies [115]. The third type structures are "craters" located on the appendages. Authors who have found the superficial structures suggest that they are important for "docking" of chlamydia organisms on the cells.

Matsumoto et al. [116] examined *C. trachomatis* D9-3 by SEM and found rosettes in the plasmatic membrane, which have eight subunits, unlike *C. psittaci* Cal 10 B which are composed of nine subunits.

- ***Specialized internal structures.*** It is believed that the inner content of EBs consists of hexagonal units of size 10 nm. In IBs are found spiral-like internal membranes that are associated with the outer membrane [115].

Data analysis for the morphology and fine structure shows that this main part from the biology of chlamydia organisms is of very important theoretical and practical significance, and the efforts of many researchers are dedicated to this problem. Light microscopic and electronic microscopic period gave the main morphological characteristics of the pathogen in experimental biological systems – CC and CE. Studies on the fine structure of Chlamydia organisms are still in its initial stage. The study of the morphology of chlamydia in clinical and pathological materials of sick people and animals, i.e., *in vivo*, is also an important issue, as the morphologic data for this group of microorganisms are derived mainly on model CE and CC and are insufficient for their overall objective characteristic.

3.6 Mode of Development of Chlamydia in Chicken Embryos and Cell Cultures

The issue for the mode of development of the chlamydiae is controversial and debatable. Bedson and Bland [7], based on light microscopic observations, postulated the so-called developmental cycle, i.e., a series of consecutive morphological changes of chlamydia in their development in the cell cytoplasm, including binary division.

Today's knowledge on the morphology and the development of these microorganisms depend on EM observations. A number of authors interpret their results in accordance with the postulates of the pioneers of chlamydia ology [117–126]. In other studies, the binary division of chlamydia has not been established [21, 23, 100, 124–126]. Prominent researcher Schachter [127] believes that there is no full understanding of the mode of development of chlamydiae.

Schematically represented, the hypothesis of the development cycle of chlamydia is as follows. The chlamydial infection begins with the attachment of the EB of the agent to the cell wall. The EBs bind to receptors on susceptible cells and enters them by endocytosis and/or phagocytosis. The EBs fall into vesicles (cell endosomes) of the host where RBs reorganize. The Chlamydiae inhibit the fusion of endosomes with lysosomes and thus oppose their intracellular destruction. The entire intracellular cycle of chlamydia occurs in the endosomes. The reticulate bodies replicate by binary division and turn into EBs. The generated chlamydial inclusions can contain from 100 to 500 "descendants." At the end of the cycle, the cells of inclusions of *C. psittaci* were lysed, and for *C. trachomatis* and *C. pneumoniae* inclusions are pushed out from the cells by reverse endocytosis.

The opposite hypothesis – of the virus-like mode of development of chlamydia without binary division – was formulated by Terskih [21]. In the first stage of infection (6–14 h) in the cytoplasm of the infected cells (within the Golgi apparatus) appear as structureless formations forming "viroplast" (matrix), which is induced by the chlamydial EBs. At this point, a specific synthesis of RNA, proteins and antigen occurs, but the culture has no infectivity. During the second stage (14–22 h) from the viroplast are gradually released large immature chlamydial forms – IBs. During this stage intensive synthesis of DNA and RNA and of the protein components of the chlamydiae is carried out. In the third stage (24–48 h), IBs, because of their rigid membrane, gradually reduce their size by dehydration and reach the final stage of development – to the size and structure of EBs. This process is asynchronous, which means that immature IBs and mature EBs [21] are observed simultaneously. In support of her observations Terskih [21] points out two main arguments: (1) in contrast to the general biological pattern to divide only biologically mature forms, many researchers believe that EBs (mature) arise from IBs (immature) and (2) the binary division is always preceded by a squeeze between the two daughter cells, such as bacteria. Therefore, it is hardly possible to make chlamydiae an exception to the general biological rule of dividing only mature forms and, moreover, that there is no credible evidence of binary division [21].

A team of authors has launched another vision of the development of the chlamydiae, which opposes the above hypotheses, namely that their development cycle is similar to the cycles of two representative of gamma Proteobacteria – *Coxiella burnetii* and *Legionella* [128].

Proceeding from the contradictions and lack of solution to the problem, we set ourselves the task of exploring the mode of reproduction of chlamydiae – through the so-called cycle of development or peculiar morphogenesis. For this purpose, we used different strains of chlamydia that cultured in CEs [23, 99, 100, 141].

According to our hypothesis, the chronology of the events is as follows. *Adsorption of chlamydial bodies on the cells. Phagocytosis.* The initial stage of the interaction of chlamydia with the cells is insufficiently studied. It is known that the chlamydial agent enters the cell by phagocytosis [19, 127]. *Eclipse.* This moment of the interaction between the agent and the cells has been studied in a one-step development cycle [129]. It was found that after adsorption of strain B 557 in L – cells, the chlamydial infectious titer decline. The level of the cell-associated chlamydial infectivity descends well below the titer of the starting infectivity. Only after 36–40 h a rapid upsurge of infectivity follows due to the newly formed chlamydia. In our opinion, this phase of the chlamydial development can be described as chlamydial eclipse or semi-eclipse when the infectiousness is still not lost completely, but reduced to a minimum. Of the chosen model we can see that in the first 4–5 days, the chick embryos survive, i.e., the chlamydial infection *in ovo* has an incubation period, which to a certain extent is due to the chlamydial eclipse. *Chlamydial matrix with initial chlamydial vacuoles.* Avakian et al. [130] have reported for viroplast within the Golgi apparatus in the cytoplasm of cells from a cell line from human amniotic cells after infection with chlamydiae. The authors observed this matrix in the period from 6 to 14 h after infection. In our investigations into the cytoplasm of the infected endodermal cells of YS, in the endothelial cells of blood vessels, as well as in certain blood cells, we observed the formation of specific chlamydial matrixes and matrix structures. Chlamydial matrixes affect the whole cytoplasm. They are varied and often characterized by elevated osmiophily, thickening ergastoplasm, mega ribosomes, and particularly with the appearance of various chlamydia-induced membrane structures. The cytoplasm of infected cells is vacuolated too characteristic by acquiring a foamy appearance. The vacuoles have different sizes. They are limited by the large ribosomes, which are arranged in the form of a crown on the outer layer of the three-layer membrane. The membrane of these specific chlamydial vacuoles has a thickness of 10 nm. The internal content of the vacuoles is similar to the surrounding matrix. We call these vacuoles, "vacuole-precursors" or "initial vacuoles." Besides clearly shaped vacuoles, rounded clusters of ribosomes are detected – "ribosomal initiation centers." On the periphery of such ribosomal clusters can be seen separate segments of the membranes of the future vacuoles characteristically with a circular arrangement of ribosomes. In the zone of vacuolated cytoplasm are found various forms of chlamydia – single or multiple.

Vacuolating cytoplasm is the first characteristic morphological sign of the infection with chlamydia of the cells. This stage of the development of chlamydia organisms can be called the stage of the *multivacuolized chlamydial matrix.* Here it is the reproduction of the different chlamydial cells.

Studies of Matsumoto and Manire [103], Kramer and Gordon [133], Stokes [131], and Kordova [132] attest to the significant role of the membrane system in the chlamydial infection. In these studies, the authors make mention of including membranes, empty vacuoles, and other chlamydia-induced membrane formation in the infected cells.

Autonomous predominantly – DNA or predominantly – RNA the synthesis of the chlamydial cells in the initial vacuoles. In the multivacuolized chlamydial matrix, different morphological forms of chlamydiae are present simultaneously. There are EBs, condense bodies, intermediate forms and reticular bodies. Clearly visible is the connection of the different morphological forms of Chlamydia with the initial vacuoles. These morphological data give grounds for believing that the morphogenesis of the individual chlamydia cells occurs simultaneously and independently in multivacuolized chlamydial matrices. In the individual initial vacuoles, DNA or RNA synthesis takes place predominantly, with the result that are formed chlamydial cell with different amounts of DNA and RNA, respectively chlamydial cells with different genetic maturity. Despite the large volume of our ultrastructural studies, nowhere is binary division or any cycle of development and transformation of one form into another observed. Our results are consistent with the data presented by Terskih [21] and other researchers for lack of binary division in chlamydiae. Some authors [103, 140] presented electronic microphotographs as examples of binary division, but we believe they can be interpreted as a moment of accidental merger or division of closely lying initiation vacuoles, launches two new chlamydia cells. Our opinion on the autonomous synthesis of chlamydial bodies is supported by some data on the chemical composition of chlamydiae. For example, it is known that there is considerable variation in the composition of the EBs and RBs. Tamura [134] found that EBs of the chlamydial agent of meningopneumonitis contain approximately the same amount of RNA and DNA, while RBs have three-fold to four-fold more RNA than DNA. Moreover, Tamura and Iwanaga [135] found in the IBs the three types of RNA. It is also known that chlamydial agents possess enzymes for DNA and RNA synthesis and may themselves synthesize nucleic acids of exogenous precursors.

Becker and Yael [136] and Tanaka [137] investigated DNA-dependent RNA polymerase in the EBs and IBs of *C. psittaci* and *C. trachomatis*. They found that the homogenized EBs released RNA polymerase that is able to synthesize new DNA. The IBs also contain a DNA-dependent RNA polymerase, which synthesizes new RNA *in vitro*. All these data certainly testify to the chemical and synthetic autonomy of the various forms chlamydiae. Most likely, said chemical autonomy underlies the autonomous predominantly-DNA or RNA synthesis, respectively morphogenesis of the different chlamydial bodies [23, 99, 100, 141].

Regardless of diametrically opposed ideas about the mode of development of Chlamydia organisms, there exists a unidirectional view of the uniqueness of the mechanisms associated with it and the reproduction of these infectious agents.

Despite the undoubted advances in the characterization of the morphology of chlamydia organisms and their way of development at the ultrastructural level nowadays, a comprehensive deepening of studies of these agents at the molecular level is necessary. Rahman and Belland [142] indicated that understanding the cycle of chlamydia has been greatly advanced by the availability of complete genome sequences, DNA microarrays, and modern cell-biology technics. The authors point out that the complex interaction between the chlamydial pathogen and the host cell requires coordination of relevant biological functions. These include interactions of membrane structures and protein complexes, processes which depend on metabolic and physical factors. An important condition is the identification of genes involved in the different phases of the chlamydial developmental cycle. This could be achieved by using a whole-genome microarray [142]. The above surveys and the overall study of Chlamydia organisms at the molecular and cell biological level will contribute to a fuller understanding of the chlamydial diseases and updating of fighting them.

3.7 Degenerative Changes in the Infected Cells

The morphological picture of the infected cells is characterized by a number of changes in the cytoplasm and nucleus. In the cytoplasm are observed very specific chlamydial structures – osmiophilic matrices, multivacuolized chlamydial matrices, chlamydial membranes, and many different forms of chlamydial bodies. At the same time, the cell organelles in the cytoplasm greatly change, degenerate, or disappear. The nucleus is osmiophilic, pyknotic, or disintegrated as a result of kariolysis. So infected chlamydia cells have strong signs of degeneration and destruction.

We observed the multiplication of chlamydia not only in the endodermal cells of the YS, but also in the endothelial cells of the blood vessels of it. These observations shed light on the pathogenetic mechanisms of the inflammatory edema and hemorrhage in the YS of CEs infected with chlamydia, and also explain to some extent the pathogenesis of capillary pathia in a number of chlamydial infections.

3.8 Morphological and Morphogenetic Features of Chlamydia in Clinical and Pathological Materials of Infected Animals and Humans

The first section of this chapter presented the morphology and mode of development of chlamydia cultured in CEs (*in ovo*) and CCs (*in vitro*). The second part of the chapter will examine morphological and morphogenetic features of the chlamydial agents in clinical and pathological material from sick animals and humans, i.e., *in vivo* [108, 141]. Undoubtedly, this question is interesting in view of the expected comparisons, analogies, and differences.

3.9 Light Microscopic Examinations

For microscopic demonstration of *Chlamydia* spp. we prepared the smears, or impressions of various clinical, pathological, and experimental materials: placentas, cotyledons, uterine-vaginal secretions, internal parenchymal organs, fibrinous deposits on the serous membranes, exudates and more. The preparations were stained by the methods of the Macchiavello, Stamp or Giemsa.

The light microscopic methods were also applied as part of the methodological complex used for the indication and identification of chlamydia in their primary isolation and cultivation into CEs and CCs. In these studies, samples were stained by the above methods and used the staining procedures as well, with Lugol's solution or the method of Ziehl-Neelsen. B histologic sections were found characteristic inclusions of *Chlamydia abortus*. They are described in Chapter "Pathological Morphology".

The materials from animals having pathology of pregnancy (abortions, premature births, births of dead or non-viable offspring) – placentas, cotyledons and genital discharges, are particularly suitable for this type of examination. In preparations (Macchiavello, Stamp or Giemsa) observed *C. abortus* in the form of polymorphic cells (oval, spherical, rod, or intermediate), stained red or red–violet in different shades that clearly stood out of the blue–green or blue background (Figures 3.1 and 3.2).

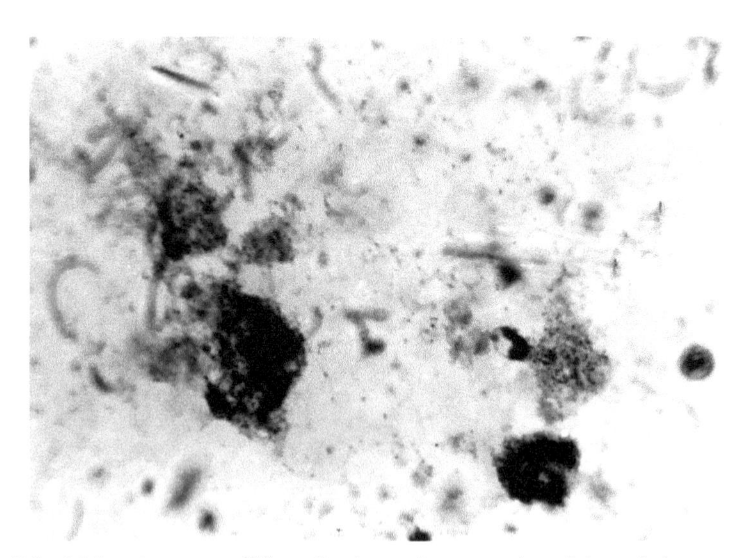

Figure 3.1 Light microscopy. *Chlamydia abortus* in preparation of aborted placenta of sheep with spontaneous chlamydial infection. Staining by Macchiavello. Magnification 500×.

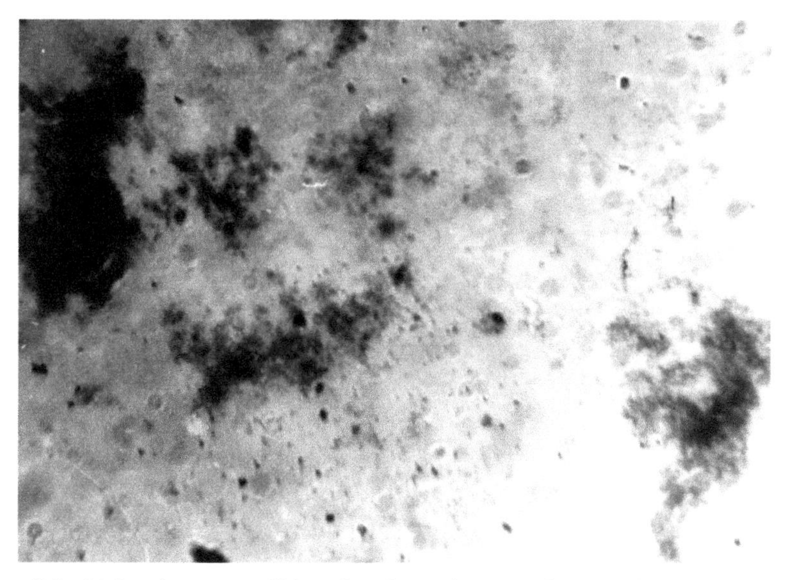

Figure 3.2 Light microscopy. *Chlamydia abortus* in smear from cotyledon of a goat with stillbirth as a result of spontaneous chlamydial infection. Staining by Stamp. Magnification 500×.

Elementary cells showed an affinity to basic dyes. After staining by Macchiavello or Stamp, EBs possessing respectively bright-red and red color were observed. In Giemsa preparations the EBs were stained in blue–violet color, and the larger the IBs were reddish-purple. Chlamydia was found intracellularly in the cytoplasm of the infected cells as single bodies distributed diffusely or formed groups of different sizes. Extracellular chlamydial bodies were also discovered, arranged individually or in the form of small groups.

The sizes of chlamydial bodies ranged from 250 to 350 nm and from 400 to 800 nm and larger.

In parenchymal organs of the fetuses chlamydial bodies were also detected, but generally the concentration thereof was lower compared to the cotyledons and placentas.

Chlamydia psittaci was clearly visualized in impression preparations from infected tissues of spleens, livers, and air bags of birds after Macchiavello or Giemsa staining (Figure 3.3). We saw the causative agent colored red (Macchiavello) or dark purple (Giemsa).

In the biological experiments on the isolation of mammalian and avian chlamydiae in white mice, guinea pigs, and rabbits, the pathogen is usually

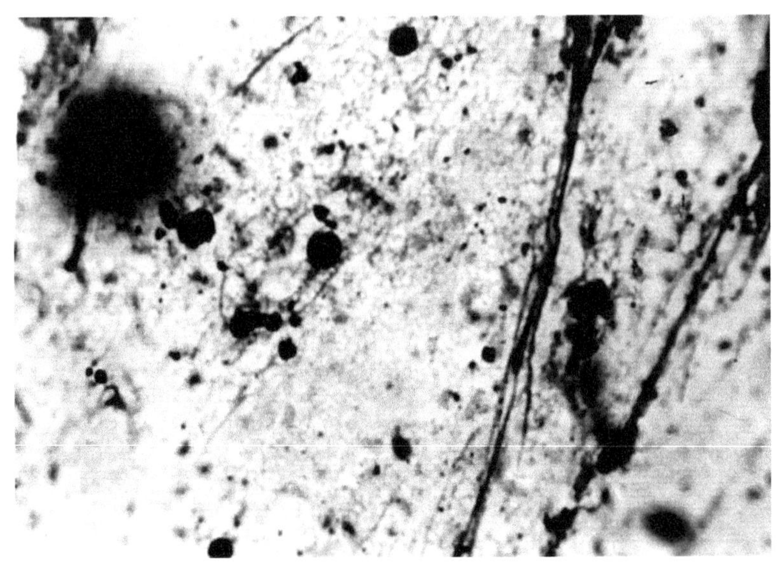

Figure 3.3 Light microscopy. *Chlamydia psittaci* in impression preparation from the spleen of a goose affected by spontaneous chlamydiosis. Staining by Macchiavello. Magnification 500×.

established in significant concentrations in the spleen. Following implemented experimental infections of these laboratory animals with virulent chlamydia and development of expressed infection, the agent can be found in the liver, too. Chlamydial bodies were observed in preparations of outer surfaces and sections of the organ, as the accumulation and their location within cells and beyond them were unequal.

The fibrin deposits on seroses of the spleen, liver, and pericardium of laboratory animals were suitable for LM-observations, where significant amounts of the agent were often found. In preparations of the peritoneal fluid of these animals, we observed chlamydial cells at concentrations of 10^1 to 10^3.

The cytological examination of direct cervical and urethral smears from people with urogenital and joint chlamydial infections and sexually acquired reactive arthritis (SARA, syndrome Reiter) used as part of the complex etiological diagnosis of these diseases. Preparations stained by Giemsa contained basophil cytoplasmic inclusions of a different color: light blue (in large, wrap-around nuclei of epithelial cells) and pink–red (smaller). The color of the inclusions depends upon their maturity – in predominance of IBs has a blue

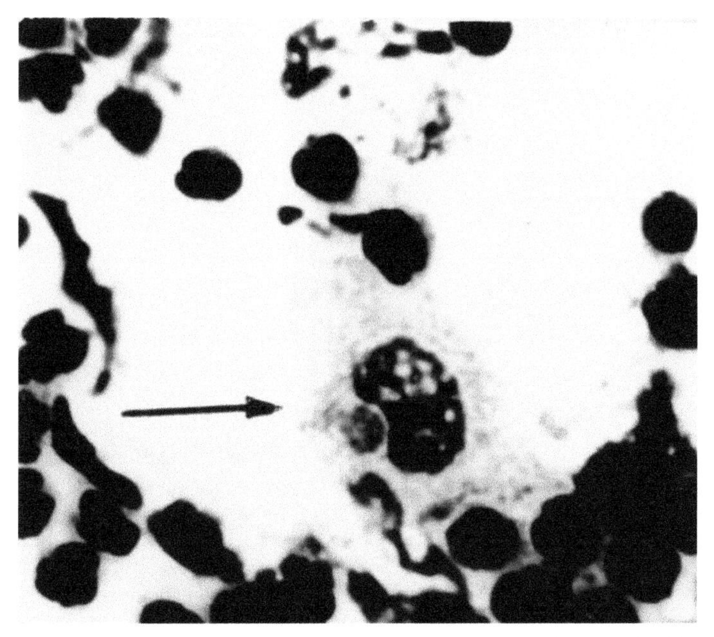

Figure 3.4 Light microscopy. Intra-cytoplasmatic basophilic inclusion of *Chlamydia trachomatis* in the epithelial cells of the cervical secretion in chlamydial endocervicitis. Staining by Giemsa. Magnification 500×.

tint, and in case of prevalence of EBs, it is pink. Overall, the cytological picture of cervicitis is characterized by a rich finding of chlamydial inclusions than with urethritis in men. Most often compact, uniform in color and homogeneous inclusions were found in the epithelial cells, in the form of a crescent or a cap (Figure 3.4).

A little more transparent and smaller acidophilus inclusions with unclear boundaries were there. The preparations of urethral secretions (NGU, PGU, SARA, Reiter's Syndrome) were rich in epithelial and epithelioid cells. We observed basophilic cytoplasmic inclusions of *C. trachomatis*, which differ in number, size and density (Figure 3.5).

3.10 Electron Microscopic Examinations

The studies in this direction were conducted on clinical and pathological material from sick animals spontaneously infected with *Chlamydia abortus*, *Chlamydia suis*, and *Chlamydia psittaci* and on people affected by infection with *Chlamydia trachomatis*.

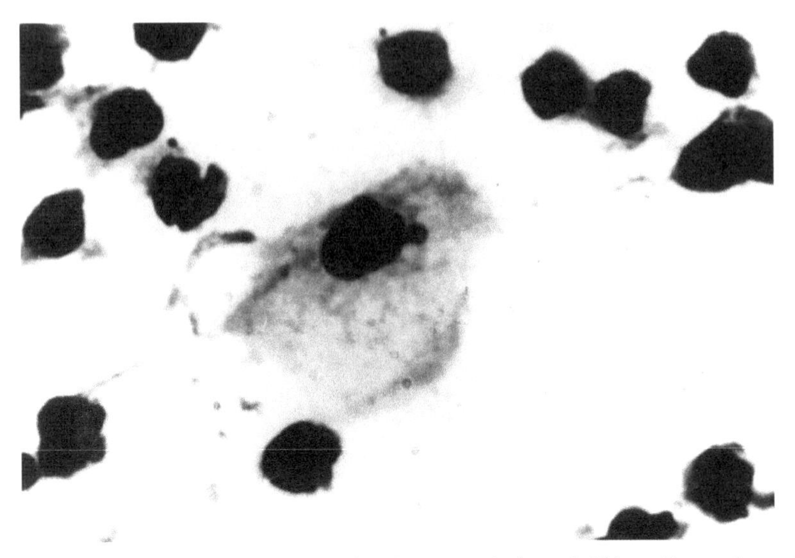

Figure 3.5 Light microscopy. Reiter's Syndrome. Inclusion of *Chlamydia trachomatis* in the cytoplasm of an epitheloid cell of urethral secretion. Staining by Giemsa. Magnification 1000×.

3.11 Ultrastructure of Mammalian and Avian Chlamydia

In ultrathin sections of placentas of sheep, goats, and cows with pathology of pregnancy, high concentrations of EBs and IBs of *C. abortus* are detected (Figure 3.6).

Chlamydial cells are often in the form of inclusions in the cytoplasm of epithelial cells of the placental tissue and cotyledons. Some of these distinct inclusions resemble "pavements" by tightly clinging chlamydial particles. Commonly, the cotyledons show very high accumulations of EBs, IBs, and intracytoplasmic inclusions.

The EBs are spherical or oval particles having a size of 250–300 nm. They are coated with a three-layer membrane with a thickness of 10 nm. The outer and inner layer of the membrane are osmiophilic and electron-dense, and its medium layer is osmiophobic and electron-transparent. The internal content of EB consists of two components: a web-like plasma and low electron density and ribosome-like granules with a size of about 15 nm and nucleoid with high electron density and eccentric positions (Figures 3.6 and 3.7).

The elementary little body is the mature full genomic infectious form comprising two nucleic acids – DNA and RNA.

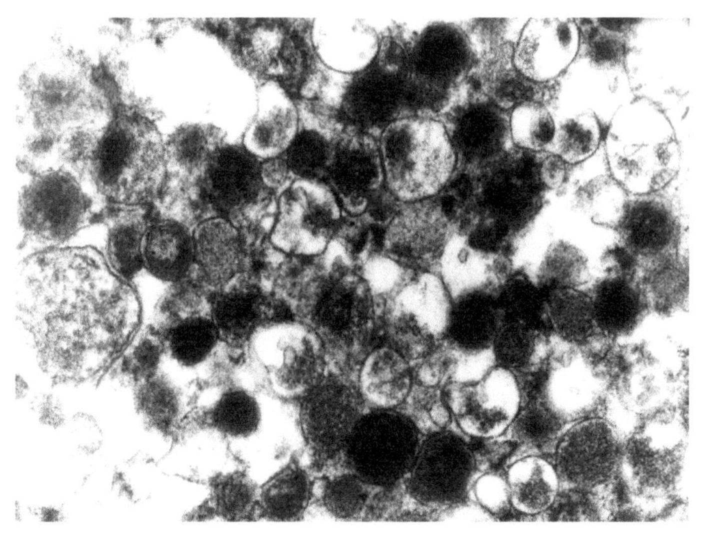

Figure 3.6 Direct electron microscopy. Ultrathin section of aborted ovine placenta. Cytoplasmic inclusion of *Chlamydia abortus*. Magnification 35,000×.

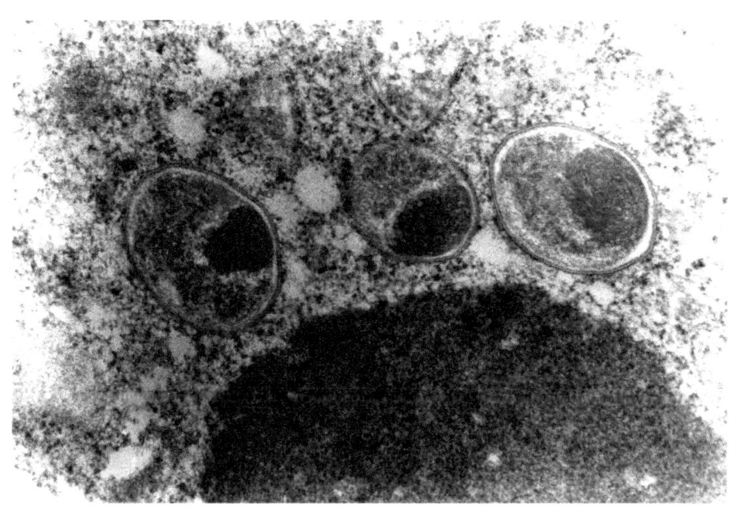

Figure 3.7 Direct EM of ultrathin section from cotyledon of aborted sheep. Elementary bodies of *Chlamydia abortus*. Magnification 80,000×.

The IBs are collectively indication of other morphological types of particles in the chlamydia population with a lower degree of maturity. In descending order according to the degree of their infectious maturity, they

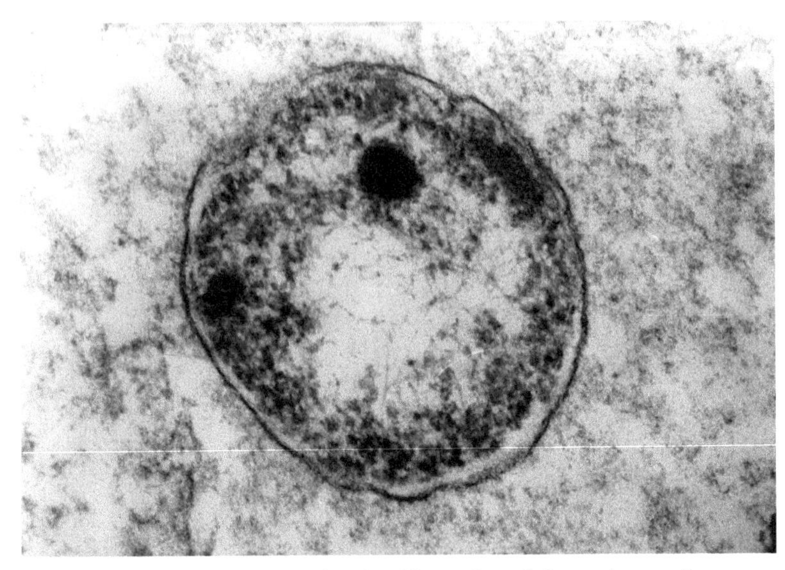

Figure 3.8 Electron microphotograph. Ultrathin section of sheep placenta from a case of a stillbirth. Condense body of *C. abortus*. Magnification 100,000×.

are divided into CBs with an electron-dense center, similar to the nucleoid of EBs and reticular bodies (RBs) having only plasma with granular-fibrilar structure and moderate electron density (Figures 3.6 to 3.8). The dimensions of the RBs and CBs are considerably larger than those of EBs – from 400 to 800 nm – 1 micron. The shape of these particles is oval and they possess a membrane similar to that observed in EBs. The IBs contain primarily the RNA and less DNA in a ratio of 4:1.

In negatively stained preparations of suspensions from cotyledons and placental tissues of domestic ruminants affected by abortion and related conditions, we found the presence of EBs and IBs with morphological titers reaching 10^8–10^9 (Figures 3.9 to 3.12).

In the presented electron microphotographs the typical morphology of *C. abortus* is seen using the method of negative staining, namely chlamydial bodies resembling blocky plastic spheres, dissolved bags or dropped rubber balls. The internal content of chlamydial bodies is slightly mobile and freely wrapped. Note that the massive presence of EBs in cotyledons and placentas, their typical EM morphology, and the technological speed of the procedure determine the reliability and accuracy of negative staining as a method for express demonstration of chlamydial infection in specified disease states.

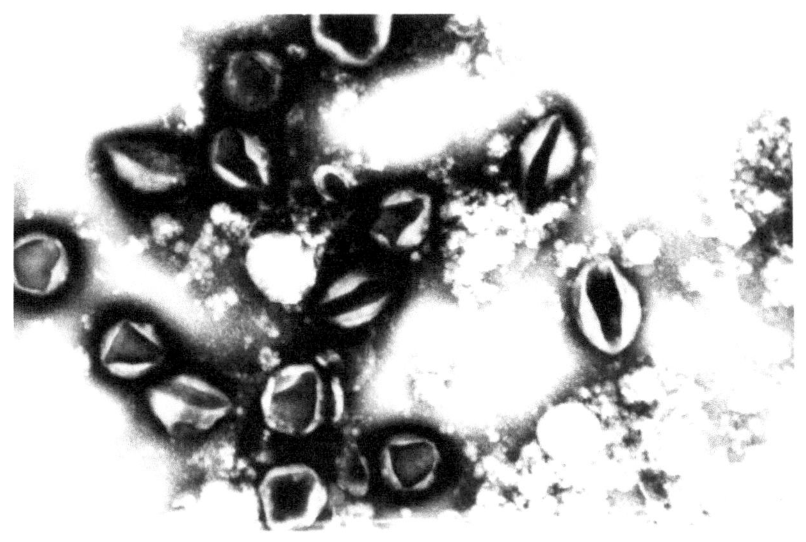

Figure 3.9 Direct EM. Negative staining of *C. abortus* in suspension from the cotyledon of an aborted sheep. Multiple chlamydial bodies. Magnification 35,000×.

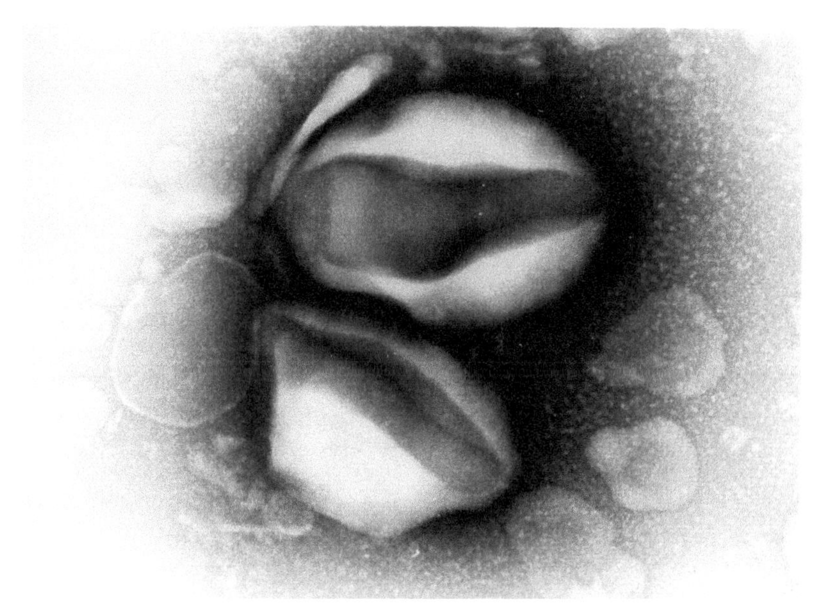

Figure 3.10 Electron microphotograph. Negative staining of concentrated and purified suspension of cotyledons of aborted sheep. Typical morphology of the EBs of *C. abortus*. Magnification 120,000×.

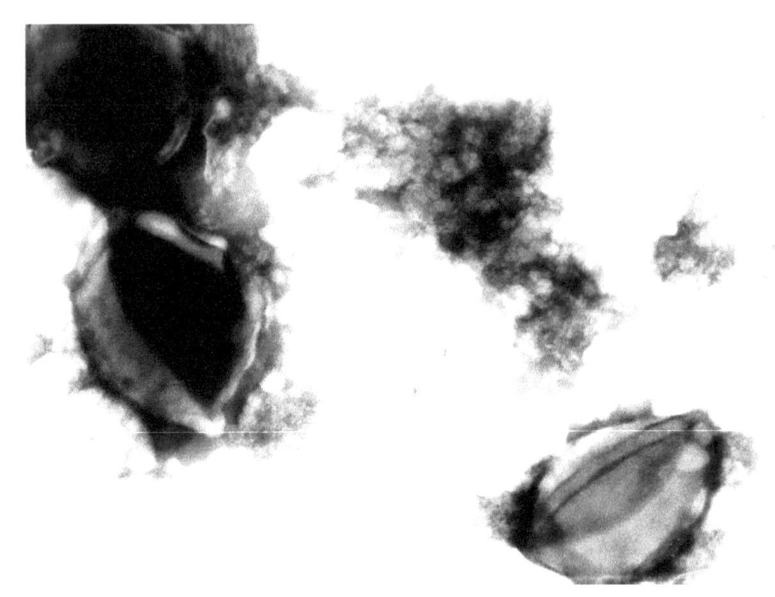

Figure 3.11 Direct EM of negatively stained suspension from the placenta of a goat that gave birth to a dead kid. Chlamydial elementary bodies. Magnification 60,000×.

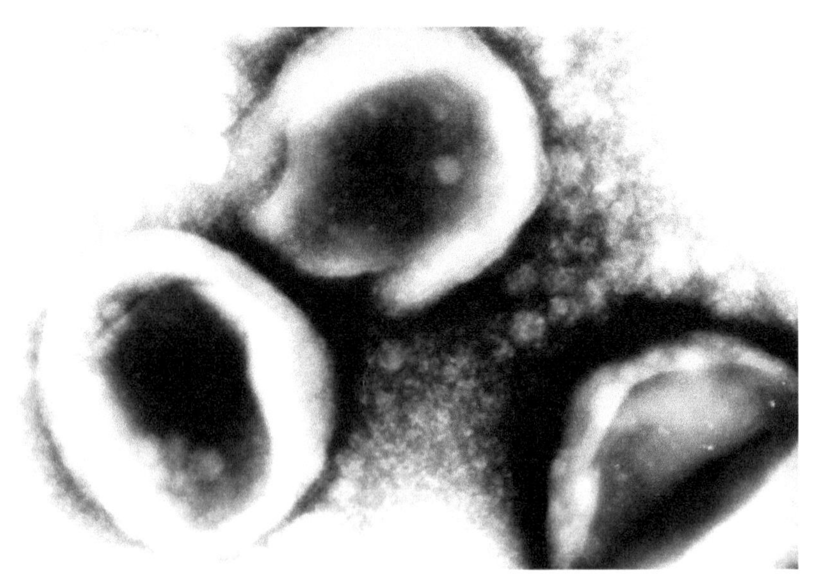

Figure 3.12 Negatively stained suspension of bovine placenta after the birth of an unviable calf. Chlamydial elementary bodies. Magnification 120,000×.

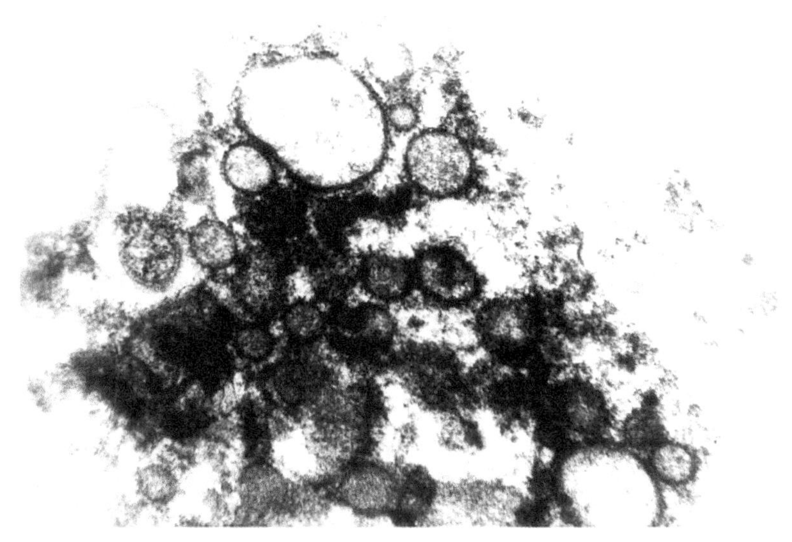

Figure 3.13 Ultrathin section from the spleen of a parrot affected by psittacosis. Chlamydial vacuoles, reticular and condensed bodies of *Chlamydia psittaci*. Magnification 20,000×.

We have found that the direct EM is reliable for the indication of *Chlamydia suis* in inflamed lungs of pigs which died as a result of chlamydial pneumonia. Similarly, the pathogen was found in placentas of swine affected by chlamydial abortions. The morphologic features of infection with chlamydiae in these conditions were similar to those described for ruminants.

Studies of ultrathin sections and negatively stained preparations from spleens, livers, and lungs of waterfowl, pigeons, parrots, and canaries showed accumulation of *C. psittaci* in the positive cases in these organs with typical ultrastructural characteristics, namely presence of EBs, RBs, CBs, intermediate forms, and other specific morphological features (Figures 3.13 and 3.25).

Along with the typical EBs, CBs, and RBs, in the placentas and internal parenchymal, different aberrant forms of the chlamydial bodies, miniature particles, and specific membranous formations are seen.

3.12 Ultrastructure of *Chlamydia trachomatis*

Our studies were focused on sexually transmitted infections with *C. trachomatis*, representing an important medical and social problem.

In direct causal connection with them are part of the sexually acquired reactive arthritis (SARA) and oculi-urethral-synovial syndrome (Reiter's Syndrome), as well as the chlamydial infections in newborns.

We examined ultrathin sections of cervical and urethral biopsies of men and women and lung necropsy material from deceased children.

In the chlamydial cells of the species *C. trachomatis*, pronounced polymorphism is also observed. The EBs are spherical particles with a diameter of 200–250 nm which possess a full-value genome that determines their infectivity and they comprise DNA and RNA. The main morphological characteristics of EB is the presence of a three-layer membrane and inner contents comprising nucleoid and chlamydial plasma. The EBs of *C. trachomatis* are osmiophilic and have high electron density. Their three-layer membrane consists of inner and outer osmiophilic electron-dense layers, separated by a middle electron-transparent layer. The nucleoid is compact, osmiophilic and electron-dense. As a genetic apparatus, the nucleoid contains both nucleic acids. The second component of EB plasma, consists of very small (2–3 nm) ribosomes having high osmiophily and moderate electron-optical density. In some cases, also small granules having a polysaccharide nature are observed.

In negatively stained preparations, the EBs have a characteristic appearance of a burst and flat rubber balls.

As in ultrathin sections and negatively stained suspensions, the EBs show highly specific morphology which is completely different from the morphology of the bacteria. Another typical morphological distinction of EBs is their compact nucleoid that differs from the widespread fibrous nucleoid in rickettsiae and bacteria. The ultrastructure of EBs is typical. It serves as a certain marker of morphological indication and identification of the pathogen. The relative share of EBs of *C. trachomatis* in clinical specimens (*in vivo*) is not as large, as in model infections *in vitro* in CCs. The share of IBs in clinical diseases is much larger as compared to that of the EBs.

Chlamydia trachomatis IBs are also of two kinds: CBs and RBs. The IBs (400–800 nm) are considerably bigger than EBs. Their shape is oval, but of great variety. The outer membrane has a thickness of 7 nm, and the plasmatic membrane – 5 nm. An electron-dense center, similar to the nucleoid of EBs, and a net-like plasma is observed in CBs, while the RBs have a net-like plasma only.

The IBs contain mainly RNA and less DNA in a 3:1 ratio. Due to their genetic incompleteness, in view of the low DNA content, IBs are less infectious.

In the plasma of IBs, there is a multitude of tiny ribosomes. An important morphological peculiarity of IBs is their low osmiophily and lower electron density. The greater rigidity of IB membranes than that of EB is very characteristic. In *C. trachomatis* also found that along with typical EBs and IBs, a number of chlamydial aberrant forms, miniature particles, and chlamydial membranes are detected.

In Figures 3.14 to 3.19, EM are illustrated – morphology of *C. trachomatis* (*in vivo*) in biopsy materials of patients with chlamydial cervicitis, urethritis and Reiter's Syndrome. Chlamydia organisms in the urethral biopsies were found in the epithelial, epithelioid, and macrophage cells. In cervicitis the main localizations of *C. trachomatis* were the cylindrical and the cubic epithelium as well as macrophages.

DEM in the lung necropsy revealed infection with *C. trachomatis* in two children aged 2–3 months who died with a clinical diagnosis of "viral infection."

Figure 3.20 shows that the cytoplasm of the displayed alveolar epithelial cell is vacuolated and filled with EBs and IBs of *C. trachomatis*. The cellular organelles are missing. The nucleus is in a state of pyknosis, kariorexis, and kariolysis. The nucleoplasm is enlightened and the chromatin is located

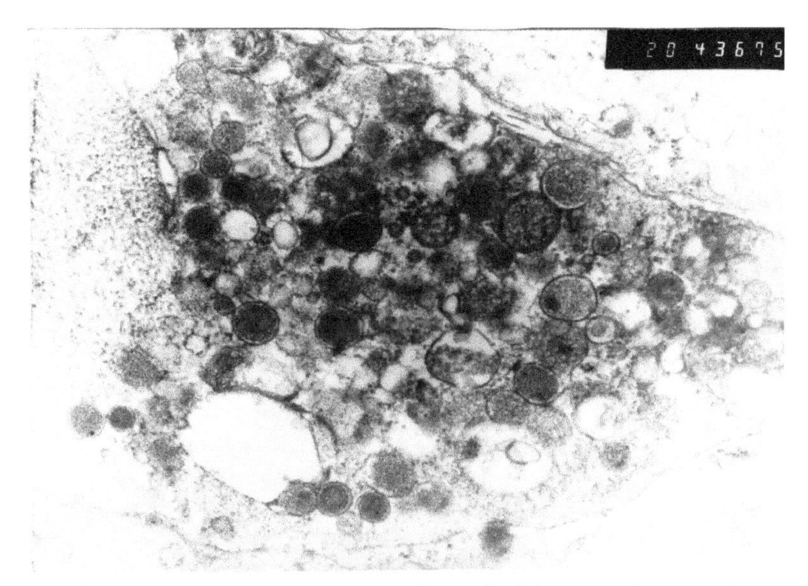

Figure 3.14 Direct electron microscopy. An epithelial cell in urethral mucosa of a man with urethritis. Numerous EBs, RBs, and CBs of *Chlamydia trachomatis*. Magnification 40,000×.

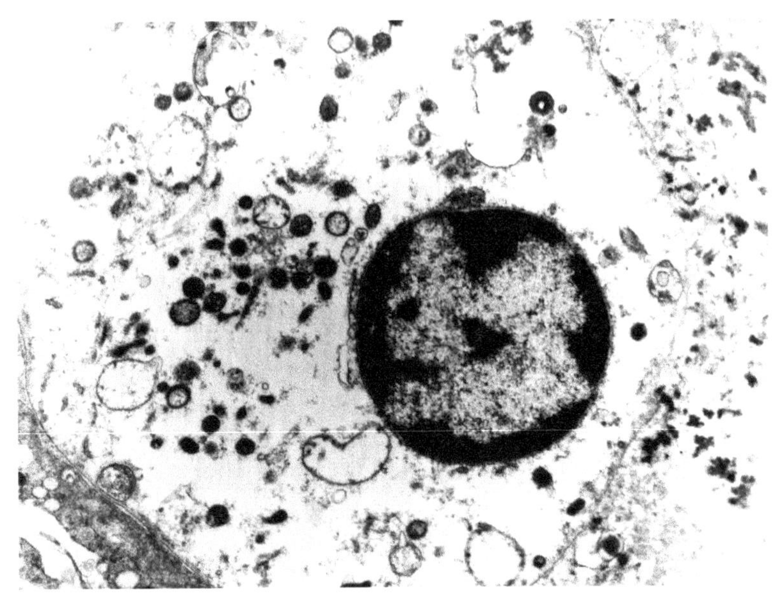

Figure 3.15 Direct EM. Ultrathin section of cervical biopsy of a woman with endocervicitis. Multiple bodies of *C. trachomatis* located in the cytoplasm of an epithelial cell. Magnification 6,000×.

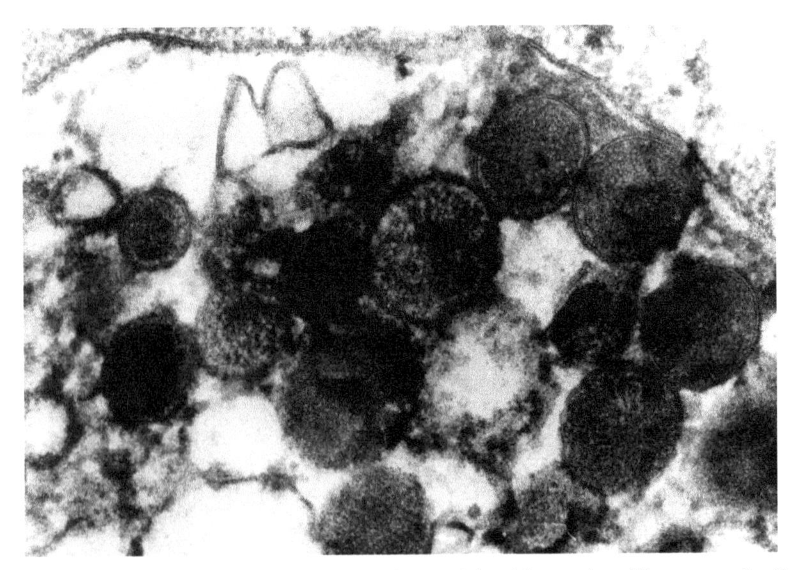

Figure 3.16 Reiter's Syndrome. Urethral biopsy. Ultrathin section. Elementary bodies of *Chlamydia trachomatis* with sizes ~200 nm. Magnification 100,000×.

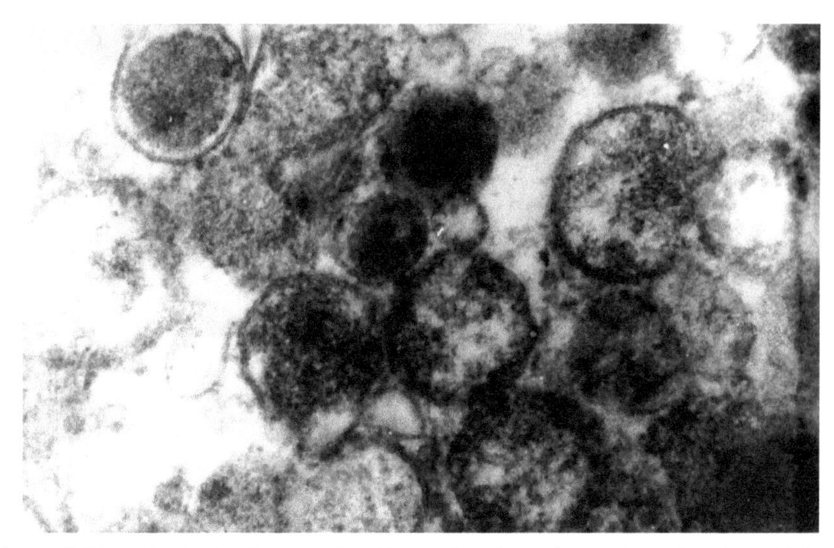

Figure 3.17 Reiter's Syndrome. Ultrathin section from the urethra of male. Initial and elementary bodies of *C. trachomatis* in the cytoplasm of an epithelial cell. Magnification 50,000×.

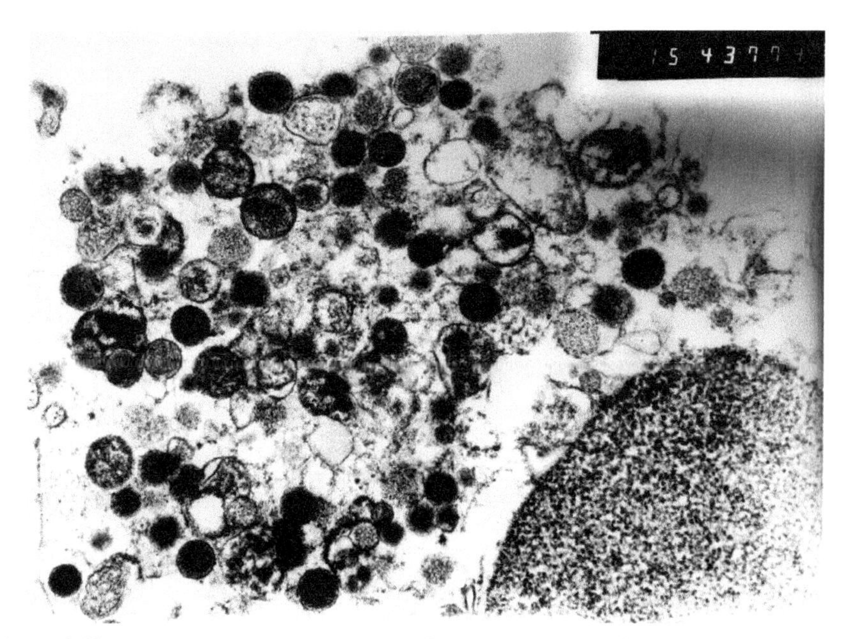

Figure 3.18 Urethral biopsy of a man with NGU. Intracytoplasmatic inclusion of *Chlamydia trachomatis* in epitheloid cell. Magnification 15,000×.

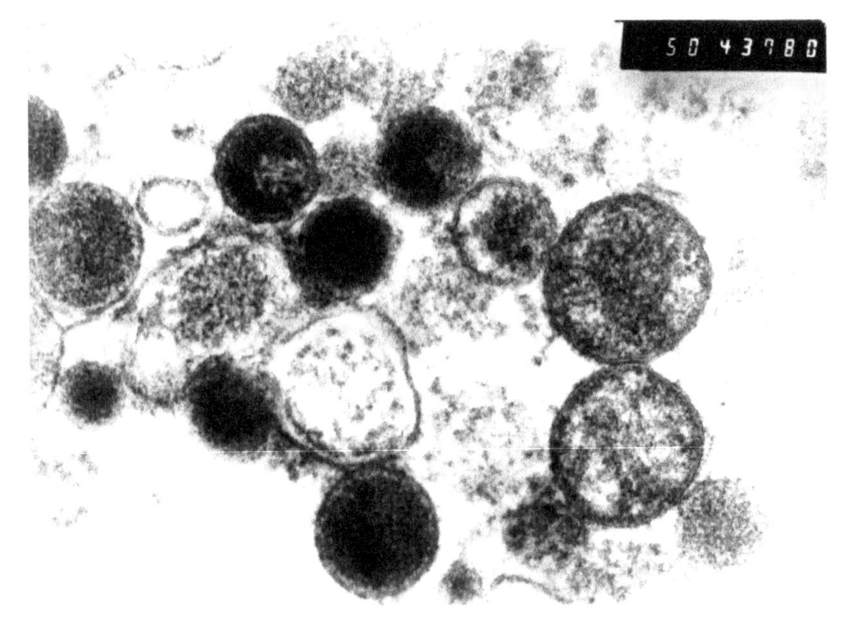

Figure 3.19 Fragment of Figure 3.18. Morphology of elementary and condense bodies of *C. trachomatis* in epithelioid cell from urethral mucosa of a man. Magnification 100,000×.

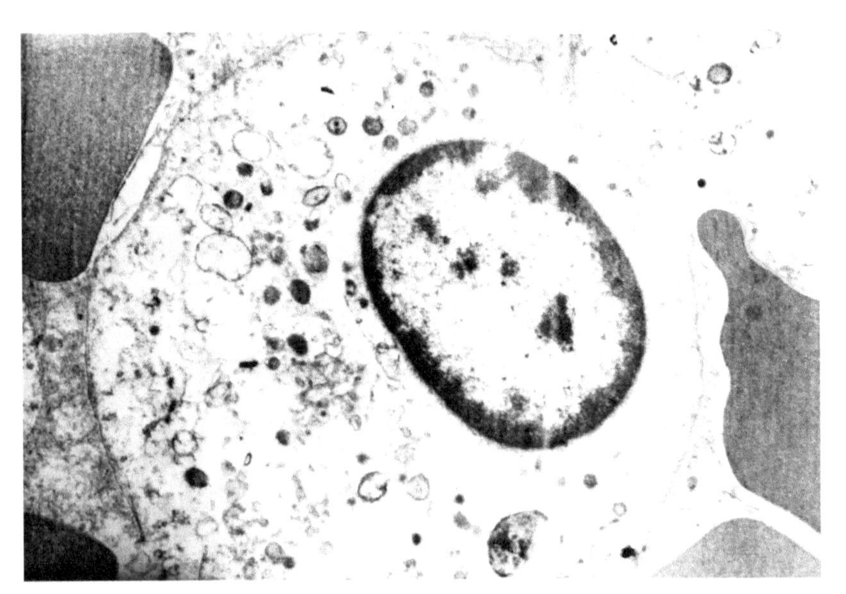

Figure 3.20 Direct EM. Ultrathin section. Alveolar epithelial cell from lung of a child with clinical diagnosis "viral pneumonia." In the cytoplasm are observed numerous cells of *C. trachomatis*. Magnification 5,000×.

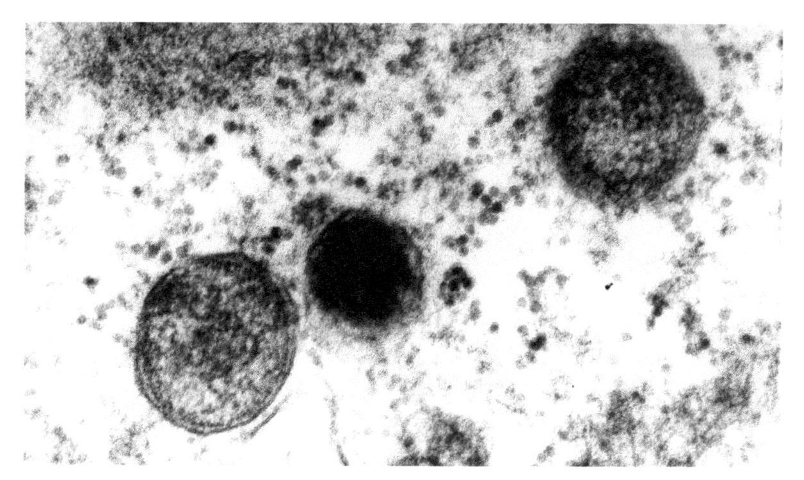

Figure 3.21 Direct EM. Ultrathin section. Reticular body of *C. trachomatis* in the cytoplasm of epithelial cell from the lung of a deceased child. Magnification 50,000×.

marginally. The typical ultrastructure of the second fatal clinical case with a baby is presented in Figure 3.21.

3.13 Comparative Analysis of the Electron Microscopic Morphology of *Chlamydia psittaci* and *Chlamydia trachomatis*

The data for the comparison are given in Tables 3.1 to 3.3. It is seen from the tables that there exist differences between *C. psittaci* and *C. trachomatis*. On comparing the morphological features, differences in eight points in EBs have been found, in six points in CBs, and in six points in RBs. Most significant are the smaller size of *C. trachomatis* and its more pronounced spherical shape. *C. psittaci* is marked by its more compact nucleoid, eccentric situation, and the higher osmiophily and electron density of EBs. Ribosome-like granules of *C. psittaci* are larger than those of *C. trachomatis*. The differences stated are best expressed in chlamydial populations from clinical and pathological materials from animals and humans – the targeted object of the present study, i.e., in the natural environment of the development of such micro-organisms (*in vivo*).

The differences between the two *Chlamydia* species are less markedly expressed in CCs (*in vitro*) and in CEs (*in ovo*).

Table 3.1 Comparative ultrastructural data of elementary bodies of *Chlamydia psittaci* and *Chlamydia trachomatis*

Ultrastructure	*C. psittaci*	*C. trachomatis*
Size	250–350 nm	200–250 nm
Shape	Spherical or oval	Spherical mainly
Membrane:		
• General structure	Three-layered	Three-layered
• thickness	10–12 nm	10 nm
Internal content:		
• Nucleoid	Compact eccentric	Diffuse or compact
• Plasma	Differentiated	Undifferentiated
• Ribosomes	Size 5 nm	Size 2–5 nm
• Electron density	Very high	Moderately high
• Osmiophily	High	Moderately high

Table 3.2 Comparative ultrastructural data of condense bodies of *Chlamydia psittaci* and *Chlamydia trachomatis*

Ultrastructure	*C. psittaci*	*C. trachomatis*
Size	Some 400–500 nm	Some 300 nm
Shape	Oval	Spherical mainly
Membrane:		
• General structure	Three-layered	Three-layered
• Thickness	10–12 nm	10 nm
Internal content:		
• Nucleoid	Compact	Net-like
• Plasma	Net-like granulated	Net-like granulated
• Ribosomes	Size 5 nm	3–5 nm
• Electron density	Low	Low
• Osmiophily	High	Moderate

Table 3.3 Comparative ultrastructural data of reticulate bodies of *Chlamydia psittaci* and *Chlamydia trachomatis*

Ultrastructure	*C. psittaci*	*C. trachomatis*
Size	400–800–1 micron	400 nm on an average
Shape	Oval	Spherical mainly
Membrane	Three-layered	Three-layered
General structure:		
• Thickness	10–12 nm	10 nm
Internal content:		
• Plasma	Net-like granulated	Net-like granulated
• Ribosomes	Size 5 nm	Size 3–5 nm
• Electron density	Low	Moderate
• Osmiophily	Low	Moderate

Table 3.4 Comparative morphogenetic data on *Chlamydia psittaci* and *Chlamydia trachomatis*

Morphogenesis	*C. psittaci*	*C. trachomatis*
Adsorption	At hand	At hand
Phagocytosis	Phagosomes	Phagosomes
Matrix	Vacuolized	Vacuolized
Autonomous synthesis:		
• Vacuoles	Multitude large	Multitude small
• RNA–DNA Osmiophily	High	Moderate
Full-genomic forms (EBs)	Part of the population	Part of the population
Genome deficient forms – (IBs):	A significant part	A significant part
• Reticular	Many	Many
• Condense	Many	Few
• Intermediate	Single	Single

3.14 Morphogenetic Features of *C. psittaci* and *C. trachomatis*

The data for comparison in the morphogenesis of both chlamydial species, based on examinations *in vivo*, are presented in Table 3.4.

We believe that the main phases of the morphogenesis of chlamydia are the following:

- Adsorption of chlamydial cells in the host cell.
- Penetration into the cell by the mechanism of cellular phagocytosis and formation of phagosomes containing EBs.
- Formation of cytoplasmic vacuolated chlamydial matrix, characterized by ribosomal initiation vacuoles in the ergastoplasm and endoplasmatic reticulum, where a characteristic circular positioning of the ribosomes with the building up of chlamydial membranes is observed (Figures 3.22 and 3.23).
- Autonomous synthesis of chlamydial bodies in a multitude of chlamydia vacuoles – precursors (initiation vacuoles). In the different vacuoles are observed simultaneously IBs and EBs. In Figure 3.24, a vacuole precursor is shown, with an active autonomous nucleic acid synthesis in a epithelial cell from aborted ovine placenta. Condensation is clearly seen in the inner contents of the initiation vacuole.
- In a final chlamydial population, a small amount of EBs is observed, which have a full-value genome and a large number of IBs of a reticular, condense, or intermediate type, varying in their morphology. The IBs are genome-defective, due to the low DNA content. According to

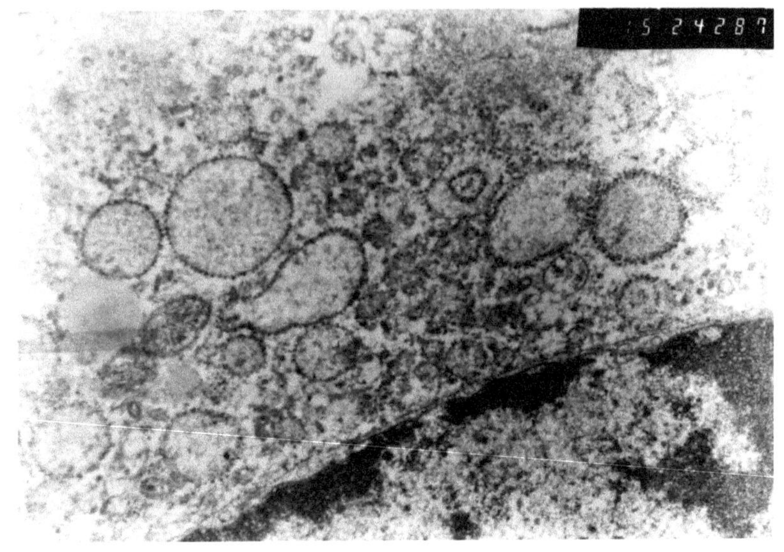

Figure 3.22 Electronic microphotograph. Ultrathin section. Placenta of sheep aborted after infection with *C. abortus*. Multiple vacuolation in the cytoplasm of an epithelial cell. The initial centers and initiation vacuoles with characteristic circular arrangement of the ribosomes. Magnification 15,000×.

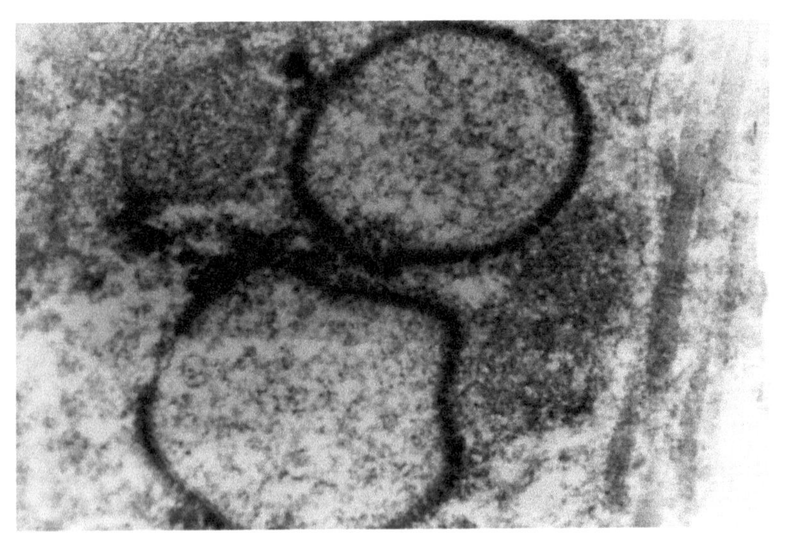

Figure 3.23 Ultrathin section of cervical biopsy material of a woman with endocervicitis caused by *C. trachomatis*. Ribosomal initiation vacuoles and formation of chlamydial membrane. Magnification 40,000×.

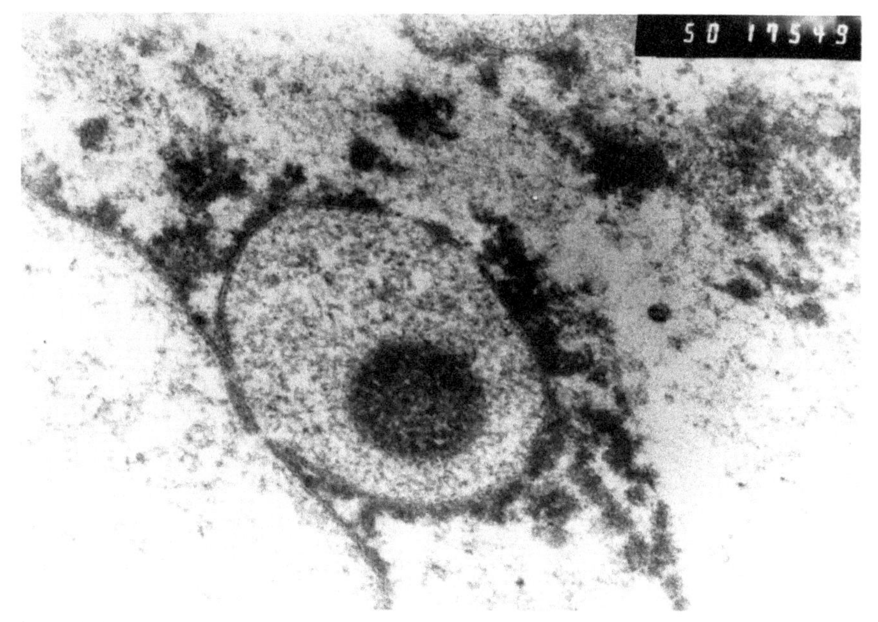

Figure 3.24 Ultrathin section from sheep placenta. Autonomous synthesis of nucleic acids in a chlamydial vacuole – precursor. Condensation on the internal contents of the initial vacuole. Magnification 100,000×.

our observation, mature chlamydial populations in clinical and experimental materials are characterized by a great variety of cytoplasmatic vacuolized matrices, IBs and EBs.

An essential feature is that the typical EBs constitute a small part of the population. We would like to note that in *C. psittaci*, the CBs, RBs, and EBs are to be found in large amounts while the *C. trachomatis* population is presented mainly by RBs and EBs, and a small number of CBs.

Another peculiarity is that *C. trachomatis* is observed mainly in the epithelial cells of the conjunctival, urethral and cervical mucosae and likewise in the lung pneumocytes of children having pneumonia.

Chlamydia psittaci also infects the conjunctivae and the lungs of animals and humans. As for *C. abortus*, its affinity towards epithelial cells of the cotyledons is most striking. In the locations mentioned above, *C. trachomatis* does not form compact inclusions so often. Conversely, the percentage of detached compact *C. abortus* inclusions in epithelial cells of sheep cotyledons is much higher.

In accordance with our own investigations on *C. psittaci* and *C. abortus*, their reproduction occurs as a unique morphogenesis. Moreover, the identity of the morphogenesis, observed on a model CEs and in the organism of affected animals, is of particular importance [22, 23, 99, 100, 108]. We found that the reproduction of *C. trachomatis* from patients with urogenital infections also occurs as a form of morphogenesis [141]. Therefore, the mechanism of reproduction in these species is the same. Irrespective of the unanimity of the morphogenesis, differences were found in the morphogenetic features of *C. psittaci* and *C. trachomatis* related to three points (Table 3.4).

In the developed stages of the infection with chlamydia, progressive impairment of the infected cells is established. As a demonstrative model for the observation and characterization of these processes, the placenta of small ruminants is especially suitable. In the affected cell areas, vast areas in the cytoplasm of epithelial cells are observed occupied by osmiophilic chlamydial matrices containing various chlamydial cells (Figure 3.25).

These matrices are too often sharply limited and demarcated from the surrounding cytoplasm with chlamydial membranes, thereby forming typical cytoplasmic inclusions. The nuclei of the infected epithelial cells are in a state of ultracytopathic effect (Figure 3.26).

The last stage of development of chlamydiae in the cytoplasm of epithelial cells is accompanied by severe degenerative processes in the cells: lack of nuclei and the disappearance of the cell membrane and the cellular organelles.

The territory of the cell is completely filled by the chlamydial matrices containing chlamydial vacuoles, reticular, condense, elementary and intermediate- bodies, aberrant structures of bodies, and miniature particles. An important feature of chlamydia *in vivo* is a frequent finding in the matrices of areas with strongly pronounced production of chlamydial membranes. Another feature of the reproduction of chlamydia is the lesions in the blood vessels of the placenta as a result of accumulation of the pathogen in the lumen of capillaries and around the capillary spaces.

3.15 Electron Microscopic Diagnostics of Chlamydiae

Over the past three decades, a number of reference reviews have been submitted on the use of the electron-microscopic methods of diagnosis of viruses and viral diseases. However, despite convincing evidence of the effectiveness of EM in the research of Chlamydiae, EM diagnostics to this day has not found a proper place in the diagnostic arsenal of these microorganisms and

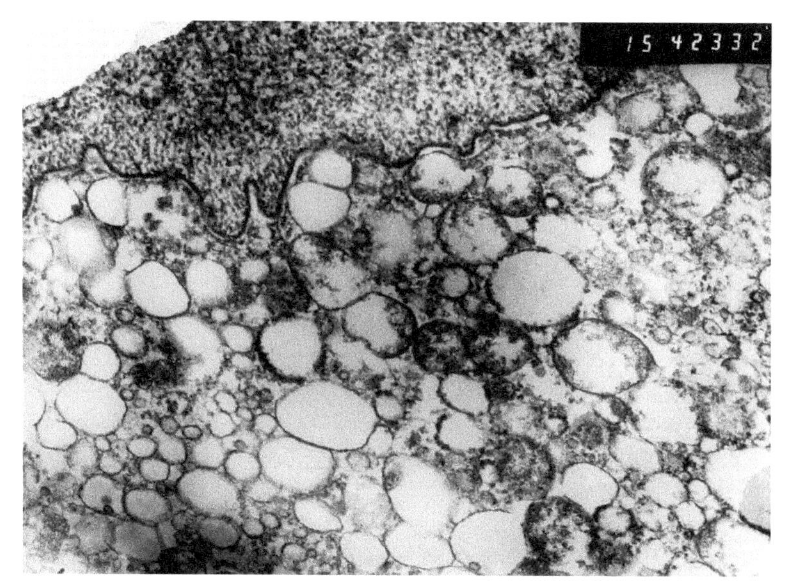

Figure 3.25 Ultrathin section of cervical biopsy of woman with cervicitis caused by *C. trachomatis*. Multi-vacuolated chlamydial matrix. Magnification 15,000×.

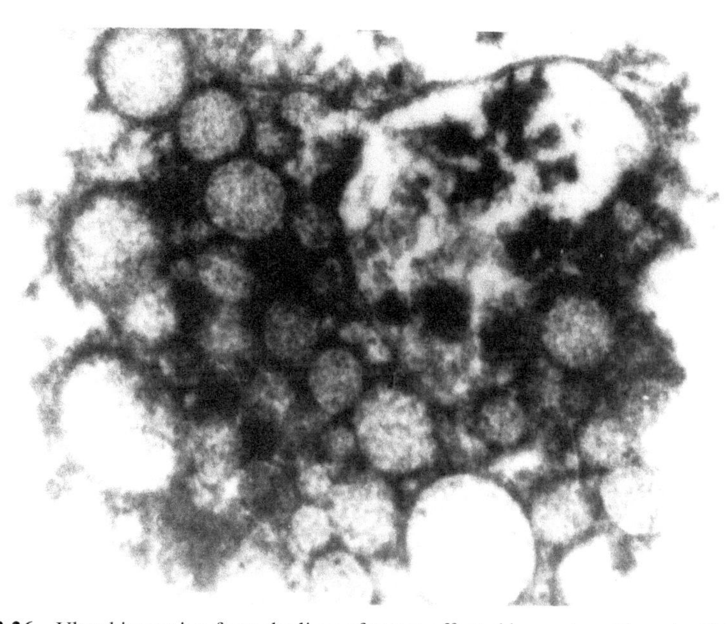

Figure 3.26 Ultrathin section from the liver of parrot affected by acute psittacosis. Chlamydial vacuoles and bodies of *C. psittaci* in the cytoplasm of a hepatocyte. The nucleus is in a state of ultracytopathic effect. Magnification 40,000×.

diseases that they induce. It should be noted that since 1930, all laboratories specializing in chlamydia research have begun using light microscopy (LM) in the morphological diagnosis of these agents on a large scale. The shortcomings of LM are well-known. Difficulties come down to the fact that not all methods for staining chlamydiae produce equal results, the dimensions of the chlamydial agents are very small, and numerous artifacts are likewise seen in the preparations. Thus, LM is an outdated, not very precise method for the morphological diagnosis of chlamydiae.

In our long-lasting work in the field of chlamydiae, we regularly apply electron microscopic methods for ultrathin sections and negative staining. Detailed knowledge of the ultrastructure of chlamydiae and reliable results obtained in the examination of more than a thousand samples allowed us to introduce electronic microscopy as a routine laboratory and diagnostic practice for these pathogens [143–149].

- *Chlamydial abortions, stillbirths and unviable offspring in sheep and goats.* As a rule, pathological materials from sheep and goats affected by abortions and related conditions contain EBs and IBs at high concentrations. The presence of chlamydial bodies as differentiated inclusions in the cytoplasm of epithelial cells of the placentas and cotyledons is very characteristic. These inclusions look like pavement formations of EBs, CBs, and RBs. The accumulation of the pathogens and the formation of specific inclusions in the tissues of cotyledons is extremely rich. We established the presence of EBs and IBs at high concentrations, reaching up to 10^9 in negatively stained preparations, prepared from 10% suspensions of placetae and cotyledons. We likewise examined parenchymal organs, livers, and lungs above all from aborted fetuses. However, only in well-preserved, fresh organs could we detect IBs mainly in the cellular cytoplasm. The organs, having sustained autolysis or going rotten, cannot be used for EM indication by both methods. We were able to detect infections with *C. abortus* in many farms in different administrative districts of Bulgaria with the aid of direct EM of placentas and cotyledons from sheep and goats affected by abortions and related conditions.
- *Chlamydial infections in cattle.* The EM diagnostics of the chlamydiae is efficient in ultrathin sections from the YS of CE, inoculated with material from aborted cows and calves suffering of polyarthritis and keratoconjunctivitis [150, 151]. Accumulations of the agent (CBs, RBs)

were detected as early as the primary inoculation with initial pathological or clinical material in 7-day-old CEs. EBs were also observed in passages carried out much later.

- *Chlamydial infections in swine.* Direct EM (DEM) proved to be a suitable method to indicate chlamydiae in pneumonia-affected lungs of pigs, having died after chlamydial pneumonia. These data were confirmed by isolation of the agent and the positive serological investigations. We showed on the zero and subsequent passages IBs and EBs accumulations in ultrathin sections of YS, infected with materials of swine, having pneumonia. The EM data were positive for the presence of IBs as of the initial inoculation for the diagnosis of swine abortion and pig arthritis and pig pericarditis by the isolation of the pathogen in CE. Later, when the strains were serially passed, EBs were also found.
- *Avian chlamydiosis.* Direct EM of pathological materials from parrots, canaries, pigeons and ducks, as well as EM of ultrathin sections of YS, inoculated with these materials, indicated a great multitude of chlamydiae accumulations in parenchymal organs and in the YS.
- *Chlamydiosis in wild murine rodents.* Chlamydial IBs were established in ultrathin sections from bronchopneumonial lesions of the lungs of the rodents *Clethrionomys glareolus* and *Apodemus flavicollis.* The Chlamydiae were intracellularly found in the cytoplasm of individual alveolar cells. The cytoplasm of these cells was vacuolized, containing chlamydial matrixes. In separate cases similar investigations were carried out with kidney tissues where individual manifestations of chlamydial infection were found. The presence of Chlamydiae in YS inoculated with lung and kidney suspensions were electron-microscopically established [152].
- *Chlamydial infections in humans.* When applying EM on clinical materials from sick persons, positive results were obtained in cases of ornithosis, uro-genital diseases (urethritis, prostatits, endocervicitis), oculo-urethro-synovial syndrome of Reiter, and pneumonia in children. In cases of ornithosis and prostatitis, the examination of negatively stained preparations of concentrated urine was efficient. Numerous EBs, CBs, and RBs were found. The direct indication of chlamydiae was likewise possible in an ultrathin section of spittum (ornithosis). In the cervical and urethral biopsies were observed different types of bodies of *C. trachomatis* and other morphological features of the reproduction of chlamydiae in the cells. Similar findings were found in epithelial cells from pneumonic lungs of deceased children.

A part of our findings established by using DEM are shown above in Figures 3.6 to 3.12 and 3.22 (ruminants), Figure 3.13 (parrot), and Figures 3.20 and 3.21 (humans).

3.16 Morphology as a Method of Indication and Identification of Chlamydial Organisms

The morphologic identification of the characteristic chlamydial inclusions, EBs and IBs is one of the four compulsory tests for the identification of a certain chlamydial strain [19].

When using LM, smears or impression preparations stained by the above-mentioned methods are examined for small, dense chlamydia cells with the color corresponding to the method of staining. The classical method for the isolation of chlamydia-infected CEs is usually the examination of the YS. Following intranasal inoculation of mice, preparations from pneumonic lungs are examined. Miscellaneous staining procedures are used for LM examinations of monolayer from infected CCs. When the isolation of the agent is done in the YS of CE, in the process of passages the chlamydia organisms cause mortality with an average time of death and specific mortality curve [46]. On the other hand, infection with chlamydia leads also to typical changes in CEs and their YE that appear to be a useful indicator of the infection.

A very valuable method for of indication and identification of chlamydia is EM. As a rule, in the early passages of the agent in a biological system, a typical morphological picture of chlamydia in LM is not found. At the same time, these samples were positive in EM. Admittedly, in the isolation and cultivation of chlamydiae, the EM-methods are particularly useful for an early indication and identification of these organisms [165–167]. also Reliable differential diagnosis of coxiellas, rickettsiae, mycoplasma, bacteria and viruses is performed by EM and mono- or mixed infections can be identified with these organisms.

The usefulness and effectiveness of the EM methods for direct detection and identification of chlamydiae in clinical and pathological materials is dealt with in other sections of the book.

3.17 Interaction of Chlamydial Organisms with the Host Cell

This question has been studied on a model CCs. When using the chlamydia with a high multiplicity of infection, the already mentioned One-step development cycle was achieved [129].

After penetration of the chlamydial agent into a cell, the infectious titer is lowered. The level of cell-associated chlamydial infection is significantly lower than the starting titer. The reproduction of the reticular cells is carried out for about 36 h. These morphological forms of the chlamydia are not infectious. The next stage is characterized by a stormy upsurge of infectivity, which coincides with the transformation of RBs in EBs [129]. In the various CCs, the average production of chlamydia is unequal: in Hela, infected with *C. trachomatis* is 35–60 units (IU); in McCoy (*C. trachomatis*) – 1000 IU; in L (*C. psittaci*) – 400–500 IU.

Several teams studied the relationships of *C. trachomatis* with human monocytes and polymorphonuclear leukocytes [153–155]. Register et al. [156] explored the interaction between EBs and marked human polynuclear leukocytes and found that the EBs were attacked by phagocytosis of leukocytes within the first hour after contact. Later, however – even after 10 h or after the destruction of leukocytes – ET retained infectivity, albeit small, sufficient for infection of susceptible cells. These data indicate that chlamydia cannot fully be disposed from the polymorphonuclear leukocytes.

Data obtained in the mouse model after infection with *C. abortus* (sheep abortion) was presented by Buendia et al. [157]. The authors indicate that polynuclear neutrophils play a significant role in the primary infection and preventing the uncontrolled multiplication of the chlamydia in the liver and spleen. Young et al. [158] indicate that the anti-chlamydia activity of human leukocytes is associated with their toxic effect on *C. trachomatis*. After 1 h of incubation, chlamydial titer is reduced by 3–3.5 lg. The role of phagocytes in the fight against chlamydia is significant, but their capabilities are restricted to a certain limit.

It is known that the chlamydiae inhibit the DNA synthesis and the protein synthesis in the host cell, leading to profound disability.

Ojcius et al. [159] characterize the cytotoxic activity of *C. psittaci* in the vacuole surrounded by a membrane during a 2-day infectious cycle. Within 1 day of inoculation, it was found that the infected epithelial cells and the macrophages died by apoptosis. The inhibition of protein synthesis in the host cell had no effect on the cell death, but blocking the entry of chlamydia or their protein synthesis prevented apoptosis. This means that chlamydia reproduction is a necessary condition for cell death. Since the apoptotic cells secrete pro-inflammatory cytokines, it is assumed that the chlamydia-induced apoptosis contributes to the inflammatory response of the host [159]. Gibellini et al. [160] come to similar conclusions in a study of apoptosis induced by *C. psittaci* 6BC and the *C. trachomatis* LGV serotype 2 in cells LLC-MK-2.

To survive in eukaryotic cells, chlamydia should resist the defense mechanisms of the host cell and acquire the metabolites necessary for their development. The membrane of the chlamydial inclusion (MCI) in the cell, which until recently did not attract much attention, undoubtedly play a key role in the relationship of chlamydia with the cells [161]. The discovery of gene-encoding proteins present in the MCI of *C. psittaci* and *C. trachomatis*, is an important development in this direction [162–164].

4

Genomic Structure, Genes, and Molecular Regulation of the Development of Chlamydiae

Over the last decade, a number of studies were carried out on the genome of the Chlamydiae and comparisons made within the group and with other organisms [168–170]. The results obtained contributed to a deeper understanding of the chlamydial metabolic and biosynthetic capabilities [168]. At this stage, however, the presented data on the genome give relatively little information that would serve to decipher the mechanisms determining the pathogenic properties of chlamydiae. Cited studies focused on the genomes of C. trachomatis (including the mouse biovar MoPn) and C. pneumoniae, only provide limited opportunities to track the evolutionary path of chlamydia. The main reason for this is supposed the too small size of the genome of these species. Apparently studies are needed on the members of the completely different *C. psittaci*, before proposing a detailed phylogeny of chlamydia [171, 172].

The picture that is revealed today is the presence of remarkably stable genomes with relatively small variations in the gene families. This gives grounds to assume that the genomic content has appeared early in the evolution of chlamydia organisms – probably before the start of their intracellular life. In addition, the spread of chlamydia on different hosts was not accompanied by large horizontal changes similar to those observed in other free-living microorganisms [171].

The genome of *C. trachomatis* is 1,042,519 bp and 894 calculated protein coding genes [168]. The genome of *C. pneumoniae* is 1,230,230 bp and has 894 protein coding genes [169]. *C. trachomatis* contains another extra-chromosomal plasmid genome of 7493 bp, whereas for *C. pneumoniae*, extra-chromosomal elements were not detected [170]. 186 genes in the genome of *C. pneumoniae* are not homologous with chain sequences of the genome of *C. trachomatis*, and the other 70 genes from the genome of *C. trachomatis* are not represented in the genome of *C. pneumoniae*.

Sequencing of the chlamydiae genome has provided new means to analyze the biology of these organisms from the molecular and structural standpoint. The five complete genome sequences published are as follows: *C. trachomatis* D [168]; *C. trachomatis* MoPn [170]; *C. pneumoniae* CWLO29 [169]; *C. pneumoniae* J138 [209]; and *C. pneumoniae* AR39 [170].

From the standpoint of pathogenesis, the most important result of these two genomic studies is the identification of a new multi-gene family of chain membrane proteins and the set of components for secretion system of type III factors that play a role in the virulence of chlamydia.

Regardless of the very great similarity in the biology of chlamydiae, the tissue tropism and disease events they produce are extremely diverse, for which the genomic analyses have no satisfactory explanation [170]. For example, *C. trachomatis* (serovariants A–C) causes the eye disease trachoma, while serovariants D–K cause a number of sexually transmitted disease syndromes. All these serovariants, united in the trachoma biovar, cause an infection localized in the mucosa of the host mucous membranes. On the other hand, serovariants L1, L2, L3 of *C. trachomatis*, classified as biovar LGV, cause systemic infection, primarily of the lymphoid tissues. In *C. pneumoniae* is also a great diversity in the diseases which are caused – from the infection of the mucous membranes of the respiratory tract (pharyngitis, bronchitis, pneumonia), to vascular infection and atherosclerosis [170, 173, 174].

A large number of genes are found in the *C. trachomatis* and *C. pneumoniae* with identified or obscure features: genes necessary for the synthesis of peptidoglycan [171]; genes encoding enzymes involved in energy metabolism [171]; genes encoding the 97–99 kD outer membrane protein [175]; a gene encoding a transporter of a specific adenosyl/methionine (SAM) [176]; genes coding for MOMP [177]; genes responsible for stress (heat-shock genes groES and groEL [178, 179]; genes specifying resistance to antibiotics [180], etc. Flores et al., [181] presented traits of a newly identified protein (Cpn1027) of the membrane of inclusions of C. pneumoniae. Two membrane thiolproteases genes – *mtpA* and *mtpB* are expressed late and encode proteins which may play a proteolytic role in the formation of OM complex [192, 202]. According Maurer et al. [203], Albrecht et al. [204], Belland et al. [192], and Nicholson et al. [191], investigations have shown that the developmental cycle of Chlamydia is marked by differential temporal expression patterns of large sets of genes, which have been characterized as "early" (EB to RB conversion), "mid" (RB replication), and "late" (RB to EB conversion) [205]. Some authors suggest that the developmental cycle is ultimately regulated at transcriptional level [193, 206, 207]. Nicholson et al. [191] indicate that the signals that trigger inter-conversion of the morphologically distinct chlamydial forms are unknown. Besides that, the mechanisms that control and regulate intracellular development are poorly understood [191]. The hypothesis for the regulation of the cycle at a transcriptional level was tested by microarray analysis, which provides a comprehensive assessment of the global gene regulation through the chlamydial development [191]. It was reported for the identification of seven cohesive gene clusters with 22% (189 genes) of the genome differentially expressed during the cycle of development. The correlation of these gene clusters with hallmark morphological events during the cycle suggests three global stage-specific networks of gene regulation. The evaluation of the DNA microarray of the stage-specific transcription has been made by comparing global gene expression profiles at different time points during the developmental cycle. These time points (6, 12, 18, 24, and 36 p.i.) represent times associated with phenotype changes, and are known to occur from early to late [191, 208].

The research to decipher the function of genes in the chlamydial developmental cycle continues.

Thompson et al. 2005 [182] sequenced the genome of the *Chlamydophila abortus* (*Chlamydia psittaci var. ovis* – causative agent of abortion in sheep) strain S26/3. In the comparison of the genome of *C. abortus* with the genomes of other members of the family, Chlamydiaceae identified a high degree of conservatism with respect to the chain sequence and to the total content of genes. The genome is composed of 1,114,337 bp circular chromosomes with a total content of G + C of 38.8%. Typical for a representative of the genus Chlamydophila, the investigated *C. abortus* has only single copies of genes 23S, 16S and 5S rRNA, and unlike the members of genus Chlamydia, which have two copies of these genes. 961 coding sequences (CDS) are identified, giving an encoding density of 88%. Of those CDS, the functional engagement of 746 is indicated based on previous experiments or database. Of the remaining 215 CDS, 110 were significantly similar to the proteins of other members of the Chlamydiaceae, and in 15, no similar elements have been found. A total of 38 tRNAs have been identified, which meets all of amino acids, with the exception of selenium cysten [182].

The circular schematic representation of the chromosome of *C. abortus* shows that different genes are located in the respective circuits (1 and 2):

of the membrane surface structures; of the central and intermediate metabolism; of the degradation of macromolecules; of the information transfer and cell division; of the degradation of small molecules; of regulators; of the pathogenicity or adaptation; of the energy metabolism; unknown genes; and pseudo-genes.

So far there is no evidence of recent horizontal transfer of the genes of *C. abortus*, including a proven complete lack of phage genes, unlike other sequenced species of the genus Chlamydophila [170, 183].

Hypothetical scheme for the molecular events during the development of chlamydia in the cell is as follows [184]:

- In the *initial stage*, chlamydia enters through endocytosis. Two suggested actions: (a) release of previously synthesized proteins; (b) preventing the fusion with lysosomes.
- *~15 min. after infection (p.i.). Transformation of EB in RB.* The EB starts the protein synthesis [185].
- *~60 min p.i. Transformation of EB in RB.* Peak transcription of gene *euo* (early-stage gene having essential significance for initiation of the chlamydial developmental cycle). The same is undetectable for 8 hours p.i. Damage to histone (HC-1 – chlamydial protein, which is thought to play a key role in the condensation of the DNA during the formation of EB). Selected activation of the transcription (probably suppressing genes in places rich in adenosine-triphosphate); likely involvement of chlamydial protein SW1-B, which is supposed to play a role in condensation/decondensation of chlamydial DNA. Transcription of the proteins of the chlamydial inclusions Inc D, E, F, G and the putative proteins of inclusions CT 128, CT 228, and CT 147 (probable role in the endosomal fusion) [185–192].
- *1–2 h p.i. Transformation of EB in RB.* Reduction of the disulfide-linked MOMP to the level of detectable monomer. Lack of CRP (cysteine-rich protein). Accumulation of MOM during the cycle. Onset of the transcription of Inc D [193].
- *2–8 h p.i. Transformation of EB in RB.* Metabolic and "housekeeping" genes, some of which are specific to the early and middle phase of the cycle [189].
- *8–16 or more hours p.i. Division of RB.* Inc-proteins. Growths in the membrane of chlamydial inclusions. Inc A partially mediates inclusion fusion in *C. trachomatis*. RB-specific proteins include p52, *C. trachomatis* TroA and p242. Lipids of the exo-glycolytic pathway intercepted for incorporation into the expanding chlamydial endosome [194, 195].

- *8–16 or more hours p.i. Division of RB.* RB have no histone-like proteins. Diffusion of DNA that enables transcription and translation of the genes.
- *8–16 or more hours p.i. Division of RB.* RB lack Omp2 and Omp3 (outer membrane proteins), leading to the formation of fragile, but permeable structure of the wall [193, 196].
- *15 or more hours p.i. Transformation of RB in EB.* Probably stimulated by external factors? Up-regulation of histone proteins. DNA condensation. Up-regulation and cross-linking of Omp2 and Omp3 CRPs, perhaps involving CT780, CT783-encoding thioredoxin disulphide isomerases and probably involved in the exchange of disulfide bonds between cysteine-rich proteins and membrane. DNA condensation on histone proteins HctA, HctB (intermediaries in the late DNA condensation in chromatin) [191, 192, 197].
- *40 or more hours p.i. Release* Productive cycle only. By lysis. Possible Exocytosis?

Molecular biological research related to the regulation of the development cycle, the control of the expression of chlamydial genes in various forms of infection and the molecular response of chlamydia in relation to thermal and oxidative stress and other external influences are presented in a number of other publications [198–201].

5

Isolation and Cultivation of Chlamydia

5.1 Introduction

The isolation of chlamydia in a variety of biological systems has been the "gold standard" for diagnosis of chlamydia infections [9, 16, 18, 19, 23, 83, 214–216]. Historically and nowadays, there are three groups of biological systems for the isolation and cultivation of *Chlamydia spp.* – chicken embryos, cell cultures, and laboratory animals.

5.1.1 Chicken Embryos

Rake et al. 1940 [217] first reported that the agent of LGV multiplies in the yolk sac (YS) of chicken embryos (CE). Over the next decades and currently the CE method for isolation and cultivation of the causative agent is a major part of the diagnostic arsenal in the chlamydiology [20–22, 24, 54, 83, 215, 216, 218].

To this day, the assessment of Meyer et al. [219–221] that CE is the perfect biological model for chlamydial isolation remains valid and current. In the first place, all known chlamydial agents multiply in the YS of 6–8-day-old CE [19, 21, 23, 222, 223]. The obvious natural adaptation ability of chlamydia for culturing in YS allows for serial passage and maintenance of the reference and other strains over the years. An important finding is the preservation of the virulence of the agent in continuous passaging in the YS [18, 21, 23, 46].

Another important feature of the method is the possibility of obtaining a large amount of infectious material, which facilitates obtaining of antigens for diagnostic purposes and experimental work with different strains.

It is essential for the successful isolation and cultivation of chlamydia in the YS of CE to ensure the proper selection of the source material, its freshness, and its timely delivery to the laboratory [23, 215, 225, 228].

In need of delaying the procedure for the isolation, the field research material from animals is placed in a transport medium. The most frequently used medium is one containing sucrose, phosphate and glutamate (SPG) supplemented with 10% fetal bovine serum and antibiotics – kanamycin, vancomycin, and streptomycin (200–500 μg/ml), gentamicin (50 μg/ml) – and a fungal inhibitor (amphotericin B) [229]. An important prerequisite when working with materials suspected of chlamydia or with isolated strains is to work carefully in view of the zoonotic risk by complying with the rules on biosafety [230–232]. Compliance with certain rules when working with clinical specimens from humans (selection of suitable samples, the time for transportation, storage, and processing) is also important in attempts to isolate *Chlamydia trachomatis*.

The sensitivity of CE to chlamydia declines with age. Inoculated CE usually incubate at 37°C, but the development of *Chlamydia psittaci* improves at 39°C. Other researchers propose that *C. trachomatis* be cultured at 39°C and at high relative humidity. CE sensitivity to chlamydia and other microorganisms is reduced in certain seasons of the year. Role in this change play physiological, nutritional, environmental and other factors in the avian flocks and individual birds. In Russia and Bulgaria, the unfavorable season for cultivation in the CE is spring, while California has seen a decrease in the sensitivity of the CE in July and August. The feeding of birds with feed containing antibiotics also leads to deterioration of cultivation in the CE [24].

Despite its large-scale application, this method has not been the object of detailed analysis. According to some authors [210–213], the time of death of chick embryos after inoculation is in an inverse relationship with the concentration of chlamydia in the inoculum. Storz [19] has pointed out that individual chlamydial strains adapted to the YS of the CE have characteristic dynamics of mortality. Depending on the concentration of the agent in the inoculum, the CE die between the third and fourteenth day after the inoculation. Storz [19] believes that, provided a sufficiently large number of embryonated eggs are used, it is possible to analyze the infectiousness of the strain by recording the average day of mortality for a given chlamydial dilution.

Our objective was to study and analyze the inherent trends in the pathogenicity for CE upon infection in the YS of 22 chlamydial strains with which 5539 7-day CE had been infected, mostly for primary isolation of the pathogen in several nosological units: miscarriages in sheep; abortion in cows; keratoconjunctivitis in calves; polyarthritis in calves; arthritis in pigs; ornithosis and orchiepididimitis in man; and chlamydiosis in wild murine rodents – Cletrionomus glareolus and Apodemus flavicollis [46]. The inoculations were carried out with optimum concentrations of the chlamydial suspensions – from 1:10 to 1:1000, depending on the height of the passage. A total of 137 passages were performed. At the isolation of the strains in the YS we registered mortality of the CE from the first to the eighteenth day after the inoculation. The embryos died specifically between the fourth and sixteenth day after the infection. Individual CE died specifically at later periods as well (seventeenth to eighteenth days). In newly isolated strains, in some cases 100% or almost 100% mortality of CEs has been detected in the first passage. This model of lethality in the zero passage and in the subsequent early passages was accompanied, as a rule, by the discovery mainly of initial bodies (IBs), while there were only individual elementary bodies (EBs), or none at all. The accumulation of IBs and EBs increased after several passages in the yolk sacs. Virulentization of the chlamydial agent for CE was to be observed in the first 4–5 passages, parallel with the adaptation of the strain to the YS. In strains S, P, and R1 (bovine keratoconjunctivitis), for instance, the titers after the fifth passage fluctuated between 10^5 and $10^{5.8}$ ID_{50}/ml for 7-day CE, in strain D (ovine abortion) – $10^{4.5}$–$10^{5.1}$ ID_{50}/ml, etc. Upon comparative investigations of the pathogenicity for CE of native non-concentrated suspensions and purified concentrated suspensions, we established that with the same amounts of the inoculums, the concentrated suspensions cause death earlier – on the fifth or sixth days. Conversely, the moderate (10^4–10^5) and lower concentrations (10^2–10^3) of the suspensions yielded mortality primarily between the eighth and the sixteenth days.

The pathogenicity of the 22 strains for chick embryos was demonstrated by the overall number of dead embryos and by the data on the daily mortality of the CE. The results indicated that 4808 of the 5539 inoculated CE had died. Furthermore, 210 (4.36%) die by the third day, while 4606 (95.64%) died between the fourth and sixteenth days. The peak of the mortality for the 22 strains was the fifth day, followed by the sixth and seventh days.

The data obtained are presented graphically in Figure 5.1.

The curve shows two peaks of mortality. The first peak is the basic one, showing that the maximum pathogenicity of the chlamydiae for CE is on the fifth day after inoculation. The second peak is considerably lower than the first one, and reflects the pathogenicity of the chlamydiae on the twelfth day after inoculation of the CE. In addition to the peak moments in the pathogenicity, the graph shows that mortality is low during the first 3 days. Our morphological and antigenic investigations showed that no chlamydiae are to be detected during that period and no antigen is being accumulated.

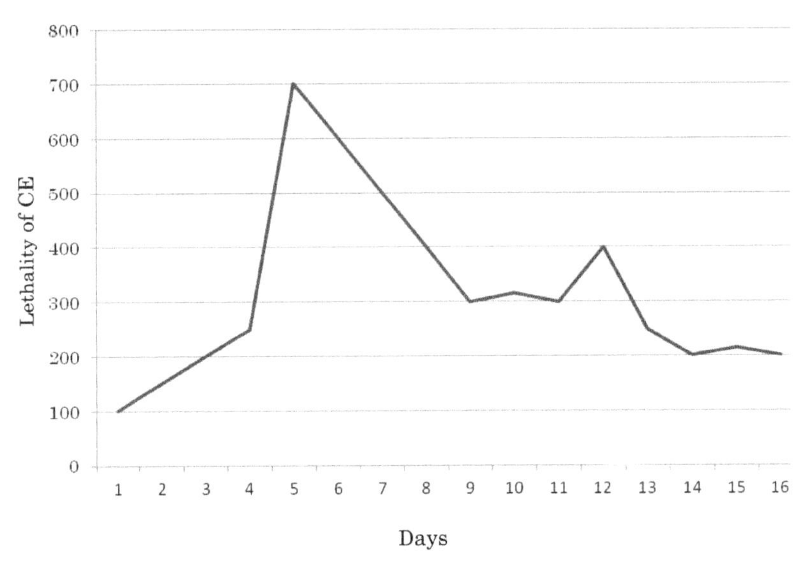

Figure 5.1 Lethality of chicken embryos upon infection with chlamydia. Summary data from 22 chlamydial strains from different animal species and from humans upon inoculation of 5539 7-day-old CE in the YS.

This signifies that during that period, mortality is not related to the chlamydial infection. A considerably lower level of CE mortality is to be established between the first and second peaks and after the second peak until the end of the observation, but the morphological and antigenic investigations have shown that chlamydial bodies and chlamydial antigen are accumulated during that period. The data presented definitely attest to the regularities in the pathogenicity of different chlamydial strains for CE. They have important practical value in the primary isolation and passages of chlamydia from clinical and pathological materials.

The infected CE show characteristic pathological changes. The membranes of the YS are hemorrhagic and thinned. The blood vessels are "injected" protruding on the wall of the YS. In some cases, gray-whitish or gray-greenish necrotic foci covered with caseous masses are detected on the surface of the YS. The yellow sacs are in a state of hyperplasia and hypertrophy. The trophoblastic villi of the YS are reduced. The yolk is liquefied and has a light yellow color. In well-adapted strains the YS are highly hyperemic. Carcasses of the CE that died from chlamydial infection show varying degrees of atrophy, hyperemia, cyanosis, petechiae on the legs, and hemorrhages in the head and body [18–24].

Infection of the chorioallantoic cavity with chlamydia is less successful than in the YS. A number of strains do not have a natural adaptive ability for allantoic cavity cultivation, so the attempts at isolation at output (zero) passage are usually negative or unsatisfactory, and usually require careful adaptation, repeated passages and advance, creating a high multiplicity of infections [23, 233]. The advantage of the method in a successful adaptation is that large amounts of chlamydiae possessing natural purity can be obtained, which are suitable for molecular, biological, immunological and morphological studies.

5.1.2 Cell Cultures

The chlamydial agents can infect more or less successfully different cell cultures (CC) – primary or permanent. There are more than 40 primary CC and cell lines in experiments on the cultivation of *C. psittaci* and *C. trachomatis*. For indication and identification of the agent, cytochemical, immunochemical, immunofluorescent, and electron microscopy (EM) methods have been used. The most widely used are the cytochemical methods for staining of infected CC by Giemsa, Gimenez, Ziehl–Neelsen, and Macchiavello, acridine orange, fluorescent-antibody staining, or other methods for detecting chlamydial intra-cytoplasmic inclusions.

For isolation of *Chlamydia abortus*, the most commonly CC used are McCoy (McCoy A – human synovial cells and McCoy B – mouse fibroblasts, BGM /buffalo green monkey cells/ and BHK-21 /baby hamster kidney). The activity of chlamydia can be strengthened after their inoculation in the CC by centrifugation (1000 to 3000g/1h, 30–35°C) [16, 235] and by chemical action on CC before or after their infection with the cycloheximide (0.5 μg/ml) in the maintenance medium, 5-iodo-deoxyuridine/IUDR/80 μg/ml for three days before infection, or emetin hydrochloride (1 μg/ml) for 5 min prior to infection. To prevent bacterial or fungal growth, gentamycin (10–50 μg/ml), vancomycin (100 μg/ml), and amphotericin B (1–4 μg/ml) were added [214, 215, 228]. Other treatments are also applied to CC to increase their sensitivity to infection: cortisone [241]; DEAE–Dextran [246]; irradiation with Co60, and others.

De Graves et al. [249] successfully cultivated *C. abortus* strain B577 isolated from ovine abortion in BGM and NCI-H292 cells and identified the agent through IF staining with monoclonal antibodies against the chlamydial lipopolysaccharide. Confluent monolayers were inoculated with *C. abortus* by centrifugation. There are a number of other examples of the successful use of the above methods to facilitate the infection of CC with chlamydia. In our laboratory practice, inoculation of concentrated and purified suspensions of *C. abortus* was carried out for infection of primary CC of CE fibroblasts (CEF) and the cell lines BHK-21 and HeLa [23]. The concentration and purification of the chlamydiae was made by using a method, including differential centrifugation, gel-filtration on Sepharose, and ultracentrifugation on a sucrose substrate [250]. The indication and identification of the agent was performed by direct immunofluorescence of preparations of cell monolayers and by EM [23, 50, 251].

For the isolation and cultivation of *C. psittaci*, cell lines BGM, McCou, HeLa, Vero (African green monkey kidney), and L-929 (mouse fibroblasts) have been used most commonly [252]. The BGM line is considered the most sensitive.

The isolation of *C. psittaci* is currently considered as a standard method for demonstrating an active infection in birds. An important methodological point is the use of fresh and well-preserved materials. It is recommended that the fecal samples and the cloacal and pharyngeal swabs be taken from live birds. After several days, these materials are supplied again to increase the probability of the proof of the infection. In the isolation and cultivation of *C. psittaci*, the aforementioned conditions are valid to enhance the efficiency of infection by centrifugation and the use of inhibitors.

In view of the zoonotic nature of chlamydiae of avian origin, efforts must be made in working with them to prevent the formation of dust and infectious aerosols and to eliminate the possibility of their inhalation [19, 53, 253].

The CC method is important for diagnosis of *C. trachomatis*, the study of its biological and morphological properties, and also for indication of the persistent infection and the effect of chemotherapeutical substances on the pathogen. For the cultivation of chlamydia, it is known that among the plurality of cell lines, only a certain part is sensitive to *C. trachomatis*. It is easier to cultivate highly virulent strains, such as from human patients with LGV. Lower virulence strains are hard to culture and have special requirements with regard to the culture medium and the functional status of the cells and their metabolism. To overcome the low biological affinity between the chlamydial agent and CC two groups of methods are applied. The first group includes the mechanical methods, which involves the application of the above-mentioned centrifugation of the chlamydial suspension on the cell monolayer. The second group consists of the biological methods. They use non-dividing cells that are in the stationary phase. These are the cells irradiated with gamma rays and treated with cytostatics, antimetabolites, polycations, or with certain types of polymers. A one-step developmental cycle of the chlamydia is achieved with these methods, which leads to completion of the production of mature EBs. Besides the inhibitors cited for *C. abortus* (IUDR, cycloheximide, emetin, and DEAE-Dextran) [234, 237, 239, 243], cytochalasin B [242], trypsin [245], and hydrocortisone [240] are also used. According Oriel and Ridgway [235] the effect of centrifugation lies in the fact that in the EBs under normal conditions, the contact between the cells and the chlamydial suspension leads to repulsion from the cells due to different electrical charges, while in centrifugation it is overcome. Mohammed and Hillary [236] offer a method for isolation and cultivation of *C. trachomatis* in McCoy cells treated with cycloheximide and polyethylene glycol and received a higher percentage of isolating the pathogen from clinical samples. Martinov et al. [105] inoculate suspensions of the cervical secretions in McCoy after pre-inhibition of the cells with IUDR and infection by centrifugation at 2800 for 1 h. After incubation for 72 hours, monolayers stained with Giemsa or with Lugol solution were assayed by LM. Kuo et al. [244] reported that rewashing Hela 229 cells with DEAE-Dextran (25 μg/ml) enhances infectivity and also may increase the size of the chlamydial inclusions.

Greene et al. [254] cultivated *C. trachomatis* and *C. muridarum* in less commonly used CC MS74 (endometrial endothelial cells) and NUVEC (human endothelial cells from the umbilical cord) and in HeLa cells. Has been established interesting biological feature, namely that infected with chlamydia cells continue to divide by mitosis and are resistant to development of apoptosis induction, although the extent of the antiapoptotic ability varied between serovars.

In general, the experimental data on the use of primary CC for isolation of *C. trachomatis* are few. Attempts have been made for culturing *C. trachomatis* in monolayers of primary human cervical CC [255]. They have been tested and some other primary CC, however, the interaction of *C. trachomatis* in CC has been studied mainly on the different cell lines. In some of our experiments with *C. trachomatis* used primary CC prepared from thyroid of sheep. The inoculations were carried out with suspensions of urethral secretions and scarified urethral mucosa of men with non-gonococcal urethritis [24].

Several teams of authors have expressed opinions that established procedures for isolation of the chlamydia are expensive and labor-intensive. As an alternative it is proposed to improve the culture media, consistent with the amino acid requirements of the agent (*C. trachomatis*, *C. psittaci*). The isolation and cultivation of chlamydia depends not only on the centrifugation and pretreatment of cells with cytotoxic drugs or other substances, but also by other factors in the culture medium. Allan et al. [247] studied the effects of certain amino acids with maturation and differentiation of morphological forms of chlamydia. When there is a decrease or lack of cysteine in the culture medium in *C. psittaci* and *C. trachomatis*, it results in a reduction of the infectivity. In another study, Allen and Pierce [248] established amino acid requirements of 4 strains of *C. psittaci* and 11 strains of *C. trachomatis*. For *C. trachomatis,* the addition of histidine to the medium is necessary. The requirements to methionine and tryptophan are different. All chlamydia need valine, as well as phenylalanine, glutamine, and leucine [248]. The amino acids are supplied in the medium with the calf serum, and it is recommended that it be a fetal calf serum. The role of reduced efficiency in the cultivation of chlamydia also has some methodological details, for example the materials from which the flasks are made. The plastic containers [256] as well as some types of tampons that have a toxic effect are unsuitable for this purpose. All these facts point to the need to seek new opportunities to improve the culture diagnosis of chlamydia.

The answer to the question about the sensitivity of the methods for isolation and cultivation of *C. trachomatis* in urogenital infections has been the object of many years of investigations. The parallel use of both methods, inoculation of CE in YS and CC, produced different results with the individual authors. In this connection, Taylor-Robinson and Thomas [437] noted that the success of a given method for the isolation of *C. trachomatis* varied in different laboratories and that a method considered as sensitive by some researchers could prove to be insensitive in the hands of others. Thus, for example, some authors believe that the CC method is some four fold more sensitive than isolation in CE. According to Ford and McCandlish [438], however, both methods yield equivalent results when used to isolate chlamydiae from NGU and RS-patients. Schachter and Dawson [20] reported that the two tecchniques, infection of CE in YS and CC, were equally sensitive when *C. trachomatis* was isolated from conjunctivitis and cervicitis cases. As regards urethritis in males, it seems that the CC-method is more sensitive [20]. Undoubtedly, the method for *C. trachomatis* isolation in CC has the advantage in the simultaneous investigation of a large number of samples. However, the difficulties linked with this method are well-known, and the preliminary treatment of CC with various means to raise their sensitivity towards chlamydiae in particular. The method for isolation and cultivation in YS of CE has the advantage that it makes possible a large number of chlamydiae to be obtained. However, the method is slower: a passage takes

15 days on an average, while at least 2–3 passages are required to accumulate sufficient concentrations of chlamydiae in order to get their LM-indication. An important methodological feature in our work should be noted here: the use of EM for an early indication and identification of *C. trachomatis* in YS, as early as the zero passage [146, 148, 165].

Information on the isolation and cultivation of the *C. pneumoniae* is scarcer and have another tissue orientation [238]. This is due to the difficulty in cultivating this chlamydial species in CC compare to *C. abortus*, *C. psittaci* and, *C. trachomatis*. Kumar [257] indicates that propagation of *C. pneumoniae* in CC has proven to be difficult. HeLA 229 cells and more recently the heteroploid HL cell line have been described as being more sensitive than McCoy for the isolation of this agent [257]. Murray et al. [96] also emphasize that the cultural diagnosis of *C. pneumoniae* is difficult. The organism does not grow in the cell lines used for isolation of *C. trachomatis*. *C. pneumoniae* can grow in the HEp-2 cell line, but this cell line is not used in most clinical laboratories. [96]. In the search for new possibilities for cultivation, it has been found that *C. pneumoniae* may be adapted to cultures obtained from human macrophages, endothelial cells, and smooth muscle cells of the aorta [258]. Later, Vielma et al. [259] reported on experiments in cultivating the agent in human aortic endothelial cells; Krull et al. [260] in human endothelial cells from other localizations; and Urima et al. [261], in human monocytic THP-1 cells and human epithelial HEp-2 cells.

Pospisil and Canderle, 2004 [262] discussed in detail the role of *C. pneumoniae* as the etiologic agent of diseases in domestic and wild animals and concluded that the isolations of the pathogen were achieved from 8 species and the strains have shown a high degree of similarity – up to 100% with a human version of the agent. Of livestock susceptible to infection with *C. pneumoniae* are the horses [263]. In Australia isolations have been achieved in diseases of koalas [264], giant frogs (Mixophyes iterates) [265], and western barred bandicoots [266], and in Africa, from another kind of frog, the African clawed frog (Xenopus tropicalis) [267], etc. In these studies, the cell line HEp-2 is primarily used, after pretreatment with cycloheximide or polyethylene glycol.

5.1.3 White Mice

Mice have long been used as an experimental biological model in the diagnosis and study of chlamydia of different origins. Mice should be obtained from colonies shown to be free of latent chlamydial infection. There have been several reports of subclinical chlamydial infections in mouse colonies [20].

A number of strains of *C. psittaci* (including current *C. abortus*) have been isolated in mice after intra-cerebral (i.c.) inoculation. This route has the advantage of not involving the respiratory tract, precluding the possibility of activating latent mouse pneumonitis [20]. It should be noted that the i.c. inoculation of mice is not as effective as the pulmonary by intranasal (i.n.) application. With few exceptions, the mammalian chlamydial strains do not affect the mice in i.c. inoculation, and so as a result of this clinical signs appear clinical signs [23, 268].

Intranasal infection of 3-week-old mice under ether narcosis with 0.02 mL of the materials is a useful method for the isolation of chlamydia from birds. Mice are particularly sensitive to the more virulent strains of human origin. The isolation of chlamydia from mammals via nasal infection of mice can often have a negative result, while in the CE it becomes successful. All known chlamydial strains infect mice, however, since their primary isolation has been on embryonated eggs and have high infectivity. They develop typical lung changes for several days after inoculation of chlamydia propagated in the YS of CEs and die 3 to 7 days after inoculation with the highly virulent strains. Mice infected with less virulent strains die between the 8th and 14th day or may survive by developing a latent infection. The lung changes were characterized by graying of the lungs, with foci of gaskets and involvement of lung hilus or the whole organ in the inflammatory process.

In the *intravenous* (i.v.) inoculation the toxin-producing (more virulent) chlamydia strains cause death of the mice as the curve of mortality has two peaks: the first, for 3–20 hours after inoculation, determined by the toxin associated with the pathogen, and a second, reflecting death resulting from the multiplication of chlamydia [152].

Most turkey, psittacine, and egret strains can be revealed by *intraperitoneal* (i.p.) inoculation. Those from chickens, ducks, and pigeons often produce enlarged spleens, but do not usually cause death. The i.p. inoculation of mice with toxigenic strains leads to infection and often to death. When used on weaker strains, the i.p route is less effective.

Oral (p.o.) and *subcutaneous* (s.c.) infection leads to protracted disease and latent infection.

Mice are a convenient model for para-clinical, pathogenetic, immunological and pathological studies [269–274].

Some authors present convincing evidence of major pathological changes in the blood vessels of people infected with chlamydia from mice. In one of the earliest studies in infections with *C. trachomatis*, Kuo and Chen [275] described a murine model of pneumonitis with four strains of trachoma and two strains from lymphogranuloma venereum. They found inclusions of the agents in the interstitial cells of the lung. The authors believed that mice can serve as a useful model for studying the pathogenesis and immunology of infections caused by *C. trachomatis*.

Harrison et al. [276] nasally infected mice with *C. trachomatis* isolated from children and registered interstitial and peribronchial inflammatory cell response early on days 4–6 with polynuclear cells and determination of IgM and IgG antibodies and death within 10–14 days. This animal model is reminiscent of infant lung infection with *C. trachomatis* in its pathological manifestation and can facilitate the understanding of chlamydia pneumonia in children. Other authors have established mouse models of cervicitis [277] and salpingitis [278] which are useful for studying the pathogenesis of genital chlamydial infections. Extensive experimental modeling of genital chlamydia infections are hiding also the agent of enzootic abortion in sheep. Rodolakis et al. [282] indicated that the murine models are very suitable for the assessment of the virulence of chlamydia isolated from ruminants.

A number of other publications over the last decade also testify to the effectiveness of white mice as an experimental biological system for the study of chlamydia. Rank [279] and Darvill et al. [280] have developed and characterized mouse models of chlamydial genital infections and described the main adaptive immune mechanisms acting in the ensuing infection. Li et al. [283] reported the mouse model of respiratory *C. pneumoniae* infection for a genomic screen of subunit vaccine candidates.

Montes de Oca et al. [281] inoculated intra-peritoneally pregnant mice Swiss OF1 aged 8–10 weeks with 106 IFU of *C. abortus* A577 and studied the development of the specific immune response against causative agents of the chlamydial abortion in sheep. Donati et al. suggested a mouse model for *C. suis* genital infection in 2015 [284] The authors applied intra-vaginal inoculations of mice with *C. suis* and *C. trachomatis*. This mouse model could be useful for comparative investigations involving *C. suis* and *C. trachomatis* species.

Various aspects of *C. pneumoniae* and the clinical condition which causes it have been studied on mouse models: pathogenesis and clinical evidence of pulmonary inflammation; the genetic relationship between chlamydia strains; virulence [211]; pathogenic properties; the role of this agent in the development of atherosclerosis, etc. [285–287]. Little et al. [288] reported on *C. pneumoniae*-induced amyloid plaques in the brains of mice similar to those observed in Alzheimer's disease in humans.

All these data show the substantial and versatile capabilities of mice as an experimental system for biomedical research of *Chlamydia spp.*

5.1.4 Guinea Pigs

Guinea pigs are also used as a biological model for the study of chlamydial infections in animals and humans. A typical and common illness for these animals is conjunctivitis with inclusions – guinea-pig inclusion conjunctivitis (GPIC) which is caused by Chlamydia caviae. Therefore, in some herds there is spontaneous chlamydial infection [289]. This infection should be excluded in guinea pigs to be used for experimental infection with chlamydia. Also, if the serum of guinea pigs with natural infection was used as a complement in complement-fixation test with chlamydial antigen, it can produce false positive results. To isolate chlamydia from clinical and pathological materials from domestic ruminants and turkeys, guinea pigs are inoculated intraperitoneally (i.p.). The infection by this method in many cases leads to fibrinous peritonitis, fever, and often death within 7 to 14 days. Guinea pigs are very suitable in attempts for a direct isolation of chlamydia from fecal samples, serving as a natural filter for these severely contaminated field materials. Infecting CE with organ suspensions of infected guinea pigs is used as a second phase of the work. For demonstration of chlamydia in pathological materials can be determined chlamydial CF-antibodies in serum of the infected guinea pigs. The direct indication and identification of chlamydia in the infected animals are carried out cytologically, by immunofluorescence or EM and polymerase chain reaction (PCR). Ornithosis human strains under i.p. inoculation led to febrile reaction after 48 hours, accompanied by anorexia and adinamiya [21]. Separate highly virulent ornithosis strains (Borg-15 strain and the agent of atypical pneumonia) cause death in guinea pigs at this kind of infection.

In the study of conjunctivitis with inclusions, it was found that this infection may have venereal character affecting female and male reproductive organs. Experimental GPIC infection of the guinea pig genital tract leads to an infection that closely resembles a genital tract infection with *C. trachomatis* both in male and female subjects [290, 291]. Mount et al. [290, 291] infected male guinea pigs in the urethra with the agent of GPIC, and the males sexually transmitted infection to females. These findings identified guinea pigs as an important experimental model for a chlamydial infection of the genital tract in humans [291–293]. De Clercq et al. [293] indicate that guinea pigs are a suitable animal model for the study of hormonal influences on genital tract chlamydial infection, since their female reproductive system closely resembles that of humans. There is a similarity in the length of sexual cycles in guinea pigs and women, and these animals and the women are both spontaneous ovulators and have an actively secreting corpus luteum [292–294]. Studies in guinea pigs and women have shown that estradiol makes these animals and humans more susceptible to chlamydial infections [295, 296].

Ocular and respiratory chlamydial infection may be useful in the study of similar diseases in humans [228].

Rank et al. [298, 299] studied humoral and cell-mediated immunity in guinea pigs. They found that both types of immunity are needed for the resolution of the chlamydial genital infection and to create immunity against re-infection. Rank et al. [297] induce pneumonia in newborn guinea pigs after intranasal infection with the agent of GPIC. Infection is characterized by enhanced respiration, radiographic evidence of pulmonary inflammation on day 6, chlamydial inclusions in the bronchial epithelial cells, and IgM and IgG antibodies with peak levels on day 12. This model chlamydial genital re-infection in guinea pigs, similarly to that in humans is short-lasting [300].

5.1.5 Rabbits

Patton et al. [301] reproduce limited acute salpingitis in rabbits by direct infection with *C. trachomatis* in the oviducts. Infiltration of the mucosa and sub-mucosa by polymorphonuclear leukocytes and lesions of the epithelial cells have been observed.

Fong et al. [302] described a rabbit model for *C. pneumoniae* respiratory tract infection and its possible role in the immunopathogenesis of atherosclerosis. Isolations of *C. pneumoniae* from the liver, spleen, and aortic arch indicate systemic infection or colonization as well as those of the lungs [302]. Another team of authors found that infection with *C. pneumoniae* accelerates the development of atherosclerosis and treatment with azithromycin prevents it in a rabbit model [303].

Iversen et al. [304] reported ocular involvement with *C. psittaci* of mammalian origin (strain M56) in rabbits inoculated intravenously. Bilateral anterior uveitis has been the major clinical symptom of the induced systemic chlamydial infection. Chlamydiae were recovered consistently by conjunctival swabbing from the fifth to the twenty-fourth day. The agent was present within the eye (viz. iris-ciliary body) in three of four rabbits killed at 15 days and in five of ten rabbits killed 60 days after inoculation. Chlamydiae had persisted in the cerebrum and joints as well [304].

The contemporary evidence on the use of rabbits in experimental work with animal and avian chlamydia is extremely scarce.

5.1.6 Cats

Kane et al. [305] have made an experimental model of human pelvic inflammatory disease in cats with the feline keratoconjunctivitis agent (*C. psittaci*). The inoculation of *C. psittaci* directly into the oviducts of eight cats produced an acute disease that was characterized by hyperemia of the tissue and polymorphonuclear leucocyte infiltration of the epithelium and subepithelial stroma. The lumens of the tubes contained exudates of desquamated epithelial cells and polymorphonuclear leucocytes. After about 30 days, the disease subsided leaving chronic inflammation in the tissue infiltrated with both polymorphonuclear and mononuclear cells. Fimbrial scarring and formation of adhesions have been apparent by 40 to 50 days after inoculation. Chlamydiae have been isolated in McCoy CC from most cats, in one for as long as 51 days after inoculation. Chlamydial inclusions have been seen in histological sections or smears of cells from the fimbriae of four of the eight cats. The peak of specific chlamydial antibodies have been on the 12th day. In this model, laparoscopy has been used for collection of specimens from both cats and diseased women [305].

5.1.7 Pigs

Pigs can be successfully experimentally infected with their natural pathogen *C. suis* to examine the pathogenesis, clinical manifestations, serologic response, and pathology of induced chlamydial disease.

Vanrompay et al. [306] demonstrated that pigs may be useful to study the pathogenesis, pathology and immune response of *C. trachomatis* genital tract infections. For this purpose, the authors used specific-pathogen-free (SPF) aged 16 weeks which was inoculated intra-vaginally with tissue-culture suspension of *C. trachomatis* serovar E, strain Bour or strain 468. As a result, an ascending infection develops, accompanied by replication of chlamydia in the epithelium of the cervix and uterus, inflammatory events, pathological changes, and humoral immune response. It is important to note that the replication of the pathogen in the test pigs was on the same specific target sites for a genital tract infection with *C. trachomatis* in women [306]. Regardless of economic considerations in the use of laboratory animals such as pigs, they have the advantage that they are physiologically and genetically closely related to humans [307, 308].

5.1.8 Primates

Monkeys showed good susceptibility to *C. trachomatis*. It should be borne in mind, however, that *C. psittaci* is endemic among lower primates. Between these two species of chlamydia exist considerable biological differences. Therefore, the results of experiments with *C. psittaci* on primates, regardless of their informational value, may relate carefully and critically to human pathology. The susceptibility of monkeys to *C. trachomatis* differs significantly. Baboons of species *Papio* and chimpanzee *Pan* species are more receptive than macaques (*Macaca* species), especially to varieties of chlamydia organisms which cause genital infections [24]. Several other species of non-human primates have been seen as potential animal models for the study of genital chlamydial infections. These include grivets (*Cercopithecus aethiops*) [312], marmosets [313], green monkeys [319], and others.

According to DeClercq et al. [293] the pig-tailed macaque (*Macaca nemestrina*) is the preferred primate model for genital chlamydial infections. The anatomy and physiology of the female reproductive tract are similar to those in humans. Their vaginal microflora closely resemble that of women, and as in women, they have a 28- to 30-day menstrual cycle. This macaque species is naturally susceptible to genital-tract infection with human biovars of *C. trachomatis*, and there is no need to pretreat the animals with, for instance, hormones to influence the infection [309–311].

Modern studies on primates as experimental models for chlamydia have a long history before. Back in 1909 Lindner [314] observed a disease similar to conjunctivitis in macaques and baboons, with inclusions in guinea pigs. The agent of this disease has caused infection of the cervix of the female animals. Darougar et al. [315] have infected the eyes of two baboons and the urethra of a third with *C. trachomatis* isolated from the rectum of the mother of a newborn affected by conjunctivitis with inclusions. Years later, Jacobs et al. [320] reproduced experimental urethritis in chimpanzees *Pan*. A chlamydial agent was re-isolated, and specific antibodies and seroconversion were found in the blood serum. Later these chimpanzees were re-infected in the nasopharynx, as a result of which they developed pharyngitis, but the serological response was weak, which was attributed to the low infectious dose [320]. Taylor-Robinson et al. [316] caused urethritis in experimental chimpanzee and for this purpose they used strains of C. trachomatis isolated from men with non-gonococcal urethritis.

In two other earlier studies, Thygeson and Mengert [317] and Braley [318] successfully infected the genital tract of monkeys and induced cervicitis.

Many other publications confirmed the usefulness of primates as an experimental model in studies of *C. trachomatis* and the similarities between the findings in the eye and especially genital chlamydial infections in monkeys and humans. Moller and Mardh [319] and Ripa et al. [312] isolated chlamydiae from diseased women with salpingitis and infected grivet monkeys with them. The latest developed acute salpingitis, pelvio–peritonitis, and also descending endometritis and cervicitis having characteristic classical clinical and pathological changes. The chlamydial pathogen was re-isolated from the monkeys in different periods of the experimental infection. Jonhnson et al. [313] produces experimental cervicitis in marmosets Conathryx after intravaginal inoculation of *C. trachomatis.*

Patton et al. [321, 322] proposed two experimental models in the pig-tailed macaque: an *in situ* model and a subcutaneous pocket model. In the *in situ* model, macaques are infected with *C. trachomatis* by cervical and/or intratubal inoculations and as a result, develop endocervicitis and salpingitis. The authors performed repeated reinfection with chlamydial agent that led to significant tubal damages and development of chronic salpingitis, and distal tubal obstruction. These consequences are similar to those seen in human pelvic inflammatory disease (PID) [323]. This model is suitable for studying the pathogenic mechanisms and treatment of PID, etiologically associated with *C. trahomatis.*

The so-called subcutaneous pocket model is established by autotransplantation of salpingeal and/or endometrial tissues into individual pockets made on the anterior abdominal wall of the macaque. Inoculation of *C. trachomatis* into the pockets produces acute infection similar to acute salpingitis or endometritis in macaques infected in the intact genital tract. From the infected pockets *C. trachomatis* can be re-isolated [322]. This model is considered to be useful for studying the dynamics of chlamydial infection and immune response.

From the foregoing it appears that experiments on primates represent a great contribution to research on sexually transmitted chlamydial infections. Shortcomings that should be highlighted include the high cost of the monkeys and complexity of work with them, and also some considerations of the ethical nature [24, 293, 324].

5.1.9 Other Animals

Laboratory and wild rats, cotton rats, susliks, Syrian hamsters, and deer mice have been used significantly less often and with varying degrees of success for the isolation, cultivation and identification of chlamydia. Different species of birds (pigeons, turkeys, parakeets, sparrows, rice birds, etc.) were used in order to study and evaluate their pathogenic properties.

5.1.10 Identification of Chlamydial Isolates

Isolates obtained from clinical and pathological materials of animals and humans, regardless of the biological system for isolation of the agent, are subject to mandatory identification to confirm their chlamydial nature [19, 23, 83, 325, 326].

The classical scheme for identification of a chlamydial agent was formulated precisely by Storz [19]:

- Identification of characteristic intracytoplasmatic inclusions, chlamydial EBs, or their earlier developmental forms;
- Analysis for the behavior of the infectious agent in indicator hosts. This should involve at least the behavior in CE after YS inoculation;
- Demonstration of group-specific chlamydial antigen in the complement-fixation test using serum with known chlamydial antibodies;
- Tests for the elimination of rickettsiae, mycoplasma, or other bacteria and viruses in the stock preparation considered to be a chlamydial agent.

Light microscopic examination of smears of placentas and cotyledons stained through the above-quoted methods is used for identification of *C. abortus* (ovine chlamydiosis). Upon sufficient experience, they give satisfactory results. Much more reliable, accurate, and demonstrative are the EM methods which detect typical chlamydial intra-cytoplasmatic inclusions and bodies [23, 147]. Fluorescent antibody tests (FATs) using a specific antiserum or monoclonal antibody may also be used for identification of *C. abortus* in smears [326]. The aforementioned methods are also used in the process of identification of the isolates in chicken embryos, CC or laboratory animals. Demonstration of chlamydial-group antigens in boiled YS suspensions or organ extract of infected mice is an indisputable proof for the identity of a chlamydial agent [19, 23]. In histopathological sections of placentas, antigen detection can be carried out using commercially available anti-chlamydia antibodies directed against LPS or major outer membrane protein (MOMP) [327]. Amplification of chlamydial DNA by PCR and real-time PCR provide alternative approaches for verifying the presence of chlamydiae in biological samples without resorting to culture. Based on the recent big progress in molecular microbiology, opportunities have been provided for acceleration of the laboratory diagnosis by direct identification of chlamydial agent in the sample, rather than a continuous method of isolation in CC and its subsequent identification. DeGraves et al. [328] and Jee et al. [329] described methods for discriminating between amplified DNA sequences originating from *C. abortus* and *C. pecorum* in cattle.

The above-described complex of tests is applied for identification of isolate *C. psittaci* (avian chlamydiosis) in CE and CC [19]. In order to avoid risky actions on isolation of the zoonotic agent from avian tissues, PCR techniques with preliminary inactivation of the respective sample can be used. Current PCR tests for detection of *C. psittaci* target the ompA gene or the 16S–23S rRNA gene [330].

With regard to the urogenital isolates of *C. trachomatis*, it should be noted that CC detection of *C. trachomatis* by LM is highly specific if a *C. trachomatis* MOMP-specific stain is used. According to the recommendations of the CDC [325], despite the technical difficulties, CC, when performed by an experienced analyst, was the most sensitive diagnostic test for chlamydial infection until the introduction of nucleic acid amplification tests (NAATs) [325, 331, 332]. There is no doubt that NAATs as a product of the technical evolution in laboratory research over recent years will contribute to the acceleration the identification of *C. trachomatis* in human urogenital infections [325].

5.2 Isolation of Chlamydia from Different Animal Species and Humans

In this section are presented isolations of chlamydial agents in our own studies on animals and humans affected by various clinical conditions.

5.2.1 Strains Isolated from Sheep

5.2.1.1 Abortions, premature births, stillbirths and non-viable lambs

In the examinations of clinical and pathological materials (placentas, cotyledons, utero-vaginal extrudes, fetal parenchymal organs) of 176 sheep originating from 81 settlements in the 9 districts, 57 (32.38%) were positive for infection with *C. abortus*. These results were received with the use of methods for the direct detection of the causative agent (DEM, LM, IF) and its isolation in YS of CE. As a rule, attempts to isolate the agent were preceded by its direct indication in the above methods. Often the examinations on the above materials were preceded or accompanied by serological tests for chlamydial antibodies. The total number of isolates of *C. abortus* was 21. The positive culture results were obtained in flocks of sheep from 21 settlements.

All isolates grew well in the YS of CE, showing a characteristic curve of lethality between days 4 and 15 after inoculation. In some cases the accumulation of chlamydial bodies in the YS cells was satisfactory even in primary isolation. In other cases, this was achieved after one to three consecutive passages. An important prerequisite for faster adaptation of a chlamydial isolate was the use of a bacteriological sterile starting material with a higher multiplicity of infection. The conclusion to be made on the basis of these studies and the many years of experience in the isolation and cultivation of chlamydiae in the YS of embryonated eggs from cases of abortions and related conditions is that these strains exhibit a natural adaptation capacity to that mode of cultivation.

The realized isolation of a given strain of *C. abortus* is demonstrated by several methods of indication and identification (LM, EM, IF, serological detection of chlamydial antigen, the pattern of infectivity of the CE in the inside yolk inoculation). Particularly demonstrative is the electronic microscopic method (Figure 5.2).

The electron micrograph of ultra-thin sections of YS show the typical ultrastructure of *C. abortus* – EBs, condense bodies (CB), and reticular bodies (RB) and other characteristics of chlamydial infection in the infected cells.

Electronmicroscopical morphology of chlamydia in negatively contrasted preparations of YS of CE is shown in Figure 5.3.

It is seen that both EM methods are characterized by a certain morphologic picture that strictly specific to each of them. The ultra-thin sections method provides us with broad possibilities to study the ultra-structural features and morphogenesis of *C. abortus* in CE and the changes in the infected cells. The negative staining method allows for rapid morphological identification of isolates and conducting systematic and accessible EM control over the cultivation of strains. In a series of experiments with different strains of abortion and other forms of fetal loss, we found a lack of natural ability adaptation of the same to cultivation in the allantoic cavity of CEs.

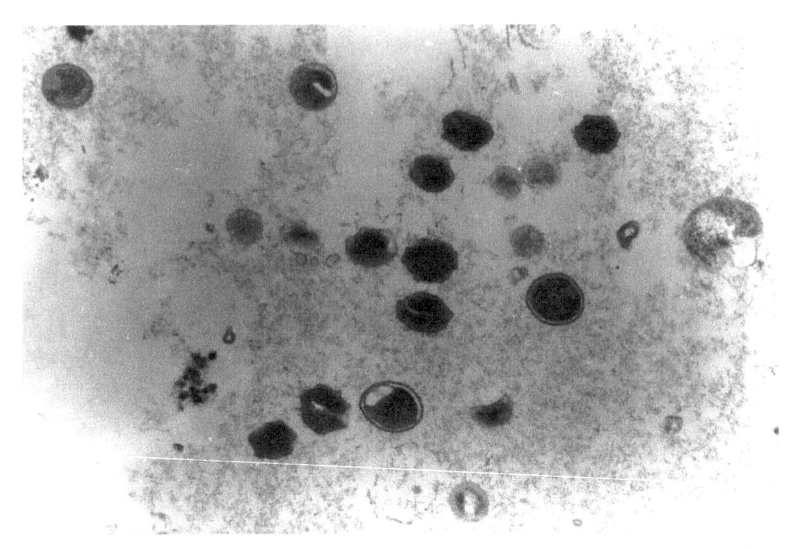

Figure 5.2 Electron microphotograph. Chlamydia abortus in ultra-thin section from yolk sac of chicken embryo. An isolate from placenta of aborted sheep. Multiple intracellularly arranged elementary bodies. Magnification 15,000×.

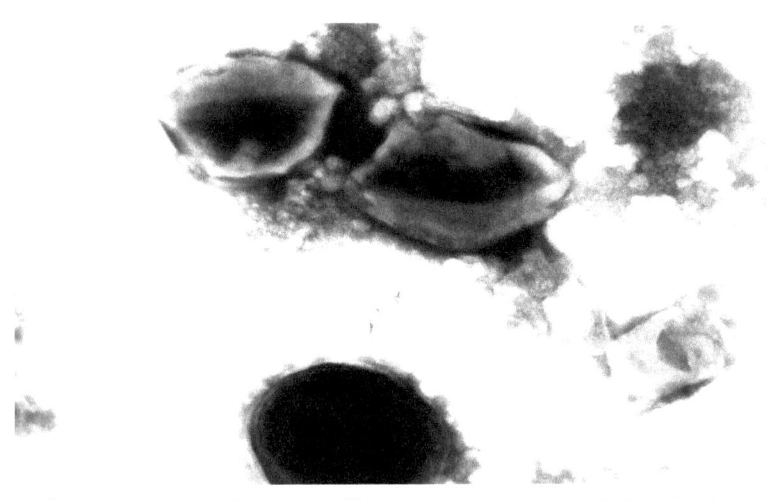

Figure 5.3 Electron microphotograph. *Chlamydia abortus* in negatively stained suspension from an isolate in YS of CE from the internal parenchymal organs of lamb born not viable. Magnification 50,000×.

More positive results were obtained when inoculating the concentrated and purified suspensions having infectious titers higher than LD_{50} log 5×10^6/ 0.5 ml daily for 6–7-day-old chicken embryos [23, 49, 54, 333].

Table 5.1 presents the results of experimental inoculations of white mice with starting field materials and three strains of *C. abortus* isolated from them.

Mice aged three weeks inoculated in ether narcosis with 0.02 ml of the suspension. The table shows that attempts to direct isolation of *C. abortus* from placental and fetal material at nasal isolation were unsuccessful. At the same time, the highly infectious strains in the YS of embryonated eggs obtained from cotyledons and placentas infected mice who developed pneumonia with typical clinical signs, pathological lesions, and mortality in some of them. The agent was discovered without disturbance in the pneumonic areas (LM, IF EM) and it was re-isolated in YS of CEs.

The intravenous route of administration of infectious suspensions, as well as from field materials and the isolated strains in YS proved to be most effective. The i.c., i.p., s.c., and oral inoculation were ineffective for

Table 5.1 Pathogenicity of *Chlamydia abortus* isolated from sheep with abortion and related conditions of pathology of pregnancy for young white mice at different ways of inoculation

Field materials. Strain in YS of CE	i.n.	i.c.	i.v.	i.p.	s.c.	p.o.
Placenta, cotyledons (outbreak GV)	(−)	(−)	(+)	(−)	(−)	(−)
Strain GV	(+)	Native susp. (−) Conc.susp. (+)*	(+) (+)	(+) 2% (+) 10%	(+)10%*** (+)20%	(+) 5%*** (+) 25%
Placenta, cotyledons (outbreak Ravno pole)	(−)	(−)	(+)	(−)	(−)	(−)
Strain Ravno pole	(+)	Native susp (−) Conc.susp (+)**	(+) (+)	(+)8% (+) 15%	(+)15% (+) 25%	(+)10% (+) 30%
Fetal organs (outbreak Chelopechene)	(−)	(−)	(+)****	(−)	(−)	(−)
Strain Chelopechene	(+)	Native susp. (−) Conc.susp. (+)**	(+) (+)	(−) (−)	(−) (−)	(−) (−)

Legend: *neurological symptoms with no mortality in 15% of mice;
**Death in 10%; Without neurological symptoms in 10%;
***Protracted course leading to latent infection;
****Lethality in 20% on the 12 to 14 day; Survival and development of the latent form at 80%.
Inoculations: i.n. – intranasal; i.c. – intracerebral; i.p. – intraperitoneal; s.c. – subcutaneous; p.o. – oral.

isolation of the agent from placentas, cotyledons and organs of fetuses. Only two strains infected mice with unequal coverage and a low-grade level of clinical manifestations. The concentration of the suspensions and the creation of a high multiplicity of infections increased the virulence of the inoculation material. The fetal parenchymal organs due to the low concentration of chlamydia were less suitable compared to placentas as material for direct isolation in mice (Table 5.1).

Three strains of *C. abortus* were isolated directly in seronegative for chlamydia guinea pigs from the feces of latently infected sheep reared in herds where they are established chlamydial abortions. The i.p. inoculations led to febrile illness, weakness, toxic-infective syndrome, and death after 10–14 days. We observed macroscopically strongly expressed fibrinous peritonitis, splenomegaly, and inflammatory-necrotic foci in the liver and lungs. The presence of the infectious agent was detected with the help of LM observations of the impression preparations of spleen and gray-yellowish taxed on the surface of the same and other parenchymal organs.

Guinea pigs proved suitable biological models for primary isolation from placentas and fetuses after i.p. inoculation, as well as for passaging of already-propagated YS chlamydial isolates.

In the experiments of culturing *C. abortus* in CC, we used two methodological approaches. The first is characterized by a pre-treatment of the CC with iodine-deoxyuridine (IUDR) to increase its sensitivity to infection with chlamydia. As a result of this treatment in the used cell line BHK-21 there was a significant accumulation of chlamydial cells in the form of inclusions in the cytoplasm of the infected cells.

The second methodological approach consisted in inoculation of CC (CEF- chicken embryonal fibroblasts, BHK-21, and McCoy). That way of infection led to a significant or strongly expressed chlamydial infection accompanied by cytopathic effect (CPE) and a large accumulation of chlamydia. The infected cells are found without any difficulties using LM (Giemsa, Gimenez), IF, and EM. Figure 5.4 shows immunofluorescence indication of C. abortus, strain GV at the 48-hour primary CC CEF infected with chlamydia in a concentration of $5 \times 10^8 - 5 \times 10^9$. We used a multiplicity of infections of 10 to 100 LD_{50} per cell in which there were no toxic and degenerative changes in the cells. Pronounced immunofluorescence-granular antigen with a specific IF light with bright yellow-greenish color found in about 75–80% of the cells.

Based on the positive experience of the infections of CC with yolk suspensions of *C. abortus*, we investigated the possibility of direct isolation of the

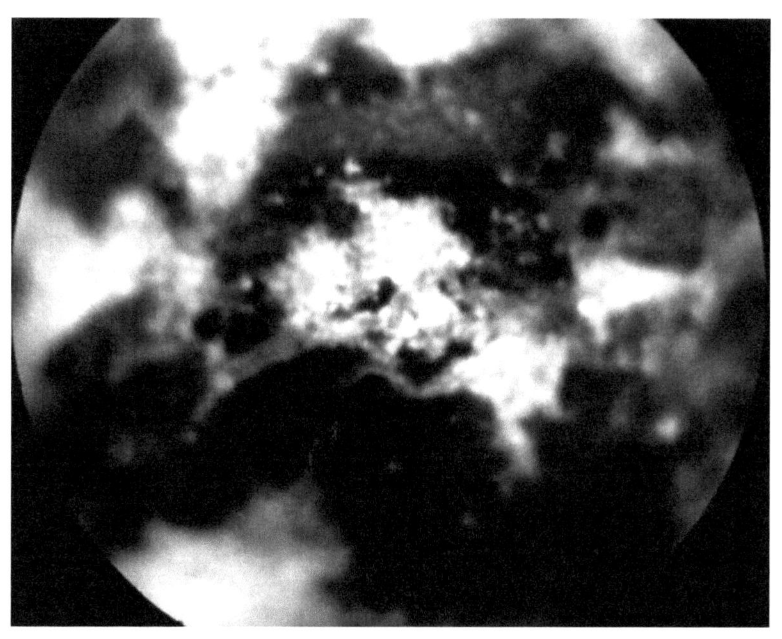

Figure 5.4 Immunofluorescence microphotograph. Cell culture CEF (48th h), infected with concentrated and purified suspension of *Chlamydia abortus* strain GV (chlamydial abortion in sheep) and stained by direct immunofluorescence-method with anti-chlamydial labeled gamma globulin. Magnification: Lens 9×, eyepiece 7×.

agent in CEF, BHK-21, and McCoy cells from field materials in chlamydial abortion – cotyledons, placentas, and parenchymal organs of the fetuses. The method was effective at compliance with the following rules:

Preliminary EM- or LM- control of the starting materials for the establishment of a high enough accumulation of EBs.

- Increasing the opportunities for penetration of chlamydia in cells by preparing a concentrated and purified suspension, or by pretreatment with inhibitors of the CC – IUDR, Cycloheximide, DEAE-Dextran and others.
- Selection of clinical and pathological materials based on the fact that the highest native concentrations of chlamydia there are in cotyledons; secondly, in the placental tissue; and third, in the fetal parenchymal organs.
- Upon conducting cultural studies in two cases, we found mixed infection – *C. abortus* and *Coxiella burnetii*. The 20 farms negative for

chlamydiosis infections were diagnosed with the other important abortifacient agent – Coxiella burnetii. Chlamydiae and coxiellas were not found in the remaining 41 settlements.

5.2.1.2 Pneumonia in sheep

From pneumonic lungs we received three isolate chlamydiae – one each from an adult sheep, lambs a year old, and adult lambs. These were animals with massive pneumonia, which were slaughtered by necessity or died due to the disease. The pathological material originated from sheep farms with serological evidence of chlamydial infection [54].

Laboratory tests were performed on two biological systems: 7-day CE and 3-week-old white mice. Direct isolation YS of CE was achieved in one case (from the lungs and mediastinal lymph nodes of lambs) even at the start / zero / passage, which was demonstrated by LM, IF and by obtaining specific antigen. The other two strains were needed for 2 to 4 passages in YS of CE to achieve a satisfactory accumulation of chlamydial EBs.

Attempts for direct isolation in mice after intranasal inoculation of suspensions of inflamed lung tissue and regional lymphatic nodes were successful with the materials of lambs and lambs a year old with respect to clinical and laboratory results – the development of a disease resulting in death after five to 10 days, demonstrating agent in the impression preparations of lung and spleen and re isolation in CE. Inoculations with the materials of sheep did not lead to clinical consequences or a positive laboratory result. Obviously the first two isolates possessed higher virulence for white mice in intranasal inoculation. All three isolates in yolk sacs, however, were pathogenic for mice in this method of inoculating the infectious agent.

To monitor the excretion of chlamydia of ovine suffering from chlamydial pneumonia, we made an attempt to isolate the agent from the feces. Thirty days after the resolution of clinical symptoms, in a fecal sample of a sheep was found the presence of C. abortus by its isolation in seronegative guinea pig at i.p. inoculation. In the other two experiments on the 50th and 75th day after the clinical recovery, we found a slight release of the pathogen on the 50th day and no release on the 75th day.

5.2.1.3 Conjunctivitis in sheep

Two chlamydial strains were isolated in YS of CE from sheep and lambs having clinical size of follicular conjunctivitis. An etiologic orientation towards making isolation attempts have given us the positive for chlamydia findings from LM investigations of smears from scarified conjunctival mucosa. We found typical chlamydial inclusions in the epithelial cells.

5.2.1.4 Arthritis in sheep

We focused on the two farms with a stationary chlamydial infection, where part of the sheep and young lambs showed clinical evidence of arthritis or polyarthritis, often accompanied by conjunctivitis.

From synovial fluids of badly affected joints we have made preparations for cytological observations. In cases of severe infection, direct detection of chlamydial inclusions was possible. Synovial fluids that were cytologically positive for chlamydia were used to inoculate KE or CC BHK-21.

The process of isolating and culturing the strain in PchAO in YS of CE is presented in Table 5.2. Summed data for the primary isolation and subsequent two passages in CE showed that the embryos died between the first and thirteenth day after inoculation, as the so-called nonspecific mortality (1–3 days) was 7.5%. The lethality as a result of the specific pathogenic effect of the agent was 92.5%. Maximum number of KE died between the sixth and ninth day. The indication and identification of the strain was achieved by LM, IF, and EM and serologically by demonstrating the group-specific chlamydial antigen in the CFT with a known anti-chlamydial serum.

The chick embryo-propagated strain was then cultured in BHK-21 cell culture, pretreated with IUDR. In parallel made inoculations of a laboratory strain R787, also isolated from the joints of sheep with arthritis [54]. The strains were grown well in a cell line and the largest accumulation of the pathogen detected in 48–72 hours (Figure 5.5).

Direct inoculation of synovium in CC BHK-21 (IUDR pre-treated) also gave a positive result, but the accumulation of the agent was less at the zero passage. This necessitated repeated passages in YS of CE to obtain higher multiplicity of infection.

Table 5.2 Isolation and cultivation of chlamydia strain in PchAO from the synovial fluid of a sheep with arthritis

	Inoculated CE	Dead CE	Days and Daily Lethality of CEs												
Passages	(No.)	(No.)	1–3	4	5	6	7	8	9	10	11	12	13	14	15
0	30	30	3	2	2	4	5	5	4	2	1	1	1	0	0
I	30	30	2	2	2	4	5	6	6	1	1	1	0	0	0
II	20	20	1	1	1	3	3	4	4	2	1	0	0	0	0
Total	80	80	6	5	5	11	13	15	14	5	3	2	1	0	0

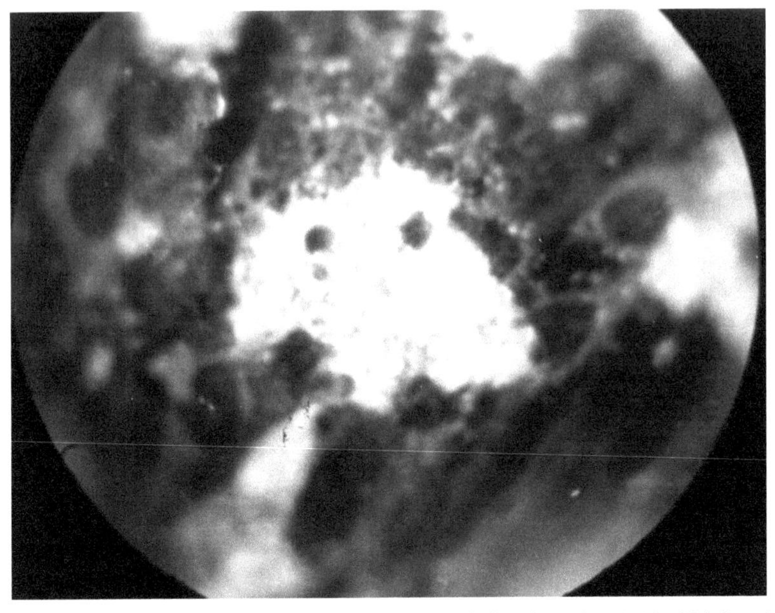

Figure 5.5 Immunofluorescence microphotography. Cell culture BHK-21 (48th hour), pre-treated with IUDR and infected with chlamydial strain PchAO isolated in embryonated eggs from a joint of a sheep with arthritis. Direct IF staining method with anti-chlamydial conjugated gamma globulin. Magnification: Lens 9×, eyepiece 7×.

5.2.1.5 Orchitis and orchiepididimytis

C. abortus strain was isolated from the semen of serologically positive for chlamydia ram showing symptoms of orhiepididymitis. The isolation was achieved in YS of CE. Other attempts to isolate the pathogen were not taken.

5.2.2 Strains Isolated from Goats

5.2.2.1 Abortions, stillbirths and non-viable kids

Five strains of C. abortus were isolated from goats in foci of chlamydial infection in five locations of three districts (SF, Pk, Lc): three from the placentas after abortion and stillbirth and one from the internal parenchymal organs of kids which were born puny, unviable, and died in the next three days [167]. The strains were still identified in the primary passage. Cultivation was very good in YS of CE.

In one of the outbreaks of abortions and related conditions (Trojan), parallel to the experiments on the isolation of the agent, conventional PCR tests of placenta, liver, and lung from a stillborn kid were performed and resulted positive for chlamydia amplification products in all the three pathological materials.

The pathogenic properties of strain Trojan, adapted for propagation in CE, investigated in newborn white mice (NWM) by i.c. inoculation and in 3-week-old white mice by six ways of inoculations [167].

I.c. inoculation of NWM caused per-acute disease and 75% mortality. Infection by the i.c. route led to nervous symptoms and lethality in about 20% of young white mice. The nasal or intravenous inoculations were also effective. Splenomegaly is established upon infection by i.p. route in the vast majority of cases and the death rate was 5%. Subcutaneous and oral administration of the agent was followed by a protracted course of disease and a mortality of less than 10%. Most often the mice survived and developed a latent form of chlamydiosis. The foregoing results show a significant extent of similarity with the data of pathogenicity of *C. abortus*, isolated from sheep with abortions and related conditions.

Strain G. Malina, isolated from the placenta of an aborted goat, tested for pathogenicity on sero-negative for chlamydia rabbits (Table 5.3). Five groups were formed, each consisting of three animals [167].

Table 5.3 Pathogenicity of chlamydial strain G. Malina isolated from aborted goats in rabbits in various methods of inoculation

Group Rabbits	A. Type of inoculum: placentas. Type of inoculation	B. Type of inoculum: strain into CE. Type of inoculation	Clinical Effect		Serological Response	
			A	B	A	B
Pregnant rabbits	i.v.	i.v.	t° >, abortions	t° >, abortions, stillbirths	1:128–1:256	1:256–1:512
Male rabbits	i.v	i.v	t° >, common signs	t° >, common signs	1:512–1:1024	1:256–1:512
Male rabbits	i.p.	i.p.	t° >, common signs	t° >	1:512–1:1024	1:256–1:512
Pregnant rabbits		s.c.		(–)		1:8–1:16
Male rabbits		s.c.		(–)		1:8–1:16

The first group included pregnant rabbits (10–15 days of gestation) that were injected twice (i.p.) during 5 days with 1 ml concentrated and purified suspension of placental tissue and cotyledons of aborted goat (A). The three animals developed fever lasting 2–3 days and aborted between the sixth and ninth day after inoculation. Their serological response was expressive – occurrence of CF antibodies on day 6 (1:8–1:16) and seroconversion (1:128–1:256) three weeks later.

The second group also consisted of three pregnant rabbits inoculated similarly (i.v.), but with a yolk suspension of the strain. The clinical effect was similar – two miscarriages and one stillbirth. Specific antibodies with similar performance but with higher titer values (1:16–1:32 – sixth day; 1:256–1:512 – after 3 weeks) were titrated in the blood serum of the rabbits (B).

The third group consisted of three male rabbits inoculated i.v. with 2.5 ml of concentrated and purified suspension from field material. The clinical follow-up showed a progressive increase in body temperature of the second to sixth day (40°C), and then started the reverse process. Delayed motor activation, pricking, rapid breathing and heartbeat, and conjunctivitis were registered during the febrile stage. The titers of chlamydial antibodies were high – 1:512–1:1024.

The fourth group consisted of three male rabbits injected at the same dose of the strain cultured in CE. The clinical consequences and the dynamics of specific antibodies were very similar to those described in the third group.

The fifth group included two pregnant rabbits and two male rabbits. Each animal inoculated s.c. with 3 ml of yolk suspension of the strain. By the fourteenth day, visible manifestations of the disease were lacking in all rabbits and serology was negative. On the 15th day we made re-inoculation with the same dose. The clinical effect was again negative, and serological response was expressed in the presence of CF-antibodies with low titers (1:8–1:16) of the thirtieth to fortieth day [167].

5.2.2.2 Pneumonia in goats

The TR90 strain was isolated in YS of CE after inoculations of suspensions obtained from pneumonic lungs of sero-positive for *C. abortus* goats. The strain adapted quickly to propagation in YS. The same was identified morphologically using light and electron microscopy and serologically through proof of chlamydial antigen. Strain TR90 was used successfully for experimental intranasal infection of goat with a view to tracking her serological response (see section serology) [167].

5.2.2.3 Arthritis in goats

Punctate synovial fluid was extracted from a goat with a mono-arthritis-affected knee and was inoculated into the YS of CE. As a result of that, specific mortality of the embryos was caused, and chlamydiae with typical morphological features (strain HR-1) were found in the separated yolk sacs. Second chlamydial strain (HR-2) was isolated from the tarsal and carpal joints of a goat slaughtered by necessity because of severely hampered movement due to arthritis. Morphological differences between the two strains were not observed. The bacteriological controls of the inocula were negative [167].

Immunological differentiation of strains C. abortus *isolated from goats, sheep and cattle affected by arthritis and miscarriages.*

We conducted experiments on the immunological identification and differentiation of chlamydial strains isolated from cases of arthritis in goats, sheep, and cattle and in parallel to the strains isolated from miscarriages in the same species. Studies performed on guinea-pig model selected as sero-negative for chlamydia. We formed six groups, the animals in each infected i.p. with one of the six strains (Table 5.4). Thirty days later, we performed new infections, but individually on three pigs from the groups with each of the remaining five strains and clinically counted the presence or absence of disease after reinfection, manifested by the typical guinea pig fever. The table shows the proportion of laboratory animals that do not become ill after reinfection with the other strains, indicating full mutual cross-immunity (bovine, ovine, and caprine chlamydia of miscarriages in the three types of animals). In CFT these strains were indistinguishable. The latter are antigenically different from chlamydia isolated from cows, sheep, and goats with arthritis who after reinfection with the abortifacient strains developed fever lasting 1–2 days. The antigenic differences were bigger among the goat strain (abortion) and the bovine (arthritis).

Thus the experiments enable the establishment of specific antigenic relationship between these strains in which a close relationship between chlamydial isolates from abortions existed on the one hand, and isolates from arthritis in ruminants on the other (Table 5.4).

These results confirm the findings of an earlier study on the close relationship between chlamydial strains of abortions in cattle and sheep and strains causing arthritis in the same species [338].

Table 5.4 Specific antigen relationships between chlamydia strains isolated from abortion and arthritis in sheep, goats, and cattle

First i.p. Infection. Strain	Second i.p. Infection after 30 days. Chlamydial Strain											
	GV – Abortion Sheep		Troyan – Abortion Goats		Sh.d. – Abortion Cows		R787 – Arthritis Sheep		HR – 2 – Arthritis Goat		GD – Polyarthritis Calves	
	Guine Pigs No.	Days of Fever No. /%	Guine Pigs No.	Days of Fever No. /%	Guine Pigs No	Days of Fever No. /%	Guine Pigs No.	Days of Fever No. /%	Guine Pigs No.	Days of Fever No. /%	Guine Pigs No.	Days of Fever No. /%
GV	–	–	3	0/0	3	0/0	3	1/0,3	3	1/0,3	3	1/0,3
Troyan	3	0/0	–	–	3	0/0	3	1/0,3	3	1/0,3	3	2/0,6
Sh.d	3	0/0	3	0/0	–	–	3	1/0,3	3	1/0,3	3	0/0
R787	3	1/0,3	3	0/0	3	1/0,3	–	–	3	0/0	3	0/0
HR – 2	3	1/0,3	3	0/0	3	1/0,3	3	0/0	–	–	3	0/0
GD	3	1/0,3	3	2/0,6	3	0/0	3	0/0	3	0/0	–	–
Control uninfected group	3	19/6,3	3	21/7	3	15/5	3	14/4,6	3	16/5,3	3	13/4,3

5.2.3 Strains Isolated from Cattle

5.2.3.1 Abortions, stillbirths and non-viable calves

We investigated clinical and pathological materials of 101 cows with abortion and deliveries of dead and unviable calves which were reared in 40 settlements of eight districts. The obtained isolates were 11 from 11 settlements in four districts (SF, Sofia-city, Pk, Kn). Isolations achieved after intra-YS inoculations of suspensions of placental tissue include seven cases of cotyledons, three cases of internal parenchymal ortgani of fetuses, and one isolate of utero-vaginal extrudes [54].

All positive culture materials were previously shown to be infected with *C. abortus* by LM, DEM or IF. Some of the farms from which the cows originate had positive serology for chlamydial infection.

Cows affected by abortions and related conditions strictly comply with the rules for intra-yolk introducing infectious material with the highest possible multiplicity, particularly in pre-detecting that the concentration of the agent is not high. This is especially true for the fetal parenchymal organs where the number of chlamydia is always less than in cotyledons and placenta. When working with fetuses, the approach to the preparation of suspensions is in the use of a large number of organs with subsequent concentration and partial purification by differential centrifugation. The strains were cultured in the YS of CE in the normal model with specific mortality between the fourth and the fifteenth days and typical lesions in the infected embryos.

Attempts for the isolation and cultivation in the chorioallantoic membrane of CEs included inoculations with the stated field materials and yolk suspensions of already isolated strains. We observed the accumulation of the agent in a low concentration using the inocula from the placentas and cotyledons and YS with good accumulation of the agent, especially after prior concentration in the above-mentioned manner. Overall, the results were not satisfactory, due to the apparent lack of natural ability for adaptation to allantoic cavity cultivation.

Attempts at direct isolation of chlamydia from bovine clinical-pathological materials were not taken. Cultivation of *C. abortus* in the cell lines BHK-21 and McCoy was successful due to the infection with strains isolated and propagated in YS of CE and prepared for inoculation in the form of concentrated and purified suspensions. In addition, in the McCoy culture we performed inhibition with IUDR and centrifugation at 2800 g for 1 h in order to increase the index of the positive diagnosis of chlamydial inclusions (infection by centrifugation).

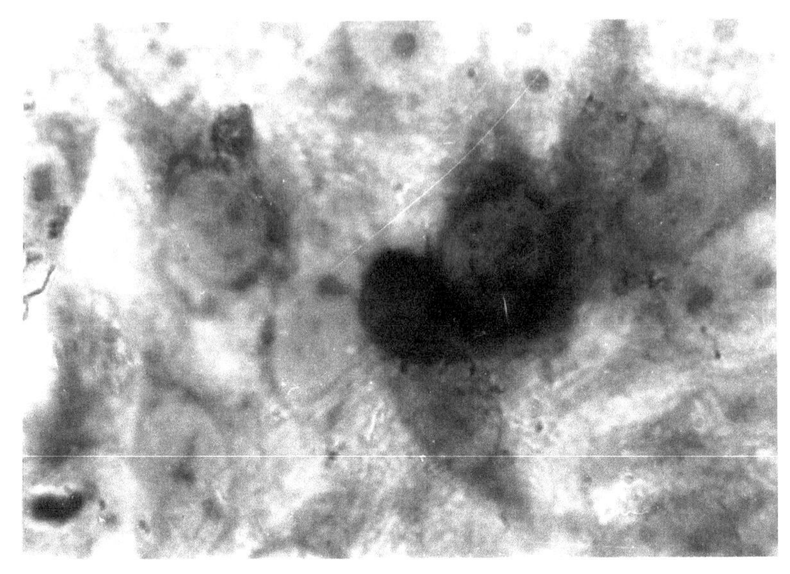

Figure 5.6 Cell culture McCoy infected with *C. abortus*, isolated from aborted bovine placenta (72 h). Intracytoplasmic chlamydial inclusions. Giemsa stain. Magnification 1250×.

Figure 5.6 shows intracellular inclusions of *C. abortus*, isolated from the placenta of an aborted cow when cultivated in CC McCoy.

The immunofluorescence indication of *C. abortus* in CC BHK-21 is shown in Figure 5.7.

As the field materials and the strains in CE had little effect on white mice at i.p., i.v., and i.c inoculations. Nasal infection following inoculations of suspensions of placentas, cotyledons, and fetal organs was also not achieved. Infection of mice was achieved with propagation in YS of CE isolates, which led to pneumonia and mortality of 25–30%.

Intraperitoneal inoculation of guinea pigs with clinical and pathological materials or with infected YS under good natural accumulation of the agent or at artificial concentration was effective. This allowed us to include a laboratory strain Sh.d. in comparative experiments with immune abortfacient and arthritic strains of ruminant animals for the establishment of specific antigenic relationships between them (Table 5.4)

5.2.3.2 Polyarthritis in calves

From spontaneous cases of polyarthritis in calves in seven settlements we isolated seven strains of *C. abortus* [23, 150]. The isolation was achieved after intra-yolk inoculations of punctate synovial fluid obtained from the knee,

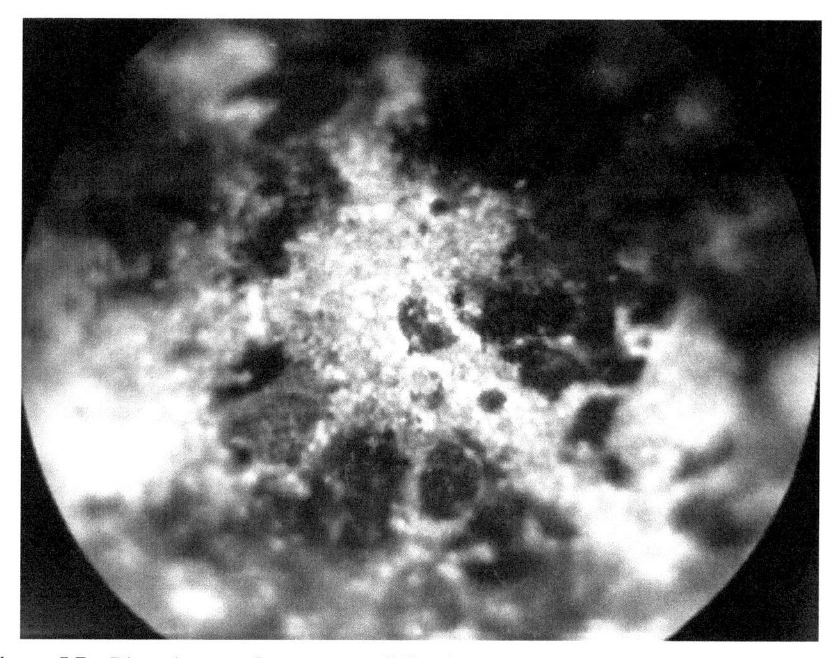

Figure 5.7 Direct immunofluorescence. Cell culture BHK-21 (72 hr) infected with concentrated and purified chlamydia, strain *Sh.d.* of chlamydial abortion in cows and stained with anti-chlamydial labeled gamma globulin. Magnification: lens 9×, 7× eyepiece.

tarsal, and carpal joints of calves aged 2 days to 12 months. The isolates were cultivated well in chicken embryos. The strain Sh., propagated in yolk sacs, was cultivated in 7-day-old mice following intranasal infection under ether narcosis. The registered mortality of 20 inoculated mice was 30%. Mice died between 11 and 20 days after inoculation. In their lungs were observed inflammatory changes. Chlamydial bodies were detected in impression preparations of pneumonic tissue stained by Stamp [23, 150].

5.2.3.3 Keratoconjunctivitis

Strain Suhodol (S) was isolated in the YS of CE from conjunctival secretions of two calves at 6 months of age exhibiting signs of acute keratoconjunctivitis [151]. The strain was passaged 19 times in the YS and reached an infectious titer of $10^{6.5}$/ID/ml. This strain was adapted easily to cultivation in the allantoic cavity of CR. For this purpose we inoculated a 50% suspension of infected yolk sacs at third level passage and inoculated 8-day CEs. After repeated passage we gathered allantoic fluid in a pooled sample, and concentrated it by centrifugation at 20,000. From the resulting pellet were prepared ultrathin sections for EM in which we found a rich accumulation of chlamydiae. Adaptation to the allantoic cavity cultivation of two other laboratory strains of outbreaks of mass keratoconjunctivitis in cattle – Rousse (P1), isolated from Ognianov et al. [336] et al. and Popovo (P) isolated from Martinov [23] – was reached similarly. It can be assumed that the tendency for allantoic cavity cultivation represents a species quality of the chlamydial strains isolated from bovine keratoconjunctivitis.

The strain propagated in embryonated eggs, strain S., was cultivated in 6-day-old mice following i.c. inoculation of 20% yolk suspension. The inoculated mice developed paralysis and 95% of them died by the 5th day. The chlamydial pathogen isolated from the brains of the mice was successfully re-cultivated in 7-day chick embryos.

5.2.3.4 Latent chlamydial infection

Bioassay on isolation of *C. abortus* from the feces of latently infected cattle in guinea pigs following i.p. inoculations finds it rather suitable because of the high degree of external contamination of such a field material. We isolated two strains of chlamydia – one in the feces of a cow with chlamydial abortion that remained a latent carrier (isolation on the 90th day after the abortion), and the second, from an asymptomatic cow belonging to a herd where chlamydiosis had been identified.

Reinhold et al. [574] emphasize that the frequently observed lack of association between infection and clinical disease in cattle has resulted in a debate as to the pathogenic significance of these organisms, and their tendency for sub-clinical and/or persistent infection presents a challenge to the study of their potential effects. However, recent evidence indicates that chlamydial infections have a substantial and quantifiable impact on livestock productivity with chronic, recurrent infections associated with pulmonary disease in calves and with infertility and sub-clinical mastitis in dairy cows. It is important to note that the existing data also suggest chlamydial infections in cattle manifest clinically when they coincide with a number of epidemiological risk factors [574].

5.2.4 Strains Isolated from Pigs

5.2.4.1 Pericarditis

In connection with the pericarditis seen in adolescent pigs in some farms, we attempted to isolate chlamydia. We investigated the pericardium and pericardial fluid of 10 diseased pigs from a farm. After inoculation of CE with a suspension of a pooled sample of said material we isolated a *C. suis* strain [47, 334] The pathogenicity of the strain for embryonated eggs at inoculation in the yolk sac is presented in Table 5.5. The table shows the process of primary isolation/ (zero passage) and the next three passages. Specific lethality of the CE was registered by the 4th to 14th day. The maximum mortality of the embryos was noted at days 6 and 12 after infection. The strain was identified morphologically by EM. EBs and IBs located in ento dermal cells of YS were observed in ultrathin sections, as well as other structures specific to chlamydial infection: multiple specific vacuolation, initiation center chlamydial membranes, and osmiophylic membranous masses. In Figure 5.8 the electron microphotograph of the isolated strain is presented.

Table 5.5 Isolation and cultivation in CE of Chlamydia suis strain of pigs with pericarditis

	Inoculated	Dead	Days Lethality of CE												
Passages	CE (No.)	CE (No.)	1–3	4	5	6	7	8	9	10	11	12	13	14	15
0	40	15	0	0	0	4	0	2	1	0	0	7	1	0	0
I	32	30	14	3	1	6	1	1	2	1	1	0	0	0	0
II	20	20	3	2	0	4	0	0	0	0	1	4	0	0	0
III	48	13	13	7	5	6	2	0	0	1	1	5	6	2	0
Total	140	113	30	12	6	20	3	3	3	2	3	16	7	2	0

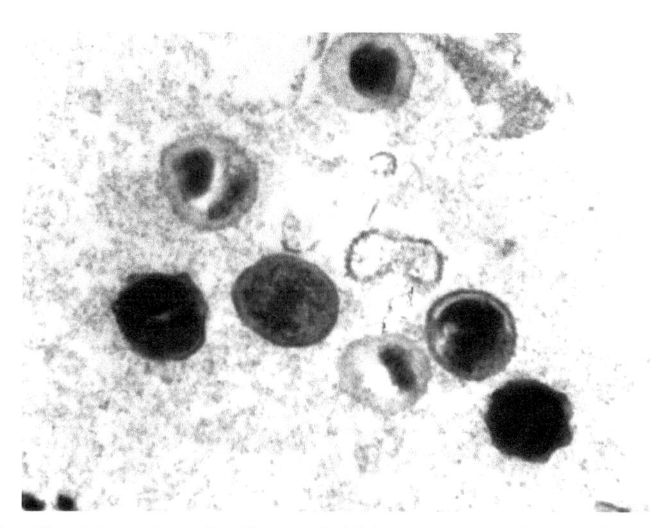

Figure 5.8 Ultrathin section of yolk sac of chicken embryos infected with the Chlamydia suis strain isolated from the pericardial fluid of pigs with pericarditis. Elementary bodies. Magnification: 30,000×.

Besides morphologically, the isolated chlamydial strain was identified antigenically. Ether-acetone was received from infected YS and purified by gel chromatography on Sepharose 2B and 4B antigens which had group-specific activity against standard anti-chlamydial sera and CF-titer of 1:80.

The isolation of chlamydia from the inflamed pericardium was performed in the age group of pigs in which pigs affected by pneumonia in the previous few years were proved virologically and serologically to have chlamydial infection [23].

5.2.4.2 Abortions

In studies in the known focus of swine chlamydiosis (Bahovitsa, Lch) was isolated strain with the same name from the placenta and fetus of aborted swine. This strain was isolated in the YS of CE separately from the two types of field materials (Figure 5.9). Other attempts at isolation have not been made.

Figure 5.9 Electron microphotograph of negatively stained preparation of YS of CE infected with *C. suis* isolated from the placenta of aborted pig. Elementary bodies. Magnification 50,000×.

5.2.4.3 Other clinical conditions in pigs

A strain of *C. suis* from a pig at 20 days of age after intra-yolk inoculation of synovial fluid from the affected knee and tarsal joints was isolated in the CE [23]. The synovial fluid from this case of arthritis was slightly hazy and contained fibrin strands.

In a joint study with Lithuanian scientists on chlamydiosis in pigs in Lithuania, a part of the comprehensive studies included the isolation of chlamydia from samples taken from the ocular mucosa, cervix, and prepuce of pigs with inflammation of these anatomical locations. In many cases, the chlamydial agent was isolated successfully in the two culture systems – YS of CE and CC McCoy – initially treated with DEAE-Dextran (90 μg/ml) and then with cycloheximide (0.9–1.0 μg/ml). The indication was achieved by the direct IF method [335].

5.2.5 Strains Isolated from Dogs

Two *Chlamydia canis* isolates (Berk-1 and Berk-2) were extracted from the YS of CE obtained from the lungs, livers, and the spleens of puppies that died soon after their birth and from the blood of an old bitch from the same farm for dogs (Berkovitsa).

5.2.6 Strains Isolated from Cats

Two strains of *C. felis* were isolated in CE from the conjunctival scarified material of cats with conjunctivitis. One case dealt with a small kennel of dogs for commercial purposes, where all the animals had persisting eye inflammation, in addition to abortions, stillbirths, premature births, and kittens having malformations over the previous two months. The second isolate was obtained from a single cat grown by another owner.

5.2.7 Strains Isolated from Guinea Pigs

A strain of *C. caviae* was isolated from a colony of guinea pigs with mass conjunctivitis, diagnosed cytologically as chlamydial. The isolation was in the YS of CE after inoculation of a suspension of ocular discharge and scarified conjunctival mucosa. With chlamydial isolate M.sv.1 having an infectious titer of LD_{50} log 10^5/0.5 ml for a 7-day KE, we reproduced the disease in guinea pigs and achieved re-isolation of the agent (see Chapter "Clinical observations").

5.2.8 Strains Isolated from Birds

Pooled results from these studies are shown in Table 5.6. The overall volume of the work included 345 samples from the clinical and pathological materials of 235 birds: parenchymal internal organs, intestinal contents, blood, and feces. In addition to the culture results, the data for the direct detection of *C. psittaci* in field samples are presented.

It covers three groups of birds: domestic, wild, and decorative [23, 53, 337]. The largest number of materials studied were on waterfowl – ducks and geese – where the average percentage of positive cases was 18.47% and had a total of nine isolates. All newly isolated strains of *C. psittaci* were obtained from samples with positive findings for chlamydia by the methods of their direct indication (LM, IF, DEM). Attempts to isolate the agent from the other materials were not made. Most isolation cases were from farms with serologically established chlamydiosis.

Figure 5.10 shows the direct immunofluorescence in primary CC CEF infected with concentrated and purified suspension of chlamydial strain isolated from the parenhymal organs of a duck affected by chlamydiosis.

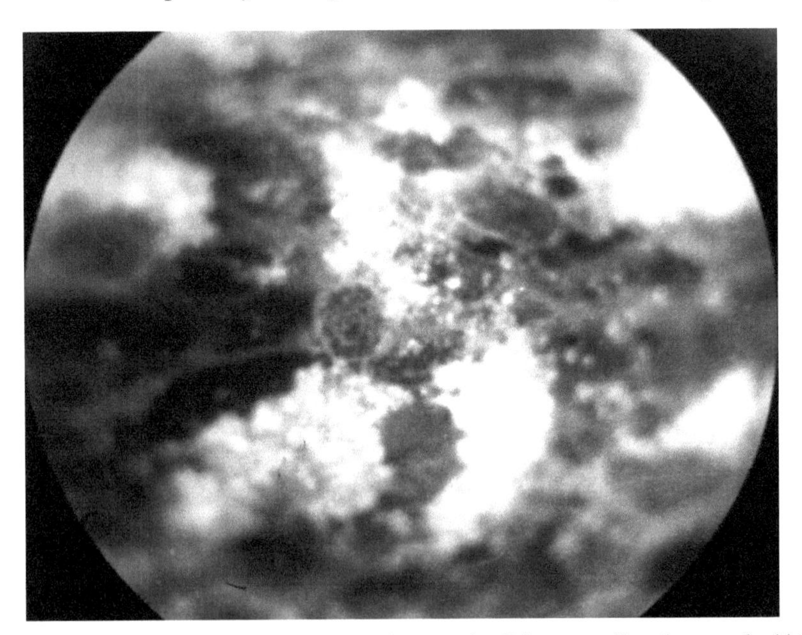

Figure 5.10 Immunofluorescence microphotograph. Primary cell cultures of chicken embryo fibroblasts (72 h), infected with concentrated and purified chlamydia isolated from duck with chlamydiosis. Direct IF staining method with anti-chlamydial conjugated gamma globulin. Magnification: Lens 9×, eyepiece 7×.

Table 5.6 Isolation and direct indication of *Chlamydia psittaci* from domestic, wild, and decorative birds

Species	District	Starting Clinical Specimens	Tested. (No.)	(+) No. /%	Demonstration of C. psittaci	
					Isolation	Direct Detection*
Domestic ducks	SF, Sofia city, Pd, Pz, Bs, Dch, Rz, Ss, VT, Pl, Rs	parenchymal organs, blood	157	29/18,47	8	LM, IF, DEM-29
Domestic geese	Rs, Bs, MT	parenchymal organs blood	23	4/17,39	1	LM, IF – 4
Hens	Sf, Pl, Rz	blood	12	1/8,33	1	–
Domestic pigeons	Sofia City, SF, Pk	parenchymal organs, blood intestinal contents	39	7/18,42	5	LM, DEM, IF – 7
Feral pigeons	Sofia City, Vd	parenchymal organs, intestinal contents	33	10/30,3	2	IF – 10
Doves (Streptopelia decaocto)	Vd	parenchymal organs, intestinal contents	4	1	1	LM – 2
Crows	Vd	parenchymal organs, intestinal contents	4	1	1	LM – 2
Pheasants	Rz, Sf	parenchymal organs	3	1	1	LM – 1
Heron	Vd	parenchymal organs	1	1	1	LM, IF – 1
Pink Pelican	VT	liver, kidney	1	1	1	IF – 1
Parrots	Sofia City, Pz	parenchymal organs, faeces	56	21/37,5	7	LM, EM, IF – 18
Canaries	Sofia City	liver, lung, spleen	7	4	1	LM, IF – 4
Finches	Sf	parenchymal organs, feces	5	5	3	LM, IF – 5

Includes all cases proven directly, including with the concomitant isolation of the agent.

The electron microscopic indication and identification of the strains of ducks, isolated and cultivated in the YS of CE is shown in Figure 5.11 [53].

A significant percentage (18.42) of positive for chlamydiosis cases including 5 isolates were received from domestic pigeons. In the majority of cases the examined single sick or dead birds were from groups or small flocks with mass clinical disease. Seven strains of *C. psittaci* were isolated from wild birds – feral pigeons, doves, crows, pheasants, herons, and pink pelicans.

From the ornamental birds were isolated 11 strains – seven of parrots, one of canary, and three of finches: zebra-like (*Taenopigia guttata*), Japanese (*Lonchura striata*) and rice (*Padda oryzivora*).

Besides CE infected in the YS, a suitable biological model for primary isolation of avian chlamydiae are white mice. In our experiments we included materials from different species of birds – ducks, domestic pigeons, parrots, herons, and pink pelicans. Intranasal inoculations with 0.2 ml were performed on 3-week-old mice under ether anesthesia. All attempts were successful, which was confirmed by the presence of inflammatory changes in the lungs

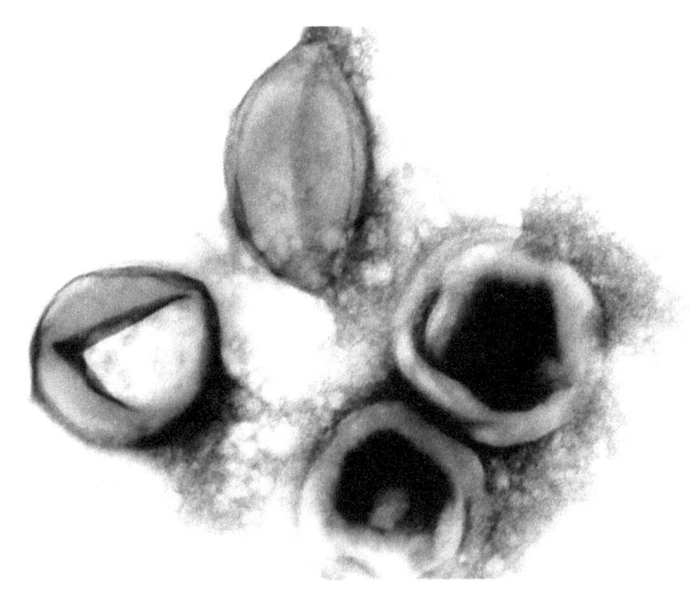

Figure 5.11 Negative staining of *C. psittaci* (Strain Trust. Pl) isolated from ducks in 7-day CE. Magnification 50,000×.

of mice with the detection of chlamydia in impression preparations made from them. The particularly suitable period to establish these findings is the seventh to eighth day. Depending on the virulence of the chlamydial agent in the material, some of the mice died between the third and seventh day in more virulent isolates and between the eighth and fifteenth day in less virulent ones. The strains from the same avian species, isolated and propagated in YS of CE to give a good accumulation of the agent, were readily cultivated in white mice through intranasal inoculation and caused pulmonary signs and clinical development with different intensities, depending on the virulence of the strain.

The experiments with white mice also included i.c. inoculations with the above-mentioned clinical and pathological materials. The chlamydiae isolated from ducks, wild pigeons, and herons showed the highest virulence, causing death of the mice within 3 to 6 days.

The remaining strains also caused infection of the mice, but there were no fatal cases. In the meninges we observed edema and strong engorgement of blood vessels. Chlamydial cells were discovered in smears from the meninges. Suspensions were prepared from the meninges, with which CE were inoculated in the YS. As a result the chlamydial agent was re-isolated.

5.2.9 Strains Isolated from Humans

5.2.9.1 *Chlamydia trachomatis*

The summary results of these investigations are presented in Table 5.7. Data refer to four groups of patients: women with cervicitis, children with inclusion conjunctivitis and/or respiratory diseases, men with Reiter's syndrome and with non-gonococcal urethritis (NGU) [24, 58, 59, 105, 148, 165].

Attempts at isolation of *C. trachomatis* from cervical secretions and scarified cervical mucosa in the YS of CE were made in 16 cases. As a result, we isolated five strains with typical morphological and antigenic characteristics belonging to chlamydia in the McCoy cell line previously inhibited by IUDR and after infection by centrifugation. We found inclusions of *C. trachomatis* in 28 of 44 samples (63.64%) after staining by Giemsa and in 18 of 38 samples (47.36%) after staining with Lugol's solution (Figures 5.12 and 5.13).

Two *C. trachomatis* isolates were obtained in the CE of male R.I., aged 47, with clinical signs of bilateral orchiepidimiditis [55]. Prior to laboratory

Table 5.7 Isolation of *Chlamydia trachomatis* from sick people

Clinical Diagnosis	Tested/ +(No.)	Isolates in CE	Isolates in CC	Other Etiological Studies of the Culturally Positive Patients	Etiologic Investigations of Mothers of Children Positive for C. trachomatis
Cervicitis	16/5	5		LM+, DEM+, CFT+, ELISA+, IF – MA+	
Cervicitis	82/46		46	LM+, DEM+, CFT+, ELISA+, IF – mAbs+	
Conjunctivitis and pneumonia baby N.T.P., 14 days *	2/1	1		LM+, CFT – not investigated.	CFT+
Conjunctivitis and nasopharyngitis baby B.V.P., 16 days	1/1	1		LM+, CFT – not investigated.	CFT+, CE+
Tracheobronchitis and pneumonia Baby M.V.V.2 months & 26 days	2/1	1		LM – not investigated CFT+	CFT+, LM+
Acute rhinitis Baby Zh.M.B. 5 months	1/0	–	–	–	not investigated
Nasopharyngitis Baby R.K.M., 2 months	1/0	–	–	–	–
Reiter's syndrome	53/29	2	27	LE+, DEM+, CFT+	
NGU	25/15	5	10	LE+, DEM+, CFT+	
Orchiepididimytis	1	1		CFT+	

Given initials and age of only positive for *C. trachomatis* children.

tests for chlamydia, he was treated with Ampicillin and Gentamycin for 10–12 days without marked therapeutic effect. The patient was tested bacteriologically for *N eisseria gonorrhoae*, *Trichomonas vaginalis*, and anaerobes and aerobes with negative results. He was found to be serologically positive

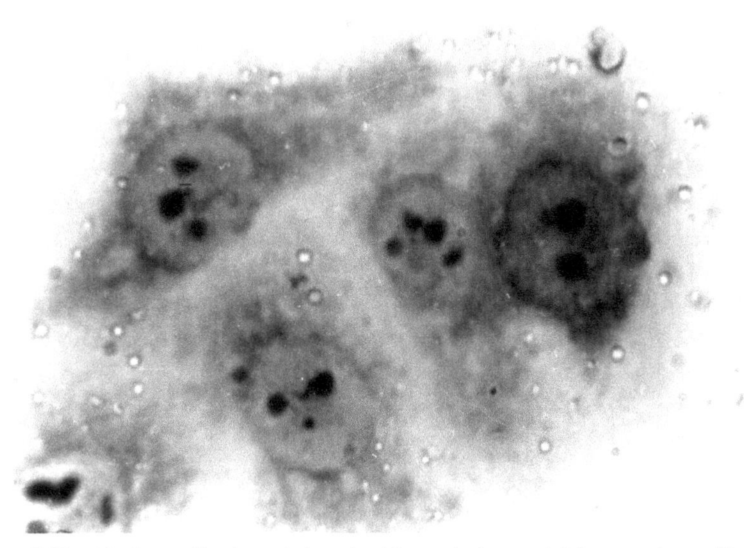

Figure 5.12 McCoy cell culture infected with cervical material from patient with endocervicitis (72 h). Inclusions of *C. trachomatis* located intracellularly. Giemsa stain, 1250×.

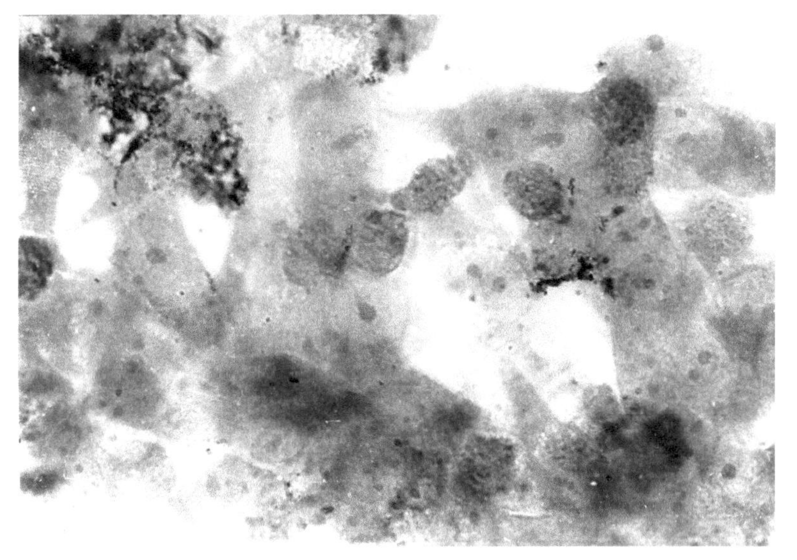

Figure 5.13 McCoy cell culture infected with cervical secretions and scarified material of a woman with endocervicitis (72 h). Iodine-positive intracellular inclusions of *C. trachomatis*. Iodine solution stain 1250×.

for chlamydia with high complement-fixing titers. The first chlamydial isolate was obtained from a blood coagulum and the second from a pooled sample of three ejaculate semen taken within 5 days.

The isolation of *C. trachomatis* in the McCoy cells from the urethral material of a patient suffering from the Reiter's Syndrome (RS) is demonstrated in Figure 5.14. The chlamydia in 27 cases was isolated in the CC (52.9% off 51 patients with RS). With such material from other two patients received 2 isolates in CE. Of the examined 25 men with NGU received 15 isolates C. trachomatis – 10 in McCoy cells and 5 in ebryonated eggs.

Figures 5.15 and 5.16 is shown the electronic microscopic morphology of the two strains of *C. trachomatis* isolated in the YS of CE of urethral materials of patients respectively with Reiter's Syndrome and NGU.

In the comparative study of the pathogenic properties of the strains *C. trachomatis* for white mice we used representatives of three groups of isolates in CE: a) D.H. (cervicitis); b) P.I. (urethritis); c) F.K. (reactive arthritis) (Table 5.8).

Table 5.8 shows that mice were inoculated in six ways. Most high performance associated with the pathogenicity has strain D.H. isolated from subjects with endocervicitis, followed by strain P.I. (NGU in man), and strain

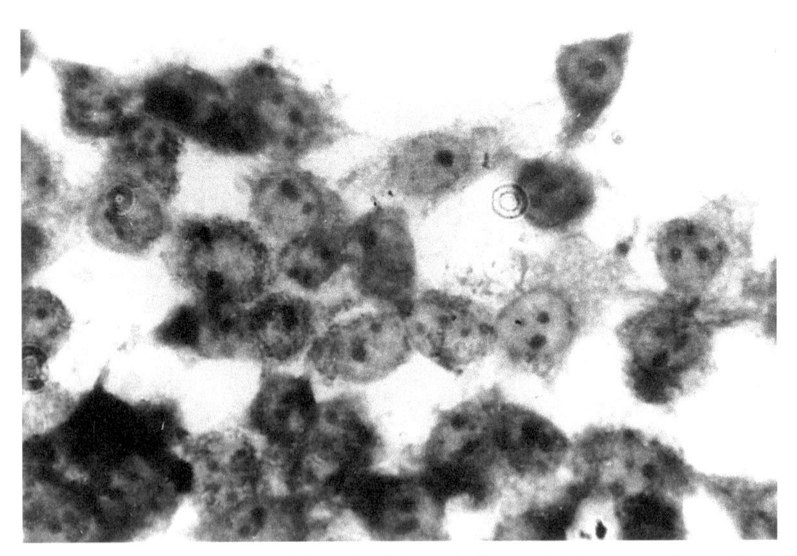

Figure 5.14 McCoy cell culture infected with urethral material from a patient with Reiter's Syndrome. Intracytoplasmatic inclusions of *C. trachomatis*. Giemsa stain 1250×.

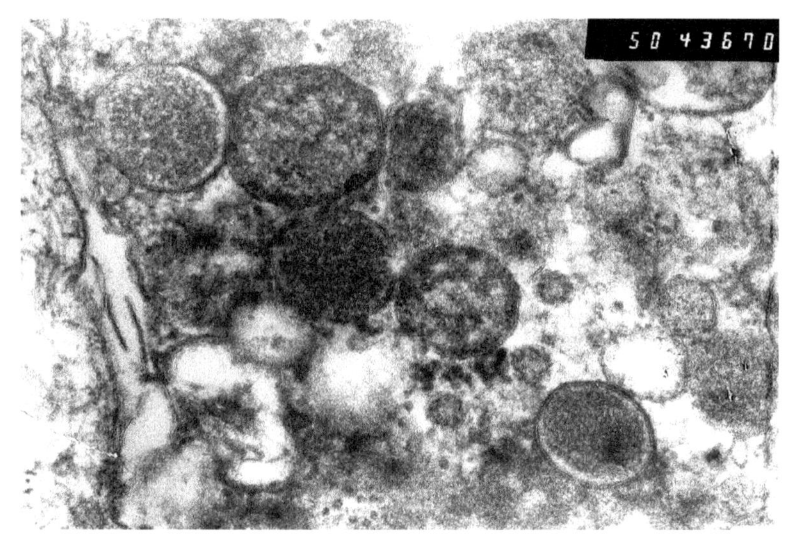

Figure 5.15 Ultrathin section of YS of CE infected with *C. trachomatis* isolated from urethral exudate of patient with Reiter's Syndrome. Elementary and reticular bodies, chlamydial vacuoles, and membranous formations. Magnification 100,000×.

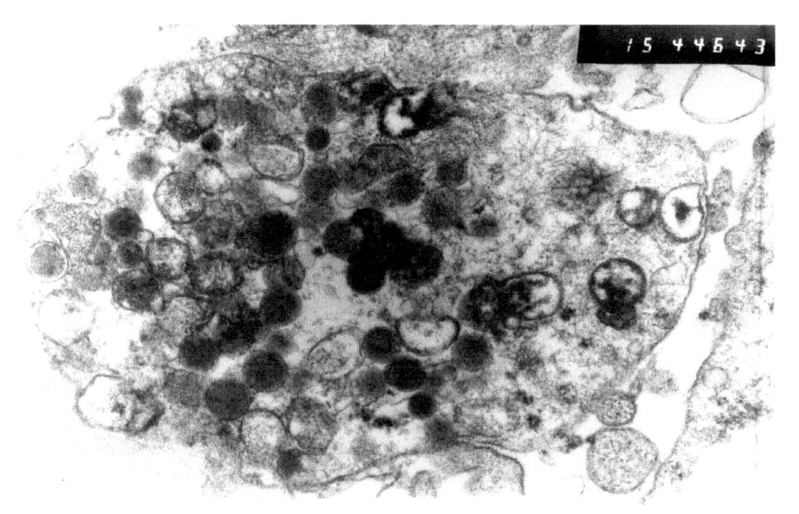

Figure 5.16 Ultrathin section of YS of CE infected with *C. trachomatis* of urethral discharge from a man with NGU. Intracytoplasmatic inclusion of *C. trachomatis* consisting of many EBs with a size of 220–250 nm and RBs with a size of ~300 nm. Magnification 30,000×.

Table 5.8 Pathogenic properties of strains of *C. trachomatis* for 3-week-old white mice

Strain	Method of Inoculation					
	i.n.	i.c.	i.v.	i.p.	s.c.	p.o.
D.H. cervicitis	(+) death in 60%	(+) death in 80%	(+) death in 45%	(+) no deaths	Protracted course leading to latency	Protracted course leading to latency
P.I. urethritis	(+) death in 30%	(+) death in 40%	(+) death in 10%	(+) no deaths	Protracted course leading to latency	(–)
F.K. -reactive arthritis in Reiter's Syndrome	(+) death in 15%	(+) death in 25%	(+) death in 5%	(–) at 90% (+) at 10% with no mortality	(–)	(–)

F.K. (Reiter's syndrome with marked reactive arthritis). The nasal route of infection proved to be quite effective, where despite the differences in the amount of mortality, the surviving mice developed pulmonary signs and macroscopically visible inflammatory changes with thickened pneumonic portions. The presence of chlamydial agent was proved in impression preparations of the lesions. Mice showed high sensitivity to the i.c. infection, leading to meningoencephalitis and significant lethality. Inoculation by the s.c. route was less effective in strains D.H. and P.I. and ineffective in F.C. and the oral administration of *C. trachomatis* had no consequences for P.I. and F.C. (Table 5.8).

5.2.9.2 *Chlamydia pneumoniae*

Chlamydia pneumoniae was isolated in the YS of CE from the sputum of a patient with atypical pneumonia. The patient was serologically positive for antibodies against *C. pneumoniae* (EIA IgA, IgG). The isolate showed good adaptation ability for intra-yolk cultivation. Other attempts to isolate the agent were not made.

5.2.9.3 *Chlamydia psittaci*

Three strains of *C. psittaci* were isolated in the YS of CEs from sick people with respiratory symptoms as a result of direct contact with birds [23, 45]. In the first case (I.I. – a girl of 4 years with acute form of interstitial pneumonia, after contact with canaries and pigeons at home), the isolate was obtained from the child's blood coagulum. In addition to the CEs, the strain was easily adapted to mice after nasal infection. The second strain was isolated from a male, K.S.K., 39 years of age with chronic ornithosis, living in an attic room adjacent to pigeons. Characteristic of the electron microscopic finding in this strain was the detection of predominantly large reticular forms of the chlamydial pathogen. The third strain was isolated from a blood coagulum taken in the acute stage of severe ornithosis in a diseased male, L.P. In this case, in addition to the isolation of the agent, positive results for chlamydia were obtained at the DEM of the ultrathin section of sputum, and in a negative-stained suspension of the urine of the patient. In the three described cases, isolation and pathogen identification was accompanied by the detection of chlamydial CF-antibodies with diagnostic seroconversion [23, 45].

5.2.9.4 *Chlamydia abortus*

Strain W.I.D. was isolated from a case of acute bilateral keratoconjunctivitis in a female due to intra-laboratory infection [23, 45]. The latter was caused by a direct occlusion of a bacteriologically sterile 20% suspension of chlamydia strain S isolated from a calf with acute keratoconjunctivitis. From the blood coagulum and from the conjunctival secretion of the sick worker, taken in the acute stage of the disease, we isolate a strain in the YS of CE, which on its cultural, morphological and staining properties was referred to as *C. psittaci* serotype 1 (now *C. abortus*). Electronmicroscopically, the isolated strain was a complex population of EBs, CBs, RBs, and intermediate forms with characteristic ultrastructural features [23, 45]. Although single, this case confirms and expands the zoonotic potential of *C. abortus*, which is known to pose a risk to pregnant women and causes sporadic cases of respiratory disease in laboratory staff, abattoir workers, and vaccine plant workers handling this agent of sheep origin [518–520]. Such a recent example is the isolation of *C. abortus* from the sputum of scientists in the field of chlamydia with a clinical diagnosis of atypical pneumonia [521]. The presence of the chlamydial agent is confirmed by molecular analysis. *C. pneumoniae*, *Mycoplasma pneumoniae*, *Legionella pneumophila* and *Rickettsia conorii* have been excluded in the differential diagnosis.

Another strain of *C. abortus* was isolated from a male, I.I.A., 28 years old and suffering from pneumonia. The patient has been in constant contact with cattle, sheep, and pigs that have had abortions and respiratory diseases [45].

6

Virulence and Pathogenicity

The problem of virulence and pathogenic action of chlamydia are immutable object of studies in previous history of chlamydia and chlamydial infections. According to the established definition, the pathogenicity is the ability to cause disease in a host. Microorganisms exhibit their pathogenic properties through their virulence, a term determining the degree of pathogenicity of the infectious agent. Determinants of virulence of the pathogen are some of its genetic, biochemical, or structural features [339–343].

Multidirectional comparative experiments were carried out with *Chlamydia psittaci* (incorporating all mammalian, avian, and human variants of the old classification) and *Chlamydia trachomatis* on a variety of mammals and birds (*in vivo*), and also in chicken embryos and *in vitro* in cell cultures.

Scientific evidence indicates that the pathogenicity is very different. For example, *C. psittaci* having high infectivity, virulence, and pathogenicity causes such infections as chlamydiosis (psittacosis/ornithosis) to birds and man and lethal pneumonias in humans, and *Chlamydophila abortus* is responsible for enzootic abortion in sheep and goats, epizootic bovine abortion and other clinical conditions in animals, and abortions in women. Because of its high virulence and pathogenicity, *C. psittaci* is in the list of particularly dangerous microorganisms, and some strains are part of the bacteriological weapons based on properties with potential for bioterrorist threat. Along with highly pathogenic strains, there are many low-pathogenic, weakly infectious chlamydiae of low virulence causing inapparent or latent infections.

For *C. trachomatis* can say the same. Strains of the epidemic trachoma and the epidemic lymphogranuloma venereum (LGV) are highly pathogenic, infectious, and virulent strains, while in non-epidemic areas of the world, the strains are low pathogenic, less infectious, and less virulent. Namely due to differences in pathogenicity differ trachoma and para-trachoma diseases. On the other hand, some authors note that the pathogenicity of chlamydia to a certain extent is specific [19, 21, 344, 345]. According to Iversen et al. [346],

C. psittaci (strain M56, the agent of epizootic chlamydiosis of muskrats and hares) was highly lethal for the snowshoe hare (Lepus Americans) following intravenous inoculation, whereas the agent was much less virulent for cottontail (*Sylvilagus floridanus*) and albino domestic rabbits (*Oryctolagus cuniculus*). Virulence appeared to be very host specific in that only strain M56 among the six chlamydiae tested was highly lethal for the snowshoe hare [346]. These and similar reports support the hypothesis of Page on peculiar "pathotypes" of chlamydia [344, 347]. In chlamydiology, the widely used terms are bio variants (biovars) and sero variants (serovars). They reflect the great diversity in pathogenicity, infectivity, and the virulence of the chlamydial agents. Although these qualities are genetically determined, there are many attempts to influence them. In routine diagnostic practice, recourse is made to passaging of chlamydia on CE as the most suitable biological system for their propagation, in order to increase their virulence and stabilize them. Back in immunological practice with chlamydia the passages are used for the attenuation (weakened) strains for use in attempts to create vaccines [24]. Some authors have tried to weaken such a particularly virulent strain *C. psittaci* as Borg, through passages in embryonated eggs in the presence of p-aminobenzoic acid [348].

Pathogenicity of chlamydia is closely associated with toxicity. Even in 1944, it was found that the agent of LGV has toxic properties associated with EBs, which perform some in their propagation in yolk sacs of CE [349]. Later, the phenomenon toxicity of chlamydia has been studied in strains isolated from birds, mammals, and human beings and have been found differences in the strength and nature of the toxic effects associated with the degree of virulence, the multiplicity of the infection, the type of host cell, and so on. Given are various explanations for the toxicity of chlamydia.

In a study of the toxicity of pathogenic chlamydia known strains Illinois and Louisiana that cause lethal pneumonia in humans, as well as in the study of other mammalian strains, it was found that the death of experimental animals has a two-phase dynamics. During the first 16–24 h, the mice die from the effects of the toxin, and to 48 h of developing infection. But other strains isolated from birds, after injection of the mice, causing mortality only by the action of the toxin [350]. Hence, the nature of the toxic effect in mice is not the same for different strains. The toxic factor is connected to the elementary bodies and is not removed by filtration or centrifugation. It destroys rapidly at 37°C or after formalin treatment, which makes it similar to the toxin of the Rickettsia or partly to the exotoxin of the bacteria. Ognyanov [37] finds that two strains of the chlamydial agent of ovine enzootic abortion have pronounced toxic properties on mice after intravenous and intraperitoneal infection. These properties are unstable and degrade rapidly when heated to 60°C or by treatment with ether and formalin. At storage (4°C), the toxic properties are lost for 48 h and at –20°C the toxic effect gradually lowering [37].

Christoffersen and Manire [351] studied the toxicity in mice of the chlamydial agent of meningopneumonitis, in conjunction with its various stages of development. They found that the intact elementary bodies of the pathogen are toxic for mice and will absorb antitoxin from hyperimmune serum. The intermediate reticulate form of the organism, however, is non-toxic and does not absorb antitoxin. It appears likely that soon after penetration of the susceptible cell, the organism loses the toxic antigen from its envelope, the antigen is missing during multiplication, and is resynthesized at maturation.

Kordova et al. [352, 353] investigated the toxicity of *C. psittaci* 6BC strain for the mouse peritoneal macrophages. In these experiments, it was shown that attenuated avirulent for mice strains are non-toxic and vice versa – the non-attenuated strains are toxic to murine macrophages. Moreover, the lysosomes are an indicator of this toxicity, namely, that they are not active when infected with attenuated strains and are highly active in infection by virulent strains.

On model L-cells, and *C. psittaci* 6BC strain was investigated so-called immediate toxicity of chlamydia at high multiplicity of infection [354]. In multiplicity, 500–1000 LD_{50} per cell death of the cells occurs between the 8th and 20th hour after infection. This effect is known as "immediate toxicity." It is delayed by ultraviolet irradiation. The heat did not affect. The mechanisms as described phenomenon consists in the inhibition of phagocytosis, inhibition of lysosomes, and ultimately to the death of the cell cultures [354]. Other authors [355, 356] also examined the importance of multiplicity of infection with chlamydia, but on a model mouse peritoneal macrophages infected with *C. psittaci* cultivated in L-cells. At a multiplicity of 100 EBs per macrophage occurs immediate cytotoxicity. If at the same multiplicity, however, chlamydia pre-heated at 56°C for 5 min, or be treated with specific antibodies, the immediate toxicity was not observed, but is established phagocytosis of the elementary bodies. The authors believe that the optimal multiplicity of infection of the macrophages is 1:1 [355, 356].

Upon modern methodological capabilities in microbiology, efforts have been made to highlight the issue from the molecular-biological point of view. Belland et al. [357] have shown that *C. trachomatis* from urogenital and ocular diseases exhibits cytotoxic activity, resulting in morphological and cytoskeletal lesions in the infected epithelial cells. Cytotoxin gene transcripts have been detected in chlamydiae-infected cells, and also a protein with the expected molecular mass has been found in lysates of infected epithelial cells. These data testify for the presence of chlamydial cytotoxin for epithelial cells and imply that the cytotoxin is present in the elementary body and delivered to host cells in a very early stage of the infection. The authors hypothesize that the cytotoxin represents a virulence factor that contributes to the pathogenesis of *C. trachomatis*-induced diseases [357].

Hosseinzadeh et al. [358] report that induced by *C. trachomatis* death of sperm is caused by lipopolysaccharide as the toxicity of EBs of the agent of serovar E is greater than that of serovar LGV. It is believed that in the elementary bodies of the pathogen, gene of cytotoxicity is present, which encodes the provision in the host cell of cytotoxin in the early stages of infection. So, this cytotoxin is one of the factors of the virulence of studied chlamydial species. The thesis of the existence of chlamydial gene cytotoxin in ocular and genital isolates of *C. trachomatis* and in the *Chlamydia muridarum* – agent of pneumonia in mice and hamsters, is supported by Carlson et al. [359] and Lyons et al. [360]. It is also known that *C. caviae* strain GPIC infects guinea pigs and encodes a single cytotoxin, and the *C. muridarum* strain MoPn infects mice and encodes three cytotoxins, arranged in tandem in the genome. All these data give the reason to believe that the toxin is an important virulence factor for mucosotropic chlamydial pathogens. An important future research goal is fuller explanation and understanding of the role and function of the toxin in mediating mucosal infection tropism and pathogenesis [359].

Issues relating to the factors determining pathogenicity and virulence of various representatives of Chlamydiaceae are constantly subject to the latest research. Rajaram et al. [361] highlight the well-known fact that the pathogenically diverse *Chlamydia spp.* can have surprisingly similar genomes. For example, *C. trachomatis* isolates cause invasive LGV, trachoma, and sexually transmitted genital tract infections, and the murine strain *C. muridarum* shares 99% of their gene content. Too intriguing is the suggestion that a region of high genomic diversity between *Chlamydia spp.* termed the plasticity zone (PZ) may encode niche-specific virulence determinants that dictate pathogenic diversity [170, 183]. Four groups of authors [170, 183, 362, 363] suggested that two PZ gene families, tryptophan synthase in *C. trachomatis* and the putative cytotoxins (tc0437, tc0438, tc0439) in *C. muridarum*, were determinants of chlamydial tropism. Recently, Newman

[364] reported studies on the characteristics of the putative cytotoxin genes in *C. muridarum*.

Bard and Levitt [365] reported that C. trachomatis *L2 serovar* of LGV *in vitro* is absorbed by 1/2 of the human monocytes and granulocytes in peripheral blood, but forms inclusions in only 0.5% of them, and they usually are digested by cells. In biopsies from human patients with active genital infections, we have found cytoplasmic inclusions in a significant percentage of peripheral macrophages [24, 146]. **As** compared, these data point to too relatively manifestation of the pathogenic properties of chlamydia depending on the particular conditions *in vitro* and *in vivo*. In confirmation of this conclusion are the data for the proteolytic cleavage of *C. trachomatis*. Hackstadt and Caldwell [366] found that the *in vitro* infectivity of the elementary bodies is unaffected by proteases. Last cleaved partially major outer membrane protein (MOMP) of the outer membrane of EBs and thus retain their ability to infect cells. But on the aforementioned data, it is seen that the digestion of chlamydia by proteases is commonplace.

The variations in the pathogenicity of the chlamydiae in its final stage are reflected in the latent infections caused by *C. psittaci*, *C. trachomatis*, and *C. abortus*. Chlamydia organisms possess important biological feature of inducing latent, persisting and inapparent course of the caused by them infections. These forms are common in chlamydiosis in birds. Martinov and Popov [45] observed chronic infections in human patients with psittacosis. The situation is similar in ruminants and pigs. Latent infection is well studied *in vitro*. Certain antibiotics, particularly penicillin, inhibit the formation of EB in CC and CE. Conversely, cytostatics, such as cycloheximide, inhibit macromolecular synthesis in the host cell and thereby stimulate the reproduction of chlamydia. Yang et al. [367] reactivate *C. trachomatis* lung infection in mice by immunosuppressive therapy with cortisone. Hanna et al. [368] studied latency in human infections with TRIC agents. They found antigen of *C. trachomatis* in conjunctival epithelial cells of healthy people and concluded that latent infection of the eyes with this agent is common. A similar opinion was expressed in other publications. For now, it is questionable whether a similar phenomenon exists in genital chlamydial infections. Schachter and Dawson [20] believe that in some patients have manifested infections and in others – inapparent. It would be wrong, however, chlamydia to be considered as part of the normal microflora of the body, even in asymptomatic infections. Rather, latent and subclinical infections are the result of the interaction of the protective mechanisms of the cell with the chlamydia, in which the multiplication of the chlamydia is weakened [20].

Mechanisms of invasion of the *Chlamydia spp*. are not sufficiently known. There are several factors of the virulence of chlamydia that contribute to the pathogenic properties of these microorganisms. In the characterization of the factors of the virulence, Clark and Bavoil [369] used *C. trachomatis* as a model. In the initial phase of the penetration of the agent in the mucous membranes of eyes, mouth and genitals play a role receptors containing sialic acid Chlamydia persist in places in the body which are inaccessible to phagocytes, T-cells and B-cells. In addition, *C. trachomatis* exists in the form of 15 different serotypes, which differ in their pathogenic properties. The serotypes cause four major human diseases: endemic trachoma (caused by serotypes A–C); sexually transmitted diseases and conjunctivitis with inclusions (serotypes E and K); venereal lymphogranuloma (L1–L3). Endemic trachoma leads to blindness, while inclusion conjunctivitis is associated with sexually transmitted form and does not lead to blindness. Another factor of the virulence is the unique cell structure of the wall of chlamydia. Research has shown that due to their cell wall, chlamydiae are able to inhibit the phagolysosomal fusion in the phagocytes. Chlamydial cell wall is gram-negative and comprises an outer lipopolysaccharide membrane, but does not contain peptidoglycan. The lack of peptidoglycan is demonstrated by the inability to detect muramic acid and antibodies directed against it. It is assumed, however, that chlamydia contain other carboxylated sugar instead of muramic acid [369]. Hatch et al. [193] indicate that chlamydia cell wall contains MOMP, packed with disulfide bonds, and cystein-rich proteins (CRPs), which may be optionally functionally equivalent to the peptidoglycan. The unique structure allows intracellular division and extracellular survival of the chlamydial organisms [193]. According to Allen et al. [370], the extracellular matrix of the chlamydial outer membrane is formed of three CRPs. They are part of the basic structure of EBs and IBs. Since chlamydia are deprived of peptidoglycan layer, these proteins are very important for the pathogenicity of chlamydia. The largest of CRPs is quoted above MOMP constituting ~60% of the total membrane protein [371]. The second largest – the outer membrane protein 2 – has a molecular weight of 58 kDa, and the third – the outer membrane protein 3 – is ~15 kDa. Bavoil and Hsia [372] disclose so-called genes secretion of type III – Type III secretion (TTS) in *Chlamydia spp*. Unlike some bacteria, the genes TTS in chlamydia are scattered in several genomic groups (clusters), and this is probably a consequence of the unique evolutionary path of these microorganisms. The authors put forward a hypothesis that genes TTS expresses proteins of virulence, which are responsible for the intracellular survival of the agent, the modulation of apoptosis, regulation of cellular transporters, acquisition of lipids from the Golgi apparatus and mitochondria, and signaling for later differentiation [372]. In connection with the TTS, Winstanley and Hart [373] introduced the term "islands of pathogenicity," representing a group of TTS genes encoding the virulence.

Schafer and Hermann [374] studied TTS systems owned by the *C. pneumonie*, strain TW183. In attempts to identify the putative virulence factors released from TTS, six genes have been demonstrated immunohistochemically – candidates for factors of virulence: CopN, cpMip, Pkn5, CpB0736, CpB0739, and CpB0856. Excretion of these genes takes place in the membrane of the chlamydial inclusions individually at different times corresponding to different phases of the infectious cycle in each of the inclusions. The positioning of those genes in the inclusions leads to the creation of a local microenvironment in the past, which enables the development, reproduction, and survival of chlamydia, as well as its interaction with the proteins of the cytoplasm into the host cell. Of the six genes (virulence factors), two – Pkn5 and CpB0739 – inhibit the modeling of the structure of the host cell.

According to Hefty and Stephens [375], little is known about gene regulatory mechanism of the main determinant of virulence – TTS system. In an attempt to clarify this question, the authors reveal 10 operons containing 37 genes associated with TTS. They consider that there are not specified so far activators or repressors which regulate specific for each phase of the development cycle of chlamydia expression of the genes of the TTS.

Specialized TTS apparatus is also proven in the genome of *C. muridarum*, which allows it to "inject" the virulence factors into the cytosol of the host cell [375]. Data supporting the role of the TTS as a factor in the chlamydial virulence are also represented in other publications [376].

Sequencing the genome of the *C. abortus* gives hopes for a better understanding of the biology of this chlamydial species and lfor lighting processes and mechanisms determining the pathogenicity of the agent [182]. Focused is attention to areas of the genome that encode proteins such as families of TMH, Inc and Pmp, considered to be strong candidates determining diverse tropism of chlamydia to host and variable clinical manifestations. The authors found an important fact that *C. abortus* lacks genes of toxicity or genes involved in the metabolism of tryptophan that genetically distinguishes them from other chlamydial species [182]. Voigh et al. [377] reported in 2012 about the first comprehensive analysis of the *C. psittaci* genome sequence. They receive valuable information about the genomic architecture of this chlamydial species and admit the ability to detect new candidate genes that shed additional light on the relationship between the pathogen and its hosts.

According to Miyairi et al. [378], the virulence of *C. psittaci* and host immunity are in a complicated relationship that is defined by the fact that the infectious agent modulates the virulence by alteration of host immunity, which is conferred by small chromosomal differences.

Carlson et al. [379] presented data on the primary role of a cryptic 7.5-kb plasmid of infectivity *in vivo*, suggesting that this plasmid is a factor of virulence, controlling (at least partially) the virulence of *C. trachomatis*, by regulating the expression of chromosomal genes. In another study, Donati et al. [387] pointed out that the recombinant plasmid protein of *C. psittaci*, pgp3, can serve as a marker of infection in cats infected with *Ctenocephalides felis* strains containing extrachromosomal DNA elements.

Between the two major patho-bio-types of *C. trachomatis* associated respectively with trachoma and sexually transmitted diseases in man, there are significant genetic and phenotypic differences. As for the differences within the group of the sexually transmitted diseases, there have not yet found reliable attributes or indicators of severity [339]. At this stage, they have been identified a number of candidates for virulence factors. These include the already mentioned putative large cytotoxin, the polymorphic outer membrane autotransporter family of proteins, TTS effectors, stress response proteins, proteins or other regulatory factors produced by the cryptic plasmid, etc. [339].

Foregoing clearly underlines the complexity of the problem of pathogenicity and virulence of chlamydial agents of different types. Among the greatest achievements in deciphering the mechanisms associated with them, there are still ambiguities and contradictions on the virulence factors and genetic prerequisites for their existence and impact. This implies continuing research efforts on the matter.

7

Chlamydial Antigens and Antigenic Analysis

Chlamydia organisms have a complicated, complex antigenic structure, for the learning of which are made many research efforts. However, a number of issues associated with chlamydial antigens and antigenic analysis remain not fully understood. To solve these problems, it is likely to contribute the progress in research on the molecular and genetic level.

The chlamydial agents are obligate intracellular parasite that elaborates antigens on its surface. These antigens are divided into genus-, species-, subspecies-, and serovar-specific determinants.

7.1 Genus-specific (Group-specific) Antigen

All members of the family Chlamydiaceae have antigenically similar glycoproteid which is stable at 100°C and possesses sensitivity to oxidation by sodium periodate. This antigen is also designated as common group-specific antigen, lipopolysaccharide (LPS) or glycoprotein complex [380]. In some publications, it is stated that the main component of the genus-specific antigen – a thermostable polysaccharide complex, contains two constituents: LPS and glycolipid (GLXA) [389, 399, 400]. The group-specific antigens are closely related to the outer membrane of the elementary bodies. In the reproductive cycle, these antigens appear a few hours before the start of the infectivity in Chlamydia culture. Using immunoperoxidase method was established surface localization of the group antigen [92]. In McCoy cells infected with *Chlamydia psittaci* and *Chlamydia trachomatis*, tested analogously and also with the IF technique, the antigen was detected in the outer membrane of chlamydia [381]. The group antigens are prepared by various methods: boiling chlamydia suspension, extraction with diethyl ether, sodium dodecylsulfate, certain acids, bases, water, and others. Precipitation of the antigen is carried out with acetone. An important point in the preparation of antigens is the use of starting chlamydial suspensions with higher concentrations of EBs and achieving a good degree of purification. For our studies, we got ether–acetone antigen in the method modified by us [382]. The essence of this modification is EM control on the strains and the chlamydial suspension serving as the raw material for the preparation of the antigen. Requirement is necessarily high concentration of elementary bodies in the starting material. When this requirement is met, the production of antigens is achieved with high titer, stability, high specificity and durability, as well as the absence of anti-complementary properties [382]. The concentrated and purified antigens prepared using a gel filtration by sepharose are also highly active and specific. Immunochemically it has been found that group-specific antigen is a water-soluble high-molecular-weight polysaccharide representing 2-keto-3-deoxy-octanoic acid with a molecular weight of 200–2000 kD [386]. This acid is critically important for antigenicity [383, 386]. It is destroyed by treatment with potassium periodate. The antigenic properties lie in the carbohydrate backbone of LPS and its specificity is determined by the number and type of linkage between residues of the sugar acid 3-deoxy-d-manno-oct-2-ulosonic acid (Kdo) [395]. *Chlamydia psittaci* contains the Kde trisaccharide and tetrasaccharide. All remaining chlamydial species contain only the Kde trisaccharide. In the serological diagnostics, the antibodies recognize these structures or partial structures thereof [395]. Reeve and Taverne [384] indicate that during the reproductive cycle, these antigens appear a few hours before the appearance of infectivity in chlamydial culture. LPS of chlamydia has the genus-specific epitope [388]. The first report for this epitope was made by Caldwell and Hitchcock [389], which used a monoclonal antibody (MAb) raised against *C. trachomatis* serovariant L2 which recognizes only LPS. This MAb has reacted with elementary and reticulate bodies of all serovariants of *C. trachomatis* and also with all eight strains of *C. psittaci* tested in the study [388, 389]. Group-specific antigens are determined mainly in direct complement-fixation test (CFT), and also in indirect CFT, reaction agglutination, reaction hemagglutination, passive hemagglutination, agar-gel immunodiffusion, ELISA, immunofluorescence, radioisotope precipitation, and allergic skin sample.

In the composition of chlamydia, another group-specific antigen is also known – chlamydial hemagglutinin. This antigen was detected first in the allantoic fluid of CE, infected with the agent of the meningopneumonitis in mice. It was later found in many other chlamydiae. It agglutinates erythrocytes of mice, hamsters, and hens. The hemagglutinin differs from the above group antigens in that it is inactivated by formalin and phenol, and partly by heat (56°C/30 min). Initially, there was speculation that the hemagglutinin is lecithin DNA-protein complex. Later, Watkins et al. [385] have identified a chlamydial hemagglutinin as LPS that agglutinated mouse and rabbit erythrocytes but not human, guinea pig, or pronghorn antelope erythrocytes. The authors found that hemagglutination by chlamydial LPS was not mediated by specific receptor–ligand interactions erythrocytes [385], but it is a property of the altered LPS-coated surface.

7.2 Species-specific Antigens

The first evidence for species specificity of chlamydial antigen is derived from Bedson et al. in 1949 [390]. Caldwell et al. [391, 392] isolated species-specific chlamydial antigen from a soluble to triton x100 fraction of purified particles of the agent of lymphogranuloma venereum (LGV). This antigen is a surface and is resistant to potassium periodate at pH 2.2–10.6, heat-labile (destroyed at 56°C in 30 min). It is also destroyed upon treatment with pronase. It is a polypeptide with an electrophoretic mobility of 0.65 and a molecular weight of 155 kD. As a result of antigenic analyses of the same and similar species-specific antigens by two-dimensional immunoelectrophoresis and IF is established similarity between the antigens of *C. trachomatis* from trachoma and urogenital infections and vice versa, it has been shown that this antigen is lacking in infections with *C. psittaci* – for example, psittacosis and other infections in animals [391–393]. Terskih [21] found the presence of species-specific antigen in EBs of psittacosis strain by CFT that reacts only with a homologous serum. The same author achieve differentiation between ornithosis and other human chlamydioses (without LGV and trachoma because of unsuccessful attempts to obtain species-specific antigens of them) [21]. Martinov et al. [394] achieved species-specific differentiation of certain animal strains chlamydia. Obtaining species-specific antigens of *C. trachomatis* strains in various clinical conditions and effected antigenic analyzes with them revealed the antigenic autonomy of chlamydiae isolated from cases of urogenital infections and inclusion conjunctivitis [244, 396, 397].

Species-specific antigens differentiate *C. trachomatis* from *C. psittaci* and are expressed on the outer membrane. Species-specific antigens contained primarily proteins that range in molecular weight from 155,000 to approximately 40,000. They are localized in the cytoplasmatic membranes of the chlamydiae. These antigens are prepared from an alkali – soluble fractions of sonicated chlamydial membranes in which the group antigen is removed. Species-specific chlamydial antigens can be demonstrated with the prior destruction of group-specific antigen with potassium periodate in a reaction neutralization, reaction neutralization of the toxicity, and CFT. For several types of chlamydia, it has been found that species-specific antigen can be demonstrated in the reaction micro immunofluorescence and precipitation in agar.

The complexity of the preparation and thermolability of species-specific antigen is not conducive to its implementation in practical serological diagnostics. From the above data, it can be concluded that so far there is no uniform method for determining the type of chlamydia by titration of species-specific antigen. Regardless of what was said, research on species-specific antigens is useful for differential diagnosis and serological classification of chlamydia organisms [23, 24].

7.3 Sub-species (Type-specific) Antigens

MAbs to outer-membrane proteins have demonstrated the presence of subspecies-reactive antigenic determinants. The type-specific antigens, similar to the group-specific, are localized in the chlamydial membrane. They are associated with the major outer-membrane protein and are secreted from infected cells as well. It is assumed that these antigens have a role in the binding of the organism to target cells and in an enzymatic process of some type that initiates endocytosis [401]. It is estimated that the molecular weights of these proteins range from 30,000 to 40,000. There is evidence that type-specific antigens of *C. trachomatis* serotype G are lipid haptens, firmly attached to proteins. It is believed that these antigens include surface peptides with molecular weight 40–43 kD [402, 403].

Wang and Grayston [398] found antigenic heterogenicity among different strains of the same chlamydial species as well as among different strains from the same chlamydial infection. This antigenic heterogenicity is due precisely to sub-species (type-specific) antigens. The first report of the isolation and purification of type-specific antigen of *C. trachomatis* type A in a cell culture BHK-21 was made by Hourihan et al. [402]. The molecular weight of the antigen is 30–32 kD.

Due to the introduction of laboratory practice microimmuno-fluorescence test, Wang and Grayston [397, 398] proved the existence of 15 immunotypes of chlamydiae among ocular and urogenital strains: A, B, Ba, C (hyper-endemic trachoma); D, E, F, G, H, I, J, K (infections of the genitals and paratrahoma); L_1, L_2, L_3 (LGV). So established type-specific antigens are also designated as serotypes of *C. trachomatis*. The A antigen is specific for the strains of trachoma in North Africa and the Middle East and Ba antigen – to the strains of trachoma among Indians in North America. Strains B and C are less prevalent. From the group of infections of the genital organs, most prevalent strains are antigens D and E. They are most often cause of urethritis, cervicitis, and inclusion conjunctivitis in adults (paratrahoma) and in newborns (blennorrhea). Next in frequency in this group are serotypes G and F. Antigens of serotypes L_1 and L_2, L_2 especially, often cross-react with antigens from other serotypes. The practical use of the system for serotyping of different strains of *C. trachomatis* showed the existence of new immuno-types of chlamydia, particularly among strains of patients with urethritis, cervicitis with Reiter's syndrome. There is a broad cross-immunological relatedness between the type-specific antigens. It is particularly pronounced in the serotypes of LGV, first L2.

Treharne et al. [426] submitted a modification of the microimmunofluorescence test for routine diagnosis of chlamydial infections. They created four antigenic pools as the first three consist of members of subgroup A Chlamydia (*C. trachomatis*): pool 1, hyperendemic trachoma TRIC agent serotypes A, B, and C; pool 2, paratrachomaTRIC agent serotypes D, E, F, G, H, I, and K; pool 3, LGV agent serotypes LI, L2, and L3. Pool 4 contained four representative isolates of subgroup B Chlamydia (*C. psittaci*).

Type-specific antigens *C. psittaci*, including avian and mammalian strains under the old classification, are also subject to investigations by various authors. Perez-Martinez and Storz [404] explore that antigenic differences of strains belonging to the series of biotypes of bovine, ovine, caprine, equine, feline, porcine, and guinea pig origin were immunotyped by an indirect microimmunofluorescence (IMIF) test. As a result, they were distinguished nine immunotypes. The authors found a close correlation between immunotypes and biotypes of the investigated strains and a pattern of either disease or host specificity could be associated with each immunotype. Andersen [405] successfully serotyped 50 mammalian and avian chlamydial strains using serovar-specific MAbs with the IMIF test.

In terms of *Chlamydophila pneumoniae,* it is known that several studies that are designed to identify specific antigens for this species during human infection have been carried out, which would be useful for the serological diagnosis. Unfortunately, no clear pattern of reactivity indicating *C. pneumoniae* infection has emerged [406]. Obviously, future studies are needed to identify new candidate antigens for developing diagnostic methods and vaccines as well as for a fuller understanding of the role of the agent in coronary artery disease. In this respect, proteomics could contribute significantly by identifying selective antigens or combinations of antigens associated with persistent infection etiologically associated with *C. pneumoniae* [406–408].

7.4 Toxic Antigens

Above we mentioned that the pathogenicity of Chlamydia organisms is linked to the production of toxin. Chlamydial toxins are formed in the process of reproduction and accumulation of the pathogen. They represent labile antigens associated primarily with the EBs. The toxic antigens are produced in different biological systems. It is believed that the toxic antigens have unquestionable role in the pathogenesis of Chlamydia.

Toxic effects of chlamydiae are demonstrated clearly on a model mice by intravenous injection of a chlamydial suspension. The concentrated suspensions of the yolk sac of chick embryos show the most pronounced toxic effects. Mice died within the first 24 h, which is not associated with the propagation of the pathogen. The preliminary immunization of mice with homologous strains or anti-toxic sera results in the neutralization of the toxic phenomena. The cross experiments for neutralizing the strains allow identifying the specific strain – toxic antigens.

The toxin combined with sera of LGV patients fixes complement [409]. According to Kiraly [410], the chlamydial toxin is species-specific and its effect is neutralized by corresponding antisera. Chlamydial toxin has not been isolated in pure form and lacks its chemical description. Gerlach [411] stated that the surface of EBs contains hepatotoxic and nephrotoxic components which have not been isolated and characterized, but they probably are related to the few proteinaceous-specific membrane antigens of the intact elementary bodies. According to other assumptions, chlamydial toxic antigen is an LPS [358, 412, 413]. Schramek et al. [413] reported serological cross reaction of the lipid A component of LPS isolated from *C. psittaci* to the analogous components of the endotoxin of *Coxiella burnetiii*. In addition to the above-cited scientific findings of Hosseinzadeh et al. [358] with respect to *C. trachomatis*-induced cell death in sperm, there is the fact that LPS negating agents have been shown to inhibit the spermatocidal properties of elementary bodies. Furthermore, it has been shown that LPS can induce sperm to generate reactive oxygen species which may be a component of the toxicity [414].

7.5 Non-specific Antigens

In the process of reproduction in the chlamydial bodies, cellular antigens are incorporated. On the other hand, in the chlamydial antigens, non-specific proteins are present. The antigens of the host cell may be incorporated such as native or modified antigens. This phenomenon was observed in the cultivation of chlamydiae in yolk sacs of chicken embryos [92, 415]. Due to the fact that the antigens of the host have been incorporated into chlamydia, using a high titre specific sera can recognize if a strain is of avian or mammalian origin. Newhall et al. [416] held antigenic analysis of surface proteins of *C. trachomatis* and found "non-specific proteins" with molecular weight 29 kD which reacted with normal human immunoglobulins.

The lipid component A of the chlamydial group-specific antigen is an unequivocal with the same antigen of the LPS of *C. burnetii*. The retention of this component in the diagnostic antigens can be the cause of cross-serological reactions. Also known are one-sided or incomplete cross reactions between chlamydial antibodies and antigens of certain Gram-negative bacteria, *Acinetobacter anitratus* (formerly *Herella sp.*), etc. There is evidence that chlamydial genus-specific antigen reacts with IgG antibodies against *Legionella sp*. Brade et al. [417] explore the LPS antigen of *C. trachomatis* and found that this antigen induced two different antibody specificities which reacted with different antigenic determinants of chiamydial LPS. The first of these antibodies cross reacted with enterobacterial Re LPS, recognizing a structure which is shared by both LPSs. The reactivity of the second antibody was restricted to chiamydial LPS. It is clear that chlamydial LPS possesses two distinct antigenic determinants, one of which is *C. trachomatis*-specific and the other of which is responsible for the cross-reactivity with enterobacterial Re-type LPS. In this process, chlamydial free lipid A plays a major role, which based on its antigenicity, cross reacts with free enterobacterial lipid A [417].

Essig et al. [418] carried out analysis of the humoral immune response to *C. pneumoniae* by immunoblotting and immunoprecipitation. They found that 100% of the Chlamydia-affected patients and 80% of the control sera obtained from adult healthy blood donors recognized a 46-kDa antigen which is non-specific for *C. pneumoniae* infection. Studies of Hunter et al. [419] have shown that certain lectins can agglutinate and precipitate the chlamydial elementary bodies, human serum and bacteria.

From the data regarding the non-specific antigenic components of chlamydia, it is evident that chlamydial agents like bacteria and viruses may contain non-specific antigens.

7.6 Antigenic Analysis

The surface antigens play a very important role in functions such as pathogenicity and virulence, infectivity and toxicity, and inhibition of metabolic and lysosomal activity of the cells. According to Schachter and Caldwell [420], the type-specific antigens bear the greatest responsibility for the biological activity of chlamydiae. They represent the final determinants of the general macro-molecular complex antigen of Chlamydia organisms. In a study of several *C. trachomatis* serotypes (B, D, G, H, and L_2) and of the *C. psittaci* meningopneumonitis, strain has been found that each of the reactions occurred with the *C. psittaci* meningopneumonitis polypeptide, indicating that *C. trachomatis* MOMPs were immunologically related; however, there is no immunological cross that the MOMPs are antigenically distinct among members of these two chlamydial species. In addition, it has been found that MOMPs of various strains of *C. trachomatis* are antigenically complex and that antigenic heterogeneity exists among the surface-exposed portions of the protein [421]. Similar conclusions have been drawn in another publication [422].

Chlamydial agents perform some different degree of antigenicity depending on the localization of the infection. In local infections, they induce low antigenicity and vice versa at systemic infections – high antigenicity. Cevenini et al. [423] analyzed the antigen specificity in urethritis in men and concluded that only patients with acute infections react with MOMP, whereas sera of patients with lighter infections ongoing react with the genus-specific antigen and also to other similar proteins. The universal nature of chlamydial genus-specific antigen is emphasized in a number of works. For example, Dhindra et al. [424] found that in the serum of different animal species, chlamydial antibodies against group antigen from *C. trachomatis* are detected.

Schachter et al. [425] perform typing of eighth ovine strains by plaque method with homologous and heterologous antisera and achieve a differentiation of two serotypes. The first involves the strains isolated from sheep with chlamydial abortion and one fecal strain of clinically healthy sheep. In the second serotype, the authors placed strains isolated in polyarthritis and conjunctivitis.

Immunochemical analysis of the antigenic determinants of the *C. trachomatis* is made with MAbs of Matikainen and Terho [427]. They found that serotype L2 has different epitopes – two separate type-specific antigens and a species-specific epitope. These three epitopes are destroyed by proteolytic enzymes. Group-specific epitopes are resistant to proteolysis, but show sensitivity to potassium periodate.

For antigenic analysis of Chlamydiae, the method for the determination of antigen specificity of the serological response by immunoblotting [423] and the method of determining the immune reactivity of the proteins by blocking them with Twin 20 [428] were also used. Ohtani et al. [429] carried out in Japan genetic and antigenic analysis of *Chlamydia pecorum* strains isolated from intestines of calves with diarrhea. The isolates have been identified by PCR, sequence, and phylogenetic analysis. The PCR-positive results were obtained when genus-specific and species-specific *omp1* genes of *C. pecorum* were targeted. Upon antigenic characterization of the isolates, it was found that their 38–40 kDa MOMP protein has strongly reacted with rabbit antiserum raised against the Bo/Yokohama strain serving as a reference for Japan. For purposes of comparison of the antigenic properties of the new isolates with those of the above-mentioned and other strains of C. pecorum have also been used immunoblot technics and cross-neutralization tests with rabbit antisera [429].

Longbottom et al. [430] reported proteomic analysis of *Chlamydia abortus* outer membrane protein fractions by liquid chromatography combined with electrospray tandem mass spectroscopy. McCoy cells have been used, infected with the known strain S26/3 – causative agent of chlamydial abortion in sheep; 79 proteins have been identified, including MOMP, 14 *Pmps*, *dnaK, OmcB*, 10 putative membrane proteins, and 15 hypothetical proteins. In five proteins, Pmps have been obtained evidence suggestive that participate directly as autotransporters or have a supporting role in this function [430].

In another advanced study, Liu et al. [431] identify *C. trachomatis* outer membrane complex (COMC) proteins by differential proteomics. As is known, the vast majority of proteins in COMC are represented mainly by MOMP and cysteine-rich proteins (*OmcA* and *OmcB*). In COMC, there are also less-abundant proteins, identification of which is hampered mainly by methodological reasons. In this study, they were used in parallel liquid chromatography-mass spectrometry/mass spectrometry analyses of *C. trachomatis* serovar L2 434/Bu EB, COMC, and Sarkosylsoluble EB fractions to identify proteins enriched or depleted from COMC [431]. In contrast to the above-mentioned well-described proteins found in abundant quantities, a variety of other COMC-associated proteins are detected that are not COMC components or are not stably associated with COMC. In addition, some novel proteins arc found, which definitely contribute to a fuller understanding of the antigenic structure of Chlamydiae and related processes and phenomena [430].

There are other contemporary works that, in addition to the classical methodology, include biochemical and molecular biological research. Summary of the literature shows that in recent years there has been significant progress in the antigenic analysis of Chlamydia organisms. However, still the antigen structure and functions of its components remain insufficiently studied. In this regard, some authors believe that one of the reasons is the poorly immunogenic properties of the chlamydiae. With these, the authors would not agree because, as noted above, chlamydia pathogens are quite different as immunogens – known are strong and weak pathobiotypes and strains.

8

Immunity and Antibodies

8.1 Introduction

Natural immunity in chlamydia does not exist. This is demonstrated by their wide distribution. In domestic mammals, more than 45 infections caused by chlamydiae are established. Kaleta and Taday [432] examine the avian host range of the chlamydial agents based on isolation, antigen detection, and serology, and presented a list of chlamydia-positive birds which contains a total of 469 domestic, free-living, or pet bird species in 30 orders. The Chlamydia-induced infections in humans are also numerous. After illness from infections caused by *Chlamydia* spp., tense immunity and resistance to re-infection cannot always be created [21, 433, 434]. With the absence of strenuous immunity, the cases of latent course of these infections can be explained. In birds, latent infection is the main. In such cases, in the blood, specific complement-fixing (CF) and agglutinating antibodies are detected. At people who are in constant contact with birds, antibodies are established in a significant percentage. It is also known that in humans with chronic or latent form of ornithosis, the CF antibodies present in a prolonged time in the blood. Increased resistance is noted in sheep after chlamydial abortion. Such sheep do not usually miscarry a second time [13, 23, 27, 37]. Rodolakis and Souriau [435] reported that immunity in pregnant ewes experimentally infected with the agent of chlamydial abortion on the 19th day after insemination protects against intradermal challenge on day 70 of pregnancy. Such sheep do not abort and do not emit chlamydiae. The authors found that immunity built after natural primary infection is less reliable at a similar challenge, as sheep give birth normally, but they all emit chlamydiae [435]. Cows aborted due to chlamydial infection, not miscarry a second time, which speaks of immunity built after the first abortion. Studies of immunity in *Chlamydia pneumonia* in calves are insufficient. In fact, it is not known whether after natural infection calves acquire immunity and which components of the immune system determine the same. Judging by the high temperature and other clinical manifestations of calves which are ill for a second time from chlamydial pneumonia, it can be concluded that there is no strong resistance to this infection after the initial illness. Relatively little is known about the dynamics of antibodies in spontaneous and experimental chlamydial infection in many species [19, 23, 436].

8.2 Cellular Immunity and Immunomorphology

Cellular (cell-mediated) immunity is an protective immune mechanism that does not involve antibodies, but rather involves the activation of phagocytes, antigen-specific cytotoxic T-lymphocytes, and the release of various cytokines and chemokines in response to an antigen [439]. There are a number of contemporary studies and data which testify that immunity to chlamydial diseases involves cell-mediated immune responses [230, 440–445].

In the preceding decades, basic evidence has been obtained about the existence and role of cellular immunity in chlamydia-induced infections. It has been found that the ornithosis agent is destroyed in leukocytes and monocytes of immune guinea pigs. Significant immunity studies on ornithosis have been carried out by Terskih [21], emphasizing the great role of local tissue immunity in this disease. According to Page [448], cell-mediated immunity is of major importance for the immunity of turkeys. Kiyazimova and Smorodinzev [446] found that the resistance of immune mice was related to the reactivity of the macrophages. Ahmad et al. [447] have examined the immunity of guinea pigs affected by inclusion conjunctivitis (GPIC) and found that it is associated with secretory IgA anti-GPIC antibodies. McKercher et al. [449] considered that the immunity to chlamydial abortion in cattle was of a cell-mediated type.

Shatkin and Mavrov [92] show that in human urogenital infections, the affected tissues are infiltrated with lymphoid cells. In a dynamic study of the cytological composition of preparations from a muco-scraped material, polynuclear leukocytes are initially found, and then lymphocyte infiltration of the affected tissues occurs. In LGV establishes delayed hypersensitivity that occurs in response to intradermal introduction of allergens. This suggests the existence of cellular immunity. Brunham et al. [450] describe lymphocyte transformation (LT) stimulated by a chlamydial antigen. By *in vitro* blast transformation of peripheral blood lymphocytes of patients with chlamydial urethritis and cervicitis, these authors study the cellular immune response to *Chlamydia trachomatis*. They found that lymphocyte blast transformation test metrics are a reliable and useful criterion for fresh infection. In control healthy subjects, LT with a chlamydial antigen is less than 3.5. In men, at a stimulation index of more than 3.5, chlamydias are isolated, but antibodies are not yet detected. In women who are usually examined late, the blast transformation index is also more than 3.5, but the pathogen cannot always be isolated; however, antibodies are detected. Cellular immune response activity is extinguished at 3–4 weeks of disease [450]. LT is also indicated in patients with Reiter's syndrome and anterior uveitis [451].

There is evidence of complete correlation between peritoneal macrophage migration inhibition and guinea pig skin allergy test as manifestations of cell mediated immunity to chlamydial antigens [452]. In mice, chlamydia was found to induce a broad antibody response in B-lymphocytes [453]. By nature, this stimulation is polyclonal. The authors [453] confirm the results of these experimental studies in mice on human patients. Levitt et al. [454] found that the hyperimmunoglobulinemia characteristic of children suffering from chlamydial pneumonia is the result of infection with *C. trachomatis*. The number of mononuclear cells is increased. B lymphocytes and monocytes of sick children are secreted extremely large amount of immunoglobulins – IgM, IgG, and IgA. Enhanced secretion of polyclonal immunoglobulins is detected in peripheral B lymphocytes *in vitro* [455].

Heggie et al. [456] investigated children with chlamydial inclusion conjunctivitis born from mothers with chlamydial cervical infection, and established lymphocyte proliferation to *C. trachomatis* and IgG antibodies in the blood serum. The authors conclude that newborn children acquire cell-mediated immunity transplacentally from infected mothers. However, in experimental conditions in mice, other authors failed to perform adequate transfer of congenital immune cells or specific antibodies to protect mice from uterine infection with *C. trachomatis* [457].

Mechanisms for cellular immunity are the subject of studies by a number of authors. There is a notion that the cytotoxic properties of the chlamydia agent are not essential for this type of immunity [458]. Antibodies are also of no relevance to the interaction of polymorphonuclear cells with Chlamydia. The opsonized with human serum chlamydia containing more than 100 mg of IgG does not stimulate more of their effects with polymorphonuclear leukocytes than non-opsonized [459]. The spontaneous cytotoxicity of blood lymphocytes to *C. trachomatis* did not also differ significantly in healthy and diseased women [460]. Robinson et al. [461] find that complement is essential for the phagocytic activity of mononuclear and polynuclear blood cells. Genital strains of *C. trachomatis* are phagocytized more effectively than the ocular strains. This dependence on complement phagocytosis is directly related to the complement activation and stimulation of chemotaxis by the chlamydial pathogen.

Huang et al. [462] believe that both T and B cell responses are involved in immunity to the agent of ovine chlamydial abortion. The authors investigate the immune responses of the ovine lymph node to this pathogen. For this purpose, they inoculate live chlamydia or control material into the draining area of the popliteal node, sequentially into the draining area of the popliteal node in two groups of sheep – immune and seronegative. In the study of popliteal efferent lymph, it was found that both categories of sheep have responded to *Chlamydia psittaci* with increased outputs of lymphocytes and blast cells. Total T cells (CDS +), helper T cells (CD4 +), cytotoxic/suppressor T cells (CD8 +), and non-helper non-suppressor T cells were maximally 4 and 7 days after challenge in immune and seronegative sheep, respectively. Proportionally, CD4 + T cells declined, CD8 + T cells increased, and T19 cells were unaltered with time after infection. Chlamydial antigens have not been demonstrated in the cells of the efferent lymph by an immunoperoxidase method [462].

In connection with cellular immunity to chlamydia, particular attention is paid to IFN-γ-producing T cells. McCafferty et al. [463] confirmed the importance of IFN-γ in controlling *Chlamydophila abortus in vivo* in a mouse model. In addition to that, IFN-γ has been identified as a clear correlate of protection in sheep by its presence in lymph of immune sheep following *in vivo* challenge with elementary bodies [464]. It is also known the ability of IFN-γ to inhibit *C. abortus* growth in ovine cells [465]. Rocchi et al. [466] emphasize that the precise nature of the cellular protective response remains to be elucidated, including the cell population(s) chieftly responsible for the production of IFN-γ. Special attention has been given in recent years to γ/δ T cells, which are an important component of the ruminant cellular immune system and able to produce IFN-γ [467]. It is very likely that multiple cell types are involved, and their interaction and cytokine profiles, as well as their effector functions, play a decisive role in the protection against ovine chlamydial abortion [466].

A study by del Rio et al. [468] shows that intragastric infection in a mouse model mimics the natural pathway of infection of *C. abortus*, the agent of OEA. In this experimental infection, IFN-γ mRNA expression was detected in the placenta just before parturition and a transient chlamydial colonization of the reproductive tract, with no excretion of *C. abortus* after parturition.

It is obvious that IFN-γ expression in the mouse placenta is associated with resistance to *C. abortus* after intragastric infection, which may be useful for a better understanding of ovine chlamydial abortion immunogenesis.

In a recent study, Poston et al. [469] point out that a major limitation in understanding the development of protective T cell memory against Chlamydia is the difficulty in characterizing low frequencies of antigen-specific T cells. With a view to a better understanding of the kinetics and phenotype of specific CD4 T cells, researchers developed a *Chlamydia muridarum*-specific T-cell receptor (TCR) transgenic (Tg) mouse. In order to better understand the kinetics and phenotype of specific CD4 T cells, the researchers developed a *C. muridarum*-specific TCR Tg mouse. Primed T cells from C57BL/6 mice immune to *C. muridarum* were stimulated with elementary body (EB)/reticulate body (RB) for 5 days and fused with BW5147 cells. Specific clones producing IL-2 and IFN-γ were harvested and screened for TCR Vα and Vβ by qPCR. Tg T cells were analyzed *ex vivo* for activation markers and cytokine production after EB/RB stimulation; 90% of peripheral CD4 + T cells express TCR transgenes, and exhibited enhanced proliferation and increased expression of IFN-γ, IL-2, and CD69, compared to controls. Adoptive transfer of Tg or WT CD4 T cells to Rag1 - / - mice rescued otherwise lethal *C. muridarum* genital infection. The authors concluded that the performance of adoptive transfer of Chlamydia-specific CD4 + T cells offers a powerful approach to characterize cellular activation, differentiation, and memory development [469].

Käser et al. [470] performs cell-mediated immunity studies in pigs to *C. trachomatis* and *Chlamydia suis* infections. After intravaginal and intrauterine inoculations with both types of chlamydia at a dose of 10–8 IFUs, Chlamydia-specific CD4 + T cells, CTLs, and gamma-delta-T cells were detected via flow cytometry while cytokine production was analyzed via multiplex. Proliferation analyses showed pronounced chlamydia-specific CD4 + T-cell response whereas CTLs and gamma-deltaT cells responded less effectively. Multiplex analyzes revealed IFNg and IL17 indicating strong TH1 and TH17 responses. This study sheds light on the porcine cellular immune response to chlamydial infection, which was initially poorly studied. It is hoped that pigs can serve as an appropriate and affordable pre-clinical animal model in chlamydia vaccine development [470].

Cellular immunity has certain immunomorphological characteristics. Qvigstad et al. [471] detected a 70% proliferative T-lymphocyte response from healthy humans and, to a lesser extent, B-lymphocytes as a result of the action of a partially purified suspension of *C. trachomatis* strain

LGV-2. Proliferation is antigen-dependent and antigen-specific. However, it should be emphasized that the HLA-antigen has a limiting effect on T-cell antigen transformation. In the cervix affected by chlamydia infection, deep cytological and histological changes have been described [472, 473]. Histological dysplasia of the cervix occurs quite often in women with glandular ectopia. Parallel to cervical dysplasia, antibodies to *C. trachomatis* are detected. Mitac et al. [473] make a detailed immunohistochemical analysis of chlamydial cervicitis and cervical intraepithelial neoplasia. In cervical biopsies infected with chlamydia, they detect squamous atypia. The discovery in histological examination of squamous atypia in the transformed areas raises suspicion of cervical intraepithelial neoplasia. Our EM data from the direct diagnosis of the elementary and initial bodies of the pathogen in cervical biopsies also testify to significant cytological and ultrastructural anomalies in the cervical epithelial tissues [146, 148]. The ultrastructural dysplasia and squamous atypia accompany the high accumulation of chlamydial bodies in the cytoplasm of the cervical cells.

Markey [474] makes a detailed comparison of the pathogens and cellular immune responses elicited by *C. abortus* and *Toxoplasma gondii* in pregnant ewes. In both experimental infections, infiltration with CD3 + T-lymphocytes was associated with the presence of lesions and was more pronounced at later time points and/or with increasing lesion severity. In lesions caused by *C. abortus*, a more pronounced lymphocytic infiltration, apparently reflecting the more pronounced inflammatory response, was found. The most numerous were T-lymphocyte subsets were CD8 + and Ûδ+ T cells. In infected chorionic epithelium have been discovered ulcerations, intracytoplasmic *C. abortus* inclusions and accumulations of neutrophils [474].

8.3 Humoral Immunity

Numerous initial studies on antibody-mediated protection (AMI) against Chlamydia and other intracellular pathogens were followed in the mid or late nineties of the twentieth century by "lulling" and neglect as many immunologists believed that the function of AMI was well understood and was no longer deserving of intensive investigations [475]. The subsequent description and use of monoclonal antibodies (mAbs) and the great advancement of science at the molecular level have led to a reassessment of the role of humoral immunity that has been placed in the shadow of cellular immunity, considered an indisputable host defense factor. Clarification of this important issue is reflected in several recent publications. For example, Lin-Xi and McSorley [476] summarize recent progress on the role of B cells during Chlamydia genital tract infections and discuss how B cells and humoral immunity make an effective contribution to host defense against important intracellular pathogens, including *Chlamydia* spp. A similar analysis on the role of B cells in humoral immunity has been made by Redgrove and McLaughlin [494]. They point out three basic features of B cells that determine their role in modulating immunity against the chlamydial infection: antibody-mediated neutralization, related to their ability to produce antibodies against chlamydial peptides [495]; an antibody-dependent cellular cytotoxicity (ADCC) mechanism that targets cells that have antibodies attached to their surface for lysis [496]; and supporting role in the formation of antibody–antigen complexes that bind to receptors on APC. The last then enhances phagocytosis and antigen presentation to $CD4^+$ T Cells [497].

8.4 Polyclonal Antibodies

These antibodies are also referred to as conventional. Infections caused by chlamydia are accompanied by the formation and presence of antibodies in blood serum and in a number of secretions. The amount of antibodies varies with the individual infections. For example, in the case of chlamydial urethritis in men, it is not particularly high. Antibody titers are higher in ascending genital infections – salpingitis, bilateral obliteration of fallopian tubes, pelvic inflammatory disease (PID), chlamydial peritonitis, and perihepatitis. The immunogenesis involves both general and local mechanisms of the immune system. When analyzing questions regarding humoral immunity, account should be taken of both its objective features and the imperfections of the serological tests for its study. In serological tests of humans and animals, antibodies are detected not in all patients and different titer values are measured. Often, antibodies are found in individuals who have had disease, and not infrequently in those who are clinically healthy. For a long time, the traditional serological reaction is the complement fixation test. In 1970, Wang and Grayston [398] developed a new microtiter indirect immunofluorescence (IIF) test, which produced many AMI data. A major function decade later in the laboratory diagnostic and research enters the enzyme-linked immunosorbent assay (ELISA).

The general indicators of the humoral response in the various chlamydial infections can be represented by immunoglobulins – IgA, IgG, and IgM. IgM class antibodies are usually detected in blood serum 1–2 weeks after infection and are present for an average of 30 days in untreated patients. These are early, fast transient antibodies. The IgG antibodies are also early. They accumulate in the serum at most 1–2 months after infection, but can remain at different levels for months and years. It should be noted that while IgM and IgG antibodies accumulate in blood serum, in the tears and cervical secretion, IgA and IgG antibodies are found. It becomes clear that IgM class antibodies are available for a short time and then disappear. They are most indicative of the diagnosis of the infection. What is more, their growth has been established during the monthlong period of their existence. But even the presence of IgM in a serum sample allows to be suspected an ongoing infection. In males with chlamydial urethritis, IgM antibodies are detected in 80%. Prolonged observations revealed seroconversion of IgG antibody titers in duplicate serum samples. According to Richmond and Gaul [477] and Taylor-Robinson and Thomas [437], serum antibodies were found in 86–88% of women affected by cervicitis. These data refer to patients in whom the chlamydia pathogen has been isolated in cell cultures. In cases where the agent was not isolated, the seropositive women were much less. The same applies to seropositive men. The greater incidence of humoral responses and higher titers of chlamydial IgG antibodies in women is explained by the significant prevalence of the oligo-symptomatic course of chlamydial infections, as well as the extensive organ-tissue areas and the high intensity of the antigenic irritation of the body's immune system. In women with cervicitis, IgM antibodies are detected in about 30%. IgG antibodies can be found in a current infection, but can also be evidence of a past infection. Establishing ongoing chlamydial infection by the classical method of seroconversion is rarely achieved. Cremer et al. [398] analyze the polyclonal nature of chlamydial antibodies. They find that there are anomalies in the antibody response when the fourfold increase in antibody does not always help in the causal diagnosis of the current infection. The reason for this is probably the serological response of the patient subject to the different proteins of *C. trachomatis*: 60–62 kDa proteins were the main that bind antibodies, while MOMP (29 kDa) responds more poorly with the antibodies [416]. Newer et al. [479] investigated the humoral immune response to membrane components of *C. trachomatis* in *in vitro* fertilization (IVF) patients. They detected a significant serological response (IgG and IgA antibodies) in the blood serum and follicular fluid against two antigens of the causative microorganism – MOMP and a recombinant lipopolysaccharide. In addition, the authors reported that they have determined the expression of human 60-kDa heat shock protein (hsp60) in the follicular fluid. These data provide further support for the possibility that a persistent upper genital tract chlamydial infection contributes to IVF failure in some women [479].

Domeika et al. [480] examined the association between humoral immunity to unique and conserved epitopes of Chlamydia trachomatis 60-kDa hsp60 and immunity to human hsp60 in women with PID. An ELISA was used to detect antichlamydial IgG and IgA antibodies, IgG antibodies to recombinant human hsp60, and antibodies to two synthetic peptides of chlamydial hsp60. Half of the patients had antibodies to human hsp60, which correlated with the presence of antibodies to the chlamydial hsp60 peptide 260–271 homologous to human hsp60. Antibodies to peptide 260–271 were associated with antichlamydial IgG and IgA. These results support the hypothesis that chlamydial infection of the upper genital tract is followed by autoimmune response to human hsp60, most likely as a consequence of an immune response to an epitope of chlamydial hsp60 cross reactive with human hsp60 [480].

The onset of chlamydial infection most often precedes the clinical manifestations, resulting in delayed demand for laboratory and physician assistance [24]. Such an opinion is also expressed by Lin-Xi and McSorley [476], namely, that the control of Chlamydia infection is hindered by the asymptomatic nature of initial infection but the consequence of untreated infection seriously threatens the reproductive health of young women.

In addition to blood serum, antibodies are found in the cervical secretions, but their titers are low. IgG and IgA antibodies with titers of 1:8 have been detected. Osse and Persson [481] study the significance of systemic and local humoral immunity to *C. trachomatis* in post-abortal pelvic infection. Chlamydial antibodies have been determined in serum and cervical secretions. No difference in the frequency of local IgA antibodies in chlamydia-positive women has been observed between those with and those without complications. The results obtained show that serum antibodies seemed to offer some protection against salpingitis in chlamydia-positive cases, whereas local antibodies did not. Such antibodies are also found in tears, milk, synovial fluid, and peritoneal exudate. In colostrum and milk of women with chlamydial infection of the lower genitalia, there are secretory IgA antibodies but no correlation is found between the level of IgA antibodies in milk and antibodies in blood serum [482]. Milk antibodies appear to be produced as a result of local immune mechanisms.

In case of re-infections, IgG antibodies are usually determined, or the titers of this immunoglobulin class are increased. In reinfection with a strain different from the first, appear IgM antibodies to the new strain and IgG antibodies to the old strain. Schachter et al. [483] examine the humoral immunity of men to chlamydial infection in order to establish specific protection against re-infection. They found that males in California have a high level of IgG antibodies: those with post-gonococcal urethritis have antibodies in 93.6%, with non-gonococcal urethritis – 83%, and those without urethritis – in 65%. However, the degree of isolation of the agent is low. Thus, the widespread prevalence of chlamydial antibodies among the male population testifies to immune protection that prevents to a certain extent the re-infections. Immunity in re-infection of the genital tract has been examined experimentally in marmoseti [484]. This study shows that the infection can persist in the presence of antibodies and that immunity is not only sero-specific or type-specific – it is polyclonal. On the other hand, antibodies are genetically dependent on the type of host of the infection and their protective effect is also genetically dependent. Thus, antibodies only best protect individuals of the same species [485]. Another sero-epidemiological study has shown that genital infections caused simultaneously by typical *C. trachomatis* genital strains and trachoma sero-variants have been detected in certain regions and human populations [486]. The polyclonal humoral immunity in chlamydia is often complicated due to mixed infections with strains belonging to different serovars.

Brunham [487] indicates that the rodent models of *C. muridarum* infection have yielded some of the most useful insights into human *C. trachomatis* immune biology. In the murine model, CD4 T cells secreting interferon gamma correlate with clearance of primary infection, and CD4 T cells or antibody correlates with resistance to reinfection [488]. The reason why there are such differences in immune protection between primary infection and reinfection is still unclear [487]. According to Moore et al. [489] and Farris et al. [490], antibody-mediated immunity depends on receptors to the constant region fragment (Fc portion) and is directed to surface molecules on the organism, suggesting a role for antibody in amplifying antigen-presenting cell functions and T cell immunity.

Mice and guinea pigs produce a strong antibody response as a result of chlamydial genital infection that is long lasting in serum but relatively short lived in genital secretions. In guinea pigs infected intravaginally with *Chlamydophila caviae*, antibody appears to be essential for immunity to reinfection. The data for mice also strongly support an important role for antibody in resistance to reinfection [298, 300, 441]. Based on experiments with guinea pigs infected in the genital tract with the GPIC agent, Rank et al.

[298] conclude that humoral immunity is essential for the recovery of female guinea pigs from this infection. In another study, Hafner et al. [498] found that immunization of guinea pigs with a chimeric peptide consisting of a variable domain IV and a region known as GP8 from the MOMP of *C. caviae* (formerly GPIC pathogen) induces good vaginal secretion antibody response, *in vitro* neutralization, and partial protection against live challenge.

The question of the mechanism by which antibodies cause resolution of the chlamydial infection in these animal models *in vivo* remains unclear. There is evidence from *in vitro* experiments that antibodies can block chlamydial infection via neutralization [491]. According to another study *in vitro* [492], antibodies may opsonize EBs and enhance uptake by phagocytes. It is possible that both mechanisms are operative *in vivo* [441].

Armitage's 2012 studies [493] show that transporters play an important role in the function of antibodies. When urogenitally infected with *C. muridarum*, both male and female reproductive tracts up-regulated expression of polymeric immunoglobulin receptor (pIgR) and down-regulated expression of neonatal Fc receptor (FcRn). Transport of IgA and IgG into the mucosal lumen is facilitated by receptor-mediated transcytosis yet the expression profile (under normal conditions and during urogenital chlamydial infection) of the polymeric and a strong correlation immunoglobulin receptor (pIgR) and the FcRn remains unknown. The author believes that these immunoglobulins play a dichotomous role in chlamydial infections, and are dependent on antigen specificity, and FcRn and pIgR expression. FcRn was found to be highly expressed in the upper male reproductive tract, whereas pIgR was dominantly expressed in the lower reproductive tract [493].

After an experimental intravaginal infection and re-infection of pigs with the *C. suis*, de Clercq et al. [499] establish inflammation and pathology in the genital tract, as well as cellular and humoral immune responses. A certain level of protection following the initial infection has been established. Protective immunity against re-infection coincided with higher Chlamydia-specific IgG and IgA antibody titers in sera and vaginal secretions, higher proliferative responses of peripheral blood mononuclear cells (PBMC), higher percentages of blood B lymphocytes, monocytes, and CD8+ T cells, and upregulated production of IFN-γ and IL-10 by PBMC [499].

In a recent publication, Khan et al. [500] reported a study of the antibody-mediated immune responses in koalas either naturally infected with *Chlamydia pecorum* or following administration of a recombinant chlamydial MOMP vaccine. In spontaneously infected coals, the authors found very low levels of *C. pecorum*-specific neutralizing antibodies between low IgG total titers/neutralizing antibody levels and higher *C. pecorum* infection load. In vaccinated koalas, it was found that the vaccine was able to boost the humoral immune response by inducing strong levels of *C. pecorum*-specific neutralizing antibodies. A detailed characterization of the MOMP epitope response was also performed in a naturally infected and vaccinated coalas using a PepScan epitope approach. Results obtained definitely showed the importance of antibodies in chlamydial infection and immunity following vaccination in the koala. [500].

One of the most recent work on immunity is that of Hagemann et al. [501], relating to humoral immune response against surface and virulence-associated *C. abortus* proteins in ovine and human abortion. It was found that aborting sheep exhibited a strong antibody response against surface (MOMP, MIP, and Pmp13G) and virulence-associated [chlamydial protease-like activity factor (CPAF), TARP, and SINC] antigens. Antibodies against the cited proteins associated with virulence disappeared within 18 weeks of abortion. Conversely, the antibodies to the surface proteins in most animals were retained after the observation period (18 months). It is indicative that experimentally infected sheep that did not abort developed antibodies primarily against the indicated surface antigens, and the duration of their storage was shorter. As is well known, *C. abortus* may cause abortions in exposed women. In such cases, Hagemann et al. [501] have found specific antibodies against the above-mentioned surface antigens. In contrast, in the control group of individuals who are at risk due to contact with infected sheep (shepherds and veterinary staff), there have been only single cases of an immune response to the antigens selected.

8.5 Monoclonal Antibodies

Casadevall and Pirotski [475] underline that beginning in the 1990s' studies using mAbs revealed new functions for antibodies, including direct antimicrobial effects and their ability to modify host inflammatory and cellular responses.

The pre-history of mAbs covers a number of studies from the eighties of the twentieth century. Fuentes et al. [502] have received 11 stable hybridoma clones by fusing myeloma cells with splenocytes from *C. psittaci* immunized mice from the F1 line. Hybridoma cells secrete antibodies that react with 12 reference chlamydial strains – five strains of *C. psittaci* and seven strains of *C. trachomatis*. Ten of these mAbs refer to genera-specific epitopes. These epitopes are chlamydial proteins [40,000 molecular weight (m.w.) components]. The localization of the various surface antigens on *C. trachomatis* was demonstrated by immunoelectronic microscopy with mAbs from Clark et al. [503]. These authors demonstrate by indirect immune-ferritin EM technique binding of antibodies to the outer membrane surface of RBs and EBs. Two different ways of coupling have been demonstrated – spot-like and diffuse. There is a direct correlation between the diffusion of antibodies to the outer membrane surface and the neutralization of *C. trachomatis in vitro*. This part of the antibodies, which are neutralized and are complement-dependent, is characterized by a diffusion location on the outer membrane surface. Other antibodies are not neutralized and spread in a spot-like type [503]. The characteristics and specificity of mAbs are studied on 19 independent hybridoma cell lines [504]. They are obtained by fusing mouse myeloma cells with lymphocytes from mice immunized with *C. trachomatis*. The mAbs produced by these hybridoma lines in the culture fluid were tested against EBs and RBs by immunofluorescence. MAbs that interact with genus-specific antigens are found to react primarily with RBs. Antibodies that interact with species-, subspecies- and type-specific antigens react equally with RBs and EBs. The physicochemical characterization of the antigens recognized by different mAbs is made by heating, pronase digestion, periode oxidation, and immunoblot techniques. The genus-specific antigen was a heat-stable, pronase-resistant, and relatively periodate-sensitive component of less than 10,000 m.w. The species-, subspecies-, and type-specific antigens were heat stable, pronase sensitive, and periodate resistant. The antibodies that detected species- and subspecies-specific antigens predominantly reacted in immunoblots with the 40,000 m.w. MOMP. The authors argue that these mAbs provide a new approach for the precise serological classification and detection of different *C. trachomatis* strains [504].

Monoclonal analysis of the genus-specific antigen is made by Caldwell and Hitchcock [389]. The mAbs are obtained by fusing mouse myeloma cells with spleen cells from mice BALB/C, immunized with formalin-inactivated EBs of *C. trachomatis* serovar. The specificity of these mAbs was determined in an ELISA against antigens of *C. trachomatis* serovars D, G, H, I, and L_2 and antigens of *C. psittaci*. The genus-specific mAb reacts with *C. trachomatis* and *C. psittaci* and is designated as L-21-6. This antibody reacted with all serovars of *C. trachomatis* and *C. psittaci* from five different animal species. The immunoreactivity of the genus-specific epitope was heat resistant (100°C, 10 min) but was destroyed by sodium metaperiodate treatment. Biochemically, this epitope is a lipopolysaccharide or glycolipid that is located on the chlamydal lipopolysaccharide. It contains 8.8%

2-keto-3-deoxyoctulosonic acid. The mAb against the genespecific antigen was examined by immunoblotting analyses against isolated LPSs extracted from *Neisseria gonorrhoeae*, *Salmonella typhimurium*, and *Escherichia coli*. As a result, the mAb L2I-6 did not bind LPS of these organisms, demonstrating that the chlamydial genus-specific LPS epitope is apparently not shared by these Gram-negative bacteria. Conversely, the polyclonal antibodies cross react with bacterial LPSs. It is clear that LPSs of chlamydia have at least three antigenic regions, two of which are close to LPSs of Gram-negative bacterial organisms, but one is an antigenic epitope for the highly specific L-21-6 mAb [389]. These findings are confirmed once again in a recent publication by Gillespie [505], which states that two mAbs are available: one is directed to the outer membrane protein and is species-specific and the other is directed to the genus-specific lipopolysaccharide and can detect all *Chlamydia* spp., but will not be able to distinguish between them.

The aggregated research data show indisputable importance of mAbs to antigenic analysis of chlamydia.

Alexander et al. [506] use the genus reactive mAb in rapid identification of *C. trachomatis*. They visualize EBs by direct immunofluorescence in genital secretions from the urethra and the cervix of 250 unselected patients. The results obtained have shown that overall the direct IF test with the genus specific mAb had a sensitivity of 90% and a specificity of 97%. In our initial studies by direct IF with the genus specific mAb to prove *C. trachomatis* in cervicitis, we used the Chlamyset monoclonal commercial Kit of Orion, Finland. Cervical secretions from 30 women, previously proven negative for gonococcal and other bacterial infections, trichomonasis and candidiasis, were examined. In 21 cases (70%), infection with *C. trachomatis* was detected [507]. The usefulness of mAbs in another IF test was tested by Forbes et al. [508] and Shafer et al. [509]. According to the authors, the mAb included in the Microtrac assay (Syva Co., California) for detection of chlamydia in cervical and urogenital samples can be used as a screening assay in parallel with cell cultures, especially in low-prevalence populations. However, it is believed that the fluorescein-conjugated mAbs for determining the endocervical infections retreat of the cell cultures by their effectiveness [508, 509]. These data are not consistent with those of Alexander et al. [506] for the high efficiency of genus-specific mAbs in the direct IF test. The IF technics are used by a number of other laboratories. mAbs in the immunofluorescence assay have been used to rapidly detect *Chlamydia* spp. in various clinical conditions [510]. Specific antigens can best be detected by immunofluorescence, particularly using mAbs [511]. Similar to those cited, there are a number of views on the advantages and benefits of using the immunofluorescence method, primarily with mAbs. Despite its usefulness, some authors have reservations about its widespread use in *C. trachomatis* human genital infections. Papp et al. [325] indicate that direct fluorescent antibody (DFA) tests should not be used for routine testing of genital tract specimens. Rather, DFA tests are the only FDA-cleared tests for ocular *C. trachomatis* infections. The procedure requires an experienced microscopist and is labor intensive and time consuming [325].

Kumar and Mittal [512] reported that they have developed a total of 12 murine hybrid clones producing IgG class mAbs to *C. trachomatis* previously isolated from the genital tract of infected women (species-specific, B-serogroup-specific, and serovar-specific). Dot-ELISA was used to check the specificity of the clones and was used to select hybridomas that produced anti-*C. trachomatis* mAb. No cross reactivity has been detected of species-specific, B serogroup-specific, and D serovar-specific anti-MOMP mAbs with other species of Chlamydiae. The sensitivity and specificity of the developed anti-MOMP mAb in the EIA for chlamydial antigen detection have been 91.3% and 94.4% for the species-specific clone, 91.30% and 98.1% for the serovar-specific, and 75.00% and 99.07% for serogroup-specific, compared to the cell culture method.

Recently, Zheng et al. [513] published data on the production of novel mAbs against a recombinant protein equivalent to the immunodominant region of CPAF from *C. pneumoniae*. It was found that the mAbs specifically reacted with the endogenous CPAF antigen of the *C. pneumoniae* strain in immunoblotting and IIF, but did not react with *C. trachomatis* strains isolated from patients with urogenital infections, neither directly with genital secretions from patients affected of acute *C. trachomatis* infection. The established high specificity of the resulting mAbs has enabled the authors to develop new screening tools for the diagnosis of early pediatric pneumonia, namely, a new IIF assay and ELISA to detect *C. pneumoniae* antigen in clinical specimens from child patients suspected of pneumonia. The development of new mAbs to *C. pneumoniae* is reported in a publication of 2017 [514]. ViroStat has introduced three new mAbs that are specific to *C. pneumoniae*. These new products do not show cross reactions with *C. psittaci* and *C. trachomatis*. They function well in the ELISA and indirect IF assay and are considered candidates for rapid test application.

Regarding the mechanism of AMI against chlamydia-induced infections, there are different views. Morrison and Morrison [515] study the humoral immunity to *C. muridarum* genital tract infection. In this model, they found that mAbs to the chlamydial MOMP and LPS conferred significant levels of immunity to reinfection and reduced chlamydial shedding by over 100-fold. These authors suggest that the mechanisms of ADCC are involved in protection. The opinion of Morrison and Morrison [515] does not support the notion that direct neutralization is a major mechanism of AMI against chlamydial infection *in vivo*. Moore et al. [496] have shown that ADCC is an effector mechanism for killing Chlamydia in mature adult hosts. It is known that neonates are born with immature cellular immune system; ADCC may not function well in the presence of maternal antibodies. Under experimental conditions, this may change when newborn mice adoptively receive immune cells from adult mice. It is believed that adoptively transferred adult maternal antibodies and T-cells, by activating their mechanisms, can by common action eliminate Chlamydia infection by ADCC [516].

Sri et al. [517] investigated the time kinetics of mAbs to *C. trachomatis* in order to determine the optimal titer. They found that the highest optical density values and best reactivity in all three hybrid clones were detected at 1:1000 at day 3. The author's interpretation of the results is that a given antibody (clone) titer correlates mainly to its concentration.

9

Serology

9.1 Serological Assays

9.1.1 Complement Fixation Test

The complement fixation test (CFT) is one of the major traditional tests for the demonstration of the presence of specific antigens or antibodies. According to the definition of the English Oxford Living Dictionary, this is a test for infection with a microorganism, which involves measuring the amount of complement available in the serum to bind to an antibody-antigen complex. Complement is a group of serum proteins that are functionally related and play an important role as components of the humoral immune system. They are included in the mechanisms of several biological functions involved in the development of the infectious process. In the course of complement activation, several proteins react in a distinct order similar to the cascade in blood coagulation. In the classical activation, the complement cascade is initiating by complexes of antigens and IgG1, IgG2, IgG3, and IgM antibodies, whereby the latter are the most efficient [527].

The main quality of the CFT is its high specificity. This specificity, of course, refers to the group membership of the relevant species of Chlamydiaceae.

With regard to chlamydiae and the nosological units and syndromes caused by them, the CFT was developed for the first time in 1935 by Bedson in human psittacosis [522]. In the following decades and so far, this reaction has been widely used in the diagnosis of chlamydial infections in animals and humans and for the titration of genus-specific chlamydial antigens [18, 23, 535]. The CFT is the first described lipopolysaccharide (LPS)-based assay. This is a reliable test for the establishment of the group specificity of chlamydial agents isolated from clinical specimens or to determine the group-specific antibodies in the sera of patients and those who have had chlamydial infections.

There are different versions and modifications of CFT. Approximately 30 years after the introduction of the test into laboratory practice, Fraser [523] made a detailed review of its modifications over the years. Typical of CFT are the many preparatory procedures, the sequence of steps in conducting it, and the availability of a number of controls. Essential for the proper functioning and effectiveness of the application of CFT is the presence of qualitative antigens and other necessary reagents and well-processed and stored sera in order to avoid anticomplementary effects. Complement fixing antigens are prepared from YSs of CEs, cell cultures, lungs, or spleens of mice. In parallel with the progress of purification techniques of chlamydia, many researchers prefer to prepare antigens from partially or highly purified Chlamydia suspensions. Significant stability and activity has the antigen, which is prepared from YSs of CEs by the method of Volkert and Christensen [524], modified from Chervonsky and Popova [525]. Further, Ognianov [37] modified the method of Chervonsky and Popova [525]. Martinov and Popov [166, 382] and Popov et al. [250] improved the method for preparation of CF antigens by controlling the strains and output suspensions by means of EM and by using purified suspensions.

Regardless of the group-specific nature of the antigen, there are data on structural differences in LPS of some chlamydial species. Sachse et al. [536] draw attention to a similar fact in *Chlamydia psittaci*, which in addition to the alphaKdo(2 → 8)alphaKdo(2 → 4)alphaKdo trisaccharide contains a structurally different trisaccharide with the structure alphaKdo(2 → 4)alphaKdo(2 → 4)alpha Kdo having lower reactivity with the (2 → 8) – (2 → 4) – linked analog, and also branched tetrasaccharide alphaKdo(2 → 8)[alpha Kdo(2 → 4)]alpha Kdo(2 → 4) alphaKdo [537, 538].

Direct CFT is widely used for the serological diagnosis of chlamydial infections in mammals. In case of suspicion of enzootic abortions in sheep and goats, the animals are examined within 3 months of abortion or parturition. Individual antibody titers and mean geometric titers are reported. For the individual animal, it is advisable to take a minimum positive titer of 1:32, which directs to a probable chlamydial etiology. Adoption of this minimum titer is necessary in order to avoid misinterpretations of low titers, which are sometimes false positive results as a result of cross-antigenic reactions between *Chlamydia abortus* and *Chlamydia pecorum*, as well as with some Gram-negative bacteria (e.g., Acinetobacter [23, 526]). This consideration is necessary, but on the other hand, it should not be forgotten that there is a possibility that this low titer may be specific based on low-grade infection with *C. abortus*. Low-value titers can also be due to subclinical enteric infection with *C. abortus*. Paired sera collected at the time of the abortion and again at least 3 weeks later may reveal that a rising CF antibody titer hat will provide a basis for a retrospective diagnosis. The test has been successfully used in other clinical conditions in small ruminants related etiologically with chlamydia, as well as in large ruminants and pigs. The group-specific nature of the test allows for the study of other animal species by requiring appropriate adaptation of the serological test according to the manufacturer's instructions and accepted variants in the country concerned.

Various variants of CFT are used for serological tests in avian chlamydiosis, each of which has its advantages that we will not discuss in detail. Andersen and Vanrompay [539] characterized CFT as a valuable and generally accepted method for the diagnosis of avian chamydiosis. The OIE [83] recommended in its recent editions a modified direct CF test for the detection of antibodies. This modified version differs from the direct CF test in that normal, unheated chicken serum from chickens without chlamydial antibody is added to the complement dilution. The normal serum increases the sensitivity of the CF procedure so that it can be used to test sera from avian species whose antibodies do not normally fix a guinea-pig complement [83].

In several studies on chlamydial infections in pigs from 1966 to 2016, CFT was successfully used [23, 334, 335, 529–531]. Szymanska-Czerwinska et al. [532] investigate the prevalence of *Chlamydia suis* in the pig population in Poland using CFT. With a comparative purpose, vaginal swabs (n = 277) selected from pigs diagnosed as positive in CFT were examined by PCR. The occurrence of *C. suis* has been confirmed in PCR in 200 cases. Statistical analysis (χ^2 test and Person and Cramer correlation coefficient) has shown that both methods were coincident in the diagnosis of *C. suis* infection in swine.

In another comparative study of CFT and enzyme-linked immunosorbent assay (ELISA) (Bommeli, for group-specific antibodies) and ELISA (Pourqieur, for antibodies against immunogenic proteins with molecular weight (m.w.) 80–90 kDa to *C. abortus*) in serology for Chlamydia in cattle, Niemczuk [533] found that CFT gives results that are significantly comparable to those by ELISA.

In addition to the above-mentioned evidence for the serological diagnosis of chlamydial abortion in sheep by CFT, the data of two other authors' groups are of interest [533, 534]. McCauley et al. [533] perform a comparative evaluation of CFT and three ELISA-based tests for the diagnosis of *C. abortus* infection in sheep. In studies involving *C. psittaci* LPS antigen, they found that CFT performed comparably with the three ELISAs: Pourquier, rOMP90-3, and rOMP90-4 with a sensitivity of 96.4%. Using ovine *C. abortus* LPS antigen, the sensitivity of the CFT is lowered to 60%. These data are obviously a reflection of the above-mentioned structural differences in LPS of *C. psittaci* and *C. abortus*, knowing that this feature has important practical implications for choosing and preparing CFT antigens. In the study under discussion [533], 100% specificity was achieved in the CFT used. Wilson et al. [534] indicate that this result is based only on examinations of field sera from New Zealand, a country free of enzootic abortion of ewes (EAE), and no analyses were performed using sera from *C. pecorum* infected flocks to asses cross reactivity with this species. The authors justify their criticism based on their own comparative studies on CFT and the same three ELISAs. They also found a similarity in the sensitivity of the four tests, but considered that the specificity of CFT could be negatively influenced by the possible cross reactions with arthritogenic and conjunctival subtypes of *C. pecorum* [534]. Walker et al. [659] examined arthritis in sheep in Australia caused by *C. pecorum* and found that CFT was found to be a reliable indicator of chlamydial positivity especially at the flock level.

In terms of human chlamydial infections, CFT is still the most commonly used assay for psittacosis and LGV, but is not sufficiently sensitive for trachoma and oculogenital infections. Indeed, in human urogenital infections, CFT retreats in its sensitivity to ELISA and microimmunofluorescence (MIF), but nevertheless, in prior good selection of patients and using CFT as part of a complex etiological diagnosis involving cytological, culture, and EM methods, it can be useful. For *Chlamydia pneumoniae*, the CFT is intermediate in sensitivity [540]. According to Mabey and Peeling [541], infections caused by LGV serovars of *C. trachomatis* tend to invade the draining lymph nodes resulting in a greater likelihood of detectable systemic antibody response and may aid in the diagnosis of inguinal (but not rectal) disease. The CFT was classically used for this purpose but could be replaced by the more sensitive species-specific MIF test (MIFT) [325, 541].

De Ory et al. [542] emphasize that CFT allows low-cost screening of serum samples for different agents including Chlamydia spp. within a single assay, and is a useful tool for the serological diagnosis of acute respiratory infections. Because CFT is technically quite demanding, time consuming, and labor intensive, an automated version was developed (Seramat). Magnino et al. [543] evaluated this system for detection of CF antibodies in ovine and caprine sera and established good correlation with manually performed tests.

The Seramat automated system represents a significant technical improvement that may enable many clinical laboratories to use the CFT as a routine diagnostic tool [542].

9.1.2 Enzyme-linked Immunosorbent Assay

In the last 30 years, the ELISA has been extensively experimented and applied to detect antibodies and antigens in chlamydial infections. Several modifications of this assay have been tested: micromethod; immunofluorescent variant; rapid version with monoclonal antibodies (mAbs); with purified elementary bodies (EBs); with solubilized EB extracts; with peptides; recombinant antigen preparations; plate-based ELISAs; and solid-phase ELISAs. Various variants of ELISAs produced in the respective research laboratories or commercial kits of different companies have been used over the years. Based on the family specificity of the Chlamydia LPS antigen, a number of laboratories used intended for the detection of *C. trachomatis* immunoassay, for the serological diagnosis of chlamydia in animals. Various ELISA tests have revealed significant variations in the sensitivity and specificity depending on the animal species, clinical and epidemiological situation, and the differing antigen loads in the various types of test samples [536]. For example, the OIE [83] points out that many of the earlier ELISA tests for *C. trachomatis* based on monoclonal or polyclonal antisera against LPS epitopes, some of which were shared with other Gram-negative bacteria, had dubious diagnostic value in avian chlamydiosis. Their use when screening individual birds is questionable, as they lack sensitivity and specificity.

Cevenini et al. [544] determined antibodies in patients with nongonococcal urethritis (NGU) with micro-ELISA, using the concentrated and purified EBs of *C. trachomatis* strain L_2 for antigen. They conclude that ELISA is a fast and sensitive assay and can be used for clinical and seroepidemiological purposes. Many other authors have also used ELISA to detect antibodies. In addition to EBs, purified RBs and extracted group antigen also serve as an antigen. These antigens determine IgA, IgG, and IgM class antibody titers in blood sera. The diagnostic value of ELISA was evaluated in comparative MIF and CFT experiments. Evans and -Robinson [545] find that ELISA is less sensitive than MIF. The mutual coincidence of these two assays is 81%. ELISA is much more sensitive than CFT. Caldwell et al. [546] – one of the authors who performed the earlier studies on the problem – gave the following comparative assessment in favor of ELISA: CFT is specific but has low sensitivity; MIF is highly sensitive but technically difficult to perform in many clinical laboratories; and ELISA is fast and does not give false positive results. Titers by ELISA were higher than those of MIF. The test is recommended for diagnosis and seroepidemiology as a simple and sensitive method. As a result of its application, a number of interesting data have been obtained on the diagnosis and prevalence of chlamydial sexually transmitted infections. Chlamydial antibodies were found to be widespread among patients in venereology and rheumatology clinics (85%), among adults with nonvenereal diseases (20%), among children (20%).

The use of ELISA for the determination of antigens in chlamydial infections is the second major strand in research with this assay. Caldwell and Schachter [546] found that ELISA determines the antigen of major outer membrane protein (MOMP) with m.w. 39.5 kDa. Jones et al. [547] applied an ELISA for the detection of *C. trachomatis* in genital secretions of 206 males with urethritis and 210 women with cervicitis. For the control, the cell culture method was used. The sensitivity of the Chlamydiazyme test was 81% and the specificity was 98%. Overall, Chlamydiazyme provided a rapid (4 h), sensitive, and specific assay for the detection of chlamydial antigens [547]. In another study, the direct detection in ELISA of chlamydial antigen in urethral and cervical swabs had a sensitivity of 92.5% and a specificity of 97.2% [548].

ELISA reveals some laboratory and practical features of antichlamydial mAbs [549]. It examined avidity and specificity of the mAbs from the culture supernatant and in ascitic fluids of mice. Using ELISA, antibodies have been found to be low specific and highly specific. Usually, the antibodies from ascitic fluid are low specific. Some of the antibodies in the cell culture fluid are also low specific. Reverse dependence between the specificity and the avidity of the antibodies is observed. Thus, avidan IgM antibodies from ascitic fluids are not highly specific [549].

The binding range of antichlamydial mAbs with genus-specific, species-specific, and type-specific antigens is broad – ranging from pH 4 to pH 9. However, binding is nonspecific at pH 4–6 [550].

Griffiths et al. [551] adapted to sheep and evaluated for serodiagnosis of EAE a recombinant LPS-based rELISA, which was originally intended for human *C. trachomatis*. In the assay, a deacylated, BSA-conjugated antigen has been used which contains two epitopes – the alpha $2 \rightarrow 4$ linked disaccharide moiety and the $(2 \rightarrow 8)$ linked chlamydia-specific epitope [552]. Although the test was not able to distinguish *C. abortus* and *C. pecorum*, it was more sensitive to initial screening than CFT [551]. Several other commercial LPS-based assays have been developed for which there are controversial and in some cases disturbing data about their sensitivity and especially specificity. This group includes an indirect ELISA (CHEKIT^{TM} Chlamydia Assay (IDEXX Labs, USA), for antibodies against *C. abortus* in ruminants. In two studies using this assay, cross reactions with field sera from sheep spontaneously infected with enteric and arthritogenic subtypes of *C. pecorum* and with sera from SPF lambs experimentally infected with arthritogenic and conjunctival subtypes of *C. pecorum* are established [534, 553]. LPS-based assay is also ImmunoComb (Orgenics, Israel), based on a solid phase with a purified chlamydial antigen attached to a plastic comb. It is known as a comfortable and easy-to-use, high sensitivity test. In our own research, we also confirmed these qualities of ImmunoComb. Jones et al. [554] compared the effectiveness of five tests for detection of antibodies against EAE and found that ImmunoComb was the most sensitive but least specific compared to two ELISAs, CFT and immunoblotting. In another study [534], the high sensitivity of the test was questioned due to the detection of a high percentage of false-positive results.

Another group of ELISA assays are based on MOMP – one of the three main immunodomnant antigens on the surface of the chlamydial EB – LPS, MOMP, and polymorphic outer membrane proteins (POMPs). Herein it refers to competitive ELISA (cELISA), the base of which is the binding of specific mAbs against the MOMP variable segments VD1 or VD2 that is inhibited by the presence of serum antibodies [559]. Vretou et al. [555] indicate that the inhibition process is dependent on the affinity of the competing antibodies and their amount, and these antibodies may recognize linear or conformational epitopes of the native MOMP. In a later study, Vretou et al. [553] made the evaluation of this cELISA assay with well-documented reference sera and a cut-off at 50% inhibition, and found a specificity of 98.1% and a sensitivity of 77.7%.

Morre et al. [557] compared three commercially available peptide-based ELISA systems for detection of *C. trachomatis* IgG and IgA antibodies with MIF assay and found that the three ELISA tests performed as well as the MIF. The authors concluded that these assays might be useful for the serodiagnosis, as they are less expensive, less time consuming, and easer to perform than the MIF assay.

The test Panclabort, commonly referred to as MOMP-P (dianoSTI, Scotland) belongs to the group of MOMP-based assays. In its base, synthetic peptides are derived from VD1 and VD2 amino acid sequences. The specificity of this assay was 95.9% and the sensitivity was 70.4% [533]. In a recent work, Menon et al. [558] developed new multipeptide ELISA that is highly specific for the detection of tubal factor infertility in women related to *C. trachomatis* ($p = 0.011$) compared to another ELISA ($p = 0.022$) and MIF ($p = 0.099$). The authors indicate that the sensitivity of the assay should be improved before clinical utility.

A third set of ELISA assays are based on POMPs [556, 560, 561]. The members of this multigene family are rather complex and immunoreactive antigens comprising antigens present in the Chlamydia organisms. Current research has shown encouraging results from the use of POMP-based ELISAs in serological diagnostics of OAE. Buendia et al. [556] performed field evaluation of ELISAr-Chlamydia, an ELISA based on a recombinant antigen which expresses part of a protein from the 80–90 kDa POMP that is specific to *C. abortus*. The test has been compared with CFT and an ELISA using purified *C. abortus* EB. The authors believe that ELISAr-Chlamydia provides the most balanced results between sensitivity and specificity, especially in flocks with no clinical OEA but reactivity to *C. abortus* [556].

Longbottom et al. [560] examined a truncated fragment of p91Bf99, also known as rOMP91B expressed as GST fusion protein with several reference sera, and found its higher sensitivity (84.2%) and specificity (98.5%) compared to CFT and cELISA. The individual fragments of the POMP90 molecule differ in terms of specificity and sensitivity. In this respect, it is distinguished the fragments rOMP90-3 and rOMP90-4, excelling other recombinant fragments of such indicators as 100% specificity and 95.7%/94.3% sensitivity, respectively. When testing the two fragments with EAE-positive field sera, rOMP90-4 was more efficient [561]. In a comparative study of sheep sera with the discussed rOMP90-3 and rOMP90-4, the commercial assay of Pourquier-POMP 80–90, and the ELISA commercial MOMP-P test, again rOMP90-4 performed best: 100% specificity and 98.1% sensitivity [533]. In a wider study, Wilson et al. [534] compared the properties of six ELISA assays for diagnosing *C. abortus* infection in sheep: rOMP90-3 and rOMP90-4; Pourquier ELISA *C. abortus* – serum B version (commercially available based on POMPs); CHEKIT *C. abortus* antibody test (commercial); EB ELISA; and soluble protein (SolPr) ELISA. The two rOMP90 ELISAs have been found to be the most sensitive and specific. Overall, the rOMP90-3 ELISA performed the best in terms of sensitivity (96.8%) and specificity (100%). It did not cross react with sera from OAE-free flocks or from animals infected with *C. pecorum*. In the latter case, the lack of cross reactivity was detected both in field sera from animals infected with *C. pecorum* and in the study of SPF sera that had been experimentally infected with various subtypes of *C. pecorum* [534]. There are other examples where commercially available serological tests have been evaluated and compared with so-called in-house tests with variable results. From the given examples for comparative analyzes, as well as long experience of specialized laboratories in different countries, see the differences – sometimes significant with respect to sensitivity, specificity, and overall effectiveness of serological tests and products. It implies the need for the introduction of commonly accepted standards, overcome the difficulties associated with the use of particulate antigens, harmonization and proper validation of the tests and reagents for their application.

9.1.3 Reaction Immunofluorescence

This reaction (RIF) is widely used in the diagnostics and scientific studies of chlamydia and chlamydia-induced diseases. Different modifications of the reaction were tested. For its great application, two main explanations can be made: (1) it is highly specific in chlamydial agents and (2) chlamydiae usually do not induce light-optical cytopathic effect in cell cultures as many viruses; therefore, their IF indication becomes the basic method. There are several essential features of Chlamydia RIF: (a) in the direct method, rodent sera give a brighter glow from human sera and guinea-pig sera; (b) unlabeled sera cannot be used for a blocking reaction to control RIF if they are diluted 1:20 and more; and (c) a drop of guinea pig complement at a 1:10 dilution should be added to improve the light intensity of the indirect RIF with human sera. Various data are available in the literature on the preparation of conjugated sera for RIF. Terskih [21] reported that when conjugating sera to the direct method, the titers of the antibodies were reduced 1000-fold. In our research on the preparation and use of fluorescent sera for the direct RIF method, we did not encounter such difficulties and we received high-titer fluorescent sera [23, 251].

Many authors have performed comparative studies of RIF with other diagnostic tests. As a result of these studies, it is known that RIF is specific, convenient, and rapid to diagnose chlamydiae. It is more sensitive to CFT and has advantages over the cytological methods by Gimza and similar stains. Stephens et al. [562] find that Immunofluorescence staining is significantly more sensitive than Giemsa staining for detecting chlamydial inclusions, especially from specimens containing low titers of Chlamydia. The sensitivity of RIF is increased by the use of fluorescein-conjugated mAbs for detection of *C. trachomatis* inclusions in McCoy cell cultures [563]. Direct IF with mAbs is also successfully used to indicate chlamydial antigens in human clinical specimens – urethral and cervical secretions from urogenital infections and samples of cases of trachoma and neonatal conjunctivitis and pneumonia. By comparing the sensitivity of RIF to the cell culture method, conflicting results are obtained. Nevertheless, RIF should be considered a rapid, practical method that can be used in peripheral hospitals and laboratories where there are no conditions for working with cell cultures.

The use of RIF for the detection of antibodies in chlamydial infections is the second main direction in the research with this test. It provides great information about humoral immunity. Wang and Graystone [398] have great contribution for its widespread use in antibody titration who developed the MIFT as an epidemiological tool for the serotyping of human *C. trachomatis* strains. The research with the MIFT Test is the basis of the proposed by Eb and Orfila, 1982 [564]; Eb et al., 1986 [565] classification of the chlamydial agents of animal origin. Andersen [566] performed serotyping of avian chlamydial strains using serovar-specific mAbs with MIFT. In a later publication, the same researcher [567] reported serotyping of 150 US isolates of *C. psittaci* from domestic and wild birds using a panel of 14 serovar-specific MAbs. A total of 93 (97%) of the Psittaciformes isolates were of serovar A; 11 (79%) of the Columbiformes isolates were of serovar B; 64% of the Galliformes isolates were of serovar D, and all the Struthioniformes isolates were of serovar E. The three Falconiformes isolates did not react with any of the MAbs to the avian and mammalian isolates and are presumed to represent a new strain [567]. The author concludes that these specific chlamydial strains are usually associated with certain types of birds and that some serovars may be unusually virulent for certain species of birds [567]. In a serological study of the prevalence of the chlamydial diseases among pigs in Italy, Di Francesco et al. [570] found MIFT as a suitable method for the assessment of co-infection by several chlamydial species.

By means of MIFT, Chlamydia trachomatis strains have been divided into 15 serotypes [504, 568]. Regarding human psittacosis, one author group assessed the available laboratory methods and expressed doubts about the effectiveness of MIFT for diagnosis of this zoonosis [569].

It is important to note that in the OEA none of the serological tests that are available can differentiate vaccination titers from those acquired as a result of natural infection [571].

9.1.4 Other Serological Tests

It is already described that CFT, ELISA, and RIF are the major serological tests for research and diagnosis of chlamydia infections. A number of other tests are also known: agglutination, indirect agglutination, EB agglutination test, hemagglutination, latex agglutination test, agar gel immunodiffusion test, neutralization of infectivity, neutralization of toxicity, counter electrophoresis, radioimmunoprecipitation, immunoperoxidase method, and intradermal allergic test. These tests, with few exceptions, for various reasons have not taken hold in the general laboratory and diagnostic practice.

9.2 Serological Examinations

This section presents the results of our own serological investigations for Chlamydia-induced infections in animals and humans affected by various clinical conditions.

9.2.1 Serological Examinations in Sheep

9.2.1.1 Serological status of Chlamydial infection in the sheep population in Bulgaria

In 1986–2007, 10,606 sheep were tested serologically for chlamydia: 5530 diseased, contact, and clinically healthy animals tested with diagnostic and prophylactic purposes and 5076 controls – clinically healthy sheep certified as seronegative for *C. abortus* – imported from abroad in a quarantine period.

The samples originated from 154 settlements in 20 districts of the country and the imported animals from the respective quarantine bases [54].

The total seropositivity for chlamydia of 5530 sheep studied was 13.3% (5530/735/13.3%). Catching over a long period of time (1986–2007) was purposeful. It was conditioned by the need to track the dynamics of the range of chlamydial infection over the years and to analyze it under the conditions of organizational and structural changes in agriculture.

Figure 9.1 shows the obtained serological results. It is seen that the studies are divided into three periods.

The seroepizootiology of the first period (1986–1991) reflects the final stage of the existence of centralized industrial animal husbandry with cooperative and state ownership. The average seropositivity of sheep for *C. abortus* during these years was the highest –19%, and the outbreaks of chlamydiosis were 75.

The second period (1992–1998) is characterized by intensified processes of degradation of the system of industrial animal husbandry, the liquidation of a large number of agrarian complexes, and the fragmentation of livestock farming and agriculture on numerous small private farms and yards.

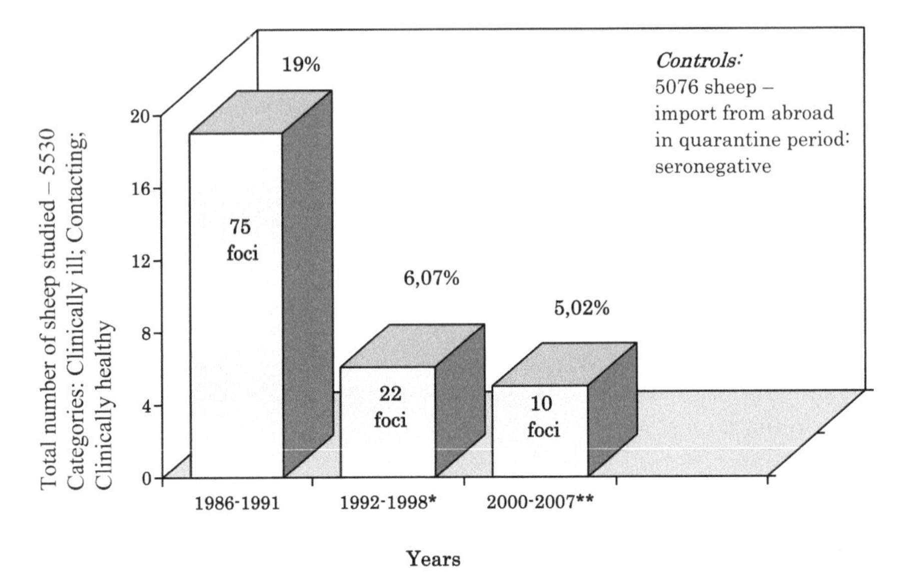

Figure 9.1 Serological examinations of sheep for antibodies against *Chlamydia abortus* for the period 1986–2007.

*No examinations were conducted in 1999.
**The period includes 2000 and 2004–2007.

The average seropositivity of sheep for chlamydia in the period was 6.07% with 25 outbreaks revealed.

In the third period (2000–2007), the small sheep breeding was fully established in the country, the number of sheep was repeatedly reduced compared to the first period, and the rearing of the animals was mainly in the private yards. Seropositive sheep averaged 5.02%, and the outbreaks of the infection –10. We will note that despite the period 2000–2007, it actually covers the last years (2004–2007) and 2000. In 2001–2003, serological studies were not undertaken.

The data presented clearly show the tendency to significantly reduce the seropositivity, respectively, the range of chlamydial infection among sheep during the second (3 times) and the third period (3.8 times). Within the studied periods, we observed variations in the size of the serologically positive reactions, depending on the category of studied sheep, the presence or absence of acute and chronic clinical manifestations, and the influence of the season (lambing campaign, postpartum period, etc.). Findings were diverse in different regions, regions, and farms. Serologically positive animals have been found in all areas studied, practically on the territory of almost the entire country.

9.2.1.2 Serological examinations in sheep with abortions and related conditions of the pathology of pregnancy

We focused on 89 farms from 19 districts where abortions, births of dead and nonviable offspring, and premature births were observed. In such flocks, placental retention was often seen in both abortions and normal births. In a number of cases, serological reactions were accompanied by microscopic, electron microscopic, immunofluorescence, and culture studies (see there).

In the serological tests of 869 sheep with abortions and related conditions, antibodies against *C. abortus* were found in 302 (34.76%). Complement – fixing titers with values $\geq$ 1:32 (1:32 – 1:1024) established in 2/3 of the seropositive animals. Positively reacted sheep were found in 65 farms from 16 districts, 21 of which for the first time. In 24 farms, the serological results were negative. The positive reactive sheep with abortions, stillborn, and nonviable lambs were <50% in 35 farms, between 10 and 40% in 28, and <10% in 2.

Our studies, carried out on a large number of serum samples from many herds, show that the serological response to spontaneous chlamydial abortions and the deliveries of dead and nonviable lambs is convincing and similar in the three clinical conditions. We should also include premature births. The detection of specific CF antibodies in the first 7–14 days after the clinical event (abortion or other forms) is common with the majority of sheep. Serodiagnosis is facilitated when at least 5–10 sheep are tested from one unhealthy herd or from different groups and flocks in a sheep complex situated at the same site. In these cases, the single detection of seropositive animals with different titer values was based on the assumption that abortions, premature births, stillborn, and nonviable offspring are etiologically related to *C. abortus*. In the study of a very small number of blood samples from sheep or single aborted sheep, for a minimum positive titer indicating probable chlamydial etiology, we assumed 1:32.

The possibilities of single serological tests to detect the participation of chlamydiae in abortions and other similar clinical forms are illustrated in Figure 9.2. The graph shows that a comparison was made between two groups of outbreaks of the specified clinical conditions. In the first group of animals from 15 sheep farms, the low titers (1:8 to 1:16) were 16%, the 1:32 titer was 14%, and the 1:64 to 1:1024 values accounted for 70% of serological responses. The GMT was 91.

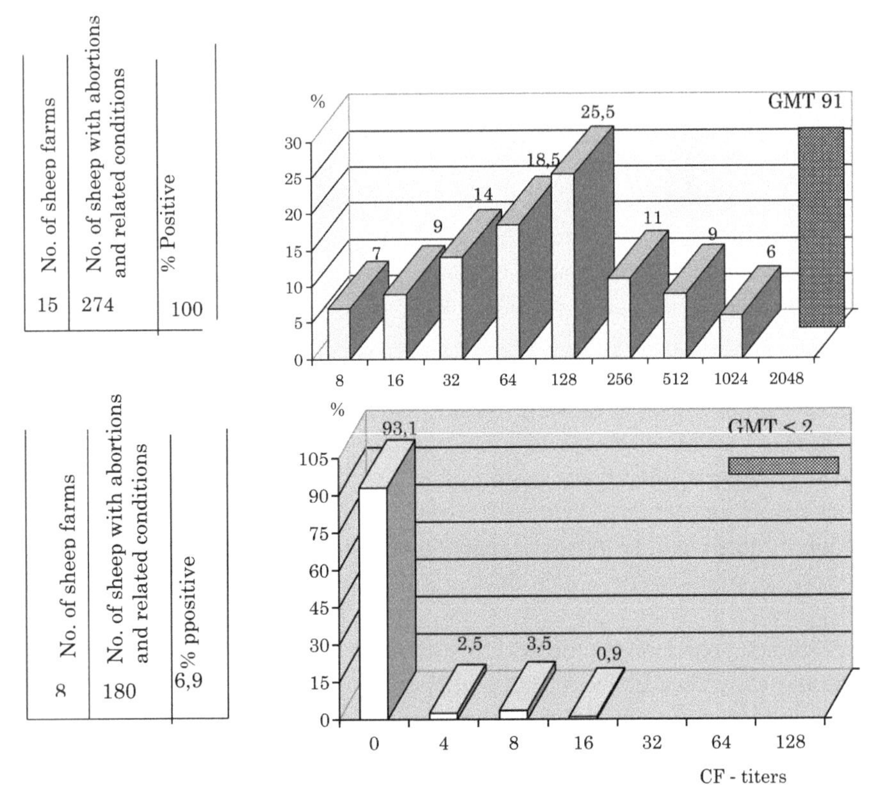

Figure 9.2 Comparative serological evidence of etiological participation of *Chlamydia abortus* in outbreaks of abortions, stillbirths, or deliveries of unviable lambs.

The opposite is the situation in the second group of sheep from eight farms. Serologically negative for *C. abortus* were 93.1% and the other 6.9% had low titers – 1:4 to 1:8 (6%) and 1:16 (0.9%). The GMT was <2.

Comparable data suggest that the first group of outbreaks may be expected to have a high probability of *C. abortus*'s etiological involvement.

The above results are typical of the two opposing possibilities of using serological methods as a means of allowing or rejecting the participation of chlamydiae in the etiology of abortion. There is also a third category in this regard – the probable partial involvement of *C. abortus* as a pathogen causing abortions, stillbirths, and other forms related to fetal loss. In such cases, it was usually groups of sheep with individual and GMTs indicative of active chlamydiosis, but significant percentages of serologically negative animals were also found. In such herds of sheep, obviously other etiological agents also played a role.

Serologic examinations of twofold blood samples within 14–21 days we applied in cases where in a group of sheep or flock in the first study near abortion (stillbirth) (1–7 days), we found predominantly low or negative titers (<1:16) of specific antibodies. For a credible titration (diagnostic seroconversion) indicating the participation of chlamydia in the etiology of these clinical cases, we recorded a fourfold or greater titer increase in the second sample compared to the titer of the first. The two time increase in titers we assumed was a dubious rise. We also noted the lack of difference in titer height in the two serum samples and the decrease of the titer in the second sample compared to the first (reverse dynamics).

In Figure 9.3, the dynamics of CF antibodies against *C. abortus* in serial serum titrations from two naturally infected and aborted sheep is demonstrated. Six studies were seen from the 10th to the 119th day after the abortion. There is similarity in the movement of titers in both animals. We recorded the maximum titer (1:1024) on the 20th day. This titer persisted until the 55th–56th day, after which a downward process began – 1:256 for one sheep and 1:128 for the second (69th day), 1:64, respectively, 1:32 for 89 day, and low values (1:8 to 1:16) until the end of the observation (119th day).

Figure 9.4 reflects the results of serological examinations of pairs of blood samples taken purposefully with a comparative purpose at different times. Eight first serum samples were titrated between 10 and 15 days after abortion,

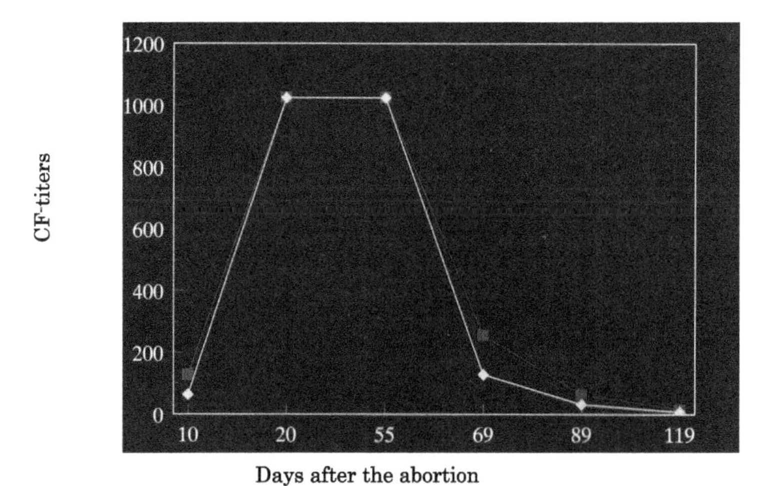

Figure 9.3 Series titrations of chlamydial CF antibodies in blood sera of naturally infected and aborted sheep.

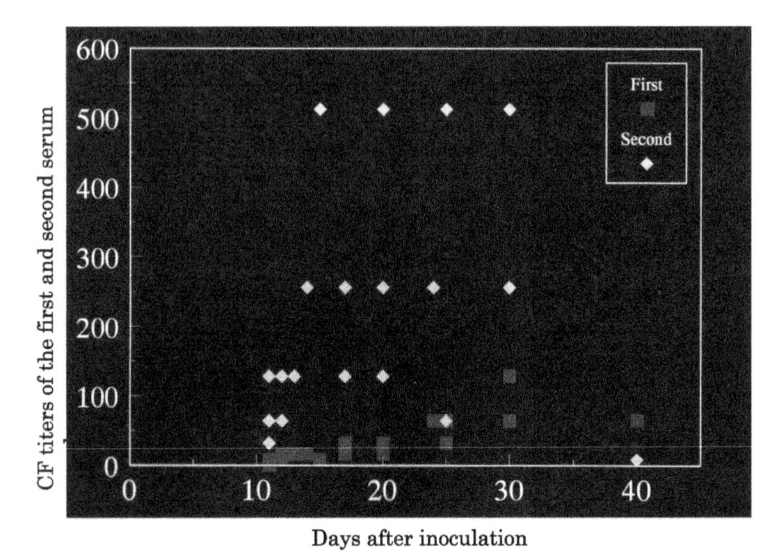

Figure 9.4 Titration of CF antibodies against *Chlamydia abortus* in twofold serum samples from spontaneously infected sheep.

two on the 17th day, three on the 20th and 25th day, two on the 30th, and one on the 40th. Antibody titers in the first subgroup (10–15 days) range from 0 to 1:32; in the second subgroup (17th day) – 1:16 to 1:32; in the third subgroup (20th day) – 1:32; in the fourth subgroup (25th day) – 1:32 to 1:64; in the fifth subgroup (30th day): 1:64 to 1:128; and at the sixth subgroup (40th day) – 1:64.

With one exception, the second serum samples taken 2 weeks later showed a fourfold and greater increase in antibody titers of values of 1:32 to 1:512. In the first sample sheep taken on the 40th day after the abortion (1:64), the retitration after 14 days showed a titer decrease of 1:16 (reverse dynamics). This can be explained by the late titration of antibodies in the first sample, most likely after the apical peak titers in an earlier period.

The data obtained from the pairs (duplicate) blood samples in sheep clearly indicate the usefulness and diagnostic value of the methodological approach used.

Another direction of our serological research is the follow-up of the antibody response in the dynamics of ewes after their experimental infection with the chlamydial ovine miscarriage agent (Figure 9.5). Experimental pregnant animals infected with the reference strain S26/3, originally isolated in Scotland, and passaged repeatedly into CEs. We used a suspension of

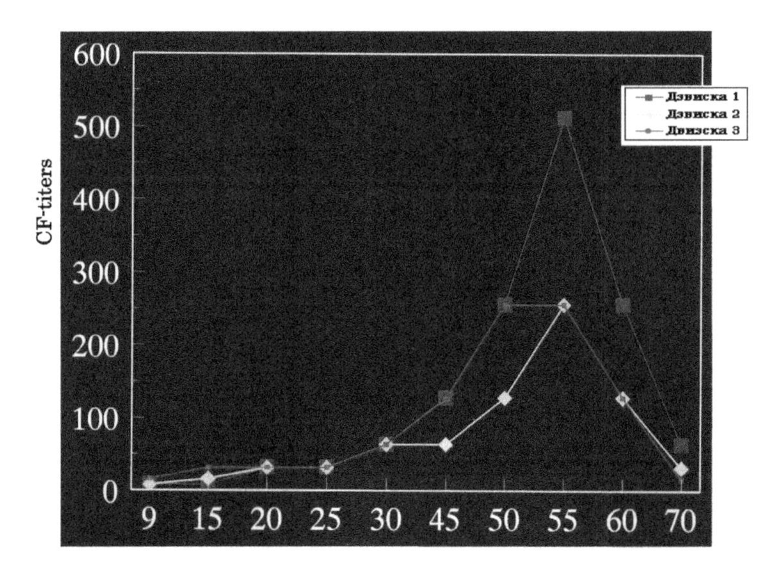

Days after inoculation
Pregnancy: ewe № 1 – 3.5 months; ewe № 2 — 3 months; ewe № 3 — 1.5 months.

Figure 9.5 Dynamics of CF antibodies against *Chlamydia abortus* in the sera of experimentally infected pregnant ewes.

the strain of infectious titer LD_{50} log 5×10^5/0.25 ml for a 7-day CE that inoculated intramuscularly at a dose of 5 ml. The three ewes responded with similar onset and movement antibody values, the peak titer in one of them being 1:512 (55th day), 1:256 (50–60th day) in the second, and also 1:256 (55th day) at the third. By the 70th day/end of the observation, we recorded a significant decrease in the titers – 1:32 to 1:64.

The studies described above are a continuation of our earlier work on Chlamydia-induced abortions in sheep carried out on samples and materials for serology and other etiological investigations from a large number of animals originating from regions throughout Bulgaria [23, 44, 49, 333, 572, 576].

9.2.1.3 Serological examination of sheep respiratory diseases

In some sheep, farms were observed respiratory diseases among sheep and lambs which in some cases were detected simultaneously with abortion and related conditions associated with fetal loss, while in others alone. The clinical picture is presented in a separate chapter.

We serologically studied 243 animals from six farms in four regions (Sf, Kn, Vd, and Lc). In 66 of these (27.16%), antibodies against *C. abortus* were detected. Individual titers ranged from 1:16 to 1:1024. For the individual groups of samples tested, the number of positive reactions was 26–100%. A greater incidence of infection was found in sheep (26.7–100%), while in the older lambs, it was between 26.7% and 50%. These data, albeit insufficiently representative of the prevalence of respiratory chlamydiosis in the country, undoubtedly indicate a significant involvement of chlamydia in the etiology of these ovine diseases in the studied holdings.

Figure 9.6 is a graphical representation of the serological data obtained. It can be seen that high titers (1:128 to 1:1024) make up 39.5%, moderate (1:32 to 1:64) – 48.5%, and low (1:16) – 12%. The GMT was 64.

The method of study paired serum samples – in the first week of clinical onset and 14–21 days later, applied in individual animals. The established diagnostic increase in titers (>4 times) undoubtedly demonstrates the effectiveness of the serological diagnosis method for chlamydial pneumonia in sheep.

The dynamics of specific CF antibodies after experimental tracheal infection of animals from three age groups (adult sheep, an older lamb, and a you 6ng lamb) is shown in Figure 9.7. In all three cases, we found a distinct serological response, but with differences in titer height and overall dynamics. The most pronounced reaction occurred in the sheep with a maximum titer value (>1:512 <1:1024) between 30 and 50 days. Followed by the older lamb

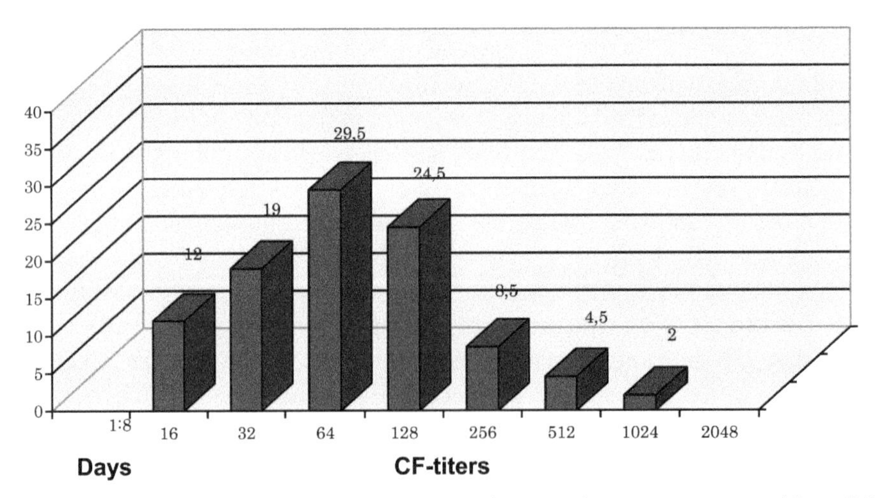

Figure 9.6 Distribution of CF titers and geometric mean titer at spontaneous chlamydial pneumonia in sheep and older lambs.

with two peaks of > 1:256 <1:512 at 30 and 45 days with an intermediate drop (> 1:128 <1:256) at day 38. At the young lamb, the indicators were relatively lower: a peak titer of 1:128 (30 days), followed by a gradual decrease to 70 days.

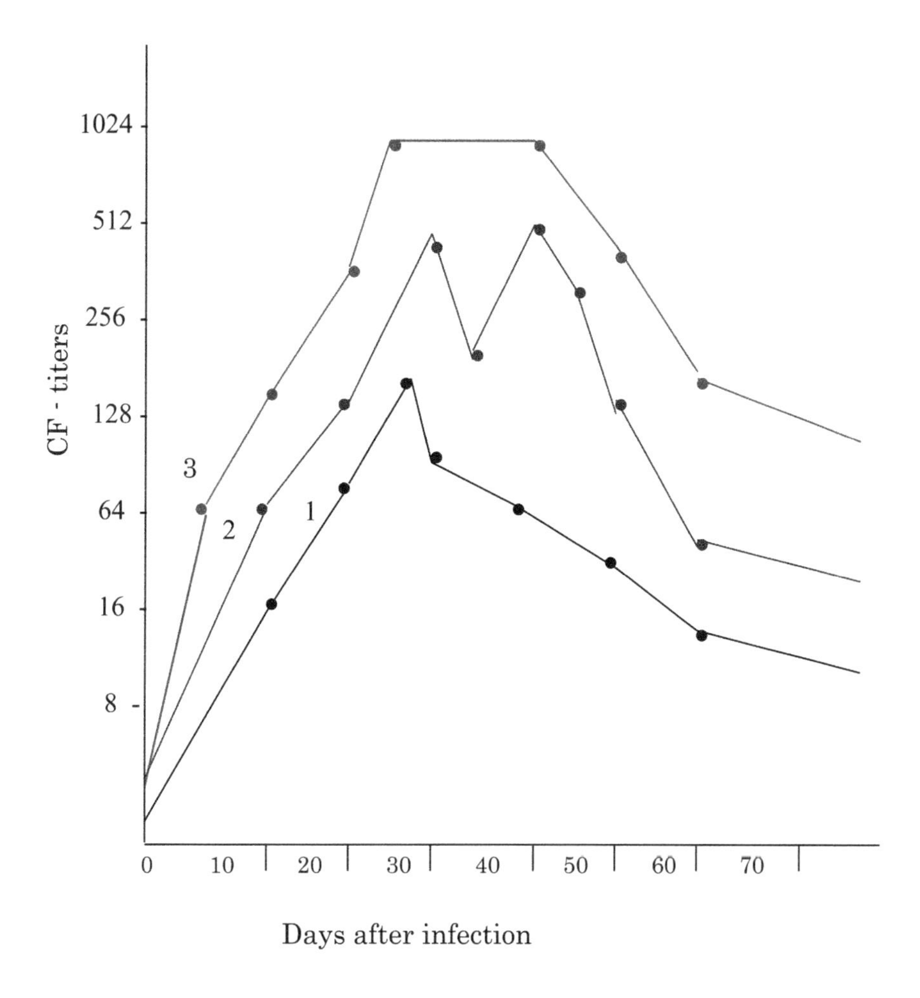

Figure 9.7 Experimental chlamydial infection in sheep. Dynamics of CF antibodies after tracheal infection with a strain isolated from the lungs of a sheep with pneumonia.

Table 9.1 Serology if rams for antibodies against *Chlamydia abortus*

Category	Examined (No.)	Positive (No. /%)	Holdings with Serologically Positive Rams (No.)
Rams from nine holdings	411	22/5.36	4
Controls: Rams imported from abroad in quarantine	213	0/0	0
Total	624	22/3.35	4

9.2.1.4 Serological examinations of Rams

We surveyed a total of 624 rams: 411 of tribal and commodity holdings in nine settlements and 213 of twi quarantine animal bases – imports from abroad (controls with certificate from the exporting country that are seronegative for chlamydia).

The results of the studies of the 411 rams indicated that 22 of them (5.36%) of four holdings – Kostenets (Sf), D. Mitropoliya (Pl), Babino (Kn), and Shumen were seropositive for *C. abortus*. Two of the mentioned settlements had proven chlamydial abortions on the sheep, and in one case – chlamydial abortions and pneumonia. The control 213 animals were seronegative for *C. abortus* (Table 9.1).

The seropositive rams in two of the farms were clinically healthy, and in the other two parts of them had orchitis or orchiepididymitis, and the rest – contacting them had no signs. CF antibody titers ranged from 1:16 to 1:128. The etiological involvement of *C. abortus* in the observed orchitis and orchiepididymitis was confirmed by isolation of the agent from seminal fluid. All the investigated rams were serologically negative for infectious epididymitis (*Brucella ovis*).

9.2.2 Serological Examinations in Goats

9.2.2.1 Serological status of Chlamydial infection among goats in Bulgaria

During the period 1986–2007, 1249 goats were tested for antibodies against *C. abortus*: 1043 sick, contact, and clinically healthy animals from 45 settlements in 14 districts and 206 bigger kids – imported from abroad in a quarantine period serving as a control [167]. The latter are certified by the exporting countries as serologically negative for chlamydiae. The total mean seropositivity for *C. abortus* of 1043 goats tested was 12.75% (1043/133/12.75%). The results obtained are shown in Figure 9.8.

Figure 9.8 shows that the studies are provisionally divided into two groups covering the periods 1986–1996 and 1997–2007, respectively. In the last period, we have divided two subgroups – first, for the years 1997–2004, and second, 2005–2007, reflecting the seropositivity for *C. abortus* of goats at a more recent period.

The first period's seroepizootiology is based on data obtained in the last 5 years of industrial livestock farming in Bulgaria where the share of goat breeding was relatively small and the first five years of the transition between the mentioned economic model and the fragmentation of the agro-livestock sector In the latter, due to a number of socioeconomic, commercial, and other reasons, which we will not analyze here, began a process of reducing the number of cattle and sheep in the country and increasing the goats mainly in the private yards. During this period (1986–1996), the total seropositivity

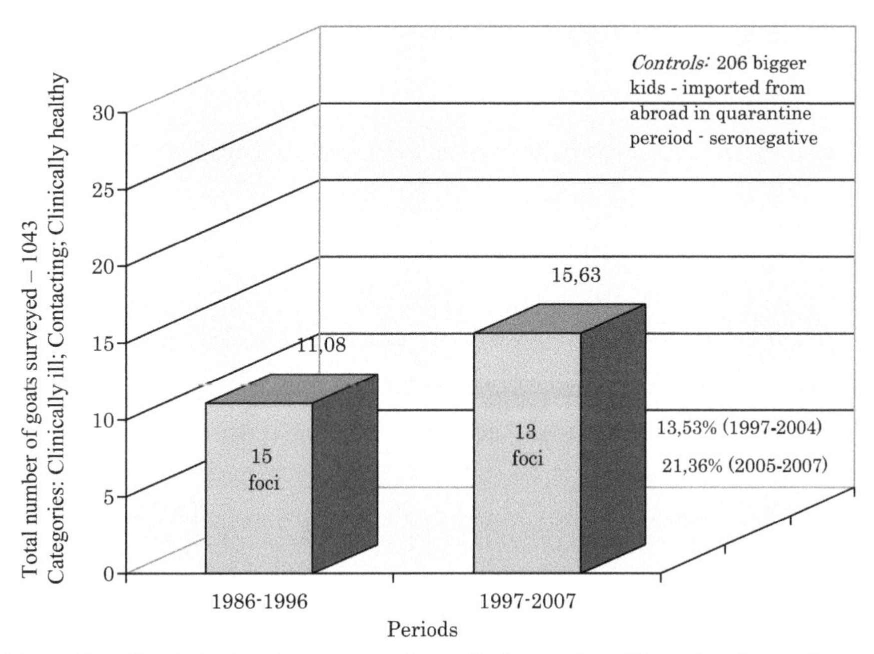

Figure 9.8 Serological testing of goats for antibodies against *Chlamydia abortus* for the period 1986–2007.

for *C. abortus* in goats was 11.08% and the new outbreaks of chlamydiosis were 15.

During the second period (1997–2007), we witnessed significantly increased number of goats in Bulgaria as a source of milk at the expense of the greatly reduced size of the national herds of cattle and sheep. The total rate of the serologically positive goats was 15.63% (13.53% for the years 1997–2004) and 21.36% for the 3-year period 2005–2007. In the next few years (2008–2011), new seropositive animals continued to be detected, with average annual rates ranging from 15 to 18%.

9.2.2.2 Serological examinations of goats with abortions, stillbirths, and deliveries of nonviable kids

In 1984, we reported serological tests of 658 goats: 167 belonging to herds having mass abortions in 6 farms and 347 control healthy animals from farms and quarantine bases for imported animals. As a result of these studies, as part of the applied diagnostic methodological complex, including also agent isolation, EM, and cytology, two outbreaks of chlamydial abortions in goats were revealed [573].

In another study in a much later period, 2009 [167], the subject to serological tests was 126 goats with abortions and similar conditions, originating from 13 settlements in 5 districts. It was about the outbreaks of these diseases, which were associated with great economic losses. Serological data on *C. abortus* infection were obtained in 26.99% of the goats surveyed (Sf, Pk, Lc, and Dc). These results were supported by the direct proving of the agent by LM, DEM, and IF and by its isolation from placenta and fetal parenchymal organs (see the relevant chapters).

The percentage distribution of titers and GMT in serologically positive goats is shown in Figure 9.9.

The chart shows that the low titers (1:16) are 2% and the average (1:32 to 1:64) – 35%. High titers (1:128 to 1:4096) account for 67% of the positive reactions. The highest titer is 1:4096 (1%), and the most common is 1:256 (27.5%), followed by 1:128 (23.5%). The average geometric titer is high – 104.

The twofold study of pairs of serum samples in goats with abortions and births of dead or unviable kids gave very good results (Figures 9.10 and 9.11).

We modified the classical sampling scheme, with the first sera targeting differently: 1, 2, 3, 5, 7, 10, 15, 20, and 30 days after the clinical event (the stillbirth or delivery of nonviable kid). In general, the height of the titers rose proportionally to the 20th day when we found the highest value of 1:128.

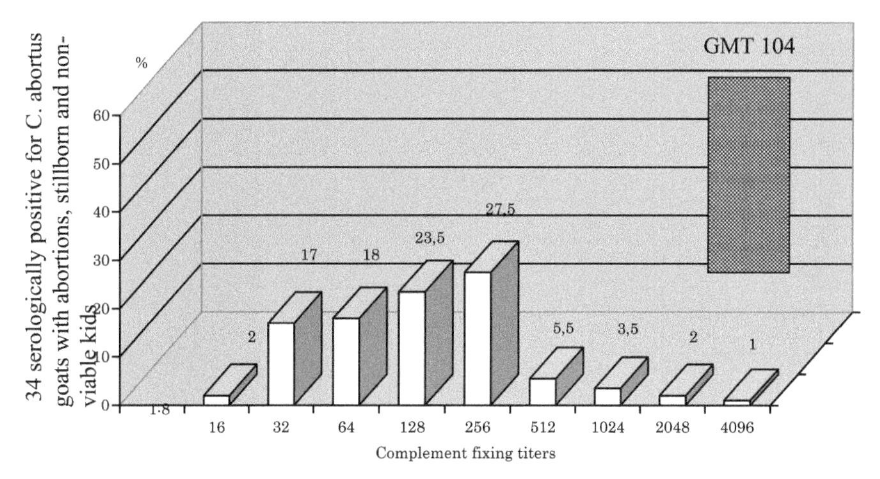

Figure 9.9 Chlamydal abortions in goats, stillbirths, and deliveries of nonviable kids. Percentage distribution of CF titers and GMT in serologically positive animals.

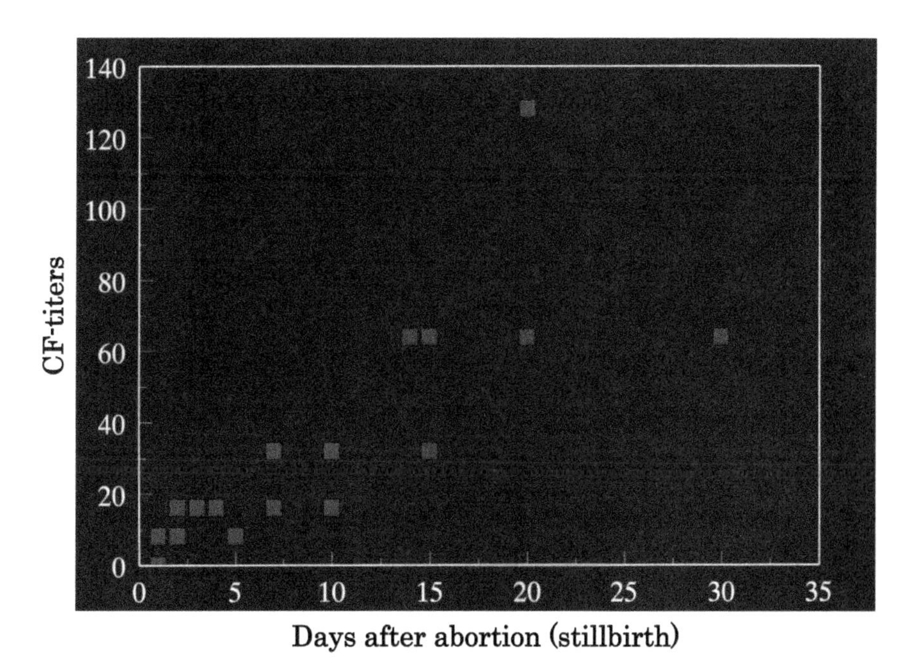

Figure 9.10 Natural *Chlamydia abortus* infection in goats with abortions, stillbirths, and unviable kids. Complement fixing antibodies in the first serum sample.

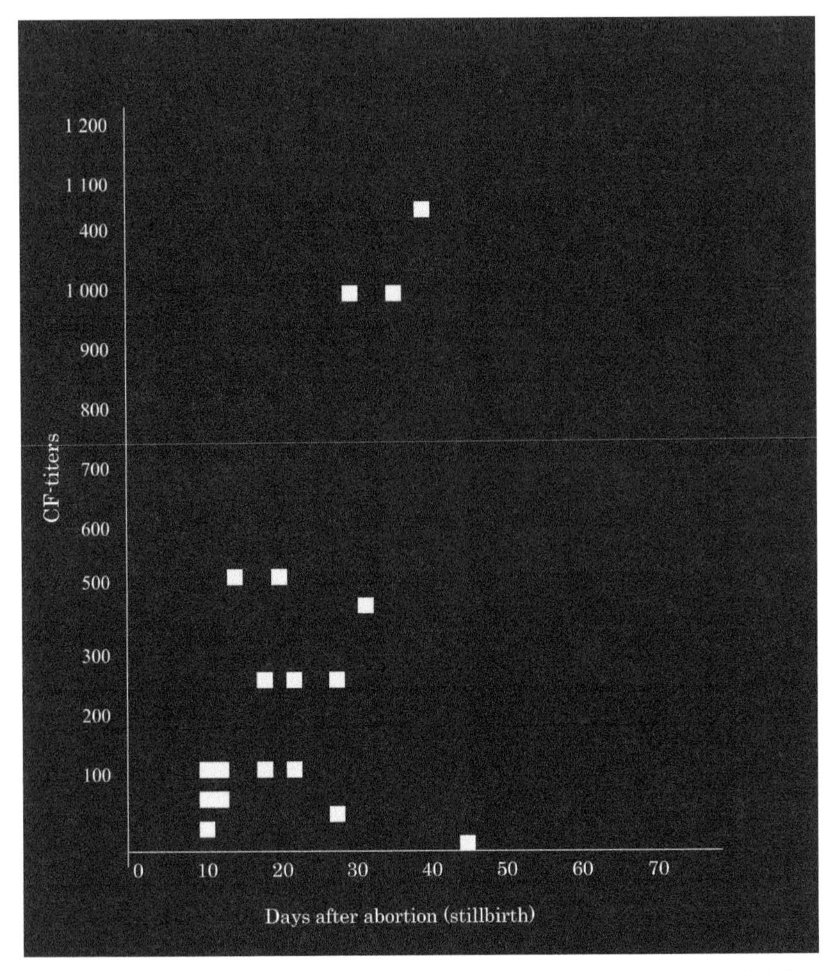

Figure 9.11 Natural *Chlamydia abortus* infection in goats with abortions, stillbirths, and unviable kids. Complement fixing antibodies in the second serum sample.

This trend was more valid for group rating of titer values. Analysis of individual samples showed that a titer of 1:8 was found in one animal on the 1st–2nd day, but also on the 5th day; 1:16 – between the 3rd and 10th day; 1:32 between the 7th and 15th day; 1:64 – 14th – 30th day; and 1: 128–20th day.

Repeated testing of goat blood samples was also carried out over extended periods, from 12th to 45th day, depending on the time of the first sample taking, so that a 10–20 day interval was provided for each of them. It can be seen that almost all samples have seroconversion – an increase of >4 times, which as the titer height is higher as the titer in the first sample is higher. The peak titer values achieved are 1:1024 and 1:512. On a first sample taken at day 30 (1:64), the retest after 15 days (45 days) showed a decrease to 1:8, apparently due to the passing of the tittle's apogee in earlier periods.

It follows from the above that the serology for chlamydia of twofold serum samples in goats with different forms of pathological birth is a reliable method for both individual and herd diagnosis of the disease.

9.2.2.3 Serological investigations in goats with pneumonia

In our studies, we observed pneumonia in goats in herds where there are etiologically proven chlamydial abortions (stillborn, nonviable offspring) or as a self-acting clinical form.

We examined serologically 131 goats with pulmonary symptoms belonging to individual stationary flocks in five settlements in different regions (Sm, Pd, Sn, and Sf). Serologically positive for infection with *C. abortus* were 59/45.04% of the five settlements. The percentage of individual CF titers indicates that the high values (1:128 to 1:1024) have a high relative share of 60.5% and the mean titers (1:32 to 1:64) – 37.5% (Figure 9.12).

The mean geometric titer is 60. These serological data are an indication of active chlamydial infection inducing the production and accumulation of specific antibodies with values giving rise to a high probability of allowing the agent's etiological involvement in pneumonia.

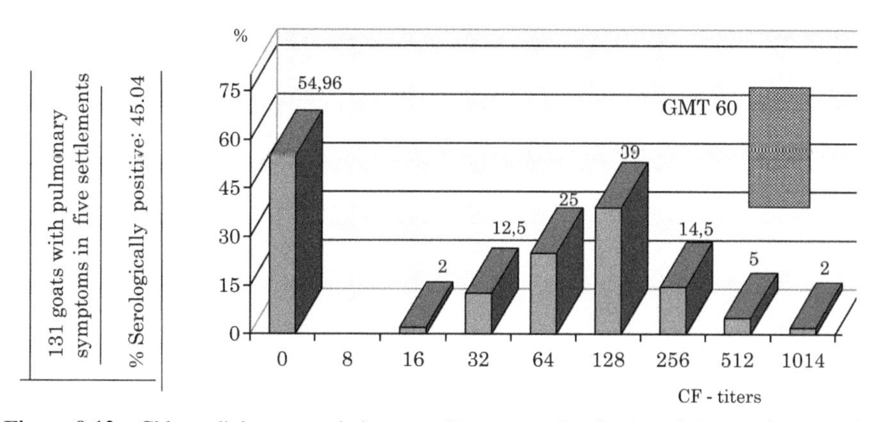

Figure 9.12 Chlamydial pneumonia in goats. Percentage distribution of titers and geometric mean titer.

Direct proof of the pathogen in inflamed lungs of affected goats in many cases confirmed serological data. Serological response of a seronegative for chlamydia goat (nonpregnant, with a birth) after experimental infection with strain TR90 isolated from the lungs of a goat with pneumonia is shown in Figure 9.13. The infection was achieved after intranasal inoculation of the concentrated and purified yolk suspension of the strain adapted to cultivation in the YS of CE.

The infectious titer of the strain was 10^6 ID_{50}/ml. Clinical follow-up revealed a pronounced febrile reaction and progressive development of interstitial lung inflammation.

The serologic response was very demonstrative (Figure 9.13). In multiple serum sample titrations between the second day after inoculation and the first third of the fourth month, antibody dynamics was detected as follows: antibody (1:8) – second day; seroconversion (1:512) – 47th day; decrease to

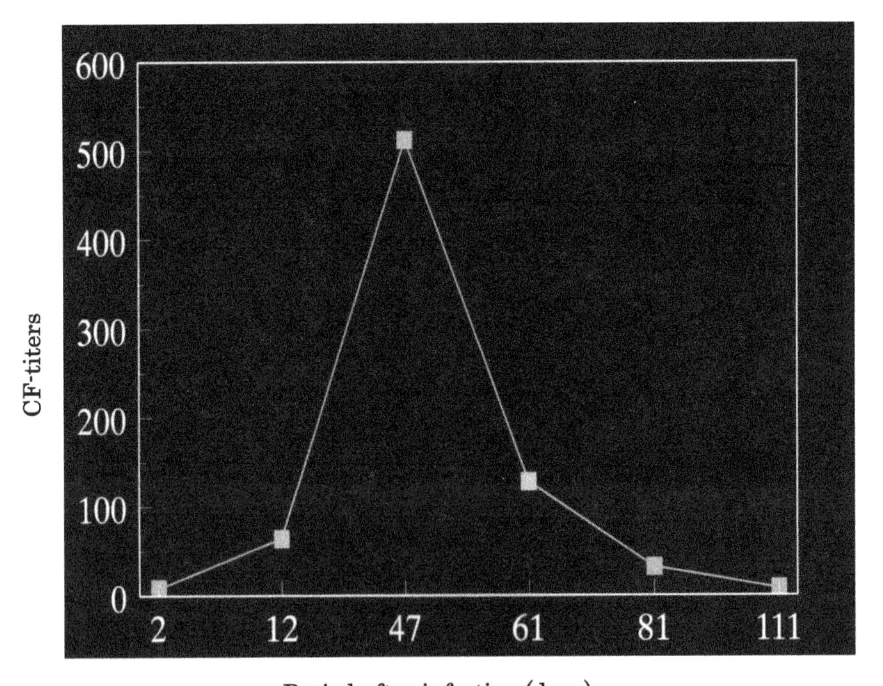

Figure 9.13 Experimentally induced chlamydial pneumonia in the goat. Dynamics of specific antibodies.

1:128 (2 months); 1:32 (2 months and 20 days), and 1:8 (3 months and 10 days) – end of observation.

The results obtained from the serological tests of goats under the conditions of natural infection in herds and in experimental chlamydiosis testify to the effectiveness of the serological method in the diagnosis and studies of chlamydal pneumonia in goats.

9.2.3 Serological Examinations in Cattle

9.2.3.1 Serological status of Chlamydial infection among cattle in Bulgaria

In 1986–2007, 17,370 cattle were serologically tested: 8181 indicating signs of disease, contact, and clinically healthy animals from different farms in the country and 9189 controls imported from abroad in quarantine in the bases that were designed for this purpose. The first category of cattle originated from farms in 118 settlements in 25 districts. The total mean seropositivity of the animals for chlamydia was 4.58% (8181/375/4.58%) [54].

In bovine animals, like small ruminants, we have adopted a targeted approach for a prolonged period of time (1986–2007). This was necessary in order to track the seroepisootiology of chlamydiosis, in particular in connection with the radical socioeconomic and structural changes in the country's agrarian sector (see Section 9.2.1).

The results are presented in Figure 9.14. It shows that 92 outbreaks of chlamydial disease among cattle (84 in 1986–1996 in 21 districts and 8 in 1997–2007 in 5 districts) were detected for the whole period from 1986 to 2007. It is noteworthy that the total average seropositivity for the first period is 7.75% and for the second – much lower (0.65%). The interpretation of this result should take into account the structure of the bovine samples tested, namely, the large number sera – 3424 of animals collected from a region (Ss) selected as clinically healthy and originating mainly from holdings without epidemiological and clinical history and laboratory evidence of chlamydial infection. All these bovines destined for export gave negative serological results. Conversely, among the other 230 animals from farms with health problems, serologically positive were found in 12 settlements from different areas. Conditional summing in a group of virtually unequal in a history and clinical epizootiological respect animals clearly affected the actual figures reflecting the seroepidemiology of chlamydioses in cattle in this period.

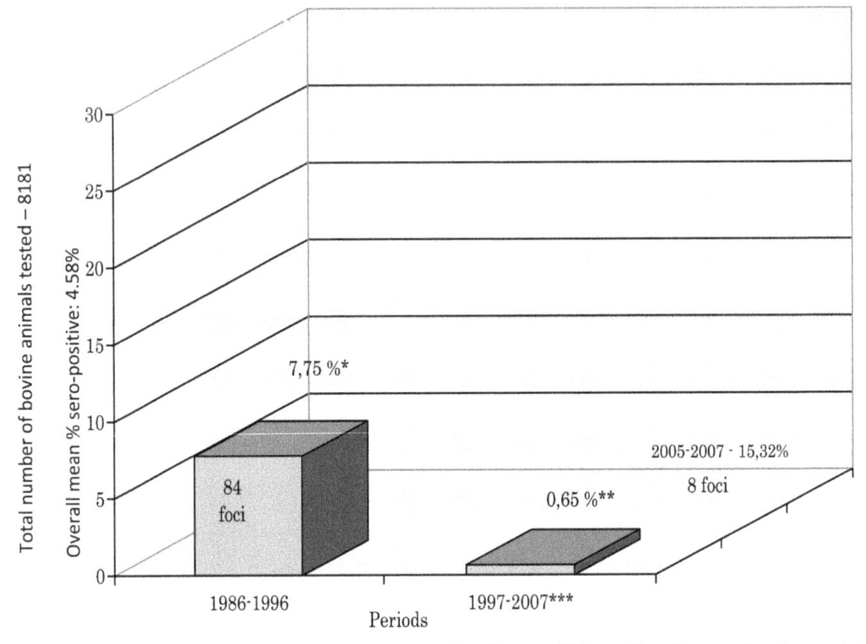

Figure 9.14 Testing bovine blood sera for antibodies to *Chlamydia abortus* for the period 1986–2007.

Legend:
*Includes serology of 4527 clinically ill, contact, and clinically healthy cattle from 106 settlements in 25 districts.
**Low % due to the structure of the test animals: (a) serology of 3424 clinically healthy bovines from one region selected for export (all negative for chlamydia); (b) serology of 230 sick animals, contact, and clinically healthy cattle from 12 settlements in 11 districts (8 outbreaks of chlamydia).
***No studies were conducted in 1999, 2001, 2002, and 2004.

Figure 9.14 also shows that data from the last 3 years (2005–2007) based on a relatively moderate number of samples – 230 indicate an average seropositivity of 15.32% in the investigated cattle from outbreaks of chlamydiae in six settlements. The other two outbreaks were discovered in 1998 and 2003.

In the study of control quarantined animals – imported from abroad, 13 seropositive cattle were found in three lots of animas from Greece, Germany, and Israel, including two bulls. All other control animals from a large number of import lots were seronegative. Thus, the final figures from the massive control studies carried out were as follows: tested 9189/seropositive 13/0.14%.

9.2.3.2 Serological testing of cows with miscarriages, stillbirths, and births of nonviable calves

The serology of cows affected by abortions and similar disease conditions covered 636 animals from 36 settlements in 11 districts. The births of dead or nonviable calves dying in the next 2–3 days accounted for around 1/4 of the loss of the fetus. Positive for chlamydiae cattle we found in farms from 20 settlements in seven districts (Kj, Sm, Bs, Rs, Vd, Sofia, and Sofia-city). The average percentage of 93 cows responded positively to 14.62%. The percentage of serologically positive cows from individual problem farms varied widely.

Table 9.2 shows that the outbreaks of abortions and related clinical conditions are divided into three groups: (a) eight outbreaks with seropositive over 50%, highest GMT, and possibly predominantly etiological involvement of *C. abortus*; (b) 16 outbreaks with positive serology of 10–49%, medium–high GMT, and possibly partial involvement of the chlamydial agent; and (c) seven outbreaks with seropositivity below 10%, low GMT, and no involvement in the etiology of the outbreaks but likely to play a role in sporadic cases of abortions, stillbirths, and deliveries of unviable calves, of course if the individual animal has a diagnostic increase in antibodies in a second blood sample or there is isolation of the chlamydial pathogen from the placenta or fetal parenchymal organs. In the latter two groups, the causes of the cases of massive fetal loss in cows should also be attributed to other etiological

Table 9.2 Serological evidence of *Chlamydia abortus* infection in abortions, stillbirths, and deliveries of nonviable offspring in cows

Seropositivity for *C. abortus* in Outbreaks of Miscarriages and Related Conditions in Cows	Number of Outbreaks	GMT in Serologically Positive Cows	Probable Degree of Etiological Involvement of *C. abortus*
>50%	8	56	A dominant part in outbreaks
10–49%	16	42	Partial participation in outbreaks
<10%	7	17.1	• Without participation in outbreaks • Participation in single cases of abortions (stillbirths and unviable calves)

factors with partial involvement (in association with *C. abortus*) or with a major etiological role (with single cases caused by chlamydia).

It should be noted that the methodological approach shown in Table 9.2 was used successfully in our previous studies on the role of chlamydia in bovine abortion, which involved 6887 animals: 2083 aborted from 103 holdings and 4804 healthy cows and heifers. The mean seropositivity in aborted cows was 19%, while in controls – 0.08% and 99.92% were seronegative [436]. If the lowest serological titers in most cases should be ignored (and therefore possible role for their appearance based on latent intestinal carriers of the agent), then the convincing titers as height and dynamics should be taken as a serious indication of possible etiological involvement of chlamydia, especially when serology is supported by positive for chlamydial infection findings in isolation and cultivation, and the direct detection of pathogen by various methods.

The percentage distribution of individual serological titers and GMT of CF antibodies against *C. abortus* in outbreaks with a massive range of infection is presented in Figure 9.15.

Part of the test animals were pregnant cows and heifers at ease or unfortunate in terms of abortions farms. Pooled results showed that of 80 test animals, 26 (32.5%) were seropositive for *C. abortus* with various titers. These data, on the one hand, shed light on the extent of infection among this category of animals and, on the other hand, are of prognostic significance for possible loss of the fetus due to abortions and similar conditions as a result of active chlamydiosis.

The dynamics of CF antibodies in natural chlamydial infection in a cow that gave birth to a nonviable calf is shown in Figure 9.16. The serological response in the cow is characterized by the occurrence of antibodies (1:8 to 1:16) on the fifth day postpartum and progressive titration up to 50th–60th day (1:256). There is a gradual decrease to medium and low values over a long period (160 days). The chlamydial etiology of this clinical case was confirmed by pathogen isolation. The data obtained indicate a certain course of the titer curve in the disease, which also allows retrospective proof of the infection in a similar clinical picture.

9.2.3.3 Serological examinations of bulls

The results of the serological testing of the bulls are set out in Table 9.3.

It is seen from the table that objects of serological examinations for *C. abortus* infection were 249 bulls of two categories: (a) breeding bulls from the Livestock Breeding Centers (LBC) in the country and (b) bulls – imports

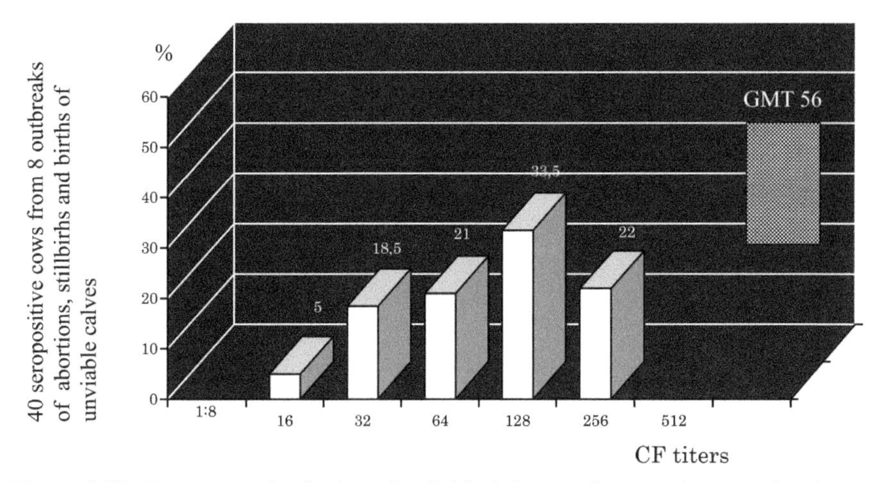

Figure 9.15 Percentage distribution of individual titers and geometric mean titer in cow abortions, stillbirths, and deliveries of nonviable calves with probable etiological involvement of *C. abortus*.

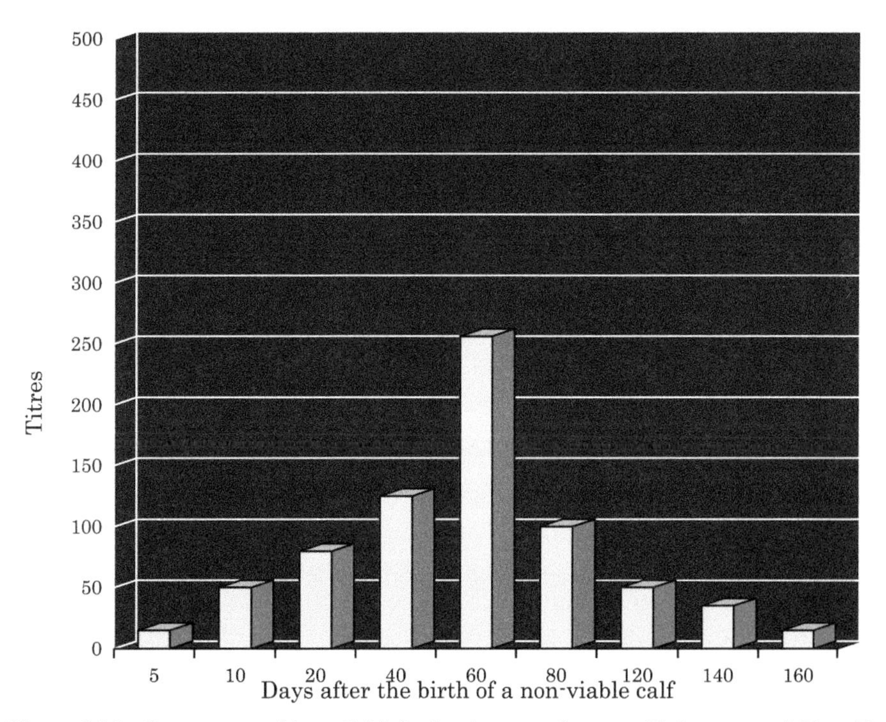

Figure 9.16 Spontaneous chlamydial infection in a cow that gave birth to a nonviable calf. Dynamics of CF antibodies against *Chlamydia abortus*.

Table 9.3 Serology of bulls for CF antibodies against *Chlamydia Abortus*

Tested Bulls (No.)	Origin	Seropositive (No. /%)
78	Livestock breeding centers	4/5, 12%
171	Imports from abroad in quarantine	2/1, 65%
Total: 249		6/2, 4%

from abroad in a quarantine period. Single seropositive animals (2.4%) were found, including 4 (5.12%) in one of the LBC (Yab) and in one group of imports from Germany (1.65%). The titers of specific CF antibodies ranged from 1:16 to 1:64/1:128. These data, part of the preventive control of breeding and sperm donor infections, show the importance of systemic serological tests for the detection of chlamydia-infected animals in LBS and border quarantine bases.

9.2.4 Serological Examinations of Buffaloes

These studies were of a limited nature. Altogether, we surveyed 31 animals – 27 offspring from three farms and 4 maquis from a farm. Chlamydial antibodies were found in seven serum samples (22.58%): in six infants with respiratory signs in one farm and one aborted malachian. The obtained data confirm our previous findings of significant circulation of chlamydial infection among some buffalo herds in Bulgaria. In those studies, of a total of 214 animals serologically positive for chlamydial infection were 49 (22.89%) [23].

9.2.5 Serological Examinations in Pigs

We surveyed 2346 pigs from 30 large pig farms and smaller farms in 16 districts [23, 47, 334, 575].

The animals belonged to different age groups and had unequal clinical status. The resulting serological results are summarized in Figure 9.17.

It can be seen that the average percentage of seropositive for *C. suis* pigs is 7.37%. During the first period (1986–1991), the positive animals were 5.15% of the farms in eight settlements in seven districts. These indicators for the second (1992–1997) and the third period (1998–2010) are, respectively, 14.61%, seven settlements, six districts and 15.5%, nine settlements, six districts. The period includes the following years: 1998, 2001, 2003, 2004, 2006, 2007–2010.

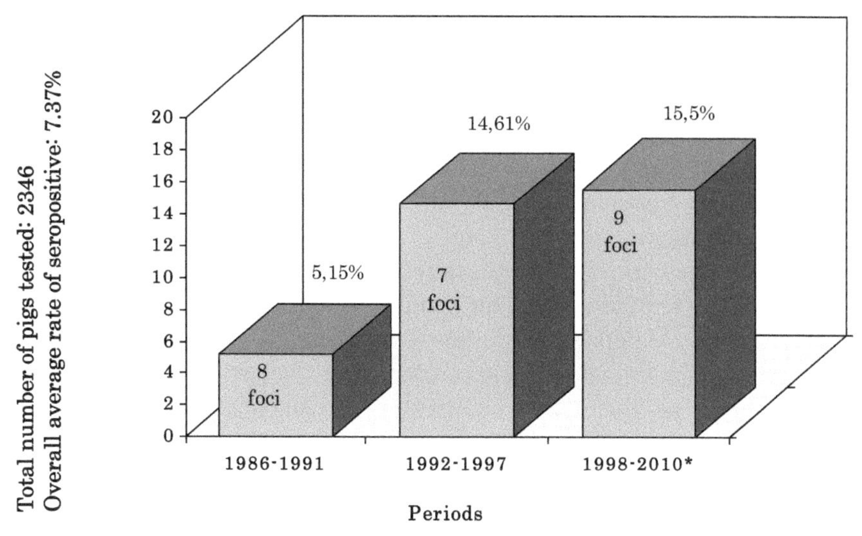

Figure 9.17 Serological investigations of pigs for antibodies against *Chlamydia suis*.

A detailed analysis of the performed tests with samples of different categories of pigs is made in Table 9.4.

Table 9.4 Investigations of different categories of pigs for antibodies against *Chlamydia suis*

Pigs Category	No. Tested	Seropositive No. /%
Sows from groups with pathology of pregnancy (abortions, stillbirths, and contact asymptomatic animals)	631	46/7.29
Pigs with reproductive disturbances	51	14/27.45
Adult pigs from groups with respiratory diseases (clinically ill; contact asymptomatic animals)	868	27/3.11
Adolescent pigs (over 3 months of age) from groups with respiratory disease (clinically ill; contact asymptomatic animals)	484	37/7.64
Boars with different clinical status (urethritis; orchitis; orchiepididymitis; clinically healthy)	161	28/17.39
Clinically healthy pigs from seamless farms	151	0

The investigations showed positive reagents at 7.64% in adolescent pigs and twice less – 3.11% in adult pigs. Seropositive adolescent pigs with such a clinical picture were found in farms in nine districts (Mt, Vr, Lc, VT, Rs, Tc, Sn, Vn, and Pd), and pigs in five districts (Mt, Tc, Sn, Vn, and Pd).

In the group of boars with urogenital pathology and male contact animals, there is a significant percentage (17.39) of seropositivity for *C. suis*, which is the second largest after the infertility pigs (Table 9.4).

Within the framework of cooperation with scientists from Lithuania, we have carried out extensive studies on swine chlamydial infections in this country [335]. For a comparative purpose, we present the main results of this study. Chlamydial infections were serologically detected in 87.5% of the country's regions. The quantitative estimate of the antichlamydial antibody load was 1.24 ± 0.59 log2, indicating predominantly chronic type of the disease. In farms (groups) with cases of abortions and stillbirths, the mean percentage of seropositive pigs was 7.29. Both aborted and contact-clinically healthy animals are included here [335]. In Bulgaria, positive responses to aborted pigs were found in five holdings of five regions (Lc, Ss, Vn, Pd, and Sf). The range of the serollogically positive animals was within 10–60% for the individual farms. The involvement of *C. suis* in the etiology of abortions was confirmed by the direct detection of the pathogen in placentas and the isolation of two strains.

The category of pigs with reproductive disorders comprises 51 animals from one pig complex (Lovech). It was about pigs with difficulty in fertilization or with permanent infertility. The resulting seropositivity score of 27.45% is indicative of the potential of *C. suis* in similar porcine pathology.

Concerning porcine respiratory diseases, we will note that chlamydia serology revealed constantly the chronic type of the disease. Of the total of 2502 pigs tested, 7.7% contained CF antibodies. This seropositivity rate is very similar to the above-mentioned 7.37% in the study of 2346 pigs in Bulgaria. And in this study, it was found to be affected animals of different age groups having clinical disease or asymptomatic (Table 9.5).

The table presents general serological data on the extent of infection in the relevant groups, including both the clinically ill and the asymptomatic pigs (left half of the table) and also the data referring only to the clinically ill animals. Among the latter, the growing fattening pigs are predominantly with pneumonia and enteritis having a seropositivity of 17.6%. In piglets suckers, there is no change in figures, while in the other two groups, there is an increase of 0.9–2%. The difference between total seropositivity (7.7%) and that of clinically ill pigs (11%) was 3.3%.

Table 9.5 Serological investigations for *Chlamydia suis* of pigs of different age and technological groups

	Total Tested Animals			Investigated Clinically Ill Animals		
	Samples Tested	Positive		Examined	Positive	
Groups	(Total No.)	No.	%	No.	No.	%
Piglets suckers	59	2	3.4	59	2	3.4
Pigs for fattening	432	21	7.6	108	19	17.6
Sows	1935	94	7.8	748	81	10.8
Boars	76	4	7.9	68	6	8.8
Total	2502	192	7.7	983	108	11.0

Table 9.6 compares the serological methods for the diagnosis of chlamydiosis in pigs.

Blood samples and clinical and pathological materials were tested for both versions of the immunofluorescence method – a direct (DIF) and indirect (IIF) from four separate farms, respectively, 1633, 614, 143, and 94 samples. Via CFT seropositivity of 13.8–20.2% was found; by DIF – 26.1%; IIF – 28.6%. High detection rates showed enzyme immunoassay (EIA) (35.7%) and MIF test – 31.9%. Especially suited was the EIA in serodiagnosis of swine genital chlamydial infections. The latter two reactions have shown a higher degree of sensitivity than CFT, but the question of their specificity is where CFT outperforms them. This question is essential, especially when it comes to diagnosing the disease in the individual animal, not for the diagnosis of the herd.

Niemczuk [529] made evaluation of Chlamydia spp. prevalence in pigs in Poland using CFT and ELISA and made a comparative statistical analysis of the results. The last (the chi-squared at $P = 0.00$) has revealed a semihigh conformity of the results obtained using the two tests. With regard to CFT, the author believes that this test can be used for serological diagnosis of

Table 9.6 Comparative results from the use of different serological methods for the demonstration of swine chlamydiosis

No. of	CFT		DIF		EIA		IIF		IIF		
samples	(+)	%	(+)	%	(+)	%	(+)	%	(+)	%	*P*
1633	259	15.8	487	26.1	–		–		–		<0.01
614	85	13.8	–		219	35.7	–			–	<0.001
143	23	16.1	–		–		41	28.6	–		<0.05
94	19	20.2	–		–		–		30	31.9	>0.05

swine chlamydial infections but recommends its prestandardization and validation under the conditions of a laboratory and with the use of swine blood sera [529].

9.2.6 Serological Testing of Horses

We tested a total of 210 horses, including 6 aborted mares, 12 stallions, 8 young horses, and 184 adult horses of both sexes coming from eight settlements in seven districts. Seropositive for chlamydia were two horses (Karlovo and Pd). This low seropositivity (0.95%) indicates a poor circulation of chlamydial infection in the study groups of horses.

9.2.7 Serological Examinations of Dogs

The CFT testing covered two periods: 1988–1992 and 2005–2011, and included a total of 503 dogs with 39 positive results (7.75%).

During the first period, we tested 408 dogs from Berkovitsa (Mt), Sofia, and Dobrich, and for the first time in our laboratory-diagnostic practice, we detected chlamydial antibodies in 12 animals from Berkovitsa and Dobrich with a total average seropositivity of 2.94%.

In the second period, reflecting the state of the problem in recent times, we studied 95 dogs from Berkovitsa (dog farm), Sofia (insulator for stray dogs), and private owners. Clinical conditions of dogs and serological results are presented in Table 9.7. The table shows the variation in the clinical symptomatology of seropositive animals. The overall percentage of positive dogs is significant – 28.42%.

9.2.8 Serological Survey of Cats

For the first time in our research, we undertook a serological study of 16 cats selected as negative for pathogenic bacteria. Clinically, cats with conjunctivitis predominated, one of which had aborted 10 days ago. Single animals had respiratory signs, breast cancer, hypotension, and increased liver function. In the rest, no clinical data were recorded.

A total of six cats (37.50%) were seropositive for *Chlamydia felis*, in five cases, the animals were affected by conjunctival inflammation, and in one case, there was abortion and conjunctivitis.

Table 9.7 Serological examinations for chlamydiae of dogs with different clinical conditions (2005–2011)

Clinical Status	No. Tested	Positive		Origin
		No.	%	
Fever state with conjunctivitis	9	3	33.33	Sofia, private owners
Gastroenteritis and nervous signs	20	4	20.00	Dog Isolator – Sofia
Mucopurulent cervicitis	8	2	25.00	Sofia, private owners
Urethritis and balanoposthitis	2	2		Sofia, private owners
Respiratory diseases	17	4	23.52	Sofia, private owners
Polyarthritis	2	2		Sofia, private owners
Toxic infectious syndrome with high urea and ESR	1	1		Sofia, private owner
Asymptomatic adolescent dogs – contact with dead puppies	9	4	44.44	Berkovitsa, a dog farm
Asymptomatic contact adult dogs from an infected farm	12	4	33.33	Berkovitsa, a dog farm
No clear clinical diagnosis	3	1		Sofia, private owners
Clinically healthy dogs – prophylactically tested	12	0		Sofia, private owners
Total	95	27	28.42	

For several cats, we underwent both serological tests for *C. felis* (CFT), *C. pneumoniae* (EIA IgG), and *C. trachomatis* (EIA IgG, and IgA) due to established infections with the last two chlamydial species in the animal owners. The results obtained with the cats were negative.

The etiology of conjunctivitis in seropositive cats was confirmed by a direct LM indication of *C. felis* in conjunctival secretions and scarified materials and by isolation of the causative agent in two cases.

9.2.9 Serological Investigations in Birds

The serological examinations for avian chlamydiosis (*C. psittaci*) comprise a large number of samples of waterfowl, chickens, pigeons, parrots, and other species [53, 337, 577–580].

9.2.9.1 Chlamydiosis (ornithosis) in ducks

We studied 5181 Mulard ducks, production of which in the early 90s of the twentieth century marked a significant growth. The samples originated from 49 farms with different capacities in 46 settlements in 19 districts. The surveys we conducted between 1991 and 2007. The results obtained are shown in Figure 9.18.

In the chart, the studies were divided into two parts: (a) 1991–1998 with samples of 2600 ducks, 15 foci of chlamydiosis, and an average seropositivity for *C. psittaci* of 13.46% and (b) for the period 2000–2007 with 2581 tested ducks, 12 new foci of the disease with an average seropositivity of 6.23%. The rate of seropositive ducks for the period was 9.86%, and for the last three years of the study (2005–2007 incl.) – 9.37%. The extent of infection in individual farms ranged from 2 to 80%.

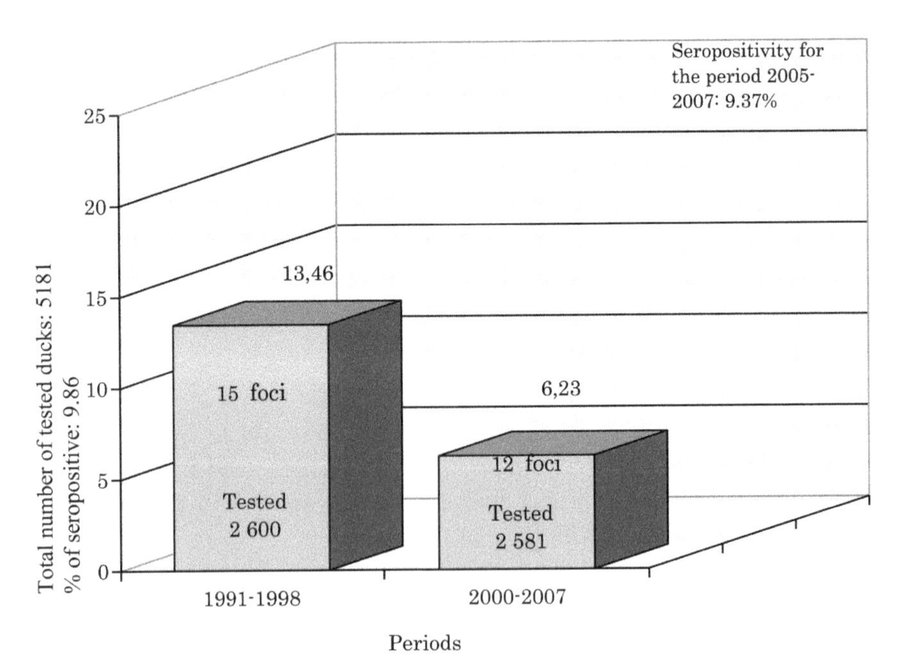

Figure 9.18 Serological tests for avian chlamydiosis in Mulard ducks (1991–2007).

Table 9.8 Serological status of *C. psittaci* infection in industrially grown ducks during outbreaks of ornithosis in humans in the 1990s

Year	No. Tested	(+) No.	(+) %	Outbreaks of Chlamydiosis
1991	534	135	25.20	3
1992	165	33	20.00	2
1993	455	105	23.07	3
1994	885	146	16.49	5

In part of the farms with large and medium capacity – Trustenik (Pl), Kostinbrod (SF), Lom, Razgrad, and others, avian chlamydiosis had an epizootic nature with high mortality. These epizootic outbreaks caused epidemic outbreaks of ornithosis among farm staff and part of the population in the settlements.

Table 9.8 shows the serologic status of chlamydial infection among Mulard ducks during the epidemic for the people period 1991–1994.

From the table, it can be seen that a total of 2039 ducks were tested for the four-year epidemic period. They originated from 19 farms in 14 districts. The mean seropositivity for *C. psittaci* was 20.54% (16.49–25.2%). We have revealed 13 outbreaks of avian chlamydiosis including five which were directly related to outbreaks of ornithosis among humans (Trustenik, Kostinbrod, etc.)

Besides the large amount of investigations of Mulard ducks, we tested serologically for ornithosis and other kinds of ducks. For example, in a study (April 2008) of 14 semiwild ducks kept at a Kostenets (Sf) breeding farm, we found that 11 (78.57%) were positive for IgG antibodies in solid phase EIA analysis. Despite the limited number of samples tested, the resulting high percentage of seropositive ducks in a herd without clinical manifestations is evidence of *C. psittaci* latent infection.

9.2.9.2 Chlamydiosis (ornithosis) in geese

Studies of chlamydial infection of geese grown industrially for meat and liver are also of interest from an economic and epidemiological point of view [83, 580, 581].

The serology of these birds in the present study covered 505 blood samples from eight farms in seven areas. The average seropositivity for *C. psittaci* infection was 28.76% and in the individual farms from 10 to 46.6%. Positively reacted geese we found in all farms surveyed.

Particular attention is paid to goose chlamidiosis data during the above mentioned epidemic period for human ornithosis in humans (Table 9.9). They

Table 9.9 Serological status of chlamydiosis in industrially farmed geese during an epidemic period for ornithosis in humans

Year	No. Tested	(+) No.	(+) %	Outbreaks of Chlamydiosis
1992	75	9	12.00	2
1993	211	86	40.75	1
1994	155	20	12.90	3

show that for three years, six outbreaks of chlamydiosis have been detected in geese with an average seropositivity of 26.07% (12.9–40.75%). Ornithosis in humans has been proven in one of the outbreaks of chlamydial infection in geese.

The serological results for avian chlamydiosis in ducks and geese were supported in a number of cases by the microscopic and virological indication and identification of *C. psittaci* in the internal parenchymal organs of dead birds.

These data clearly testify to the single etiology of chlamydiosis (ornithosis) in waterfowl and ornithosis in humans in its various forms of morbidity, including the epidemic.

9.2.9.3 Chlamydiosis (ornithosis) in hens

In the testing of 1272 hens and chickens from 11 poultry farms in 11 settlements in 8 districts, we found seropositive birds in 8 of them with an average rate of 17.13. The infection with *C. psittaci* has been proven in the regions of Pleven, Dobrich, Razgrad, Bourgas, Yambol, Pazardzhik, Sofia-region, and Pernik.

Positive responses for individual study groups ranged from 10.95% to 46.6%. The maximum range was found in laying hens from one farm (46.6%) and chickens from another farm (41.17%). Cases of epidemic outbursts of ornithosis in humans from the same village were not noted, but the etiological involvement of chickens infected with chlamydia in the sporadic cases of human disease is allowed.

9.2.9.4 Psittacosis in parrots

We surveyed 21 parrots belonging to private owners from Sofia and V. Tarnovo. Four of them (19.04%) had CF antibodies against *C. psittaci*. The serological results for psittacosis were supported by the isolation of chlamydial strains from the parenchymal organs of diseased parrots.

9.2.10 Serological Examinations of Wild Mammals and Birds

The prevalence of chlamydial infection among some wild species was investigated by serological studies of mouflons, crows, quails, and feral pigeons [53, 578, 582].

We generally examined 150 blood samples. Positive responses were found among the four studied species. The examined 46 mouflons were imported from abroad and quarantined within the stipulated deadline. In two cases (4.34%), we found chlamydial antibodies with titers of 1:64. Interesting is the high seropositivity (60%) of the studied crows. For the first time in the country, we studied quail and found positive cases at 9.09%.

The feral pigeons in the present study in the regions of Sofia-city, Pleven, and Vidin showed seropositive reactions at 25.5%, which again confirms the significant circulation of *C. psittaci* in this population and the zoonotic risk to the human population. The zoonotic significance of wild pigeons is emphasized in a number of older and contemporary publications [53, 578, 582–585].

For *C. pecorum* is known that affects domestic ruminants, pigs, and coalas. Recent serologic and molecular studies have shown other wildlife globally may harbor strains of *C. pecorum*. These hosts include ibex, buffalo, red deer, and wild boar – Burnard and Polkinghorne [1101]. The impact that these infections have is currently unclear, however [1101].

The resulting serological data are reported in Table 9.10.

9.2.11 Serological Examinations of Humans for Chlamydia spp

Summarized serology results of some contingents of sick people for chlamydial infections are outlined in Figure 9.19.

It is seen that for the period 1997–2011 serologically for *C. psittaci*, *C. trachomatis*, and *C. pneumoniae* tested a total of 21,251 people. Positive results are distributed as follows: *C. psittaci* – 6.67% of 2817 tested; *C. trachomatis* – 35.77% of 14,052; and *C. pneumoniae* – 41.00% of 4382

Table 9.10 Chlamydial CF antibodies in the blood serum of wild animal species

		Positive	
Species	No. Tested	No.	%
Mouflons	46	2	4.34
Crows	20	12	60.00
Quail	33	3	9.07
Feral pigeons	51	13	25.49

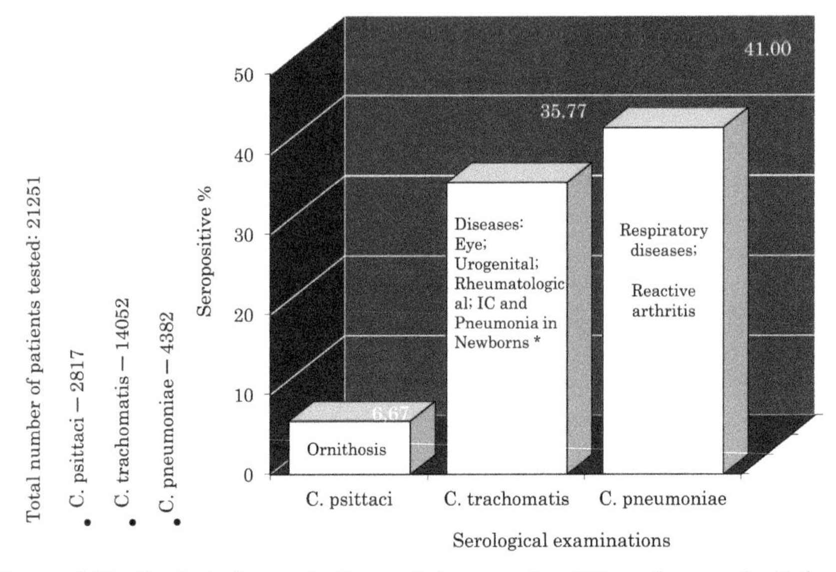

Figure 9.19 Serological examinations of humans for Chlamydia spp in Bulgaria (1997–2011).

tested. The clinical features of etiologically proven diseases are presented in a separate chapter describing clinical manifestations of Chlamydia-induced diseases in animals and humans.

9.2.11.1 Infections with *C. psittaci*

Table 9.11 shows our serological results for *C. psittaci* by years.

Of the table, it is shown that with the exception of 1997 and 1998, serology covers from 103 to 353 samples per year or midyear – 188. The seropositivity ranges from 3.86% to 10.6% – 6.67% on average for the entire period. We will note that the covered period is postepidemic in relation to avian chlamydiosis, which in the previous years (1991–1994) manifested itself with massive diseases in waterfowl birds and resulted in several epidemic outbursts among people. During this prolonged postepidemic period, the serological testings were conducted on suspected sporadic cases of human ornithosis and in contact with them asymptomatic individuals.

The serological status of the epidemic *C. psittaci* infection among humans during the Chlamydia epizootic diseases of waterfowl in 1991–1994 is shown in Table 9.12.

It can be seen from the table that from two epidemics and two epidemic blasts have been tested a total of 462 clinically ill and contacting them oligo-symptomatic or asymptomatic individuals. For the individual

Table 9.11 Serological investigations by CFT of humans for *C. psittaci* (1997–2011)

Year	Tested Blood Samples (No.)	Positive	
		No	%
1997	16	1	6.25
1998	29	4	13.7
1999	114	10	8.77
2000	121	10	8.26
2001	103	11	10.6
2002	353	26	7.36
2003	361	33	9.14
2004	259	10	3.86
2005	241	16	6.6
2006	272	13	4.7
2007	172	10	5.81
2008	201	14	6.96
2009	194	11	5.67
2010	183	10	5.46
2011	198	9	4.56
Total	2817	188	6.67

Table 9.12 Serological data on outbreaks of ornithosis in humans in Bulgaria in 1991–1994

Year	Settlement	Epidemiological Form	Serological Studies by CFT of Clinically Ill, Oligo-Symptomatic, and Asymptomatic Individuals			Source of Infection
			Examined No.	Positive		
				No.	%	
1991	Trastenick	Epidemic	199	51	25.62	Ducks
1991	Kostinbrod	Epidemic	209	69	33.01	Ducks
1993	Razgrad region	Epidemic blast	38	18	47.36	Geese
1994	Targovishte region	Epidemic blast	16	10	62.50	Geese

epidemics, seropositive cases are from 25.62% to 62.5% (average of 32%). Consequently, the average seroprevalence during the ornithosis epidemics is 4.8 times greater than that in the period after the epidemic (6.67%). A similar increase in seropositivity was found in the surveys of waterfowl during the epizootics related to human epidemics, compared to the postepidemic period.

9.2.11.2 Infections with *C. trachomatis*

Serological tests for *C. trachomatis* conducted on a large number of blood samples from individuals affected by different diseases: urogenital infections (NGU, PGU, cervicitis, and salpingitis); rheumatological problems [Reiter's syndrome and sexually acquired reactive arthritis (ReA)]; ocular diseases (conjunctivitis, iridocyclitis, and uveitis); and chlamydial infections in newborns (neonatal inclusion conjunctivitis, and pneumonia). In total, samples and materials from 14052 patients were examined. The total average seropositivity of these infections for the period 1997–2011 was 35.77% (Figure 9.19).

In detailing the serology for *C. trachomatis*, it has been found that for years the minimum number of samples tested was 275 (2007) and the maximum was 3418 (1997). The average annual number of serum samples for the whole period (1997–2011) was 936. This volume of laboratory diagnostics is significant and gives a good idea of the prevalence of infections caused by *C. trachomatis*. The range of seropositive responses by year varies from 24.9% to 41.8%. There has been a steady trend in the disclosure of seropositive patients throughout the period. Schematic separation of the 1997–2011 period into three parts – 1997–2002, 2003–2007, and 2008–2011 showed a picture reflected in Figure 9.20.

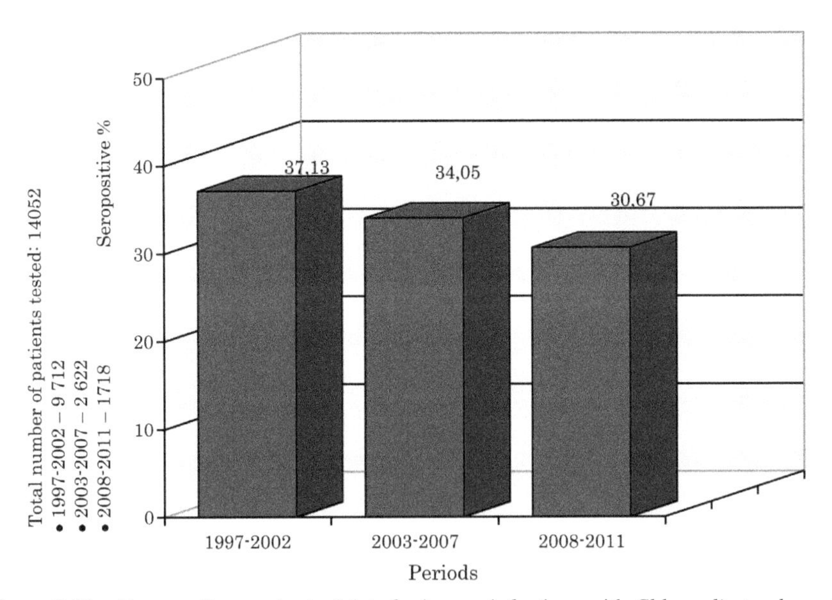

Figure 9.20 Comparative serological data for human infections with *Chlamydia trachomatis* in 1997–2002, 2003–2007, and 2008–2011.

It is clear from the graph that despite the threefold increase in the number of patients surveyed in 1997–2002 compared to 2003–2007, the seropositivity of the two periods is similar – 37.13% in the first and 34.05% in the second. During the third period of 2008–2011, there is a certain decrease in the percentages – 30.67, but overall this is also a significant percentage, which indicates a lasting trend in the seroprevalence of *C. trachomatis* infections.

In patients suspected of *C. trachomatis* infection, we generally applied the method of complex etiological diagnostics including serological tests, methods of direct indication, and identification of the agent and isolation of chlamydiae in CE and CC [24, 57, 58, 105, 148, 507]. As an example of this methodological approach, a major part of which is serology, we present Table 9.13 on cervicitis in women.

Three serological methods (CFT, ELISA, and RIF-MAbs), two methods for direct detection (cytology and DEM), and two culture methods (CC and CE) have been used. Data analysis revealed that 69.3% of the positive cases were demonstrated simultaneously by two or more methods. In statistical data processing as the underlying method, we adopted the DEM. Between the DEM and the cytological method, we found a large direct relationship (r = 0.76). The relationship between DEM and CFT is smaller (r = 0.66), whereas between CFT and CC (McCoy) is also direct and large (r = 0.85). Significant is the relationship between CFT and the cytological method (r = 0.65). The statistics give reason to work with a smaller number of complex diagnostic methods used.

Table 9.13 Complex etiological diagnosis of cervicitis caused by *Chlamydia trachomatis*

Methods	Investigated Patients (No.)	Positive (No. /%)
Cytological	113	64/56.63
CFT	113	61/53.98
ELISA	300	200/66.66
RIF – MAbs	201	120/59.70
DEM	72	51/70.83
CC McCoy:		
– Giemsa	44	28/63.64
– Lugol iodine solution	38	18/47.36
Isolation in CE	16	5/31.25

9.2.11.3 Infections with *C. pneumoniae*

Serological studies for 1997–2007 divided into three groups: (A) for the 1997–2002 period; (B) for 2003–2007; and (C) 2008–2011. During the first two periods, we examined approximately the same number of samples. The average seroprevalence rates were similar – 41.95% (1997–2002) and 45.83% (2003–2007). However, there is a certain increase (3.88%) of the positive respondents in 2003–2007. In the third period (2003–2007), there is a lower annual and for the whole 4-year period seropositivity – 30.36% (Figure 9.21).

In the analysis of these studies, the significant extent of infections with *C. pneumoniae* in the years under study was seen as the lowest percentage (22.27) recorded in 2011, and the two highest rates – 53.4 and 50.2 – were, respectively, in 1998 and 2007. The total volume of serological work is significant – 4382 serum samples (43 to 615 per year) or 292 samples per year on average. Titration of specific antibodies was performed by ELISA and indirect solid phase EIA (IgG, IgA, and IgM).

The investigations conducted were mainly in respiratory diseases – interstitial type atypical pneumonia, protracted exacerbated bronchitis, flu-like

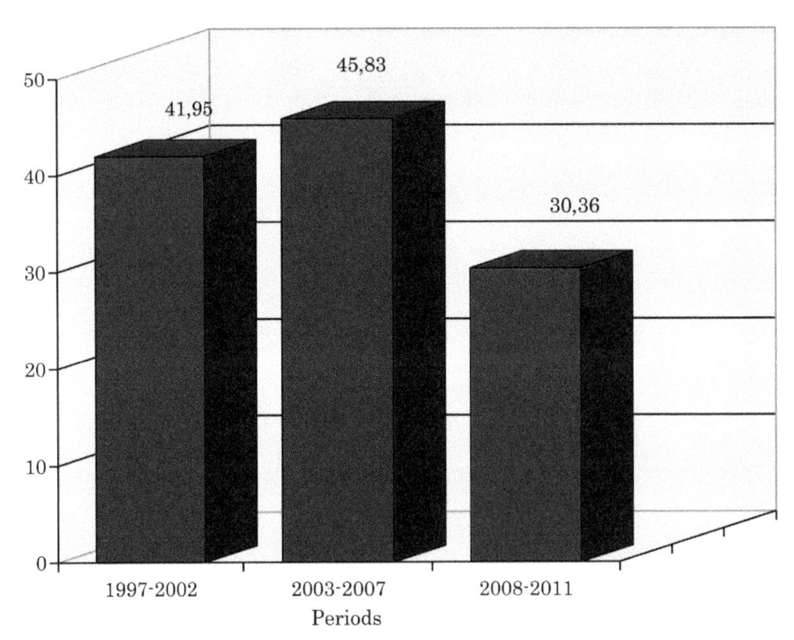

Figure 9.21 Comparative serological data for human infections with *Chlamydia pneumoniae* in 1997–2002, 2003–2007, and 2008–2011.

conditions, and prolonged subfebrility with unclear etiology. An interesting and significant piece of research is the serology of patients with pneumonia who developed bronchial asthma. As an example of the latter, we present Table 9.14, referring to some of our studies in similar cases [586, 587].

Table 9.14 Serological study in indirect solid phase EIA for the detection of IgG antibodies against *C. pneumoniae* in asthma patients

Antibodies	Newly Discovered Asthma	Chronic Asthma	Controls
Positive IgG	3	9	1
Acute infection	1	4	0
Reinfection or chronic infection	2	5	0

Table 9.15 Infectious causative agents of respiratory infections that precede reactive arthritis

Microorganism	No. of Cases	%
Streptococcus	12	38.7
Chlamydia pneumoniae	8	25.8
Adenovirus	4	12.9
Virus herpes simplex 1	3	9.7
Mycoplasma fermentans	2	6.4
Reovirus	1	3.2
Unidentified	4	12.9

Table 9.16 Results from studies of patients with reactive arthritis for carriers of antigens of HLA B7-CREG

Controls				PR-ReA No. 30		UG-ReA No. 12		EC-ReA No. 7	
Antigens		Carriers		Carriers		Carriers		Carriers	
B7-CREG	Tested	6p.	%	6p.	%	6p.	%	6p.	%
B7	1085	113	10.4	1	3.3	1	8.3	1	14.3
B13	1085	76	7.0	2	6.7	1	8.3	–	0
B22	1085	51	4.7	2	6.7	–	0	–	0
B27	1085	117	10.7	4	13.3	6	50.0***	3	42.8*
B40	1085	122	14.2	9	30.0**	4	33.3	1	14.3
B40+B13	49	–	0	–	0	–	0	–	0
B73	99	2	2.0	–	0	–	0	–	0
CW2	249	44	17.6	6	20.0	2	16.7	1	14.3

* = $p < 0.05$;
** = $p < 0.01$;
*** = $p < 0.001$.

Another specific aspect of the serologic work on *C. pneumoniae* is postrespiratory ReA (PR-ReA) [588–590]. These studies, based on a broad experimental basis, were performed in a comparative aspect with respect to other bacterial and viral agents of respiratory infections that preceded the development of PR-ReA (Table 9.15).

There is a significant etiologic relative share of *C. pneumoniae* – 25.8% – second in size after streptococci and 2–8 times greater than that of other microorganisms.

Table 9.16 provides data on the frequency of carriers of antigens from the so-called HLA B7-CREG group in PR-ReA patients and with ReA associated with a previous urogenital infection (mainly *C. trachomatis*) – UG-RheA or enterocolitis infection – EC-ReA. It is noteworthy that while UG-ReA and EC-ReA significantly increased the frequency of HLA B27 antigen challenge, in PR-ReA significantly ($p < 0.01$) more frequently than in the control population, HLA B40 antigen was detected – finding, which was reported for the first time – Dimov et al. [588]; Dimov et al. [589].

10

Detection and Differentiation of Chlamydia Microorganisms by DNA Detection Systems

10.1 Conventional PCR

The development and introduction of polymerase chain reactions in research and diagnostic work is the result of an effort to improve them by creating fast, accurate and specific tools and implementing it based on significant advances in molecular microbiology. PCR methods make it possible to distinguish different Chlamydia species and to identify the pathogen directly in clinical and pathological samples from diseased animals and humans. One of the important results of the PCR application is the possibility of avoiding in many cases the isolation of the agent in different biological models. In addition, there is evidence of greater sensitivity and specificity of the method in most cases compared to pathogen isolation.

The principle of PCR consists of *in vitro* enzyme amplification of selected nucleotide sequences limited by known sequences.

The components of each amplification reaction are as follows:

- *DNA-matrix for the process.* It results from thermal denaturation of specific DNA.
- *Enzyme Taq-polymerase* which is necessary for the process of copying the desired DNA sequence. This enzyme is thermo-stable and active throughout the amplification process. Synthesis with Taq-polymerase is best performed at 72°C.
- *Primers.* Short synthetic oligonucleotides number 14–40, complementary to the multiplying sequence-limiting regions. The synthesis of DNA at each new cycle of the PCR reaction begins with the primers. Their optimal concentration in the amplification reaction is 0.1–1 μM.
- *Deoxyribonucleotide triphosphates* (dATP, dGTF, dTTP, dTTP). They play a role as a substrate for the construction of the newly synthesized chain. In this process they are selectively included on the principle of

complementarity. An important mandatory condition is that the four types of dTTPs are present in equimolar amounts in the reaction mixture. For each nucleotide a concentration of 20–200 μM is required.

- *Reaction Buffer.* It provides the proper pH = 8.3–9 of the medium and ions required for enzyme work. The buffer composition is 10–50 mM Tris-HCl at pH = 8.3–9; to 50 mM KCl and 0.5–5 mM MgC12. Buffer components can vary qualitatively and quantitatively, depending on the requirements of the particular Taq polymerase. As additional ingredients, they use nonionic detergents (such as DMSO – dimethyl sulfoxide), bovine serum albumin (BSA), gelatin, and others.

At the beginning of the PCR is a prolonged denaturation, which aims at separating the DNA double strand. In order to amplify the desired site, repetitive cycles of a certain temperature are required, each of which involves the following three steps:

1. *Thermal denaturation at 90–97°C.* Under this temperature regime, complete separation of double-stranded DNA into single-stranded amplification matrices is ensured;
2. *Hybridization (annealing).* It is performed between the primers and their complementary regions of the matrix single-stranded DNA. This interaction takes place at a very specific temperature, which is calculated according to the length of the primer sequences used and the base composition.
3. *Synthesis (extension).* At this stage a new complementary to the template DNA strand in the direction 5 → 3 is synthesized. The most common temperature is 72°C. Its height is determined by the specificity of the enzyme used and selection of optimal temperature is essential to providing the highest activity of the DNA polymerase.
4. *Final stage.* It is common practice to keep the temperature at 72°C from 5 to 15 min after the last cycle. The purpose of this is to finish the partially elongated products. After the end of the PCR, the tubes should be stored at −20°C.

Various modifications and protocols for conducting PCR are described in the literature. A significant part of them target the *omp*A gene or the ribosomal RNA operon [76, 591–596]. Sachse et al. [536] characterize the *omp*A gene as an ideal target for diagnostic PCR as well as for intra-species differentiations assays. They believe that this is due to the heterogeneous primary structure of this gene. The latter encodes MOMP and harbors four VDs, each of which is flanked by a conserved region. Serovar-specific segments are mainly found on VD2 and VD4, unlike group-specific and species-specific antigen determinants that are encoded by the conserved regions [536]. Madico et al. [597] reported for PCR assay targeting 16S rRNA or 16 + 23S rRNA for detection of *Chlamydia trachomatis*, *Chlamydia pneumoniae* and *Chlamydia psittaci* in the old classification. This version is also available in triplex mode. The *omp*B was also successfully used as a target gene in the PCR modification developed by Hartley et al. [598]. This system is well suited to the molecular identification of Chlamydiaceae species under the revised classification, which is achieved by two methodological steps: (1) obtaining a family-specific product and (2) performing restriction enzyme analysis and/or enzyme-linked oligonucleotide assay of the amplicons. Laroucau et al. [599] proposed and successfully used a *pmp* gene-based PCR for the diagnosis of avian chlamydiosis. In its sensitivity, this assay out-performs the conventional PCR targeting 16S–23S intergenic spacer and the *ompA* locus.

The issue of the sensitivity, specificity, and validation of PCR assays is key to the reliability of the results obtained and their interpretation. The sensitivity and specificity varies in sample preparation and the performance parameters of the PCR test. Sensitivity is increased by targeting a relatively short DNA segment, using a nested procedure or using real-time PCR techniques [83]. DeGraves et al. [600] draw attention to such a methodological moment as the stabilization of DNA by the addition of suitable reagents in cases where a delay in processing the samples is anticipated. Several authors groups [594–596] have found that, in terms of sensitivity and specificity, nested PCR is equivalent to the isolation of the agent. While conducting PCR, extraordinary precautions should always be taken to prevent contamination. Apfalter et al. [601] considered the complex issue of accuracy and quality control in conducting PCR, emphasizing that the validation should cover all steps related to the study: sample preparation, pre-analytical procedures, DNA extraction, the design and implementation of the actual reaction, and reporting, interpretation, and verification of the results.

10.2 Real-time PCR

The purpose of conventional PCR and PCR in real time is identical – the production of a mixture of DNA molecules in which the concentration of the nucleotide sequence of interest is many times greater than that of all other sequences. Real-time PCR can be applied to traditional PCR applications as well as new applications that would have been less effective with traditional PCR [606]. The use of conventional PCR provides the opportunity to demonstrate the presence or absence of an infectious agent. The real PCR can determine the amount of this pathogen in the test sample. The difference in the number of DNA molecules at the start of the reaction and its end is huge. For this, from a practical point of view, counting the number of copies at the end of each cycle is replaced by calculating the rate of increase using a logarithmic function. Although theoretically the number of new copies increases in geometric progression, in practice, under PCR conditions, growth has slowed down after a certain time. This is due to several reasons: exhaustion of reagents, reduction of enzyme activity, parallel disintegration of a portion of already synthesized fragments, etc. The logarithmic increase of amplification product after each cycle is called log-linear phase. This phase is followed by a plateau where the increase in yield becomes less steep. Reporting of the results for the two methods takes place in a different phase of the amplification. Conventional PCR typically has 35–50 cycles when the plateau is reached and a high amount of amplicon is produced and it is then possible to see the product of the last cycle. In real-time PCR, it is possible to detect the increase in copy number of the product throughout the reaction. Thus, the actual differences in the initial quantities of the matrix in the different samples are taken into account. In contrast to conventional PCR, the quantitative real-time PCR is based on measurement in the log-linear phase.

The accumulation of the amplified product is monitored by measurement of the fluorescent signal generated by exonuclease digestions of a specifically annealed dual-labeled fluorogenic probe [536, 602, 603]. The generation of fluorescence signals is a result of specific binding of dye molecules, such as SYBR Green I, to double-stranded DNA [604] or by labeled oligonucleotide hybridization probes [602]. As fluorescent readings of PCR in the log-linear phase can be correlated to the initial number of target gene copies, quantitation of the number of microbial cells present in the sample can be accomplished [536]. The concentration of DNA (as copy number) is calculated from the number amplification cycles necessary to generate a fluorescent signal of a given threshold intensity [536].

Two variants of quantitative analysis are known: Absolute – which measures the exact number of copies of a given nucleotide sequence; and relative – which determines how many times more or less a given sequence is present than a reference one. The exact determination of the quantity of a given sequence is done by means of a standard curve. In order to construct this curve, serial dilutions of a sample containing previously known amounts of the same sequence are used. The fluorescence of each of the dilutions was monitored for PCR. Based on this a graph of the dependence of the illumination on the concentration of the dilutions can be constructed. In this way it is determined what concentration corresponds to the fluorescence measured in the unknown sample.

In general, in recent years, real-time PCR was established as the preferred method of molecular diagnostics, due to its ease of standardization, speed, high throughput assays and lower risk of contamination [536]. Espy et al. [607], summarizing the scientific and experimental experience of a number of authors, have shown that it has revolutionized the way clinical microbiology laboratories diagnose human pathogens. There is also reason for such an opinion with regard to infectious agents of veterinary interest [330, 593, 600, 605, 607].

Reduced problems with contamination and labor savings in real-time PCR are an advantage of this method, due to the fact that it is based on one reaction in a closed system [330, 593, 605]. Since the nucleic acid amplification and detection steps are performed in the same closed vessel, the risk of release of amplified nucleic acids into the environment, and contamination of subsequent analyses, is negligent compared with conventional PCR methods [607].

Most of the Chlamydiaceae-specific real-time PCR assays target the 23S rRNA gene [593, 600, 607]. The published by deGraves et al. [600] the real-time PCR version uses fluorescence resonance energy transfer (FRET) technology run on the LightCycler. This methodological approach contributes to the possibility of detecting chlamydial agents as both group-specific and species-specific levels in the nomenclature of four species. Geens et al. [605] developed a *C. psittaci* species-specific and genotype-specific real-time PCR which is targeting *omp*A gene and also uses the LightCycler technology. Unlike the authors cited above, in this study the species-specific test uses the SYBR Green detection method. For PCR detection of genotypes of *C. psittaci* (A, B, C, D, E, F, and E/B) individual tests have been developed [605]. In line with the trend of designing individual real-time PCR protocols for detecting the respective species of chlamydia, Pantchev et al. [330] developed separate tests for *C. psittaci* (avian chlamydiosis) and *C. abortus* (EAE-ovine chlamydiosis). These assays target the *omp*A gene as well. In order to eliminate cross-reactions between the genetically related *C. psittaci* and *C. abortus*, the authors use a minor groove-binding probe (MGM^{TM}). For the same purpose, said sample was used by Menard et al. [608], which proposed a TaqMan real-time PCR based on *inc*A gene sequence analysis of *C. psittaci* strains.

Jalal et al. [609] developed and validated a multi-target real-time PCR (MRT-PCR) for detection of *C. trachomatis* DNA. The targets for amplification in a single reaction were the cryptic plasmid (CP), the major outer membrane protein (MOMP) gene, and an internal control. The authors conclude that this assay could be used in the qualitative format for the routine detection of *C. trachomatis* and in the quantitative format for the study of the pathogenesis of *C. trachomatis*-associated diseases. An important methodological detail is that this assay has demonstrated the potential to eliminate the need for confirmatory testing in almost all samples, thus reducing the turnaround time and the workload [609].

In a recent paper Butcher et al. [610] reported a reduced-cost *C. trachomatis* (Ct)-specific multiplex real-time PCR assay evaluated for ocular swabs and used by trachoma research programs. The authors developed two low-cost quantitative PCR (qPCR) tests for Ct using readily available reagents on standard real-time thermocyclers. The assays were highly reproducible (Ct plasmid and genomic targets mean total variance of 41.5% and 48.3%, respectively) and performed well in comparison to the commercially marketed comparator test (sensitivity and specificity >90%).

De Puysseleyr et al. [611] developed a sensitive and specific TaqMan probe-based *Chlamydia suis* real-time PCR for examining clinical samples of both pigs and humans. The analytical sensitivity of the real-time PCR was 10 rDNA copies/reaction, without cross-amplifying DNA of other chlamydia species. The validation of the reaction was carried out using pharyngeal, conjunctival, and stool samples of the slaughterhouse employees as well as porcine samples from two farms where there was reproductive failure, and from a farm with clinically healthy animals. Positive results for *C. suis* have been detected in diseased pigs and in the eyes of asymptomatic humans. These results have also been confirmed by the isolation of the pathogen in McCoy cells. In addition to that described, the *C. suis* isolates obtained were tested for the presence of tetracycline-resistance gene *tet* (C) by Tet (C) PCR designed to demonstrate that gene. The *tet* (C) gene was only present in porcine *C. suis* isolates [611].

In the study of Hardick et al. [612], the existing CPN90/CPN91 targeting primer and 197-bp region of the 16S rRNA gene of *C. pneumoniae* was used with a Taqman probe (CPNTM) and the Roche Lightcylcer platform to develop a method for real-time detection of the pathogen from peripheral blood mononuclear cells (PBMCs) and respiratory specimens. The analytical sensitivity of the assay was between 4 and 0.4 inclusion-forming units

(IFUs)/PCR reaction. Overall, the assay had a sensitivity of 88.5% and a specificity of 99.3%. By highlighting the well-known advantages of real-time PCR over traditional PCRs, the authors conclude that the method they adapt is very useful in the diagnosis of *C. pneumoniae*. The authors also express the view that there are a variety of molecular-based amplification methods for *C. pneumoniae* detection and that these methods are more sensitive than culture, but have the disadvantage of being inconsistent and non-comparable across studies. According to another study [613], real-time PCR holds promise for *C. pneumoniae* quantitation in human atherosclerotic plaques. This is an assay using novel probes and FRET LightCycler technology, which is particularly suitable for the specific and quantitative detection of a low-DNA copy number in conventional PCR-negative samples.

10.3 Our Experience in the Development, Adaptation and Application of PCR Techniques

10.3.1 Normal (Conventional) PCR

Figure 10.1 shows the results of normal PCR with primers. **CTU** (ATGAAAA AACTCTTGAAATCGG) and **CTL** (CAAGATTTTCTAGAYTTCATYTT GTT) of 12 strains of chlamydiae of different origins (see legend). The resulting amplification products have a size of ~900 bp (base pairs).

We have successfully tested samples of passage materials [yolk sac (YS) of chicken egg (CE) and cell culture (CC)] and clinical-pathological materials (suspensions made from placentas and from fetal parenchymal organs) stored in a frozen state or as lyophilisates [52, 615, 616].

For comparative purposes investigated in normal PCR chlamydiae and representatives of other groups of microorganisms – Coxiella burnetii and rickettsiae (Figure 10.2).

In normal PCR for detection of chlamydia, we used the pair of primers CTU and CTL, and for Coxiella-**CB-1** (ACT CAA CGT ACT GGA ACC GC) and **CB-2** (TAG CTG AAG CCA ATT CGC C). From the results in Figure 10.2 **A2.** it is seen that the amplification products obtained in chlamydiae have a size of 900 bp, whereas in the *Coxiella burnetii* strains (Figure 10.2 **B2.**) the DNA products are about 260 bp in size. In normal PCR with DNA from strains of *Rickettsia spp* a pair of primers 190.70r Rr (ATGGCGAATATTTCTCCAAAA) and Rr 190.701n (GTTCCGTTAATG-GCAGCATCT) was used, sequenced by gene rOmpA. In amplification with

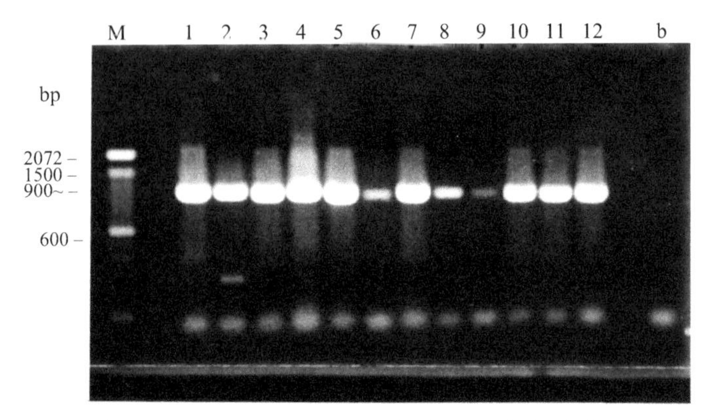

Figure 10.1 Amplification products from normal PCR with DNA of 12 strains of *Chlamydia* spp. Line M – DNA marker 100 bp; Line 1 – strain P-1 (chlamydial keratoconjunctivitis of cattle); Line 2 – *Chlamydia psittaci* – psittacosis; Line 3 – *Chlamydia abortus* – ovine abortion; Line 4 – SP4 / 78 – Chlamydial pneumonia in sheep; Line 5 – *Chlamydia abortus* – epizootic (chlamydial) abortion in cattle; Line 6 – *Chlamydia abortus* (goat abortion); Line 7 – guinea pig infected with *Chlamydia abortus* – sheep abortion; Line 8 – *Chlamydia psittaci* – ornithosis canary; Line 9 – Chlamydial encephalitis in buffalo calf; Line 10 – strain D (*Chlamydia abortus* – ovine abortion); Line 11 – *Chlamydia psittaci* – ornithosis chickens; Line 12 – Angelina strain, *Chlamydia trachomatis* cervicitis; Line B – blotting sample without DNA (negative control). **Markings:** 2072 bp, 1500 bp, ~900 bp, 600 bp – markers.

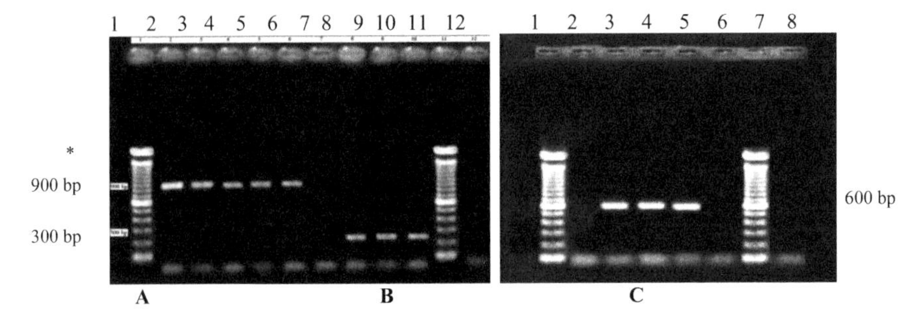

Figure 10.2 Amplification products from conventional PCR with DNA from strains *Chlamydia* spp., *Coxiella burnetii,* and *Rickettsia* spp. **Lines 1 and 11** – DNA marker 100 bp; **A. Lines 2–7** – 900 bp amplifies from: 2. *Chlamydia trachomatis* M (urethritis); 3. *Chlamydia psittaci ornithosis* (canary); 4. *Chlamydia psittaci* CP3 (ornithosis pigeon); 5. *Chlamydia psittaci* GA (abortion in sheep); 6. *Chlamydia psittaci* T-1f (abortion in goats); 7 and 12, negative controls (samples without DNA); **B. Lines 8–10** – 260 bp amplifies from: 8. *Coxiella burnetii*, Henzerling (reference human strain); 9. *Coxiella burnetii* M-44 (per person); 10. *Coxiella burnetii* (aborted cow); 1 and 7 DNA marker 100 bp; 2 and 6 negative controls (samples without DNA); **C. Lines 3–5** – amplifies with a size 590 bp of: 3. *Rickettsia rickettsii*; 4. *Rickettsia conorii*; 5. *Rickettsia sibirica*; 8. Blank. ****Indications*** 900 bp, 600 bp, 300 bp markers.

these primers, DNA products were synthesized with a size 590 bp from both genomic DNA of *R. rickettsii*, and by *R. conorii* and *R. sibirica* (Figure 10.2 **C2.**).

The results obtained show that conventional PCR is a suitable method for differentiation to a microorganism level, but is not suitable for intra-species differentiation between individual members of the same genus – *Chlamydia*, *Coxiella* and *Rickettsia*. For this purpose, we have adopted a different methodological approach, namely the use of other specific PCR techniques, described below.

10.3.2 REP-PCR in Chlamydiae of Different Species and Nosological Origin

Studies with REP-PCR (Repetitive-element-PCR; Repetitive extragenic palindromic sequence polymorphism-PCR) conducted with 32 chlamydial strains from domestic animals with a variety of clinical conditions (sheep, goats, cattle, buffaloes, pigs, dogs), residential and ornamental birds, experimental animals (guinea pigs), wild murine rodents, people affected by *C. trachomatis* infection – cervicitis, urethritis, Reiter's syndrome [52, 614, 616] (Figures 10.3 and 10.4).

We used primers **REP 1R-I** (III ICG ICG ICA TCI GGC) and **REP 2-1** (ICG ICT TAT CIG GCC TAC)

Figures 10.3 and 10.4 show the electrophoresis of REP-PCR amplified products from the examined chlamydia. All tested strains of sheep origin had 21 different bands (fragments) ranging between 400 bp and 2100 bp (Figure 10.3 – lines 8–15). Some of these bands were more intense than others. Strains GV (line 13) and P (line 14) were used in our vaccines PM-1 and PM-2 (EAE). They have 4 identical bands of 750 bp, 1200 bp, 1600 bp Strains A22 (Figure 10.3 – line 11) and S26/3 (Figure 10.3 – line 12) used in our vaccine PM-3 (EAE). These two strains have four identical bands with sharp intensity (450 bp, 1200 bp, 1500 bp, and 1600 bp) and five identical bands between 550 bp and 800 bp at lower intensities. These four vaccine strains have more or less similar matrix DNA profiles (fingerprints) with the other chlamydial isolates of sheep (Figure 10.3 – lines 9, 10, 15, 16). They have some common fragments with strains isolated from birds and dog (Figure 10.3 – lines 1–8) and strains of *C. trachomatis* (Figure 10.4 – lines 8–11) isolated from humans (600 bp) as well, and chlamydiae isolated from pigs (Figure 10.4 – lines 12, 13).

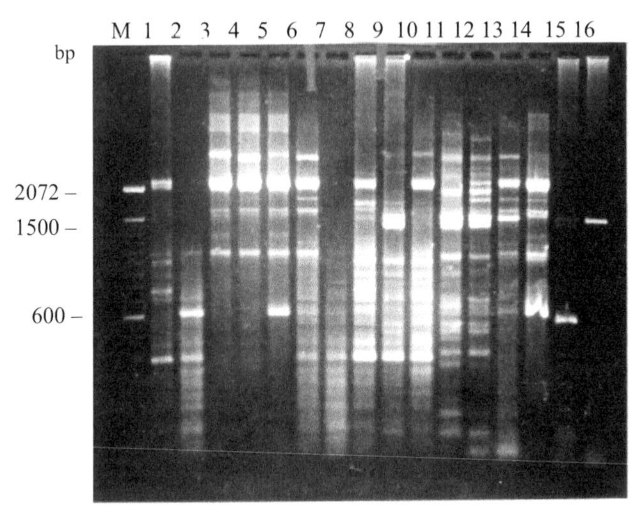

Figure 10.3 Gel I. Electrophoresis of REP-PCR amplification products of 32 chlamydial strains of mammalian, avian, and human origin. **Legend – Gel I. *Lines***: M – m w marker 100 bp; 1 – Strain P2, psittacosis in parrot; 2 – Strain P3, psittacosis in parrot; 3 – CP3/2A7, ornithosis in pigeons, United States; 4 – GR9/B6, ornithosis in ducks USA; 5 – VS1/EB, psittacosis in parrot, United States; 6 – keratoconjunctivitis in dog; 7 – canary ornithosis; 8 – chicken ornithosis; 9 – OA, chlamydial abortion in sheep; 10 – SP4/78, chlamydial pneumonia in sheep; 11 – A-22, chlamydial abortion in sheep; 12 – S26/3, chlamydial abortion in sheep; 13 – GV, chlamydial abortion in sheep; 14 – P, chlamydial abortion in sheep; 15 – German strain, chlamydial abortion in sheep; 16 – H574, chlamydial abortion of in sheep.

In all strains that we examined, we found the presence of different polymorphic bands (fragments) typical of each individual strain. This fact is indicative of the sensitivity and capabilities of REP-PCR for intra-species and intra-strain differentiation of chlamydia.

Reproducibility of the REP-PCR method is excellent, as the same results were obtained with multiple amplifications of different DNA preparations [52, 614, 616].

The UPGMA cluster analysis and the REP-PCR Jacard coefficient allowed the construction of dendrograms (Figure 10.5).

The results showed >80% similarity between the above four vaccine strains and the remaining sheep chlamydial isolates and >60% similarity between the chlamydial isolates from other animal species (Figure 10.5). The group analysis of the four human strains of C. trachomatis showed a completely different profile for these strains (Figure 10.6).

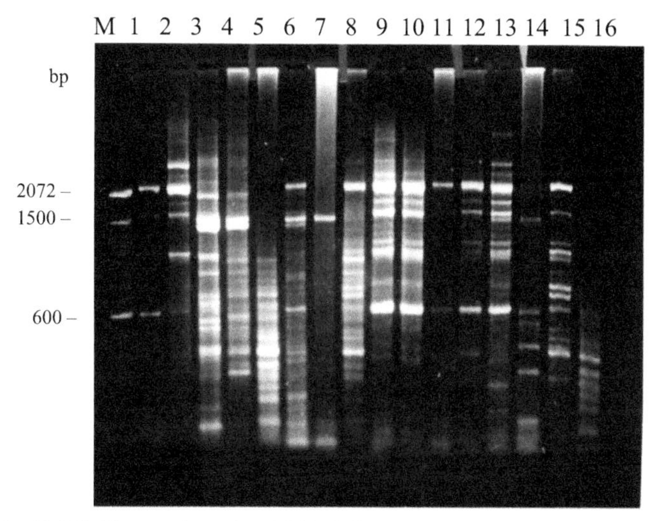

Figure 10.4 Gel II. Electrophoresis of REP-PCR amplification products of 32 chlamydial strains of mammalian, avian, and human origin. **Gel II. *Lines*:** M – m w marker 100 bp; 1 – Veliko Tarnovo, polyarthritis calves; 2 – C, chlamydial keratoconjunctivitis in cattle; 3 – P, chlamydial keratoconjunctivitis in cattle; 4 – EBA 4/78, epizootic abortion in cattle; 5 – strain 1, chlamydial abortion in goats – WT; 6 – T1r, chlamydial abortion in goats – placenta; 7 – T 1f, chlamydial abortion in goats – fetus; 8 – strain Angelina, *Chlamydia trachomatis* cervicitis; 9 – strain KC, *Chlamydia trachomatis* cervicitis; 10 – strain B3, *Chlamydia trachomatis* urethritis; 11 – strain VB, *Chlamydia trachomatis* Reiter's syndrome; 12 – strain fetus pig; 13 – chlamydial abortion swine; 14 – strain of wild murine rodent *Chlamydia glareolus*; 15 – guinea pig; 16 – strain chlamydial encephalitis buffalo calf.

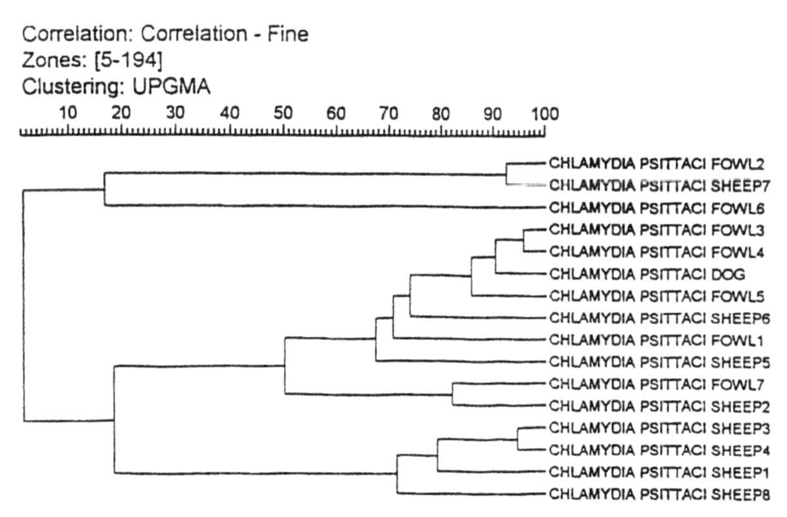

Figure 10.5 Dendrogram of chlamydial strains from mammals and birds. REP-PCR. UPGMA.

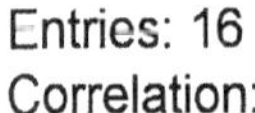

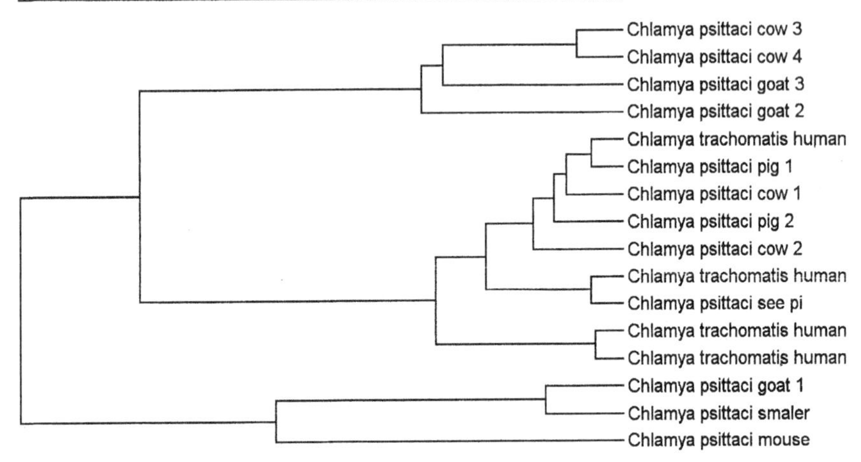

Figure 10.6 Dendrogram of chlamydial strains isolated from different animal species and humans. REP-PCR. UPGMA.

10.3.3 REP-PCR. Comparative Analysis of Strains *Chlamydia* spp., *Coxiella burnetii*, Brucella and Representatives of α-1 (Rickettsiae) and α-2 Subdivisions of Proteobacteria

In the study, we used 36 strains grouped according to their origin (genus) in five main groups: Chlamydia, Coxiella, Rickettsia, Brucella and Proteobacteria. In REP-PCR amplification with primers REP 1R-I and REP-2-I DNA profiles were observed composed of a different number of fragments of a certain size (Figure 10.7).

In four of the five tested strains of chlamydia (lines 1–4 and 36), the number of fragments is between 9 and 23 such as 2 are common to all of them. In each species, however, occur one or more specific fragments shared with some of the other species, which determines the genetic relationship between them.

Looking at the profiles of *Chlamydia trachomatis M* and *C. trachomatis W* (lines 1 and 2), it is clear that they are composed of 23 fragments, 14 of them being the same size.

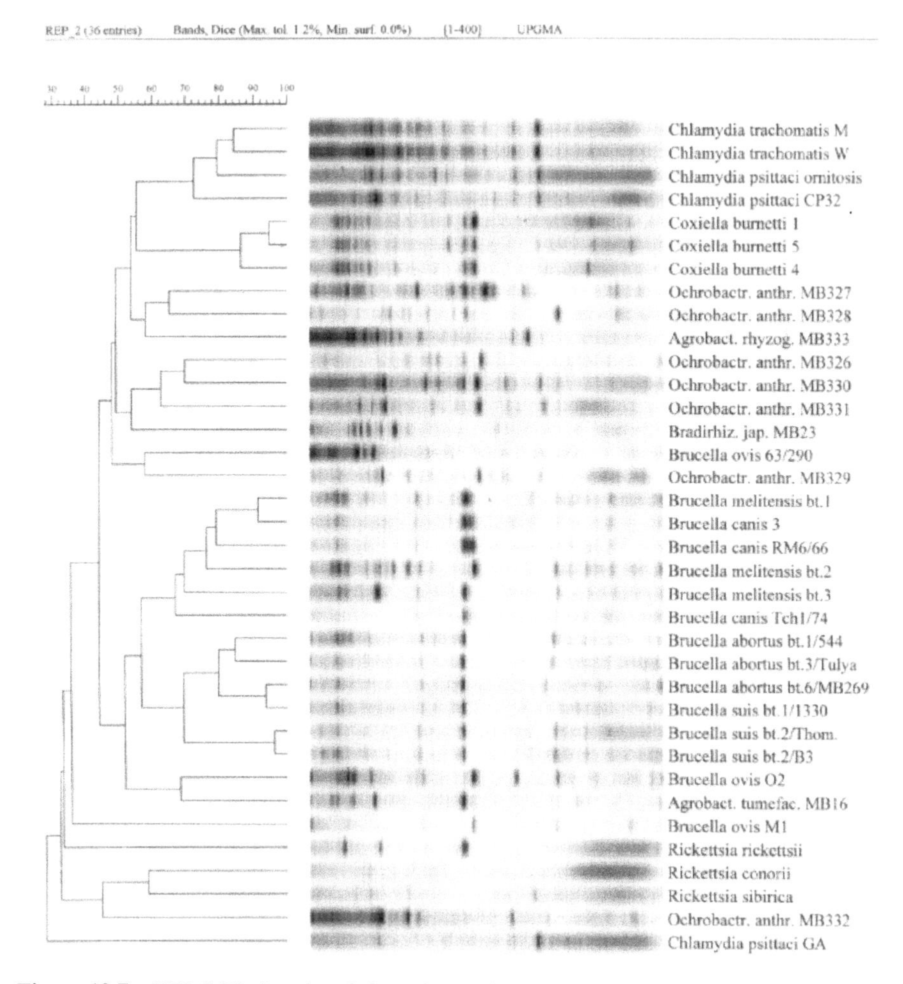

Figure 10.7 REP-PCR. Results of clustering analysis of strains Chlamydia, Coxiella, Rickettsia (α-1 subdivision of Proteobacteria), α-2 subdivision of Proteobacteria, and Brucella, with computer UPGMA program.

The *Chlamydia psittaci* strains (lines 3 and 4) have DNA profiles composed of 17 and 19 fragments, of which 5 have the same size. *C. psittaci* ornithosis (line 3) has 12 common fragments with *C. trachomatis M* and nine with *C. trachomatis W*. The *C. psittaci* CP32 DNA profile shows the presence of four common *C. trachomatis M* and three with *C. trachomatis W*. In turn, *C. psittaci* GA (line 36) has only two identical fragments with the other *C. psittaci* and one with *C. trachomatis* located at different levels of DNA profiles due to their different sizes.

The profile of the three strains of *Coxiella burnetii* (lines 5–7) is composed of 18–20 fragments, of which 11 are of the same size. This speaks of conservatism in their genome. *Co. burnetii 4* and *Co. burnetii 5* have two identical fragments.

In the strains Rickettsia (lines 32–34) pronounced polymorphism is observed – their profile is composed of eight to nine fragments, and only one of them is of the same size in all three tested strains. *Rickettsia rickettsii* has one common fragment with *Rickettsia conorii* and also one common with *Rickettsia sibirica*, located on different levels. *R. conorii* and *R. sibirica* have two common fragments.

The fragmentary profile of the Brucella species contains from nine to 22 fragments; as presented in the current study, 14 strains have only one common fragment. The analysis by species shows that the *Brucella abortus* group has a DNA profile consisting of 15–18 fragments, of which 13 are of equal length. The *Brucella canis* DNA profiles contain between nine and 17 fragments, of which seven are the same size, and for *Brucella ovis* – from nine to 20 fragments, with seven being the same. *Brucella melitensis* strains have DNA profiles composed of 18–22 fragments, of which six are of the same size. *Brucella suis* strains have 13–16 fragments, with only two being the same. *Br. abortus* has the most conservative profile.

When included in the study, 10 strains of α-2 subdivision of Proteobacteria differences are observed between 14 and 30, which do not show fragments of the same size, i.e. the representatives of this group exhibit the greatest polymorphism.

In the present study, the different Brucella species (lines 15, 17–29, 31) and the bacteria from the α-2 subdivision of Proteobacteria (Ochrobactrium, Agrobacterium, and Bradirhiz. Jap) (Lines 8–14, 16, 30, 35) which has been shown to generate very close DNA profiles in REP-PCR (Tcherneva et al., [617, 618]) are included for comparison. Analysis of the strain groups so selected showed that α-2 proteobacteria are genotypically closest to Rickettsia (lines 32–34), then to Brucella (lines 15, 17–29, 31) and Chlamydia (lines 1–4) but the farthest from Coxiella.

The *C. psittaci* GA DNA profile (line 36) is closer to α-2 bacteria than to the other chlamydia.

The *Co. burnetii* profile (lines 5–7) is most conservative and distinguishes at the DNA level from the other strain groups. Notwithstanding this conservatism, the genetic analysis of the *Co. burnetii* strains investigated shows the presence of the same size fragments with Proteobacteria – from one to five; with Ochrobactrium (lines 8–9; 11–13; 16 and 35); and two with Agrobacterium (lines 10 and 30) and Bradirhiz. Jap. MB23 (line 14). *Co. burnetii* shows five identical fragments with *Brucella ovis* O2 (line 29); four with *Br. abortus*, bt. 1, 3 and 6 (lines 23–25); four with *Br. suis*, bt. 1 and 2 (lines 26–28) and two with Br. canis (lines 18, 19 and 22). *Co. burnetii* 1, 4, and 5 have three identical fragments of *R. coronii* (line 33), one with *R. rickettsii* (line 32), R. sibirica (line 34), and Chlamydia trachomatis (lines 1 and 2).

Received REP-PCR profiles can distinguish the types of microorganisms between them, but also show a complex of fragments characteristic of individual genera. At the same time, the presence of a different number of fragments of the same size with the representatives of the individual species and genera of microorganisms speaks of their closest or more distant genetic similarity.

When visualizing the resulting matrix DNA profile (fingerprint), it is difficult to read all the DNA fragments obtained for a particular strain. After the introduction of the computer analysis, this became possible.

In our dendrogram construction study, we used the UPGMA method. The dendrogram obtained by the REP analysis, according to the arrangement of the studied strains on the electrophoretic field, groups the strains into eight main groups (Figure 10.8). Some of the groups include representatives of the same genus and species, and others – different species of the same or different genera, according to the existing classification of the species.

Based on dendrogram analysis, the similarity between the chlamydia strains used in the study was 85% for *C. trachomatis* (lines 1–2), and 72% and 79% for *C. psittaci* ornithosis and *C. psittaci* CP32 (lines 3–4), respectively. An exception is observed with *C. psittaci* GA (line 36), which has a 30% similarity to the other chlamydiae, and phylogenetically stands a long way from them.

In the *Co. burnetii* group (lines 5–7), the human *Co. burnetii* 1 strain (Henzerling) is phylogenetically much closer to *Co. burnetii* 5, a terrain strain isolated in Bulgaria from an aborted cow (95% similarity), than to the human isolate *Co. burnetii* 4 (86% similarity). To the other groups of strains, *Co. burnetii* strains show a genetic distance of 2% to 65%. *Co. burnetii* 1 and 5 are genetically the least distant from *Br. suis* bt. 2 (2%) (Lines 28–29), *Br. abortus* bt. 6/MB269 and *Br. suis* bt. 1/1330 (3%) (lines 25 and 26), and *Brucella melitensis* bt.1 and *Br. canis* 3 (lines 17 and 18). For *Co. burnetii* 4, the genetic distance is smallest with *Br. abortus*, bt. 1 and 3 (Lines 23–24) and *C. trachomatis* (lines 1–2) – 2%.

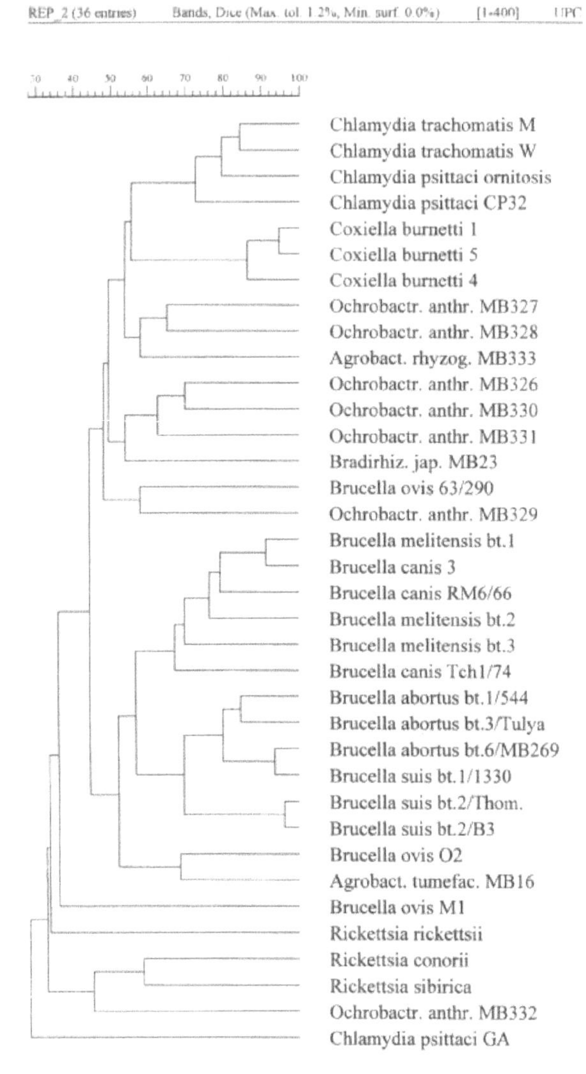

Figure 10.8 Dendrogram of strains Chlamydia, *Coxiella burnetii*, Rickettsia (α-1 subdivision of Proteobacteria), representatives of α-2 subdivision of Proteobacteria, and Brucella. REP-PCR. UPGMA.

Rickettsia, characterized by their polymorphism, exhibited less genetic similarity – 36% in the reference human strain of *R. ricketsii* (line 32) causing Rocky mountain Spoter Fever; and 58% between *R. sibirica* (line 34) causing the North Asian spotty typhus – Strain "Netzetaev," isolated from the blood of a sick person and *Rickettsia conorii* (line 33), responsible for the Mediterranean spotted fever – Kardzhali-1 strain, isolated from an *R. sanguineus* tick. Consequently, the last two strains phylogenetically stand on one branch.

Brucellosis has been found to be very closely related at the DNA level, and individual species are not always strictly distinct. *Br. melitensis* bt.1 and *Br. canis* 3 (lines 17 and 18) are phylogenetically in a single clone, with the similarity between them being 88%, and for *Br. abortus* bt.6/MB269 and *Br. suis* bt.1/1330 (lines 25 and 26) – 92%. *Br. suis* bt.2/Thom. and *Br. suis* bt.2/B3 exhibit 97% similarity (Lanes 28–29). On their DNA profile, some of them also come close to Proteobacteria α-2 subdivisions. For example, *Br. ovis* 63/290 and *Ochrobactrum anthropi* MB329 (lines 15 and 16) exhibit 56% similarity and *Br. ovis* O2 and *Agrobacterium tumefaciens* MB16 (lines 29 and 30) – 64%.

The presented data show that computer analysis makes it possible to accurately determine the locations, structure and interconnections between the various micro-organisms that could contribute to the discussion of possible changes in the classification of micro-organisms.

10.3.4 ERIC-PCR

The method Enterobacterial repetitive intergenic consensus sequence-based PCR (ERIC-PCR) employs ERIC chain sequences (short inter-spatial repeatable DNA chains) having common characteristics with repetitive extragenic palindrome chains (REP) but differing in the number of chains: REP chains – 38 nucleotides; ERIC chains – 162 nucleotides.

The studies were carried out with primers ERIC-1R and ERIC-2 (Figure 10.9).

Experiments with ERIC-PCR revealed the presence of a 300-bp band common to all strains tested. This band can serve as a marker for chlamydia in the diagnostic process.

We found that compared with REP-PCR, the method ERIC-PCR is less sensitive to the chlamydial vaccine strains. For these strains, ERIC-bands are around 19. Only five of them show sufficient intensity for comparative purposes. Two of these fragments (300 bp and 850 bp) were found in the strains A22 and S26/3 (lines 11, 12) and five fragments (400 bp, 500 bp, 700 bp, and 1400 bp) can be seen in strains GV and P (lines 13, 14).

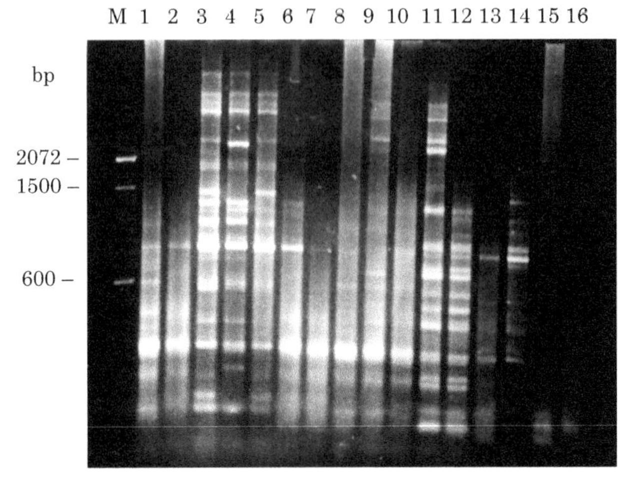

Figure 10.9 ERIC-PCR amplified products of 16 chlamydial strains of mammalian and avian origin. **Lines:** 1 – Strain P2, psittacosis parrot; 2 – Strain P3, psittacosis parrot; 3 – CP3/2A7, ornithosis pigeons USA; 4 – GR9/B6, ornithosis ducks USA; 5 – VS1/EB, psittacosis parrot USA; 6 – keratoconjunctivitis dog; 7 – canary ornithosis; 8 – chicken ornithosis; 9 – OA, chlamydial abortion of sheep; 10 – SP4/78, chlamydial pneumonia; 11 – A-22, chlamydial abortion of sheep; 12 – S26/3, chlamydial abortion of sheep; 13 – GV, chlamydial abortion of sheep; 14 – P, chlamydial abortion of sheep; 15 – German strain, chlamydial abortion of sheep; 16 – H574, chlamydial abortion of sheep.

In Figure 10.10 is shown a dendrogram of the above 16 strains, and also of 16 other strains of chlamydia from different mammals and birds, as well as *C. trachomatis* from human patients.

10.3.5 RAPD-PCR

In the method RAPD-PCR (Randomly amplified polymorphic DNA analysis) were used primers **P5** (CGG CCC CGG T) and **OPL** (04 GAC TGC ACA C).

In one of our experiments with primer P5, we included 12 strains of chlamydia isolated from bovine (keratoconjunctivitis, epizootic abortion), buffalo (encephalitis), ovine (abortion, pneumonia), goat (abortion), parrots (psittacosis), canaries, chickens (ornithosis), and from a person with endocervicitis (Figure 10.10).

It can be seen that the DNA profiles of the investigated chlamydial strains are similar but not identical. They are composed of a varying number of fragments (8–23).

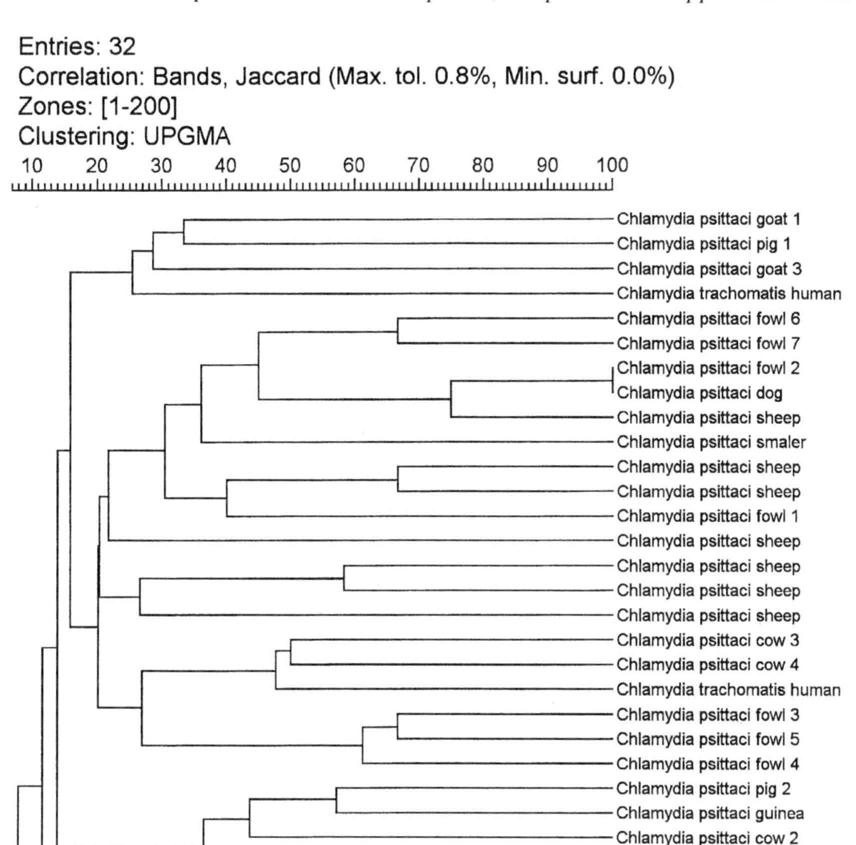

Figure 10.10 A dendrogram of 32 strains of chlamydiae isolated from various mammals, birds and humans. ERIC-PCR. Computer program UPGMA.

In the experiments with primer OPL 04 we included the same strains of chlamydia and four strains of brucella as controls: *Br. suis, Br. canis, Br. abortus, Br. ovis* controls (Figure 10.12).

Legend: Line M – DNA marker 100 bp; **Line 1** – strain P-1 (chlamydial keratoconjunctivitis in cattle); **Line 2** – *C. psittaci* – psittacosis; **Line 3** – *C. abortus* – ovine abortion; **Line 4** – SP4/78 – Chlamydial pneumonia in sheep; **Line 5** – C. abortus – epizootic (chlamydial) abortion in cattle;

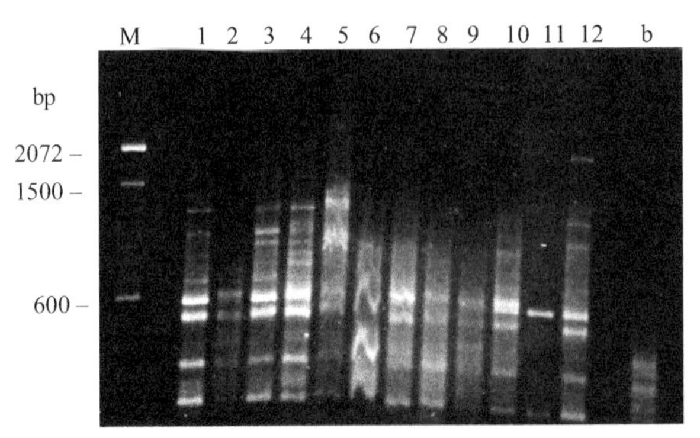

Figure 10.11 Electrophoresis of RAPD-PCR amplification products of 12 strains of mammalian, avian, and human chlamydial strain origins. Primer P5. **Line M** – DNA marker 100 bp; **Line 1** – strain P-1 (chlamydial keratoconjunctivitis of cattle); **Line 2** – *C. psittaci* – psittacosis; **Line 3** – *C. abortus* – ovine abortion; **Line 4** – SP4/78 – Chlamydial pneumonia in sheep; **Line 5** – *C. abortus* – epizootic (chlamydial) abortion in cattle; **Line 6** – *C. abortus* (goat abortion); **Line 7** – guinea pig infected with *C. abortus* – sheep abortion; **Line 8** – *C. psittaci* – ornithosis canary; **Line 9** – Chlamydial encephalitis in buffalo calf; **Line 10** – strain D (*C. abortus* – ovine abortion); **Line 11** – *C. psittaci* – ornithosis in chickens; **Line 12** – human strain Angelina, *C. trachomatis*, cervicitis.

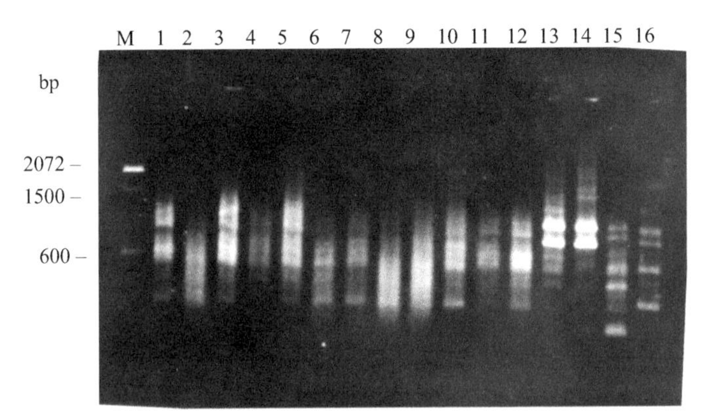

Figure 10.12 Electrophoresis of RAPD-PCR amplification products of 12 strains of mammalian, avian, and human chlamydia origin and 4 strains of *Brucella* (control group). Primer OPL 04.

Line 6 – *C. abortus* (goat abortion); **Line 7** – guinea-pig infected with *C. abortus* – sheep abortion; **Line 8** – *C. psittaci* – ornithosis canary; **Line 9** – Chlamydial encephalitis in buffalo calf; **Line 10** – strain D (*C. abortus* – ovine abortion); **Line 11** – *C. psittaci* – ornithosis chickens;

Line 12 – Angelina human strain, *C. trachomatis* cervicitis; **Line 13** – *Br. suis*; **Line 14** – *Br. canis*; **Line 15** – *Br. abortus*; **Line 16** – *Br. ovis*.

In experiments with primer OPL 04, we found similarities in the fragments of chlamydial strains, as well as differences that we will not discuss in detail. Differences were also found in the fragments of the brucella control group, both with respect to chlamydia and within the group.

In another trial, we performed a comparative RAPD analysis of five strains of chlamydia, three strains of rickettsiae, three strains of coxiellas, and two strains of bacteria.

In both experiments, the results obtained with primer P5 were visually more distinct, so we used them to compile dendrograms structured with the PHYLIP computer version 3.5c.

In Figure 10.13, we present a dendrogram of chlamydia and control groups of microorganisms. The RAPD analysis shows a similarity of the *C. trachomatis* strains tested in 93% (Lanes 1 and 2); to *C. psittaci* CP32 and *C. psittaci* ornithosis in 96% (lines 3–4); and the similarity between the two groups was 88%. *C. psittaci* GA and *C. psittaci* T-1f (lanes 5 and 13)

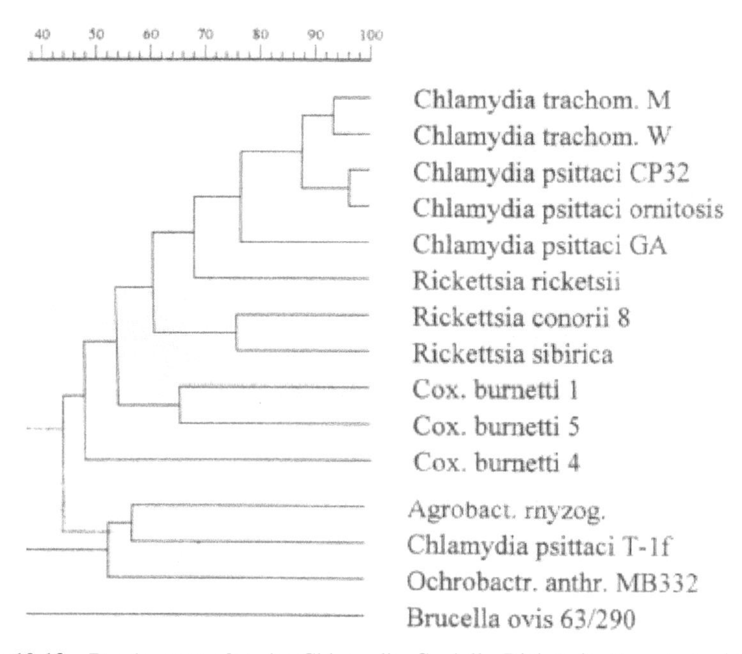

Figure 10.13 Dendrogram of strains Chlamydia, Coxiella, Rickettsia, *Br. ovis*, *A. rhizogenes* and *O. anthropi*. RAPD-PCR. Computer PHYLIP, version 3.5c.

phylogenetically are farther away from the other chlamydia species, with the similarity being 77% and 56%, respectively. *C. psittaci* T-1f (line 13) exhibits greater similarity to Agrobact. rhizogenes, with which they are on a branch of the phylogenetic tree (line 12).

When comparing these results with the REP-based dendrogram analysis (Figure 10.8), there was less similarity – 85% in the first two strains and 79% and 72% in the second two. The REP-dendrogram of *C. psittaci* CP32 and *C. psittaci* ornithosis are of different genetic distances, whereas in the RAPD-based dendrogram, the same. The similarity of *C. psittaci* GA to the other chlamydia is also much weaker – only 30%, whereas in the RAPD assay it is 77%. Consequently, the RAPD analysis is more suitable for building dendrograms to determine the genetic distance between the Chlamydia strains – lines 1–5, 13 (Figure 10.13).

The same is observed when comparing the dendrograms of the Rickettsia strains. In the RAPD analysis, the percentage similarity between *R. conorii* and *R. sibirica*, which are phylogenetically equally spaced (Lines 7 and 8), is greater (75%) than with PCR analysis (58%). The similarity of *R. rickettsii* with the other strains was 68% for RAPD and 36% for REP analysis (line 6). The greater sensitivity of RAPD due to the higher resolution of the method with regard to Chlamydia and Rickettsia makes it suitable for determining the genetic similarity between the representatives of these two genera.

In RAPD analysis of strains, *C. burnetii* can be seen (Figure 10.13) that *C. burnetii* 1 and *C. burnetii* 5 (lines 9 and 10) in the dendrogram are also of the same genetic distance, but the similarity between them is lower (64%) than in REP-analysis (95%). The same applies to *C. burnetii* 4 (line 11), which exhibits 47% similarity to the other coxiellas in RAPD – and 86% in REP-analysis. Therefore, REP analysis is better suited for building dendrograms to determine genetic similarity in strains of this pathogen.

Ochrobactrum anthropi MB329 and Brucella ovis 63/290, used as control strains in REP-analysis, are on one branch of the dendrogram with 56% similarity, while in the RAPD-analysis, they have a different distance and exhibit a 52% and 28% similarity, respectively.

10.3.6 Other PCR Techniques

During the studies on the incidence of swine chlamydiosis, as part of the diagnostic complex used, was adapted and applied PCR according to the originally proposed protocol by Pollard et al. [335, 619]. The test targets rRNA genes. It is designed to detect and differentiate *C. trachomatis* and *C. psittaci* (old nomenclature) in laboratory samples of infected McCoy cells [619]. In our experiments for the detection of chlamydiae by PCR, we infected chicken embryos and McCoy cells with clinical and pathological materials from diseased pigs. In the test were used short nucleotide sequences named *C. psittaci* (CMOMPM / CPSMOMPC), *C. trachomatis* (CMOMPN / CTMOMPC), and *C. pneumoniae* (CMOMPN / CPNMOMPC. The results showed that out of 83 tested clinically sick pigs, 32 (38.55%) by PCR evidenced *C. psittaci* (old nomenclature). Under the conditions of the experiments, this result proves that the PCR is 1.6 times more sensitive than the CC method. In addition, the applied PCR technique makes it possible to differentiate these three chlamydial species in various biological samples after inoculation into CC or embryonated eggs [335].

Real-time PCR for detection and differentiation of *C. psittaci* and *C. abortus* [620]. As is known, this PCR variant allows for amplification, detection and quantitation of the target agent simultaneously.

We have adapted a method for detecting and differentiating strain representatives of the above two species in different baseline experimental and clinical and pathological materials. A number of isolates in CE, CC, and mice in a frozen or lyophilized state, as well as suspensions of direct field materials (placentas and fetal organs from aborted sheep and cows) have been investigated. The isolation of Chlamydial DNA was performed with the NucleoSpin Tissue Extraction Kit (Macherey-Nagel, Germany).

The ABsolute™ QPCR Rc OX Mix (Thermo Fisher Scientific Inc.) was used to perform real-time PCR amplification containing all components for rapid and sensitive amplification – Thermo Strat DNA polymerase, optimized reaction buffer, dNTRs, ROX – passive reference dye to normalize the results obtained, except sample and primers. Primers and probes were used, manufactured by Canadian Alpha DNA. The nucleotide sequences are shown in Table 10.1.

The real-time PCR amplification was performed on a StepOnePlus™ real-time PCR system using the following parameters: pre-denaturation for 15 min at 95°C and subsequent 45 cycles consisting of denaturation, primer hybridization, and extension of the amplicons. Denaturation is performed for 15 s at 95°C, and hybridization and extension for 60 s at 60°C. The results obtained confirm the capabilities of the method for application across a wide range of experimental and field materials. Some of the samples were known reference strains of certain diseases, for example EAE, or strains proved to be chlamydial by prior application of a complex of classical etiological

Table 10.1 Nucleotide sequences of primers and probes used in real-time PCR

	Primers and Probes	Nucleotide Sequence (5'–3')	Amplicon Size (bp)
Chlamydia psittaci	CppsOMP1-F	CACTATGTGGGAAGGTGCTTAC	76 bp
	CppsOMP1-R	CTGCGCGGATGCTAATGG	
	CppsOMP1-S	FAM-CGCTACTTGGTGTGAC-TAMRA	
Chlamydia abortus	CpaOMP1-F	GCAACTGACACTAAGTCGGCTACA	82 bp
	CpaOMP1-R	ACAAGCATGTTCAATCGATAAGAGA	
	CpaOMP1-S	FAM-TAAATACCACGAATGGCAAGTTGGTTTAGCG-TAMRA	

methods. We applied the same approach directly to field materials. In both cases, we obtained a molecular biological confirmation of the chlamydia positive results. These results have important practical implications for the possibilities of real-time PCR in the direct diagnosis of Chlamydial abortions in animals. In some mammalian and bird samples, in addition to a single infection with one of the two agents, a mixed infection with *C. psittaci* and *C. abortus* was also established [620].

10.4 DNA Microarray Technology

The DNA microarray-based technology is a modern methodological approach with great capabilities in terms of verification of the exact nucleotide sequence of a genomic target region, and accurately distinguishing the true positive amplification products from false. These features of the method allow for precise identification of the infectious agent and lead to a significant increase in the specificity of the diagnosis. The potential of a microarray-based approach for identification and typing has been demonstrated in a number of bacterial pathogens [621–625]. The method is also effective for microarray analysis of microbial virulence factors and rapid diagnosis of agents of bacteremia [626, 627]. An important advantage of the method is the ability of DNA samples to be simultaneously assayed by a large number of probes, which may be derived from different genomic regions or from a polymorphic gene segment.

Generally, the equipment for conducting DNA microarray-based technology in large high-tech labs is expensive. The most common are the DNA array systems designed for gene transcription and expression analysis. Such systems include customized high- or low-density arrays on glass slides or nylon membranes, sophisticated multichannel fluorescence readers used for detection and purpose-built hybdridization devices. The latter are needed to ensure defined and reproducible conditions for the formation of target DNA duplexes [628].

In 2005, Sachse et al. [628] reported the development of a microarray assay for Chlamydia and *Chlamydophila* spp. Unlike the equipment and other technological requirements described above, this research group adopts a more cost-effective approach by basing its system on the commercially available ArrayTubee (AT) platform (Clondiag Chip Technologies, Jena, Germany) for processing low- and high-density DNA arrays. It involves *in situ* synthesized or spotted DNA chips of 3 mm $\times$ 3 mm size, which are assembled onto the bottom of 1.5-ml plastic micro-reaction tubes. Another change is to perform hybridization and signal processing in a simple and fast way available to laboratories with standard equipment, thus saving the need for hybridization chambers. The signal amplification is performed by enzyme-catalyzed silver precipitation. The authors designed the hybridization probes on the basis of the most variable window approach, which identified species-specific nucleotide polymorphisms in a region of generally high sequence similarity. From a multiple-sequence alignment, a highly discriminatory segment in domain I of the 23S rRNA gene has been identified. The optimized variant of the chip carries 28 species-specific probes for all nine species of chlamydia, a total of six genus-specific probes for chlamydia and the former genus *Chlamydophila*, five probes identifying the closely related *Simkania negevensis* and *Waddlia chondrophila*. There are also four hybridization controls (consensus probes) and a staining control (biotinylated oligonucleotide) [536, 628]. Target DNA is prepared by standard extraction and consensus PCR using a biotinylated primer. The ArrayTubee assay for chlamydiae has proved to be very appropriate as it provides unique species-specific hybridization patterns for all nine species of the family Chlamydiaceae. The latter are processed by the Iconoclust software (Clondiag) [536].

The AT microarray assay developed by Sachse et al. [628] was the subject of a validation study conducted by Borel et al. [629]. Two main results were achieved. First, the method has been successfully applied for the direct detection of chlamydiae in clinical tissue samples. Second, the sensitivity of DNA microarray assay was equivalent to that obtained by real-time PCR [630]. These results and the AT test qualities highlighted above have given the authors a reason to raise the question of the possible introduction of an alternative standard in chlamydia diagnosis involving a combination of AT microarray assay and real-time PCR [536].

The AT microarray assay also has another useful application. This test proved to be suitable for rapid, sensitive, and reproducible genotyping of *C. psittaci* strains and can be used in the routine diagnosis. In many experiments of genotyping of chlamydial strains from culture and clinical samples, the DNA microarray assay has proved to be highly consistent with the data from PCR-RFLP typing and serotyping. In some cases, the microarray test is superior in its sensitivity compared to the two mentioned genotyping tests [631].

11

Clinical Picture

11.1 Clinical Observations in Domestic Mammals

11.1.1 Latent Chlamydiosis

The phenomenon of latent, inapparent chlamydial infection in animals has long attracted the attention of researchers in the field of chlamydia research and of practicing veterinarians [18, 19, 54, 632–636].

In sheep and cattle, latent infection with *Chlamydia abortus* is common. In goats also we found a latent form, but the percentage relative to sheep and cattle was approximately 3:1. This form was revealed by two diagnostic approaches: a) serological; b) direct detection of the agent in placentas of clinically asymptomatic animals that had normal births and in the feces (LM, EM, IF, CE).

Our widespread observations of a large number of domestic ruminants in large territories give grounds to subdivide the latent form of chlamydial infection as follows:

- latent chlamydiosis in herds with etiologically proven clinical forms of the disease;
- latent chlamydiosis in herds without clinical manifestations that are in immediate or close proximity to other infected herds and farms;
- latent chlamydiosis in flocks where the infection is first proven.

In the active foci of chlamydial infection with many sero-positive ruminants and proven clinical cases in an earlier deployed and convalescent stage, we found both coexisting in latently infected animals. The latter are related to two categories:

- constantly asymptomatic;
- suffering from a clinical disease but subsequently remaining inapparent chlamydia carriers [54].

In the group of latently infected ruminants from farms adjacent to chlamydiosis foci, there are potential risks of emergence of clinically manifested forms based on the introduction of infection with highly virulent strains and systemic re-infections with them.

We find that the latent form of chlamydial infection is not static. Its disclosure was a signal for further targeted clinical observations and sero-epizootiological studies in the herd or farm, and a key element in this activity was the identification of factors that may be potential activators of latent infection and its transformation into a clinical disease. In this regard, we focused with particular emphasis on the pregnancy, birth, and postpartum period, the increased sterility rate, as well as on the emerging competitive infections [54]. Our opinion on this issue, derived from our own independent observations, is similar to that expressed by Reinhold [633] that according to current knowledge, the clinically inapparent chlamydial infections in cattle manifest clinically when they coincide with additional risk factors.

Over the last decade data have been accumulated on the etiological role of *Chlamydia pecorum* and the clinical manifestations associated with it in farm animals. In a recent publication, Walker et al. [635] emphasizes that *C. pecorum* infections have also been associated with a sub-clinical disease, highlighting the lack of knowledge of its true economic impact on livestock producers. Several research groups presented data on the importance of sub-clinical infection with *C. pecorum* in dairy cattle, which may have a negative effect on the performance of the corresponding herd. Here are conditions such as subclinical low-grade vaginitis (DeGraves et al. [637]); fertility disorders (Wehrend et al. [638]); increased somatic cell counts (Biesenkamp-Uhe et al. [639]); reduced body weight gain (Reinhold et al. [632]), potentially reducing growth rates by up to 48% (Poudel et al. [640]); reduced milk yield (Kemmerling et al. [641]).

11.1.2 Chlamydial Abortions, Premature Births, Stillbirths and Non-viable Offspring

Abortions and related conditions in ruminants are some of the most frequently and extensively studied chlamydia – induced diseases. This concerns in the first place the sheep, wherein the disease is referred to as Ovine chlamydiosis, Enzootic abortion of ewes [EAE] or Ovine enzootic abortion [OEA]) and is caused by *C. abortus*. OEA causes serious economic losses in many sheep-rearing areas of the world, especially where flocks are closely congregated during parturient period [23, 27, 28, 37, 642, 643].

In sheep flocks where the infection first appeared, abortions usually erupted in the last month of pregnancy and covered a different percentage (up to 30% and more) of pregnant animals. We also registered earlier abortions – around the middle of pregnancy. Some sheep aborted without previous signs, others had vaginal discharge, some showed anorexia, ugliness, hyperthermia, and bedding. Placentas were grossly inflamed.

A part of the group of aborted animals recovered quickly without any disturbance. As a complication, placental retention is detected in 5 to 30–35% of cases, with subsequent endometritis or severe lethal septic metritis.

Premature births were usually observed 4 to 8 days before the time. In special care (warm rooms with good hygiene, artificial feeding, etc.) some of the newborns survived.

The sheep showed similar precedent signs in the cases of delivery of full-term stillborn lambs and weak lambs that generally fail to survive beyond 48 h. Such non-viable lambs died with signs of acute pulmonary inflammation and respiratory and heart failure. In case of multiple births in an infected sheep, one can observe delivery of offspring consisting of one dead lamb and one or more weak or healthy lambs [54].

In **goats,** the clinical picture is similar. In a relatively higher percentage compared to sheep we watched more expressive signs preceding the abortion and higher incidence of stillbirths [54]. An important fact in connection with chlamydial abortion in goats is that the long-suspected zoonotic potential of caprine *C. abortus* (from Giroud et al. [647], then referred to as "agent of the psittacosis group"), is confirmed by the work of Pospischil et al. [648], which proved abortion in a woman caused by the chlamydial abortion agent in goats.

Useful indicators for the clinical diagnosis of chlamydial abortion in small ruminants are late abortions and macroscopic picture of diffuse necrosis in fetal membranes, which should be distinguished from that caused by *Toxoplasma gondi* (cotyledons only). It is also mandatory for the differential diagnosis to distinguish the characteristics of other abortifacient infectious agents from macroscopic pictures: *Coxiella burnetii*, Salmonella, Listeria, Brucella, and Campylobacter. The information obtained, although useful, is of an indicative nature. The etiological diagnosis that is reached through the appropriate laboratory tests is crucial.

The outbreak of an epizootic of chlamydial abortion in **cattle** is sudden. The cows aborted without signs of general disease. Most abortions are usually observed between the 7th and 9th months. Earlier abortions are also registered (fifth–sixth month). In most cases, we found placental retention with

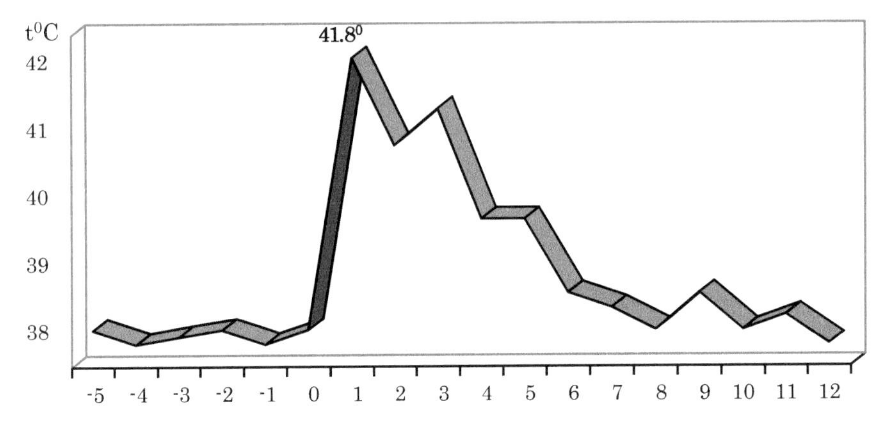

Figure 11.1 Febrile reaction of pregnant heifer after experimental intravenous infection with a *Chlamydia abortus* isolate from a stillborn calf.

subsequent inflammatory processes in the genital tract, which often led to reproductive problems. In the births of unviable calves, the latter died a few hours after the birth or in the first 1–2 days after it. Milk secretion of cows with abortions and related conditions decreased.

We conducted an experimental infection of pregnant heifer after intravenous infection with *C. abortus* strain isolated from parenchymal organs of a stillborn calf (Figure 11.1).

The graph shows that the internal body temperature is normal in the pre-infection for 5 days. Intravenous inoculation of a virulent *C. abortus* strain of bovine origin resulted in a rapid febrile reaction over the next 24 h with a peak temperature of 41.8°C. After a certain decrease on the 2nd day, a second peak of 41.4°C was recorded on the third day. The further course of the temperature curve was in the opposite direction – progressive to normalization on the 12th day. The final clinical outcome of the experimental infection was abortion in late pregnancy [54].

Overall, data on the abortifacient role of chlamydial infections in **buffalo** cows is rather limited. In three large buffalo farms in Bulgaria, Martinov [23] has seen *C. abortus* miscarriages of buffalo cows in their first pregnancy. Abortions have occurred during the last two months of pregnancy. It is interesting to note that one of the farms is a place of an old outbreak of buffalo calves chlamydial encephalitis [38]. Abortions were not preceded by clinical signs. In some cases, we found placental retention, which, when retarded, resulted in inflammatory manifestations of varying intensity. Placental lesions were similar to those in cows. In preparations of cotyledons and placental tissue, chlamydia was detected using LM, EM and direct IF. A lot later, Greco et al. [669] published data on epizootic abortion related to infections by *C. abortus* and *C. pecorum* in water buffalo (*Bubalus bubalis*). During an 11-month period, pregnant heifers suffered an abortion rate of 36.8% between the 3rd and 7th month of pregnancy. Antibodies to Chlamydiaceae were detected in 57% of the aborted cows, and in 0% of the overtly healthy cows used as controls. The serological studies were supplemented by a nested PCR assay, with 3 positive results for chlamydia from 11 tested vaginal swabs and three out of seven aborted fetuses also tested positive. In two cases there were co-infections by *C. abortus* and *C. pecorum* and in one case, a single infection with *C. abortus*. Sequence analysis of the amplicons confirmed the results of the nested PCR [669]. There is a suggestion that chlamydia also plays a role in reproductive disorders in buffaloes, but this remains to be proved [669].

11.1.3 Chlamydial Respiratory Diseases

Sheep have pneumonia with varying severity of clinical course–acute, sub-acute, and chronic. Acute and sub-acute forms affect both lambs and adult sheep. The clinical onset in a number of cases is sudden with chills and increased internal body temperature (41–41°C). Animals have rapid breathing, dry cough without nasal secretions, tachycardia, and delayed rumination. When listening to the lungs intensive vesicular respiration and bronchial breathing were established [54].

The sub-acute form appears with the same symptoms but with weaker intensity. Chronic chlamydial pneumonia is common. It is characterized by prolonged development, cough, rapid breathing of the bone-abdominal type, shortness of breath, and progressive weight loss.

We also observed several cases of a peracute form of chlamydial pneumonia. Clinically, strong mucosal hyperemia and severe shortness of breath were observed. Death occurred within a few hours, less often – after 1–2 days.

In ovine herds with active chlamydiosis, careful clinical examination revealed mild forms of respiratory chlamydial infection that could remain unrecognized due to inadequate experience and lack of etiological studies [54].

Under natural conditions, chlamydial pneumonia is often observed when feeder lambs from different herds are collected in a total herd for fattening in feed lots or irrigated pastures [644]. Clinical signs in these cases range from moderate to pronounced, with heavy mucopurulent nasal discharge, frothing, and suffocation, leading to death. A secondary bacterial infection also often interferes, which further aggravates the clinical course [19, 644].

Under the experimental conditions after intranasal or intratracheal inoculation with the chlamydial agent of ovine pneumonia, Dungworth and Cordy [645] achieved infection of lambs which react with the rapid increase of the temperature during the first 1–2 days. Animals remain febrile over the next two days, and after a gradual decrease in temperature values, normalization is reached on the fifth–sixth day. During the fever stage, the lambs were abnormal and anorectic and had respiratory distress and dry cough [645]. The same authors have shown experimentally that ovine pneumonia can also be caused by the EAE chlamydial agent. This is confirmed by Ognianov [646], which reproduces rapidly developing extensive pneumonia in one–month-old lambs after intratracheal inoculation with the EAE agent. Infected lambs have responded by raising the temperature and developing pneumonic lesions. In support, albeit indirectly, of this data is the fact of detecting chlamydial pneumonia in flocks where the OEA has already been established [646].

Figure 11.2 is a graphical representation of the temperature reaction in lambs, hog, and sheep experimentally infected with chlamydial isolate M-2 from the lungs of sheep with pneumonia. The infectious suspension of the strain with a titer of $10^{6.5}ID_{50}$/ml, propagated in the YS of CE, inoculated

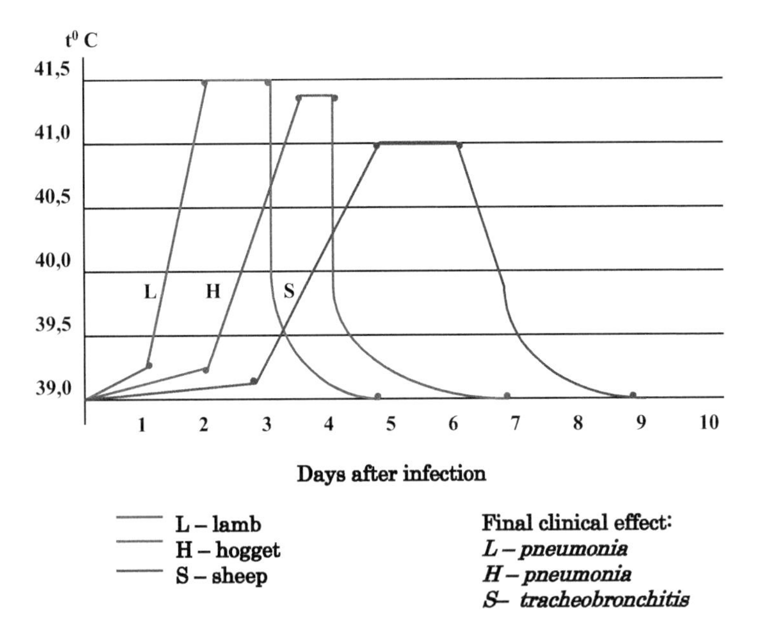

Figure 11.2 Experimental chlamydiosis in lamb, hog, and sheep. Temperature response after intratracheal inoculation with the chlamydial agent of pneumonia in sheep.

tracheally. The experimental animals – previously shown to be serologically negative for chlamydia – reacted with an increase in body temperature in the range of 41°C–41.5°C as early as after 24 h in the lamb, one day later (48th hour) in the hog, and two days later (72 h) in the sheep. After keeping the maximum temperature values for 24–36 h, their progressive reduction began to fully normalize after about 6–7 days. Animals had a worsened general condition, both in the febrile stage and after it: food failure, depression, bruising, stiffness, bedtime, and cough. Palpable tenderness of the chest, auscultatory – dry wheezing, increased vesicular breathing, and bronchial breathing were some of the finds. The final clinical outcome of the experimental infection was pneumonia in the lamb and the hog, and tracheobronchitis in the sheep with a tendency to develop lung inflammation until the end of the observation.

In **goats** from different age groups, respiratory chlamydiosis manifested in different levels of severity and clinical development forms – tracheitis, bronchitis, and pneumonia of the alveolar and interstitial type. When raising a single animal in the private yard and timely adequate healing measures, the prognosis was most often favorable. In caprine flocks, the possibility of aerosol infection in many animals often resulted in massive respiratory diseases of enzootic or epizootic nature. In some herds, chlamydial abortions and respiratory chlamydiosis were simultaneously discovered [649]. Clinically ill animals observed in the acute and convalescent stages with varying degrees of manifestations. Some goats developed heavy respiratory syndrome with bilateral involvement and adverse prognosis due to generalized septic toxicity, respiratory and cardiac insufficiency.

Respiratory chlamydial infections in **cattle** were found to be significantly less frequent than in small ruminants. They mainly affected calves up to 6 months of age with mild, moderate, or severe pneumonia with acute, subacute, and chronic courses [54]. Respiratory disorders may affect upper airways as well as the lower respiratory tract and result in pneumonia with the typical signs (nasal secretions, dry hacking cough, and dyspnea) [633]. In the absence of targeted search, some cases remained unrecognized and were diagnosed accidentally in slaughter.

Buffalos. Gupta et al. [665] reported a case of chlamydial pneumonia in a buffalo calf. The diagnosis was based on microscopic changes in the lung, the demonstration of organisms in lung sections and the experimental isolation of chlamydial organisms in guinea pigs. Soon after, Dhingra et al. [666] confirmed the affinity and the pathogenic potential of chlamydia in relation to buffaloes by isolating the pathogen from the lungs of buffaloes with pneumonia. Another publication reported isolating a *Chlamydia psittaci*-like agent from the wild African buffalo (Rowe et al. [667]). In 2000, Magnino et al. [668] detected chlamydial infection in water buffalo calves in southern Italy. The infected animals clinically manifested recumbence, depression and limb paralysis, but fever was not observed.

11.1.4 Chlamydial Conjunctivitis and Keratoconjunctivitis in Sheep

Data from the literature suggests that a characteristic of the disease is the follicular nature of the inflammation of the conjunctival mucosa. The disease is described in animals of different ages and conditions – lambs, pregnant sheep, and animals that have recently given birth, and the scale of the affected being is different – from single or small groups to enzootics and epizooties. With sheep involvement reaching up to 90% [20, 650–652]. For example, Cello [650] reported mass follicular conjunctivitis during the lambing period. Storz et al. [651] have found in sheep herds from Colorado, United States incidences of chlamydial keratoconjunctivitis reaching 90% and ranging in severity of clinical picture. In our studies of natural infection in some herds we observed conjunctival inflammations with one-sided and more frequent bilateral involvement of the eyes among fattening lambs and sheep. The illnesses were single, group, or massive. Often concurrent with conjunctivitis, some animals had arthritis or polyarthritis. In active outbreaks of chlamydial abortions and related conditions, some of the sheep showed ocular and joint symptoms.

The clinical picture is characterized by hyperemia, chemosis, and edema of the conjunctival mucosa, seborrheic secretion, and photophobia. Lymphoid follicles, especially of the lower and third eyelids, are enlarged and in many cases are confluent. In severe cases, reddening of the conjunctiva was strongly expressed in a dark shade. Some of the animals developed intestinal keratitis with deep vascularization. The so-called pink eye, a popular name, especially in the United States, is an infectious keratoconjunctivitis that is most commonly caused by *Chlamydia psittaci ovis* (old nomenclature) and *Mycoplasma conjunctivae* in sheep and goats [653]. Pink eye tends to occur as an outbreak in a flock or herd. Sheep and goats raised under intensive conditions are most commonly affected. Upon close examination, the membranes of the eye appear red and inflamed. The eyes become cloudy or opaque. Ulceration may develop. The condition is painful and may affect one or both eyes. A serious complication is the development of temporary blindness in affected animals or permanent blindness in most severe cases. Pink eye is usually treated with antibiotics that are injected into the body or placed directly in the eye [653].

The present conditions, resulting in a series of revisions and re-classifications, the name *C. pecorum* was established for one of the species in the genus Chlamydia in the family Chlamydiaceae [79, 84]. Therefore, we are currently talking about the etiological involvement of *C. pecorum* in keratoconjunctivitis and polyarthritis in young sheep and cattle [635]. It has been mentioned above that besides chlamydia, the other major cause of keratoconjunctivitis in sheep and goats is Mycoplasma conjunctivae. In this regard, it should be emphasized that a mixed infection with the two agents is possible, in which the interspecies synergism phenomenon may occur (Gupta et al.) [654].

Table 11.1 presents our results from the experimental infections we performed on ewes – seronegative for chlamydia in a pre-conducted study. It can be seen that four chlamydial strains isolated from sheep affected by conjunctivitis, arthritis, abortion and pneumonia were used for conjunctival inoculations. Chlamydial conjunctivitis was reproduced with the strains of the analogous disease and with the chlamydial arthritis strain, all of which were more pronounced in the first case. This undoubtedly speaks of a similarity in the pathogenic properties of the two strains. Conversely, inoculation with the chlamydial abortion strain was ineffective, and the chlamydial pneumonia strain caused a barely noticeable conjunctival reaction [54].

The described experimental infections have allowed us to make some clinical and etiological parallels in chlamydial conjunctivitis. Clinical findings – follicular conjunctivitis complicated with keratitis, following infection with these strains, matched positive etiological results – direct cytological evidence of the agent in scarified conjunctival material and pathogen isolation. Regarding the serological response, we note that single inoculation does not induce the production of detectable CF antibodies, but specific antibodies have been identified after two-fold inoculation. (Table 11.1).

For treatment, the usual antibiotics for chlamydia-induced diseases, of the tetracycline line, are administered – topically, orally, and parenterally. Sub-conjunctival antibiotic injection is also part of the therapeutic approach.

11.1.5 Chlamydial Polyarthritis in Sheep

The first description of the disease was made in the United States, where multiple epizooties were observed in different states, mostly in lambs [655–657].

Table 11.1 Clinical response and etiologic findings in ewes after experimental infection with chlamydiae of sheep origin

Ewe No: Strain	tC° until 14 days	Photophobia	Follicular Conjunctivitis	Keratitis	CE	LM	CF Antibodies	
							One Inoc.	Two Inoc.
No 1; Strain SM-Chlamydial conjunctivitis in sheep	+	Strongly pronounced	Strongly pronounced	Strongly pronounced	+	+	–	+
No 2; Strain PchAO Chlamydial arthritis in sheep	+	Moderate	Moderate	Slightly pronounced	+	+	–	+
No 3; Strain GV Chlamydial abortion in sheep	–	–	–	–	–	–	–	–
No 4; Strain M1 Chlamydial pneumonia in sheep	–	Slightly pronounced	Hardly noticeable conjunctivitis without follicles	–	–	–	–	–

In a recent publication, Constable et al. [658] states that ovine chlamydial polyarthritis is a common disease in feedlots, on farms, and flocks on ranges in the United States with a morbidity rate that can reach 75%, but the mortality rate is low – about 1%. Later, chlamydial polyarthritis was found in Australia, New Zealand, the UK, and Continental Europe [54, 659–662]. Virulent chlamydiae including *C. pecorum* can be isolated from the joints and eyes of affected lambs – Jelocnic et al. [660] and Polkinghome et al. [661]. Usually, larger lambs up to 6–7 or 9 months of age and, more rarely, adult sheep are affected. Watt [662] wrote that *C. pecorum* is a well-known cause of polyarthritis in lambs especially fast-growing weaned prim and British breed lambs. It affects mainly lambs up to about six months of age. We observed this disease alone or in herds where chlamydial abortions and conjunctivitis were found. It affects a joint (arthritis) or multiple joints (polyarthritis). Clinical infection was manifested as pain, warming and swelling of the affected joint, stiffness, gentle stepping, lameness, and changes in appetite (Martinov [54]). The affected lambs are depressed and reluctant to move. Many would lie down and would not resist being caught [19]. According to some authors, chlamydial polyarthritis responds rapidly to oxytetracycline [662], while others find that treatment with this antibiotic has a variable effect, and may result in residual effects of the infection, most often poor growths and ill-thrift. [658]. All that has been discussed so far indicates the importance of chlamydial polyarthritis in sheep from a veterinary, medical, and economic point of view.

11.1.6 Chlamydial Arthritis in Goats

In the United States, where chlamydial polyarthritis of sheep has long been established and well researched, there is a notion that chlamydial arthritis in goats also occurs, especially in herds that have experienced outbreaks of chlamydial abortion. As concerns and recognition of goat diseases continue to develop in the United States, chlamydia as well as other organisms may be identified as causes of arthritis in goats [663]. The element of uncertainty in this view obviously points to the need for further research into the issue. The opinion of Nietfelt [664] is also in this spirit, namely that chlamydial polyarthritis is a well-known disease in sheep, but the documentation of chlamydial arthritis in goats is lacking, despite the fact that it is sometimes cited as a cause in the species.

We advocate that the etiological involvement of chlamydia in goat arthritis is possible. In Chapter 5 of this book, we reported on obtaining chlamydial isolates from joints of goats suffering from arthritis. The disease was observed in young goats and adult goats raised in stationary herds or in private yards, but collected daily for grazing in a herd. The big joints were mainly affected. In the acute phase of the infection, fever, intrusion, fatigue, tempering, and swelling of affected joints, gait change were noted. In advanced, abandoned and untreated cases, the inflammatory phenomena underwent degenerative changes (arthrosis) leading to reduced motor function, immobilization, and long-term lameness.

11.1.7 Chlamydial Polyarthritis in Cattle

The disease is also known as transmissible serositis. The pathogenic potential of chlamydia to calf joints has long been known [22, 150, 670–673]. This disease affects cattle of all ages, but calves 4–30 days old are affected more severely. The clinical syndrome observed in experimental and natural infection is similar. Affected calves move reluctantly and have reduced appetite. Grace and lameness appear. Fever and diarrhea have been observed in some animals. The joints of most calves are enlarged and painful in palpation. Animals prefer to lie with passively stretched limbs. Despite intensified antibiotic therapy, most calves die 2 to 10 days after the first sign emerges.

11.1.8 Chlamydial Encephalomyelitis in Cattle and Buffaloes

The disease is known mostly by the name sporadic bovine encephalomyelitis (SBE), and its synonyms are Chlamydial encephalomyelitis, Buss disease, and transmissible serositis. The causative agent is *C. pecorum* biotype 2. The disease mainly affects calves <6 months old and rarely older cattle. Sporadic cases and outbreaks can occur within individual herds. Morbidity ranges from <25% to 50%. The overall average mortality rate for all categories of cattle can approach 30% and is highest in calves. At the outbreak of SBE in the former Czechoslovakia, mortality was 100% [675]. In the absence of treatment, deaths in the early stages of the disease are frequent. The first description of the disease in cattle was made by McNutt [674] in the United States, and the isolated agent was later identified as Chlamydia. Since then, SBE has been registered in a number of states in the United States, as well as in Canada, Europe, Australia and other countries. In 2016, the first disclosure of SBE in New Zealand [676] was reported, which involved 40 of 150 crossbred Friesian dairy calves between 1 and 3 months of age, exhibiting a range of neurological signs. Within a month, 13 (32.5%) of 40 clinically sick calves died. In Bulgaria, Ognyanov et al. [38] reported on chlamydial encephalomyelitis in buffaloes.

The incubation period according to some sources is 14–15 days [22], while others [676] – 6–30 days. In bovine animals, the clinical onset is manifested with indigestion, anorexia, and fever (40.5–41.7°C). Increased salivation, diarrhea, weight loss, and difficulty in breathing are observed – Stiffness and knuckling at the fetlocks are the first neurological signs to appear. Animals develop dis-coordination – stepping cautiously and falling or having difficulty in overcoming minor obstructions. Often they move circularly. Gradually, symptoms of paralysis develop. The hind legs are more severely affected. Bovine animals get stuck. Often, especially in the terminal stage, opistonus and nystagmus are observed. Severe clinical cases end up fatally on day 4–14, but in some cases death occurs – up to 30 days later. There is also a mild form of the disease in which animals heal. Sometimes SBE runs inapparent.

Buffalo disease occurs with similar symptoms. They mainly affect the buffalo calves. In addition to acute, chronic, and inapparent forms, an over-acute form of the disease is also noted. Mortality was 25% [38].

Differential diagnosis is required with respect to the encephalitis form of IBR, Aujeszky's disease, rabies, malignant catarrhal fever and listeriosis. Toxicology tests must be negative.

11.1.9 Chlamydial Mastitis in Cattle

Studies on the possible role of chlamydia in mastitis in cows have a long history. The first attempts at experimental mammary gland infection date back to 1955 when Matumoto et al. [678] caused severe acute mastitis in cows after intracisternal inoculation of a chlamydial isolate from an EAE case. Later, acute but transient mastitis was due to inoculation of the SBE agent. Other authors also used chlamydiae of ovine or bovine origin [679, 680]. In the experiment of Mecklinger et al. [679] the intramammary of six lactating cows were infected. Depending on the dose administered and the virulence of the strain, the clinical signs occurred after 1 to 5 days. The affected quarters were enlarged, hyperemic, and sensitive. Milk secretion decreased, and milk had a strongly changed amber-yellow color and contained thick fibrin threads. There was a worsening of the general condition – the temperature rose to 40.9°C on 2–5 days, followed by a decrease. More depression and anorexia were reported. The cell content of the milk increased significantly 24 h after infection. During the most marked clinical signs in sediment established chlamydia. Clinical parameters were normalized 10–30 days after the onset of the infection, but the milk production did not recover completely. The authors found chlamydial CF antibodies with a maximum titer of 1:280 on the 20th day after infection, which is followed by a rapid decline in titers [679]. Several other examples of experimentally induced mastitis in cows may be mentioned.

The issue of the role played by chlamydial agents in mastitis under natural conditions has long been the focus of researchers. According to Storz [19], in light of the experimental findings, it is conceivable that mastitis may be induced in lactating animals following an episode of chlamydemia.

Blanco [683] first described chlamydial mastitis in cows in Spain due to natural infection. After purchasing 25 cows from the Netherlands, 17 of them developed acute mastitis, accompanied by functional disorders of the mammary gland and a strong decrease in milk secretion within two months. In Germany, Wehnert et al. [680] reported chlamydial etiology mastitis in a large farm with 1,820 cows. In the studies of 45 milk and tissue samples from the mammary gland and supramammary lymph nodes, the authors isolated chlamydia 39 times. The significant pathogenicity of these isolates has been demonstrated by experimental reproduction of the disease in cows. Applied antibiotic therapy was ineffective [680].

It is now known that two chlamydia species, *C. abortus* and *C. pecorum*, are routinely detected in cattle. It is assumed that the majority of *Chlamydia* spp. infections in cattle are endemic, low-level chronic infections that do not cause obvious clinical disease, but significantly reduce performance in fertility, mastitis, and neonatal health and growth, the three most important areas of bovine production [681]. Biesenkamp-Uhe et al. [639] emphasize that in dairy cows the presence of chlamydiae was significantly associated with subclinical mastitis and increased somatic cell counts (SCC) in milk. The influence of natural *Chlamydia spp.* infection on the health of the ruminant mammary gland was studied by Ahluwalia [682]. To characterize mastitis, caused by the natural *Chlamydia spp.* infection of dairy cows, 17 dairy cows in the second or higher parity were sampled for 20 weeks after parturition. Serologically, anti-*Chlamydia spp.* IgM antibodies were found in the blood of all cows, indicating endemic chlamydial infection. Twelve (70%) of the cows were PCR-positive for *Chlamydia spp.* in any of the samples, with 11 cows positive only in vaginal cytobrush specimens, 1 positive only in milk samples, and 2 cows in both types of specimens [682]. The authors use a multivariate logistic stepwise regression model with which they found that cows with a chlamydial colonization of the mammary gland had significantly higher SCC and lower milk protein. In the presence of systemic chlamydial infection, a high degree of association is found with increased milk yield and reduced milk protein. These data represent confirmation and further characterization of both the influence of asymptomatic localized and systemic chlamydial infection on the health of the mammary gland of dairy cows and their milk production [682].

11.1.10 Intestinal Chlamydial Infections in Ruminants

The first evidence of chlamydial agents in the feces of clinically normal calves, goats, and sheep was obtained in the 1950s [684–686]. Subsequently, similar findings have been established in a number of geographic areas of the world. Later, it became clear that chlamydia is also isolated from the feces of calves suffering from diarrhea or pneumo-enteric syndrome [687, 688]. Experimental infections are usually inoculated orally with no colostrum-fed newborn calves. Clinical manifestations following infection are different depending on the type of strain used and its multiplicity [22]. The age of experimental animals is also important. When infected with *C. psittaci* serotype 1 (old nomenclature) colostrom-deprived, twenty-four-hour-old calves were observed with fever, leukocytosis, and diarrhea with watery, mucoid, and bloody appearance, leading to dehydration and often to death [689, 690].

In spontaneous infection gastroenteritis of any severity of clinical course is observed. In some cases, only mild transient diarrhea is found, and in others – stubborn sneezing and hemorrhagic diarrhea, accompanied by fever, leukocytosis, dehydration, and death. The calves are hardest hit. Often, pneumo-enteric syndrome is detected. In bulls, the stomach is not affected, but only enteritis develops. Ehret et al. [691] establish chlamydiosis in a large beef herd with 2915 animals. Clinical manifestations and lesions involving the intestinal, respiratory, nervous, skeletal, reticulo-endothelial, and urinary systems were observed in chlamydia-infected calves. Invariably at autopsy in chlamydia-positive cases, there was some degree of a fibrinous inflammatory process present [691].

The asymptomatic intestinal chlamydial infections in cattle are widespread. The only indication of the presence of the infection is the isolation of the chlamydial agent from the fecal samples [22, 692, 693]. The introduction of PCR diagnostics and commercially available ELISA assays have contributed to the more frequent detection of asymptomatic chlamydial infections in cattle [536]. Poudel et al. [640] emphasize that low-level, asymptomatic chlamydial infections in cattle are essentially ubiquitous worldwide. Furthermore, asymptomatic endemic *C. pecorum* infections reduce growth rates in calves by up to 48% [640]. In a recent publication, Li et al. [693] emphasizes that among the bovine chlamydial agents, *C. pecorum* dominates in the feces and is the endemic intestinal species in these farm animals.

Current knowledge has shown that unlike cattle, in sheep and goats, the intestinal chlamydial infection is not manifested with clinical signs. After the first publications on Omori et al. [685] on goats, and Kawakami et al. [686] and Dungworth & Cordy [645] on sheep, studies by Storz and Thornley [688], showed that intestinal chlamydial infection is natural and long-lasting. The sheep they studied have shed the agent intermittently for three years. The main localization of chlamydia in the gastrointestinal tract was the cecum.

Ranks and Yeruva [695], analyzing available data on gastrointestinal chlamydial infections in animals, make meaningful and useful comparisons, analogies, and assumptions about the problem of persistence in human genital infection with *C. trachomatis*.

For over five decades, data on the etiological involvement of chlamydia in porcine infectious pathology have been collected. Today's knowledge suggests that in pigs, *Chlamydia suis*, *C. abortus*, *C. pecorum* and *C. psittaci* have been isolated. After the changes in taxonomy, *C. suis*, i.e., the former porcine serovar of *C. trachomatis*, has been identified as the major agent causing chlamydioses in swine [702]. Approximately twenty different clinical forms and syndromes associated with chlamydial agents can be listed: conjunctivitis, pneumonia, pericarditis, polyarthritis, polyserositis, periparturient dysgalactiae syndrome, MMA syndrome (mastitis, metritis, agalactia), pseudo-membranous or necrotizing enteritis, reproductive disorders, notably abortion in sows and urethritis, epididymitis and orchitis in boars, vaginal discharge, return to estrus, increased perinatal and neonatal mortality in piglets, stillbirth, mummification, and delivery of weak piglets. Chlamydiae have also been isolated from inapparent intestinal infections in apparently healthy pigs. Despite the impressive nosological spectrum of swine chlamydial infections, some authors still consider that the impact of Chlamydiaceae on animal health on pig farms is controversial because an inconsistency seems to exist between the obviously high prevalence of chlamydiae in clinically normal swine herds [697–701]. One reason for this conception is the relatively rare publication of reports of acute swine chlamydioses. According to Longbottom [702] the porcine chlamydial infections should be considered as widespread, but they are often under-diagnosed. A similar opinion is expressed by Schautteet and

Vanrompay [703], namely that Chlamydiaceae infections are often unnoticed because they are not routinely performed in all veterinary diagnostic laboratories and Chlamydiaceae are often found in association with other pathogens, which are sometimes more easy to detect. More and more current data are accumulating that swine chlamydiae are more etiologically proven than previously thought and that their economic significance is undeniable. The economic significance of swine chlamydia is determined by mortality losses, necessity of mass slaughter due to massive pneumonia, delayed pig growth, reproductive disorders, and inadequate number of offspring.

11.1.11 Chlamydial Pericarditis in Pigs

Guenov [42] described a fibrinous pig pericarditis caused by "ornithosis virus" as early as 1961. 20–40-day-old pigs affected with growth retardation, decreased vitality, decreased appetite, and other signs of general malaise were affected. In the study by Martinov et al. [47], the disease was predominantly among growing pigs aged 30–70 days. In some cases, younger (20–22 days) and older pigs (80–100 days) were infected. The clinical picture of the animals with pericarditis is not strictly indicative. We observed a decrease in appetite to full anorexia, stinginess, and slowing of growth. Breathing is usually difficult. In some pigs, the temperature is slightly elevated. These signs are different. Often they were highly exposed and the animals were dying. In some cases, clinical symptoms were complicated by the development of pneumonia. Typically diseased animals were steered to the slaughter out of necessity, thereby leading to the discovery of typical pericarditis macroscopic changes (see Chapter 12).

11.1.12 Chlamydial Abortions, Stillbirths, and Deliveries of Unviable Offspring in Pigs

These clinical conditions were investigated by several research groups [334, 335, 704, 705, 707–709].

Abortions occurred in the second half of the pregnancy, often 1 to 2 weeks before the time of delivery. They occur suddenly, without prior symptoms. In pig holdings where the chlamydial infection first appears, abortions are predominant, whereas in farms where the disease is stationary, stillbirths and deliveries of non-viable pigs dying in 1 to 2 weeks predominate. Bortnychuk [708] reported chlamydial mass infections in a number of farms for fattening pigs in the Kiev region. Affected were all age groups. At the first onset of infection, 30–50% of young female pigs were aborted. In the following years, abortions declined, but predominantly stillbirths and deliveries of weak offspring, of which 40–50% died in the next 2–3 weeks. Other live pigs sometimes have conjunctivitis, arthritis, and central nervous system disorders. The newly introduced pigs in the infected farms suffered from pneumonia, enteritis, urethritis, and orchitis.

11.1.13 Chlamydial Vaginitis in Pigs

Sows or younger female pigs are affected. The external sexual organ is red and swollen. There is leakage of vaginal secretions with serous or mucus-purulent character. In colposcopy, the vaginal mucosa is edematous and hyperemic.

11.1.14 Chlamydial Polyarthritis in Pigs

The first report of isolation of the "transmissible agent", which is suspected to be a chlamydia from cases of arthritis and pericarditis in pigs in the United States, is dated from 1955 – Willigan and Beamer [713]. The disease was found in pigs of different age groups [23, 334, 672]. The affected young piglets became febrile and anorectic and may develop nasal catarrh, difficulty in breathing, and conjunctivitis [672]. Martinov [23, 334] isolated chlamydia from the inflamed knee and tarsal joints in 10–20-day-old piglets. The synovial fluids were slightly darkened and contained fibrin threads. Affected animals displayed signs of general malaise. In 1969–1970 chlamydial strains were isolated in Austria from numerous pigs with polyarthritis, polyserositis, pneumonia, enteritis, or conjunctivitis, from cases of abortions, and from pigs with inapparent intestinal tract infection [710, 711].

11.1.15 Reproductive Disorders in Sows

The reproductive concerns for sows have been in difficulty in breeding, temporary or permanent infertility, occurring usually after abortion, and related conditions. Also affected are pigs with chronic chlamydial disease and asymptomatic latent infected animals. *C. abortus* is mainly linked to reproductive failure. *C. suis* is found to be involved in the return to estrus – a condition with early embryonic death in more than 50% of the pregnant sows [699, 700]. Yaeger, M.J. [714] expresses some skepticism about the role of chlamydia in swine reproductive disorders. He believes that the important role of Chlamydiaceae in the reproductive problems of these animals is not clear regardless of the publications on the issue, mainly by European authors. Gross lesions have not been described. Following experimental infection of pregnant sows with an ovine abortigenic strain, chlamydia was identified in placental tissues, accompanied by placentitis, but reproductive consequences were not demonstrated [714].

11.1.16 Chlamydial Conjunctivitis in Pigs

In 1963, Pavlov et al. [715] reported infectious keratoconjunctivitis in young piglets caused by a chlamydial agent. After experimental infection, acute conjunctivitis with serous secretion and photophobia, fever, depression, and anorexia occurred between the 4th and 10th day. The cornea where the opacities were observed was also affected. In the epithelial cells of the conjunctival mucosa, chlamydial inclusions have been demonstrated. From the conjunctival discharge, the chlamydial pathogen was re-isolated and passed serially in chicken embryos [715]. In another publication relating to the spontaneous chlamydial infection in the pig, are described conjunctivitis with muco-purulent nature [716]. Several other publications have reported conjunctivitis and keratoconjunctivitis in pigs of different range of severity and varying severity of clinical course [703, 717–719]. Different factors relating to the occurrence, storage and spread of chlamydial agents among susceptible pig populations are discussed. The possible differences in the appearance of ophthalmic pathology in the extensive or intensive rearing of pigs are pointed out. For example, Becker et al. [718] consider that intensively kept pigs are predisposed to chlamydia-associated conjunctivitis. Such an opinion was expressed by Englund et al. [719] and Schautteet et al. [720] are of the opinion that intensive pig production systems may predispose pigs to the transmission and maintenance of the infection, thus increasing the infectious burden and risk of disease in the pig. In this connection, the authors establish a correlation between conjunctival and intestinal samples with a high degree of *C. suis* infection and the presence of clinical signs [719].

The possible etiological involvement of the four types of chlamydia having the potential of agents affecting the pigs is also studied. Rogers and Anderson [717] found the presence of the *C. trachomatis*-like organisms (old nomenclature) in the study of ocular pathology and in later work (1999) with such agent-induced experimental conjunctivitis in gnotobiotic pigs [717]. The studies of Becker et al. [718] on pig conjunctivitis in Switzerland and Germany have demonstrated the dominant presence of *C. suis*. These researchers reported high prevalence of *C. suis* in the eyes of Swiss (79%) and German (90%) ocular symptomatic pigs. Obviously, in pig conjunctivitis there is an etiological complex that, besides chlamydial species, can include other infectious agents, resulting in a complicated etiology, possible pathogenic interplay and clinical picture of mixed or secondary infections. In such cases, it is particularly informative to conduct molecular diagnosis with the use of species-specific nucleic acid amplification tests (NAATs) such as real-time PCR and DNA microarray, detecting the ribosomal intergenic spacer and domain I of the 23S rRNA gene [330, 628].

11.1.17 Chlamydial Pneumonia in Pigs

Despite the regular mention of pneumonia as part of the clinical spectrum of swine infections caused by Chlamydiaceae, the literature on this disease is scarce. This is probably due to the fact that chlamydia has a relative share in the etiology of the disease known by the common name enzootic pneumonia and which undoubtedly creates difficulties in the diagnosis and interpretation of the findings.

Hunter and McMartin [723] reported experimental pneumonia in pigs after intratracheal inoculation of the isolate of *C. psittaci* (28/68) from ovine pneumonia. The observed clinical signs included transient pyrexia at 24 h followed by increased respiratory rates and inappetance which lasted for a further 48 h. Clinical abnormalities found confirmation in histological findings in the lungs where at the 24-h they found acute exudative reactions and proliferative changes prevailing after 10 days. The authors note that while variations in the concentration of inocula were reflected by corresponding increases and/or decreases in gross lung damage, clinical signs, and histological reactions were unaltered. Chlamydial organisms were recovered only from the lung tissues [723].

The pathogenic potential of *C. suis* for the porcine respiratory system was established by experimental aerosol infection of conventional raised colostrum-fed pigs [628, 697]. As a result of the infection, the pigs developed acute pulmonary inflammation and severe lung function disorders. Signs included increase in body temperature of the pigs above 40°C for at least five days post infection, serous nasal discharge, dry cough, and severe dyspnea accompanied by wheezing, shortness of breath, and breathlessness. The duration of the clinical manifestations was 7 days after the infection.

Our own clinical observations were made under the conditions of a natural chlamydial infection in pigs, pre-proven etiologically by agent isolation, direct indication of the same via EM and IF, and serology [23, 334]. Pulmonary inflammation of the alveolar and interstitial type was mainly observed in adolescent pigs (over 3 months). The diseases were manifested with coughing, wheezing, and rapid breathing. We also found bronchopneumonia with a distinct and strong cough with high tone. During auscultation humid wheezing and crackles could be heard. In interstitial pneumonia, the cough was muffled, and the wheezing was dry. Severe cases ended lethally due to progressive dyspnea and toxic myocardiopathy and failure. The more moderate acute cases and sub-acute conditions were with favorable prognosis.

11.1.18 Chlamydiosis in Boars

Clinically, urethritis, epididymitis, and orchitis are found in boars infected with chlamydia [707, 724, 725]. The prostate gland can also be affected. Authors from several European countries have reported that as a result of chlamydial infection in boars, inferior semen quality is detected in animals from fattening pig farms [699, 700, 707, 712, 720, 721, 724]. These are disturbing findings, especially because of the suspected possibility of sexually transmitting part of the swine chlamydial infections. The investigation of Kauffold et al. [726] revealed intestinal infection of boars as *C. suis* was the most prevalent, while *C. abortus* was less prevalent and *C. pecorum* occurred sporadically. Among the boars are also found latently infected animals.

11.1.19 Intestinal Chlamydial Infections in Pigs

Enteric chlamydial infections of swine with *C. suis* are frequent. There is evidence of isolation or direct detection of the chlamydial agent of the intestines of swine having diarrhea in both natural and experimental infections [110, 727–732]. Overall, however, prevailing notion that intestinal chlamydiosis in pigs is often subclinical. In addition to *C. suis*, spontaneously infected pigs partially yield *C. abortus*, which is confirmed by the experimental infection of gnotobiotic pigs with this chlamydial species [110, 728, 731, 733]. According to Szeredi et al. [729], intestinal chlamydia affects about 30% of the finishing pigs. In suckling pigs, the percentage is approximately two times lower (15%) – Zahn et al. [728]; Nietfeld et al. [730].

Guscetti et al. [732] reported experimental enteric infection of two groups of gnotobiotic piglets after intra-gastric inoculation of *C. suis* S45 strain in

5×10^6 inclusion forming units (IFU) and 5×10^4 IFU, respectively. Animals inoculated with the first dose showed most severe clinical outcome – diarrhea with watery feces containing flecks of undigested curd, anorexia, weakness, less activity, decreased appetite, and marked body weight loss at three days post infection – mean 200 g. The most pronounced signs were observed between the second and third days. The piglets of the other group (5×10^4 IFU) showed less severe but more protracted symptoms. It is important to note that the reported experimental infection was achieved after the challenge of gonobiotic piglets with a strain isolated from pig without clinical disease. A similar result was achieved earlier by Rogers and Andersen [734] after infecting weanling pigs at 21 days of age with another strain of *C. suis*, also originating from a pig, which did not show any clinical signs. These results clearly indicate the enteropathogenic potential of *C. suis* from infected but asymptomatic pigs, which has an important epidemiological, clinical, and histopathological significance.

Intestinal chlamydia infection, even in cases of low-grade or asymptomatic course, results in pronounced histopathological changes that we will not discuss in this chapter. These lesions may increase the sensitivity of the intestinal tract to other agents with enteropathogenic activity and lead to clinical manifestations, mainly diarrhea resulting from mixed infection [732]. Pospischil and Wood [727] consider that in such a case there was a synergistic effect between Chlamydia and Salmonella.

Hoffmann et al. [735] collect fecal samples of 636 fattening pigs from 29 farms in Sweden. Using real-time PCR and DNA-microarray analysis they were screened for chlamydiae. All farms were positive for Chlamydiaceae with 94.3 and 92.0% prevalence in fecal swabs of the first and second time-points, respectively. The authors find a correlation of chlamydial positivity to diarrhea; *C. suis* inhabited the intestinal tract of almost all examined pigs, implying a long-term infection. Mixed infections with *C. suis* and *C. pecorum* were common, with a substantial increase in *C. pecorum* positivity at the end of the fattening period, and this finding was associated with ruminant contact [735].

11.1.20 Latent Chlamydiosis

This is a frequently discovered form of a well-balanced host-chlamydial agent relationship. Latently infected asymptomatic pigs of different ages are found both in farms and herds with etiologically proven chlamydiosis with clinical manifestations, and also in the absence of clinical signs. The chlamydial organims are excreted by the latently infected host and thus can be transmitted to new hosts under appropriate conditions [19]. The latent form under certain predisposing conditions – stress, transportation, worsening care and feeding conditions, and competing infections can be transformed into a clinically manifested disease.

11.1.21 Prevention, Treatment and Control of Chlamydioses in Pigs

Control measures include isolation of the infected pig farm, thorough disposal of aborted fetuses, placentas, maternal fluids, and carcasses, best by burning, disinfecting with 1–5% lysol, 3–5% carbolic acid, 3% sodium hydroxide, or 2% chloramine with an exposure of not less than half an hour [22]. Another 70% Isopropyl alcohol is used, 1:1000 dilution of quaternary ammonium compounds, 1:100 dilution of household bleach or chlorophenols [703]. In general, regular maintenance of good hygiene in animal housing is of great importance for the prevention and control of chlamydial infections. Shcherban et al. [707] recommend the radical method of slaughtering all aborted pigs and boars with urethritis, accompanied by bleeding from the urethra and the pigs with pneumonia and arthritis.

Traditionally, antibiotic treatment is performed with tetracyclines. As an alternative, macrolides, especially erythromycin and quinolones, are used in the first place enrofloxacin. The use of these latter is necessary when the chlamydial infection is caused by a tetracycline-resistant *C. suis* strain [721]. In 2005, Pollmann et al. [722] published interesting and promising data on the potential of a probiotic strain of *Enterococcus faecium* (NCIMB 10415) to reduce carryover infections from naturally occurring Chlamydiaceae-infected sows to newborn piglets. For the needs of practical veterinary medicine and pig-breeding, EU-licensed feed supplement for pigs containing this probiotic strain is used.

At this stage, there is no immunoprevention vaccine for swine chlamydial infections.

11.1.22 Clinical Observations in Horses

Chlamydial infection of the horse genital apparatus leads to reduced reproducibility, pure ejaculate quality, mild salpingitis, and occasional abortions [737]. Abortion was also reported Henning et al. [738], Szeredi et al. [739] and other authors. Previous abortion signs are usually absent or expressed in mild discomfort. Isolation of chlamydiae was achieved from the fetal membranes of the aborted equine fetuses that were modified to varying degrees. Another localization of the chlamydial infection is the respiratory tract. Chlamydia-related rhinitis, acute respiratory conditions and pneumonia [740–744], as well as conjunctivitis [740, 744] have been described. McChesney et al. [746] reported chlamydial polyarthritis in a foal, which later found confirmation in other publications. According to other sources, the equine biovar infection of *Chlamydia pneumoniae* plays an etiological role in young foals of bronchopneumonia: affecting the brains of horses and mules with hepatoencephalopathy; polyarthritis; eye inflammation [736, 745]; hepatitis; and fatal cases of encephalomyelitis [633]. Previous and recent studies have accumulated data suggesting a role of *C. psittaci* and/or *C. abortus* in recurrent equine airway obstruction as trigger factors of inflammation or indicators of severe disease [633, 747, 748]. Theegarten [748] finds similarities between the horse sickness and human chronic obstructive disease.

11.1.23 Clinical Observations in Dogs

There is a belief that chlamydial infections in dogs are uncommon despite the etiological data obtained during the years in some clinical conditions and syndromes. Without underestimating the possibility of a higher natural biological resistance of this species to Chlamydiaceae, there seems to be an inertia that leads to a neglect of the problem and a lack of sufficient number of targeted studies.

The chronology of the issue of chlamydia in dogs begins in the 1950s when French researchers Giroud et al. [749] and Groulade et al. [750] described canine distemper-like systemic disease occurring in acute, subacute, and chronic forms. Particularly severe was the course of the disease in younger animals. Clinical signs included fever, inflammation of the respiratory tract, gastrointestinal disorders (vomiting, diarrhea) and peritonitis. The chronic form lasts from 1 to 2 months and is characterized by varying appetite and appearance of skin lesions along the legs and abdomen. The dogs became emaciated. Treatment with tetracycline antibiotics had an excellent effect. High titers of chlamydial antibodies were found in the blood sera of dogs. A similar disease with regard to the clinical picture, pathological findings, and serological response was induced by Maierhofer and Storz [751] after experimental infection of dogs with the chlamydial agent of ovine polyarthritis. Besides the above-described features, the authors report muscle and joint soreness and incoordination. One dog had polyarthritis.

The chlamydial pathogen was re-isolated from the brain, lungs, liver, intestines, and synovial fluids [751].

In one of the cited publications [749] encephalitic syndrome in dogs was attributed to chlamydiae based on the isolation of the pathogen in CEs, and the presence of chlamydial antibodies in the blood of affected animals. Contini [753] observed a similar encephalitis syndrome in a dog and isolated a chlamydial agent with which it had inoculated mice via the intracerebral pathway. Induced chlamydial infection in mice has been manifested by cerebral and pulmonary lesions.

An interesting case was established by Fraser et al. [826], relating to a dog which had been in direct contact with the environment of the premises of an aviary, where as a result of avian chlamydiosis, there were approximately 33% mortality of a total of 300 farmed budgerigars. The dog had developed a severe respiratory symptomatology and a high titer of chlamydial antibodies had been established. Chlamydiae were isolated from the feces of two other dogs that were in contact with the sick dog. Three other reports also speak of dogs acquiring chlamydial infection from an avian sources [761–763].

In some of the early publications, data were reported for a high seropositivity rate for chlamydia in studied dogs – 43.75% [754] in Romania and 50% [755] in the former Yugoslavia. In the first work, the dogs were exposed to infected ewes from flocks with EAE. In the second work, the antibodies were detected in the sera of asymptomatic dogs. Recently, Cheng et al. [761] presented data on the prevalence of Chlamydiaceae in 442 pet dogs in Shenzhen, China. The average positive rate was 6.11%, and out of all 38 breeds of dogs tested, 14 were positive.

Individual reports describe the involvement of chlamydia in the ophthalmic pathology in dogs – conjunctivitis in South Africa [756] and superficial keratitis in Germany [757]. Krauss et al. [1108] identified *C. psittaci* infection in a case of conjunctivitis in a dog in Germany. If left untreated, conjunctivitis can create serious complications, such as infections in the cornea, lids, and tear ducts. Since conjunctivitis is often the first obvious symptom of a serious disease (such as glaucoma) it should not be ignored.

Arizmendi et al. [758] collected thoracic fluid by needle thoracentesis from a 5-month-old male Bull Mastiff dog who was referred to them with a history of 5 days of pyrexia and shifting leg lameness. The etiological diagnosis was done by isolation of the chlamydial agent in L-929 fibroblasts and CE and serology with seroconversion of CF-antibodies. It was assumed that the dog contracted a highly pathogenic chlamydial strain from a turkey.

A systemic disease associated with pyrexia, lymphadenopathy, and arthropathy of several joints of the appendicular skeleton in a dog was described in South Africa by Lambrechts et al. [760]. The chlamydial agent has been demonstrated directly in the joint fluid by means of LM, IF, and PCR. However, the origin of the infection could not be traced.

Sako et al. [759] reported the presence of viable chlamydiae in canine atherosclerotic lesions using an immunohistochemical technique with anti-*C. psittaci* polyclonal and anti-*C. pneumoniae* monoclonal antibodies, EM and PCR. These results supported the conclusion that the organism may be an active factor in the pathogenesis of canine as well as human atherosclerosis [759].

In etiologically confirmed cases of chlamydial infection in dogs, we recorded the following clinical conditions: conjunctivitis; rhinitis; polyarthritis; gastroenteritis; mucopurulent cervicitis; urethritis (Martinov, 2010, unpublished data). For example, two males and females of one owner had respectively urethritis, balanopostitis, and cervicitis. On another occasion we observed that partners of a male dog with chlamydial urethritis aborted or gave birth to dead puppies. In a previous study in Bulgaria [336] a single case of chlamydial keratoconjunctivitis was reported in a dog owned by a military unit, which was manifested by unilateral pronounced fibrinous keratoconjunctivitis, hyperemia, and edema of the conjunctival mucosa.

The data presented suggest that dogs are susceptible to chlamydia infection, which is demonstrated with a significant range of clinical manifestations. Nevertheless, a number of diagnostic problems, pathogenesis, epidemiology, actual spread, and zoonotic significance of canine chlamydiae are evident. The clarification and possible solution of these issues requires further research efforts.

11.1.24 Treatment of Dogs

Treatment is at the discretion of the practicing veterinarian. Oral tetracyclines (doxycycline, tetracycline) lasting at least 1 week, which may reach 3–4 weeks, are administered in systemic chlamydial infections. Chlamydia is also sensitive to Azithromycin. For local infections (conjunctivitis), solutions are used topically to cleanse the eye before applying medication and eye drops or ointments. In certain cases, corticosteroids are also used to reduce inflammation and infection. Treatment regimens are determined by the severity of clinical disease and individual animal's response to specific treatment.

11.1.25 Clinical Observations in Cats

According to one of the unofficial subdivisions of infections in cats caused by *Chlamydia felis*, this group includes: Pneumonia; conjunctivitis; rhinitis; heavy exhaustion and unilateral serous ocular discharge [736]. Other authors combine some of these signs with unspecified clinical manifestations in the group term Upper Respiratory Infection (Chlamydia) in cats [764]. The latter includes: Sneezing; watery eyes; discharge from the eyes; coughing; difficulty breathing; runny nose; lack of appetite (anorexia); fever; and pneumonia, if left untreated.

Cello [765] has described severe catarrhal conjunctivitis in cats, the cause of which has been successfully cultivated in mouse lungs and in yolk sacs and chorioallantois of hen's embryos. Intra-cytoplasmic inclusions containing elementary bodies (EBs) and other types of bodies and structures of Bedsonia (Chlamydia) were detected in the epithelial cells of the scarified conjunctiva of naturally or experimentally infected animals. Under natural conditions it was discovered as isolated cases in animals brought for the treatment of unilateral purulent conjunctivitis. The untreated animals usually easily develop similar changes to the other eye within one week, but not always. The disease may become endemic if it appears in catteries and may last for one year. New animals introduced into infected groups invariably develop the disease. Kittens born from infected cats, and sometimes from convalescent ones, can develop severe conjunctivitis when their eyelids open. In the conjunctival epithelial cells of such animals, chlamydial EBs are detected. Experimental instillation of 2 drops of yolk suspension containing 10^{-2} ILD_{50}/ml in the conjunctival sac of healthy cats causes severe inflammation of the conjunctiva after 3–5 days and an increase in body temperature. Instillation of a suspension of infected YS into one of the kittens' nostrils causes conjunctivitis, but not generalized respiratory infection. EBs were found in large quantities on the third day after infection and in smaller quantities until the second week. Some of the infected animals showed clinical signs up to the 90th day after infection. Reinfections of fully healed cats are also possible. In such cases, short-term mild conjunctivitis usually develops, but there are exceptions in this regard [765]. Sykes [767] emphasizes that in some cats, clinical signs can last for weeks despite treatment, and recurrence of signs is not uncommon. Untreated cats may harbor the organism for months after infection. According to another source [766], signs usually develop within a few days after infection, beginning as a watery discharge from one or both eyes. Due to the discomfort, affected cats may hold their eyes partially closed.

As the disease progresses, severe swelling and reddening of the conjunctiva may be seen and the discharge changes from a watery to a thicker yellowish substance. Some infected cats still show very mild sneezing, nasal discharge, mild fever and lethargy [766]. In addition to the features described above, chemosis and prominent follicles have been identified on the inside of the third eyelid in more severe cases [767]. Keratitis is rare.

The chlamydial pneumonia in cats (Feline Chlamydiosis; Pneumonitis) usually develops in association with the more common chlamydial conjunctivitis and rhinitis. It is caused by *C. psittaci*. The treatment is with appropriate antibiotics [768]. Generally, lower respiratory tract signs such as cough and shortness of breath are rare and may be absent [767]. Work on other clinical diseases in cats in which were found chlamydia, are single. Dickie et al. [770] reported chlamydial infection associated peritonitis in cat, and Hargis et al., [771] chlamydial gastritis in twelve cats. At this stage it is not clear whether *C. felis* causes reproductive diseases in cats [767].

11.1.26 Treatment of Cats

And in cats, tetracycline-class antibiotics are the drug of choice in the treatment of chlamydia-induced diseases; in ocular chlamydiosis, tetracycline eye ointment, topically applied 2–3 times a day [772, 773]. However, local treatment does not always lead to the clearing of the infection and recurrent infections are possible. Observations show that oral administration of doxycycline at a dose of 5 to 10 mg/kg q 12 h for 3–4 weeks is effective and results in the resolution of clinical signs in the majority of cases. Gruffydd-Jones et al. [774] applies a dose of 10 mg/kg q 24 h. The onset of clinical improvement is usually 1–2 days after taking the antibiotic. In the case of larger groups, such as cat colonies, antibiotic treatment of all animals is required to eradicate the infection. Such a positive experience was reported by Dean et al. [775], who monitored the response of *C. felis* to doxycyclin treatment by PCR. In some cases, treatment of 6–8 weeks is required. As a rule, antibiotic therapy should be continued for about 10–14 days after the clinical signs resolve.

According to Sturgess [776], for cats that do not tolerate doxycycline, a suitable alternative is amoxicillin-clavulalan administered in a 1-month course. Doxycycline surpasses Pradofloxacin by efficiency [777]. It is recommended for use instead of Enrofloxacin, as the latter gives ocular side effects. Other antibiotics and chemotherapeutics have been tested. Azithromycin is not sufficiently effective [777]. Chloramphenicol, penicillin and sulfonamides are inappropriate.

In limited studies, we demonstrated chlamydial infection in single cats with conjunctivitis (Martinov, 2009, unpublished data). One case involved a newly aborted cat from a small kennel in which all animals had persistent conjunctivitis, and in the previous 2 months there had been abortions, stillbirths, and newborns with malformations. In one cat of another owner, we found bilateral conjunctivitis and eosinophilic granuloma complex. Combined topical and oral treatment with tetracyclines and corticosteroid supplementation in the more severe cases was effective.

11.2 Clinical Observations in Birds

11.2.1 Chlamydiosis in Waterfowl

In ducks in Europe and the United States, often grown industrially in large farms, chlamydial infection (ornithosis) is common and is responsible for significant economic losses. The analysis of the evidence suggests that in the 1960s there were reports of the disease in ducks seen in various aspects. From a clinical and epidemiological point of view, the facts of major morbidity and mortality, particularly among young ducks [35, 779, 780] are essential. A number of examples can be mentioned in this respect. For example, 600 (30%) of a total of 2000 ducklings died in one farm in former Czechoslovakia [780]; in Bulgaria, out of a total of 7701 ducklings on a farm, 4081 (53%) died [36], etc. In the same early period, there were separate reports that chlamydial infection in some flocks of ducks was rather uneventful [781]. Summarized clinical evidence of chlamydial disease in ducklings from the earlier mentioned period indicates that they are weak, have watery diarrhea, and unbalanced gait. Common findings are bilateral inflammation of the conjunctiva and cornea and the discharges from nasal and auricular orifices with serous or mucopurulent type. Death occurs in convulsions. The growth of the surviving ducklings slows down. There are signs of muscle atrophy [19, 22].

According to a recent publication [736], the most characteristic clinical signs of chlamydiosis of the ducks are serous or pustular nasal and ocular discharge and also encrusted eyes and nostrils.

Our own experience with chlamydiosis on waterfalls dates back to the early 1990s to the present. It is related to complex observations and studies of the disease in a number of duck and geese farms, covering large masses of waterfowls grown industrially for the purpose of obtaining large liver and meat [53, 337, 579, 580]. The clinical course of the chlamydiosis in **ducks** was severe in some outbreaks and moderate in others. Most susceptible were young ducks up to 2 months of age. They showed depression, anorexia, unbalanced gaits, and tremors. Some birds had nasal discharge, conjunctivitis, and respiratory signs. The intestinal excretions were loose or watery with a greenish color. The progression of the disease led to weight loss, weakness, emaciation, exhaustion, and muscle atrophy. Ducks often died with convulsions. The mortality rate was 35% for the most affected flocks. In other outbreaks, the dominant clinical course was moderate, but still the mortality rate was high – 23.5%. Along with the acute cases, sub-acute, chronic and inapparent forms were observed.

Clinical observations in flocks of **domestic geese** showed similarity to those described in the ducks, but generally the signs in younger birds were moderate, and in adult geese – mild and transient. The asymptomatic form was common in adult geese [53].

11.2.2 Chlamydiosis in Hens

In hens, the acute form is rare. It is characterized by fibrinous pericarditis and hepatomegaly. Affected birds are depressed and show lack of appetite, rhinitis, conjunctivitis, swollen eyelids, and reduced egg production. In young chickens are described meningoencephalitis. In some cases, the incidence may vary from 10 to 50%, but the mortality is much lower than in ducks. Quite often, *C. psittaci* infection is subclinical and asymptomatic [53, 580].

11.2.3 Chlamydiosis in Turkeys

Chlamydiosis in turkeys often occurs in an acute form with significant lethality, especially in young birds. Meyer [219], in one of the first descriptions of the disease, recorded a mortality rate of 20–30% in some herds. The incubation period ranges from several days to 2–3 and more months. As a rule, young birds are the most common and most severely ill. Symptoms of respiratory disease, indigestion, bruising, fever falls, depression, conjunctivitis, cyanotic wattles, yellow-green watery feces, extroverted cloacae, and decreased egg production are observed. [22, 219].

Page [782] estimates that in the case of infection of turkeys with toxigenic strains of the causative agent, 50 to 80% of them in the herd show clinical signs – anorexia, cachexia, hyperthermia, yellow-green gelatinous excrements, and a sharp decrease in egg production. Sometimes the disease occurs with predominant meningeal and encephalitic symptoms. The mortality rate fluctuates from 10 to 30%. In infection with less toxigenic strains, 5 to 20% of turkeys show signs of anorexia and stinging green feces. Mortality ranges from 1 to 4% [782].

11.2.4 Chlamydiosis in Decorative Birds

Parrots and other psittacines. In parrots and other psittacines, the characteristic clinical signs are as follows: Conjunctivitis and keratoconjunctivitis; diarrhea, anorexia, and droopiness; lethargy; weight loss; dehydration; ruffled feathers; dyspnea, rales, coryza, and sinusitis; reduced egg production [736]. Mohan [784] points out that in pet birds the most frequent clinical signs are conjunctivitis, anorexia and weight loss, diarrhea, yellowish droppings, sinusitis, biliverdinuria, nasal discharge, sneezing, lachrymation and respiratory distress. In many birds, especially older psittacine birds, clinical signs may not be detected. Nevertheless, they may often shed the agent for extended periods. According to our observations, symptoms of depression, weakness, shortness of breath, and diarrhea with greenish feces were recorded in parrots from pet shops, zoos, and private homes affected by psittacosis. Some of the severely sick birds died. In other cases, oral administration of Doxycyclin two times daily in therapeutic doses for 2–3 weeks was effective and clinical recovery was achieved. In cases of continued chlamydia carriage and shedding, the therapeutic course was prolonged [53].

Finches. Finches are one of the many bird species that are susceptible to infection with *C. psittaci* [432, 567, 785–788]. Dustin [786] indicated that it reached the isolation of chlamydia from the droppings of both clinically ill and clinically healthy finches, especially if they are kept in households in which clinical cases of chlamydiosis occurred in psittacine species. According to the same author, chlamydiosis should be suspected in passerines with recurrent respiratory disease, especially if they are exposed to the already mentioned psittacine birds. Active disease outbreaks are intermittent, and infectious rates of less than 10% of the at-risk population are typical [786].

In our study, we demonstrated the disease etiologically in the study of Zebra finches, (*Taeniopygia guttata*), Japanese finches (*Lonchura striata domestica*), and Gouldian finches (*Chloebia gouldiae*, *Erythrura gouldiae*), grown in a common room with a total number of birds of 70. Clinical signs showed 15 Zebra and Japanese finches. Adult birds were asymptomatic. In some pairs, 17–20 days after hatching, we observed yellow-colored feces with normal, softened, or loose consistency. The diarrhea lasted for about 10 days. The appetite was gradually lost. The general condition of some of the finches was getting worse and they were dying. The remaining birds overcame the disease. Baytril 10% (Enrofloxacin) 0.5 ml/liter of water, permanently applied at the first signs of the disease in drinking water for 3 days, was not effective. The Cycloverft (Oxytetracyclin) administrated after the etiological diagnosis at a dose of 3 g/liter of water, 7–14-day course, had a positive effect manifested in decreased morbidity and reducing the signs, but until the period of observation, the disease was not fully terminated.

11.2.5 Chlamydiosis in Pigeons

The epidemiological and zoonotic significance of chlamydiosis in pigeons is indisputable. The disease affects both wild and domestic pigeons. Clinically, one-sided or double-sided conjunctivitis with eyelid swelling, rhinitis with serous or sero-purulent secretion, respiratory disturbances, reduced flying ability, cyanotic skin coloring, and depression are observed. Anorexia, diarrhea, and transient paralysis have also been established. In adult pigeons, sudden death has been reported. The lethality is higher in young birds. Chlamydiae are isolated from the eyes of sick pigeons. Too often, the chlamydiosis of the pigeons is inapparent or chronic with less pronounced signs, most commonly transient diarrhea. Such pigeons, however, are carriers and shedders of the chlamydial agent [22, 539, 580]. Shedding of the organism occurs in the feces as well as in conjunctival and respiratory secretions, often intermittently and without clinical signs, which makes it difficult to assess the risk of transmission of *C. psittaci* to other animals and humans [582]. The shedding of chlamydiae is exacerbated by trigger factors such as stress, concurrent infections, infestation, overcrowding, breeding, deficiency, or lack of food [539, 789].

11.3 Clinical Observations in Guinea Pigs

Chlamydial infection is one of the most common causes of conjunctivitis in guinea pig populations, which is also known as guinea pig inclusion conjunctivitis (GPIC). It is caused by *Chlamydia caviae*. The most susceptible to infection are young guinea pigs, especially those 1–2 months old. Other clinical forms of chlamydia in guinea pigs include rhinitis, lower respiratory-tract disease, and genital infections: urethritis and cystitis in males and cystitis and salpingitis in female guinea pigs [297, 298, 791]. Abortion is also noted. Chlamydial infection on a subclinical level may also occur.

The most common clinical form is conjunctivitis, which occurs in mild, moderate, and severe form. In the lighter form, conjunctival hyperemia, slight yellow-white discharge, and chemosis are observed. For the severe form of conjunctivitis, profuse ocular purulent exudate is characteristic [767, 790]. Intra-cytoplasmic chlamydial inclusions are formed in the conjunctival epithelium, which can be demonstrated in LM after Giemsa staining. Diagnosis is also successful through PCR. Oral therapy with doxycycline (5 mg/kg, bid for 10 days) is efficacious. Due to the short-term immunity, re-infections are possible [790]. In other clinical forms of chlamydiosis, the antibiotic courses are more prolonged.

Below are presented our observations on GPIC in natural and experimental infection with *C. caviae* (Martinov, 2008, unpublished data). In a colony of guinea pigs aged 4–6 weeks, we observed inflammation of the conjunctival mucosa of varying intensity. In the less severe cases, we found mild hyperemia, weak conjunctival secretion with serous, catarrhal or serous-mucosal character and white-yellowish color. In moderate forms, these signs were more pronounced. The severe forms – a small number compared to the above – exhibited a strong hemorrhagic conjunctival swelling of the eyelids and heavy purulent exudate, which literally stuck the eyelids together. Some guinea pigs had corneal clouding (keratitis).

The etiologic nature of the disease was demonstrated by LM of scarified conjunctival material, whereby cytoplasmic inclusions of chlamydia were detected in epithelial cells. From conjunctival secretions containing desquamated cells, we isolated the M. sv-1 strain of *C. caviae*.

The results of the experimental challenge of guinea pigs by conjunctival inoculation with the specified strain with infectious titer LD_{50} log 10^5/0.5 ml for 7-day-old CE and a dose 0.5–0.6 ml are shown in Figure 11.3. It shows the sequence of occurrence of the individual clinical signs, their duration and the outcome of the disease. We found that keratoconjunctivitis developed tends to self-restraint, with eye recovery being recorded between the third and fifth weeks.

11.4 Clinical Observations in Humans

11.4.1 Human Infections Caused by Chlamydia Psittaci with Avian Origin

Initial knowledge of human diseases caused by *C. psittaci* indicated their occurrence, spread and epidemic manifestation as a result of exposure to

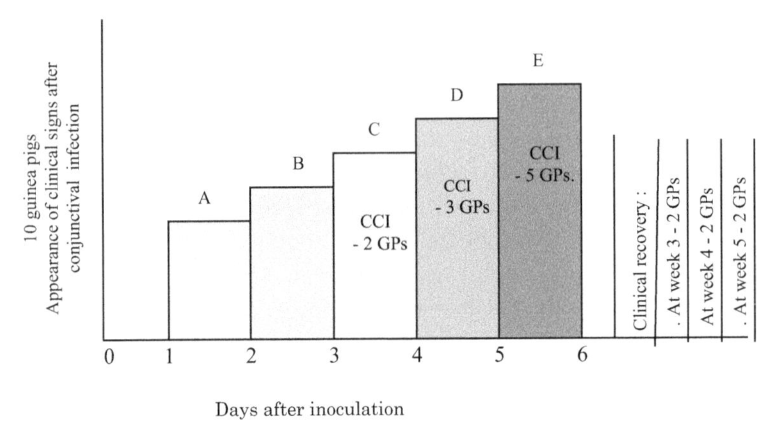

Legend:
A) Hyperemia. Persists 3–4 weeks;
B) Strengthened conjunctival secretion: • Serous-catarrhal secretion – lasts 7–8 days;
• Slurry-purulent secretion – lasts for 9–11 days;
C) Follicular hypertrophy. Detected for 2–3.5 weeks;
D) Pannus. An average of 2.5 weeks is observed;
E) Surface keratitis. In 2 GPs – 1 week; In 4 GPs – 2 weeks; In 4 GPs – 3 weeks;
CCI – Cytoplasmic Chlamydial Inclusions

Figure 11.3 Dynamics of clinical manifestations in experimentally induced chlamydial guinea pig inclusion conjunctivitis

pet psittacine birds. Later, it became clear that a number of non-psittacine species are also pathogenic to humans. Along with the expansion of poultry farming in industrial proportions, human illnesses have emerged, including epidemic outbreaks, especially in connection with the mass production of ducks, geese, and turkeys. Thus chlamydiosis (psittacosis–ornithosis) has acquired the character of an occupational disease [21, 581, 779, 780, 783, 793].

The incubation period prior to the onset of the disease in individuals infected with *C. psittaci* is from 5–14 days. The severity of this disease is different. In some cases, the disease has an inapparent nature. In other cases, a generalized disease with severe pneumonia develops. Before antimicrobial agents were available, 15%–20% of people with *C. psittaci* infection were reported to have died. Deaths nowadays are rare [792]. Patients with clinical manifestations show abrupt onset of fever, chills, headache, malaise, and myalgia. The most common is a nonproductive cough accompanied by breathing difficulty and chest tightness. Rash, enlarged spleen, and pulse-temperature dissociation (fever without elevated pulse) are less common.

The auscultatory findings are not always convincing. Lobar or interstitial infiltrates are detected in radiography. In the differential diagnosis, one should consider infections with *C. burnetii*, *C. pneumoniae*, *Histoplasma capsulatum*, *Mycoplasma pneumoniae*, Legionella species, and respiratory viruses such as influenza [792, 794]. Infection with *C. psittaci* has been reported to affect organ systems other than the respiratory tract, resulting in conditions including endocarditis, myocarditis, hepatitis, arthritis, keratoconjunctivitis, encephalitis, and ocular adnexal lymphoma. [795–801]. In a recent publication, it was reported that multi-locus sequence typing has identified an avian-like *C. psittaci* strain involved in equine placental inflammation and associated with subsequent human psittacosis [1109].

11.4.2 Clinical Data for Chlamydiosis Epidemics in Humans Associated with Outbreaks of Avian Chlamydiosis in Waterfowl

Clinical diseases of humans related to the ornithosis of ducks in Bulgaria (Trastenik, Pl) were manifested with a toxic-infectious syndrome, bronchopneumonia, or only signs of the upper respiratory tract. Two of the cases ended with death. [53, 577, 803]. Some of the affected patients experienced a flu-like condition in mild, moderate and severe form.

In the Kostinbrod (Sf) epidemic, also associated with ornithosis epizootic in ducks, slight tiredness and delayed mobility of the patients, rapid and difficult breathing, febrile reactions, respiratory signs, abundant perspiration, and clinical and radiological evidence of interstitial pneumonia were observed. Similar clinical findings were found among workers in several other farms for ducks and geese in the regions of Lom, Razgrad, Ruse and others.

Most human cases of infection associated with goose chlamydiosis occurred in the form of epidemic outbreaks (Rs-district, 1993; Tst-district, 1994). The clinical characteristics in humans were similar to those described above, but the majority of human cases were milder [53, 579].

11.4.3 Sporadic Morbidity from Ornithosis in Humans

Over the last decade no epidemic outbreaks of ornithosis have been reported in humans in Bulgaria as a result of transmission of the infection from infected duck and goose farms. Sporadic cases occur primarily as a febrile illness, often complicated by the inflammation of the lungs. Probably, one of the reasons for the lack of epidemics in humans is the improved organization of waterfowl farms and the effective prevention and control of the disease as a result of the experience gained in previous years.

Other sporadic cases have been diagnosed and confirmed etiologically in persons who have been in contact with diseased ornamental birds or pigeons. There have been cases of contamination after cleaning of houses and premises heavily contaminated by feces of resident pigeons. This is the case described by Martinov and Popov [45] concerning a 39-year-old man with chronic ornithosis as a result of prolonged pigeon contact characterized by prolonged sub-febrility, pain in sternum, and almost complete antibiotic insensitivity. Most of the illnesses occur with fever and pneumonia, mainly of the interstitial type.

For the treatment of patients diagnosed during epidemics or sporadic cases, tetracycline antibiotics are successfully administered.

11.4.4 Chlamydiae of Mammalian Origin and their Importance for Human Health

The issue of the role of mammalian chlamydiae as a disease agent in humans has long been the focus of virologists, bacteriologists, and infectionists in the field of veterinary and human medicine. Historically and nowadays a number of authors suspect infected mammals as a source of chlamydial infections for susceptible human populations. Enright and Sadler [804], in a serological investigation of slaughterhouse workers and veterinarians, detected in many cases CF-antibodies with rising titers to the sporadic bovine encephalomyelitis virus (chlamydia). The authors suggested that this was related to chlamydial infection derived from cattle and sheep. In 1955, Barwell [805] reported a laboratory infection of a male with the agent of EAE. Giroud et al. 1956 [806] consider that there is an etiological connection between pneumonia and abortion of sheep, goats and cattle in Congo and abortion in women. The reasons for this are the isolation of several pathogens with the characteristics of chlamydia from the placenta of women and lungs of the aborted human fetuses. Meyer [807] reported intra-laboratory infection, as a result of which SBE's chlamydial agent was isolated from a patient's blood. According to the data published by Prat [808], Fiocre [809], and Sarateanu et al. [810], cattle suffering from chlamydial bronchopneumonia have infected humans who have developed chlamydial infection. In 1961, Carrere et al. [811] revealed large outbreaks of EAE in sheep and goats in France, as well as a high rate of chlamydial sero-reactors among people who have been in contact with diseased sheep and goats. Horsch et al. 1978 [through 812] reported intra-laboratory infection in Germany with a bovine chlamydial isolate, clinically manifested by a febrile condition, anorexia, and acute lung inflammation. Martinov and Popov [45, 812] have observed a case of a bilateral keratoconjunctivitis in a female laboratory worker in Bulgaria, caused by the direct fall of a yolk chlamydial suspension from a strain isolated from a calf with keratoconjunctivitis. In addition to rapidly emerging acute local manifestations (irritation, severe redness, edema), the disease was accompanied by common signs – increased body temperature to 39.1°C, pain in the heart area, and general fatigue. From the blood serum of the patient taken in the acute stage of the infection the chlamydial strain has been re-isolated, identified antigenically and morphologically by EM, and the serial titrations of CF antibodies have showed a certain dynamics with a 4-fold increase in titers.

One of the well-known questions about the zoonotic significance of mammalian chlamydia is the role of *C. abortus* (EAE) in abortion in women. Several author groups published data on this issue (McKinlay et al. [813]; Johnson et al. [814]; Buxton [815]; Wong et al. [816]; Longbottom and Coulter [817]; Walder et al. [818]; Janssen et al. [819]). The most vulnerable are women involved in sheep breeding and female veterinary surgeons who are exposed to aborting sheep. McKinlay et al. [813] described two patients with severe chlamydial sepsis in pregnancy as a result of exposure to OEA. In another of the first publications on the subject, Wong et al. [816] described a case with a sheep farmer's wife who had been assisting with lambing and developed an influenza-like illness in the 28th week of pregnancy. The woman developed acute placentitis and after five days of malaise she spontaneously delivered a stillborn infant; she became acutely ill during the immediate postpartum period with septic shock, acute renal failure, and disseminated intravascular coagulation. The etiological diagnosis was made serologically by testing the patient and by the isolation of the chlamydial agent from the placenta, fetal lung, and fetal heart blood [816]. The analysis of the other reports shows that human infection has a similar clinical picture of acute influenza-like symptoms followed by abortion. Walder et al. [802] reported for a case of *C. abortus* pelvic inflammatory disease. In some cases there are severe complications of renal failure, hepatic dysfunction and extensive intravascular coagulation resulting in death.

Pregnant women are also at risk from exposure to goats from herds where there are chlamydial abortions and related conditions [648, 820]. Pospischi et al. [648] reveal a chlamydia-related abortion storm affecting 50% of the goats during the 2000/2001 lambing season. At that time, a pregnant woman (pregnancy week 19/20) who had been in contact with abortive goats developed a severe generalized infection and aborted. In her placenta *C. abortus* was shown by immunohistochemistry and PCR. Two years later, Meijer et al. [820] reported a similar case of a pregnant woman who developed a severe rapidly worsening influenza-like illness caused by *C. abortus* after indirect contact with infected goats resulting in a preterm stillbirth. The woman fully recovered after treatment with doxycycline. These testimonies clearly indicate the zoonotic risks for pregnant women, especially in the rural ares in direct and indirect contact with sheep and goats affected by chlamydial abortions and related conditions.

As is known, *C. suis* is endemic in a number of pig holdings in different countries as well as among wild boars. However, the zoonotic potential of this pathogen, phylogenetically closely related to *C. trachomatis*, is still insufficiently researched and underestimated [703, 821]. According to Rohde et al. [824] not enough is currently known about the pathogenesis and zoonotic relevance of bovine or porcine chlamydioses, as there are virtually no reliable data on cases of zoonotic transmission or on the possible modes of transmission to man. In the study of De Puysseleyr et al. [821] *C. suis* has been revealed by real-time PCR in 45 (45%) of the 100 tested rectal swabs from pigs. The agent has been also found in the air and contact surfaces as well as in eye swabs of two asymptomatic employees of 12 (16.6%) tested by both real-time PCR and culture. None of the human *C suis* isolates contained the *tet*(C) gene determining the presence of tetracycline resistant (Tc^R) *C. suis* strains. [822]. It should be point out that exposure to Tc^R *C. suis* strains poses an additional risk for pig handlers. All tested human serum samples in the above study were serologically negative for *C. trachomatis* [821]. In another recent study of De Puysseleyr et al. [823], similar diagnostic approaches in pigs from nine Belgian pig farms and farmers were used: isolation of the pathogen and a *C. suis*-specific real-time PCR. In addition, the chlamydia isolates were tested for the presence of the *tet*(C) resistance gene, and the farmers' samples were examined using a *C. trachomatis* PCR. As a result, chlamydial DNA was found in eight of the nine farms surveyed, and eight of nine farmers were positive in at least one anatomical site. In three human samples, two rectal and one pharyngeal *C. suis* isolates were identified. All human blood samples were serologically negative for *C. trachomatis* [823].

Despite the existing uncertainty about the actual zoonotic potential of *C. suis*, it is worthwhile to go back to an earlier study by Zatulovski et al. [528], which examines workers, veterinary staff, and other employees in large pig farms where chlamydial infection has been detected. In one of them, with many cases of chlamydial pneumonia in pigs, 160 people were working and 29 (18.1%) of them were positive when tested by direct and indirect CFT. Of the 28 people working in another pig farm with mass abortions and stillbirths, 9 (32.1%) proved to be seropositive. The authors found antibodies against chlamydial antigen most often in persons who are in direct contact with animals – the caregivers (23.4%) and the veterinary staff (22.7%). The high importance of the direct contact with chlamydia-infected pigs is evidenced by the low rate of positive sero-reactors (3%) among people who are not in direct contact with the animals. In most of the people who are seropositive to chlamydiae (35–24.8%) myocardiopathy, kidney diseases, and hepatocholangitis have been found.

At present, reports of transmission of chlamydial infection from dogs to humans are too scarce. Sprague et al. [825] established *C. psittaci* genotype C in four bitches from a small dog-breeding facility in Germany by examining nasal, pharyngeal, and conjunctival swabs using IF, CC, PCR, and sequencing of ompA amplification products. Clinically, the bitches showed severe respiratory distress, recurrent keratoconjunctivitis, and decreased litter size (up to 50% stillborn or non-viable puppies.) Chlamydial infection was also proven in two girls living in the same premises. The authors believe that the infection of dogs followed the introduction in the premises of a parrot and two canaries. While the interpretation of the case is in the direction of infecting girls from dogs, in our opinion the possibility of infecting children directly from the birds is not excluded. Indeed, the disease in dogs was persistent, suggesting that dogs may be reservoirs of infection with this type of chlamydia. The question remains whether or not the infected dogs shed the pathogen [825].

The third chlamydial species (except *C. psittaci* and *C. abortus*), which is considered to be zoonotic, is *C. felis* [817, 828, 829]. Several publications state that *C. felis* has been identified in sick people who have had prolonged contact with infected cats. Here is referred to follicular conjunctivitis [829, 830, 839]; functional disorders of the liver [831]; endocarditis and glomerulonephritis [832]; and atypical pneumonia [833]. Hartley et al. [835], using molecular classification techniques, found that pathogens isolated from a 39-year old HIV-positive patient with chronic non-trachomatis chlamydial conjunctivitis and from one of the patients' cats were indistinguishable from *C. felis*, and provided convincing evidence of transmission from cat to man. In this case, a prolonged course of doxycycline was required to eradicate the infection. However, some authors consider the issue of the zoonotic significance of feline chlamydiosis to be controversial and insufficiently convincing.

According to Browning [836], while there is evidence that *C. felis* may occasionally cause keratoconjunctivitis in humans, there is little evidence suggesting it can cause serious systemic disease or pneumonia. Tasker [837] points out that there is no epidemiological evidence for a significant zoonotic risk of *C. felis*. Reports are rare, attributing this pathogen to serious systemic disease or atypical pneumonia (Reinhold [633]). Indeed, zoonotic infections associated with *C. felis* are rarely published. One possible explanation for this fact is that they do not appear frequently, but are more likely to be inadequately investigated, implying that they are underdiagnosed [835, 838].

Concerning horses (*C. abortus, C. pneumoniae*), the zoonotic potential is unknown.

Data on transmission of chlamydia from infected guinea pigs to humans is too scarce. Lutz-Wohlgroth et al. [840] found *C. caviae* by PCR in the eyes of a man who has worked with about 200 infected guinea pigs. As a result, the patient had a serious ocular discharge.

No evidence for zoonotic transmission of *C. muridarum* has been published [841].

11.5 Infections Caused by Chlamydia Trachomatis

11.5.1 Chlamydial Urethritis (Non-gonococcal Urethritis – NGU) in Men

It runs in acute, sub-acute and chronic form. Criteria for this classification is not the time for development or the onset of the infection, but the clinical picture itself and the objective clinical and para-clinical findings. In this sense, the specified degrees of NGU are not mandatory and it may be straightforward as a chronic one.

The acute (catarrhal) urethritis has an incubation period of 1–2 weeks. It begins violently with pain, burning, and frequent dysuria. Sometimes the unpleasant sensations pass through the entire urethra (Dimitrov et al. [843]; Dimitrov [844]; Siboulet et al. [846]).

When viewed, the orifice is red and swollen. Exudate is almost uninterrupted. The microscopic preparation demonstrates an abundance of neutrophilic leukocytes without the presence of gonococci or other bacterial agents. In this case, the cytological findings are rich, and cytoplasmic inclusions in epithelial cells can be detected after Giemsa staining. In addition to cytology, direct EM, IF with mAbs, and PCR are successfully administered. Acute chlamydial urethritis (if any) rapidly resolves. Clinical symptoms change. Exudate from the urethra decreases and produces a mucopurulent or mucosal character. Subjective complaints decrease. The process goes into a chronic stage.

The chronic urethritis is the most typical and frequent form of chlamydial inflammation of the urethra. Usually the acute and sub-acute phase of the urethritis is absent at all. There is only mucopurulent urethral discharge, which disappears after urinating for several hours and then reappears. The microscopic finding of such an exudate contains about 10–15 neutrophils per field and epithelial cells in which the chlamydial inclusion is found. Gradually, the process can move from the front to the posterior urethra by developing total urethritis. The most typical and sometimes the only symptom is the so-called morning drop that disappears after morning urination and appears the next morning. This minimal urinary exudate in the study may be positive for chlamydia. These manifestations and signs are very difficult to see for the investigator. Sometimes he does not pay enough attention to the patient and thinks he suffers from a venerophobia. The persistence and sustainability of chronic NGU and its insignificant but overwhelming signs with frequent remissions and recurrences can continue for years, and most do not tend to self-heal. These are the most common causes of sexual neurosis in labile patients and are of great social importance.

The data for the duration of the chlamydial infection in men with urethritis are presented in Figure 11.4. It shows that the duration of the disease ranges from a few days to several years. As a complication of NGU, prostatitis, vesiculitis, orchitis, orchiepididimitus, and paraurethritis may develop, which were etiologically diagnosed in our extensive research on the problem [24, 842–844].

11.5.2 Post-Gonococcal Chlamydial Urethritis (PGU)

It is a separate nosological unit and occurs after curing gonococcal urethritis [20, 24, 57, 848–852]. It is characterized by a urethral discharge that persists after the destruction of the gonococcus in the patient. This is in fact a manifestation of a mixed urethral infection, in which the treatment of gonococcal infection does not eliminate the added chlamydial infection. Our observations show that the subjective complaints of patients, as well as the clinical picture of PGU, depend on what treatment has been used to treat gonorrhea (Martinov et al. [57, 847]). Belli et al. [853] indicate that when some quinolones are used as a single dose therapy in gonococcal infections, they are unable to eradicate co-existing *C. trachomatis* and prevent PGU.

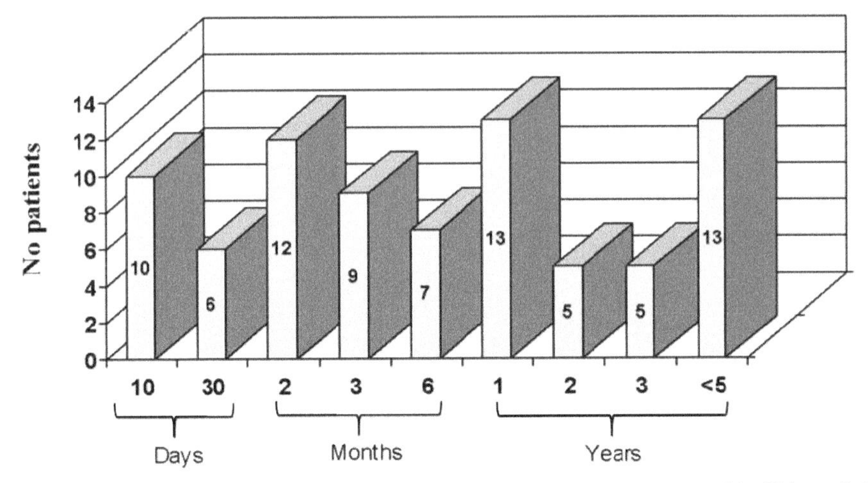

Figure 11.4 Duration and Dynamics of Chlamydial Infection in Men with Chlamydial Urethritis.

However, when ciprofloxacine was administered at a dose of 750 mg twice daily for 4 days, the mixed *C. trachomatis* infection was eradicated in 60% of the patients, and the PGU rate had been reduced from 35% to 12.8%. Obviously, the duration of quinolone treatment is essential as a prerequisite for a possible reduction in the percentage of chlamydial PGUs and even for its prevention. The same approach should be taken in attempts to treat PGU with other potentially effective antibiotics and chemotherapeutic agents.

We have paralleled the symptoms of two types of classical treatment of gonococcal infection – kanamycin and tetraolean (the trade name of a combined preparation containing tetracycline and oleandomycin).

Kanamycin does not at all affect chlamydia. This fact determines the picture of PGU after kanamycin as acute or sub-acute. The purulent discharge characteristic of gonococcal infection stops, but passes without a pause in a new type of discharge with mucous nature – during the day, scarce, and in the morning, harder. A number of inclusions of *C. trachomatis* were detected in Giemsa preparations. There was no remission of the disease after treatment with kanamycin. Thus, the picture of gonococcal infection is switching to a picture of acute or sub-acute chlamydial urethritis.

Tetraolean affects, albeit poorly, chlamydia because it contains the tetracycline component. Therefore, very often between the onset of PGU and the healing of the chlamydial infection was seen a "bright" asymptomatic period. Typically, 7 to 10 days after treatment with tetraolean, chlamydial urethritis of the chronic type was initiated, characterized by mucinous exudate of the morning drop type, oligo-symptomatology and wavy development. In the exudate, a small number of dense basophilic inclusions of *C. trachomatis* can be found.

11.5.3 Chlamydial Prostatitis

Chlamydial prostatitis is one of the most common complications of urethritis caused by *C. trachomatis*. According to Mavrov and Cletnoy [854], 46% of the extensively investigated patients with chlamydial urethritis develop inflammation of the prostate gland. Other sources also indicate high percentages (18–56%) of *C. trachomatis*-induced prostatitis [855–859]. Urethro-prostatitis actually develops, which is the most common phenomenon of ascending chlamydial infection [24]. Mardt et al. [862] introduced the term "urethro-prostatosis", which largely clarifies the nature of the developing inflammation, as the prostatic capsule and paraprostate space are very often involved.

Chlamydial prostatitis is of the diffuse type and generally runs chronically. The clinical symptoms are varied: itchy urethra, starting from the perianal region, urinary urgency, feeling cold and softening of the glans penis, anal and perianal itching, dull backache, subjective feeling of incomplete bowel movements, discomfort in the rectum, disturbances erection, short duration of intercourse, and reduced pleasure in ejaculation [24]. The classic clinical test method is the shading of the prostate gland to give the prostate secretion. The latter was examined under a microscope as a native preparation without staining under glass roofing. The discovery of the 10–12 leukocytes per field indicates an inflammatory process. In some cases are detected 30–60 and more leukocytes per field, and even "pavement" of leukocytes. When leukocytes are disposed in groups, rather than diffusely throughout the microscopic field, then it refers to the so-called "Insular prostatitis" – a concept introduced by A. Bonev, 1984 (in [24]).

Different methods were used to diagnose: direct IF, MIFT, solid phase EIA for local IgA antibodies, CC, PCR, LPS-antigen, EM, and in situ hybridization. Weidner et al. [860] discuss the laboratory diagnosis of chlamydial prostatitis and point out two major problems: (1) diagnostic urethra may reflect only urethral contamination and (2) prostate biopsy specimens from the gland may also contain urethral material. The ejaculate has the same limitations, and an ideal test for detection of Chlamydia species in ejaculate specimens is not available yet [860]. According Wagenlehner et al. [861], especially in prostatitis, the exact role of *C. trachomatis* is still under debate because of the technical difficulties in localizing the pathogen to the prostate. Despite these considerations aimed at improving laboratory diagnosis, there is no doubt about the relationship between chlamydial prostatitis and prostatitis developing as a complication. At this stage, it remains a controversial issue – whether prostatitis is due to the direct action of the chlamydial pathogen or the development of inflammation of the prostate gland is result of a possible specific allergic or autoimmune response of this organ toward the toxicity of chlamydiae.

11.5.4 Chlamydial Vesiculitis

Vesiculitis is an acute inflammation of seminal vesicles and ejaculatory ducts. The inflammation of these parts of the male genital system is determined by the proximity, the general blood supply, and the communication link between the urethra, the prostate gland and the seminal vesicles. Usually the inflammation of the prostate and seminal vesicles occur simultaneously. This disease state is referred to as prostatovesiculitis. Often the latter is of a chronic type. The symptoms of vesiculitis include fever, polyuria, terminal hematuria, and painful erection and ejaculation [863]. In the zone above the symphysis there are two-sided painful spasms extending to the perineum. In a rectal study, one or two painful and dense seminal vesicles are felt. Ejaculates from such patients are altered in terms of viscosity, color, consistency, and contain many dead sperm cells [24]. We will note that in 8 cases of chlamydial prostatitis or uroprostatitis we have experienced both inflammation of gl. Cowperi. In the follow-up of two of these patients, they were palpated, and the glands Cowperi were found to be thickened and painful [864].

11.5.5 Chlamydial Epididymitis, Orchitis, and Orchiepididymitis

Chlamydial infections are the cause of many inflammations of the testes and their appendages. Nilsson et al. [857] emphasize that epididymitis appears to be an established complication of chlamydial urethritis, especially when urethritis is complicated with prostatitis. Grant et al. [1111] found that in 59 of the patients with an etiologically proven C. trachomatis infection, the most frequent clinical symptom was one-sided epididymitis (44%), followed by prostatitis (15%) and urethritis (13%). According to these authors, in many cases the epididymitis is not accompanied by urethritis, but develops itself. Ostaszewska et al. expressed such a view expressed [867]. However, the ascendant pathway of epididymitis is considered to be most likely as a consequence of chlamydial urethritis (Oriel and Ridgway [235]; Martinov [23]). We advocate that if there was no urethritis at the time of the development of the epididymitis, this speaks of the sequence of the processes of climbing the infection or of oligosymptomatic urethral inflammation [24]. In favor of the perception of the ascending urethro-prostate-epididymal pathway of the chlamydial genital infection, there are arguments such as the near anatomical location, the communication between the urethra and the epididymis, and the very similar histological structure of the epithelium.

The clinical picture of the epididymitis is characterized by swelling, enlargement, and soreness of the epididymis. The scrotal pouch above it is red, warm and stretched. Other signs such as enlarged lymph nodes in the groin and a lump on the testicle may be observed. Sometimes the patients are febrile. The overall condition is worsened. After about a week, the process may resolve or move into chronic. Since epididymitis may lead to reduced fertility, antibiotic treatment using tetracyclines should be administrated to counteract this condition [857]. According to the recommendation of the CDC, doxycycline 100 mg orally twice a day for 10 days is suitable for this purpose. Alternatively, a single dose of azithromycin 1 g orally may replace doxycycline to cover *C. trachomatis* [868]. Where necessary, depending on the severity of the disease and the individual response of the patient, the duration of the treatment course may be extended. In mixed infection – *C. trachomatis* and gonorrhea, to the above treatment is inserted Ceftriaxone 250 mg IM in a single dose [868, 869]. When confronting male partners with chlamydial epididymitis, specific antibodies are detected and *C. trachomatis* isolated from the cervix in high percentages. Orchitis and orchiepididymitis include the testicle which becomes red and enlarged in the acute stage of chlamydial infection. Martinov et al. [55] reported a bilateral orchiepididymitis, a severe and persistent disease with unclear etiology, an exclusion of other etiological agents (*Trichomonas vaginalis, N. gonorrhoae*, aerobes and anaerobes), and a lack of effect of ampicillin, gentamicin, and sulfonamides. After the etiological diagnosis (chlamydial pathogen isolation and serology in dynamics), the patient responded very well to targeted therapy with tetracyclines in prolonged courses.

After the acute orchitis process has resolved, the size of the testicle can be greatly reduced and atrophied. The combination of atrophied testicle, obliterated epididymis, and seminal canal is a sure prerequisite for pathological abnormalities in the insemination capacity of the patient [24, 857]. Chlamydial epididymitis and orchi-epididymitis are one of the most common causes of pathological changes in seminal fluid composition, and for oligospermia and azospermia.

11.5.6 Paraurethrits

This is a clinical condition that is a complication of chlamydial urethritis, which affects the paraurethral strokes. Urethroscopy occurs in the form of erythematous macules, in the middle of which can be seen the openings of the urethral crypts. With efficient treatment of urethritis, the described paraurethral changes resolve. In some cases, however, they remain as reservoirs of chlamydial infection, the cause of which is insufficient drainage at the bottoms of the para-urethral strokes.

11.6 Chlamydial Infections in Women

Since pioneering work on sexually transmitted chlamydial infections, a large number of publications have made clear that *C. trachomatis* is responsible for a wide range of genital diseases and complications in women. These include cervicitis, vaginitis, bartholinitis, paraurethritis, endometritis, salpingitis, oophoritis, parametritis, pelvic inflammatory disease, placentitis, Fitz-Hudge-Curtis syndrome (FHC, perihepatitis), abortions, stillbirths, and infertility.

Chlamydiae are transmitted mainly through sexual intercourse. The clinical manifestations of urogenital chlamydia are very diverse, which determines several forms of the disease. Most often chlamydial infection occurs in subacute, chronic and persistent forms, less often – in the type of acute inflammatory processes [875]. The causes of widespread prevalence of chlamydial genital infections in women are complex and include a number of factors such as the nature and specificity of the infectious agent, the susceptibility and behavior of human populations, the particularities of the clinical course, the difficulties in diagnosis, often ineffective treatment, numerous complications, etc. All these questions have been and will be the subject of future thorough research by specialized scientific communities.

In this section of the book, we will only attempt to briefly describe some of the clinical features of these medically and socially significant diseases.

11.6.1 Chlamydial Cervicitis

Usually the uterine cervix is affected by chlamydia in the form of cervicitis or endocervicitis with a wide range of subjective complaints and clinical and colposcopic findings. The relationship between clinically manifested cervicitis and *C. trachomatis* is shown in a number of publications.

For example, Bruncham et al. [870] isolates the agent in 20 out of 40 women with mucopurulent cervicitis, and Paavonen et al. [871] found an even higher percentage (76.5%) in the study of 465 patients with the same type of cervical inflammation. In the United States, Schachter et al. [872] reported isolation of *C. trachomatis* in 26% in cervicitis and 47% in cervical erosions. Hobson et al. [873] reported for multiple isolates of chlamydia (63%) of women with hypertrophic cervical erosions. Hare et al. [874] associated Chlamydia with lymphocytic follicular cervicitis in 38% of 93 patients.

It should be stressed that there are a number of cases where women have a hidden course of chlamydial infection and they have no complaints, secretions from the genital tract, or hyperemia of the mucous membranes. Such women are carriers of chlamydiae and major distributors of infection sexually.

Our research group (Popov et al. [24], Martinov et al. [105], Dimitrov [844]) carried out studies in 118 women, previously selected as negative for gonococci, aerobes and anaerobes, Trichomonas, and Candida. *C. trachomatis* infection was detected in 113 patients by a complex etiological diagnostics involving eight methods – cytology, serology (ELISA, CFT, IF-mAbs), DEM, Isolation (CC McCoy – Giemsa, Lugol), and CE.

Our data presented in Table 11.2 testify to the variety that exists in clinical complaints [24, 105, 844].

An anamnestic survey of 62 patients with chlamydial infection showed that the disease duration ranged from 4 days to 8 years. In 29 patients, the duration of the disease was from 2 to 8 months, and in other 25 – from 1 to 8 years. Therefore, unrecognized and untreated chlamydial cervicitis are characterized by significant durability and do not exhibit a tendency to self-healing. Data on the persistent nature of chlamydial cervicitis, in the absence of re-infection, were presented in the works of Rees et al. [876],

Table 11.2 Subjective complaints of 72 patients with etiologically proven chlamydial cervicitis

Symptoms	Number of Patients	%
Constant discharge	20	24.36
Periodic discharge	2	2.36
Discharge with color and smell	18	23.40
Pain	6	7.8
Bleeding in intercourse and manipulation	11	11.62
With more than one complaint	5	6.1
No complaints	10	24.36

Table 11.3 Colposcopy of 72 patients with etiologically proven chlamydial infection

Colposcopic Findings	Sick Women	
	Number	Percent
1. Endocervicitis	26	36.11
(A) follicular	15	20.83
(B) granular	3	4.16
(C) banal	8	11.11
2. Exocervitcitis	3	4.16
3. Macular colpitis and cervicitis	3	4.16
4. Ectropion and ectopia	13	18.05
5. The glandularis area	5	6.94
6. Follicular colpitis	1	1.38
7. Ovula Nabothi	1	1.38
8. Ectopia and exocervicitis	6	8.33
9. Ectopia, exocervicitis, and Ovula Nabothi	2	2.77
10. Ectopia, endocervicitis, and Ovula Nabothi	2	2.77
11. Ectopia and Zona glandularis	3	4.16
12. Exocervicitis and Zona glandularis	4	5.56
13. Dysplasia of the cervix	3	4.16
Total	72	100.00

McCormack et al. [877], and other authors. On the other hand, re-infections and super-infections with untreated sexual partners are also important for long-term maintenance of the infection.

The results of colposcopic studies of 98 patients with chlamydial infection are presented in Table 11.3.

A variety of findings are seen, with endocervicities of three types (36.1%), followed by ectropion and ectopia (18.05%) and Zona glandularis (a zone of transformation between exocervix and endocervix). Follicular endocervicitis is characterized by raspberry-like, closely spaced slightly transparent follicles. In several cases with a lengthy period of time, we observed granular endocervicitis, in which nodular formations were found which were equal in size to the follicles. In the study of patients with ectropion and ectopia, vascularization of the cervix, vesiculation of the glands, and easy bleeding in mechanical irritation was observed. These and other changes met alone or simultaneously in different combinations (Table 11.3).

The treatment is performed with tetracyclines, macrolides, and quinolones on non-uniform schemes according to different publications. Described are treatment in 7-day, 14-day and even 21-day courses in separate stubborn and complicated cases. The CDC [878] provides Recommended

Regimens as follows: Azithromycin 1 g orally in single dose or Doxycycline 100 mg orally twice daily for 7 days; the alternative regimes include: Erythromycin base 500 mg orally four times a day for 7 days or Erythromycin ethyl succinate 800 mg orally four times a day for 7 days, or Levofloxacin 500 mg orally once daily for 7 days, or Ofloxacin 300 mg orally twice a day for 7 days.

11.6.2 Chlamydial Vaginitis

Chlamydial vaginitis (colpitis) is a rare disease. The reason for this lies in the histological structure of the vaginal epithelium – a flat multilayer with a tendency for twitching, which is not a suitable field for the development of chlamydia infection. Chlamydiae here can only be retained in the epithelium of some serous glands located in the vaginal mucosa. According to some authors [875], it can occur with hormonal disorders, accompanied by structural and morphological changes in the mucous membranes. In women with hysterectomy and hormonal predisposition, the isolation of chlamydia and the detection of chlamydial antibodies occur against a background of mixed infection.

Chlamydial colpitis data in a "pure" form is rare. In our practice, we have only observed two cases of chlamydial macular vaginitis without additional flora [24]. Mixed infections with bacterial flora have been reported in a number of publications [879–881]. Gonococci, staphylococci, colibacteria, proteus, anaerobic non-spore-forming bacteria, Candida, and others have been detected. Hanna et al. [881] indicate that the isolation of *C. trachomatis* in mixed infections is more frequent at a higher (alkaline) pH. More often, the secondary process develops during maceration of the vaginal mucosa caused by secretions from the upper genitalia. Its clinical manifestations depend on the infection that has joined [875]. The relationship between Chlamydia and *T. vaginalis* is more particular. These protozoans have the ability to phagocyte chlamydia. The combination with *C. trachomatis* makes the trichomonasis macular colpitis more stubborn and lasting [24]. Oriel and Ridgway [235] and Osborne et al. [882] reported chlamydial vaginitis in newborns and children most likely acquired during childbirth in direct contact with infected maternal birth paths. While the latter is an epidemiological conception, the explanation for child susceptibility seems to be related to the peculiarity that the vagina of individuals before puberty has an unformulated flat epithelium that is more vulnerable to chlamydial infection.

11.6.3 Chlamydial Urethritis, Urethral Syndrome, and Paraurethritis in Women

In the etiologically proven cases of urethritis, the clinical picture is characterized by dysuria, itching, urination soreness, frequent urgent need for urination, urethral discomfort, and small colorless urethral discharge. Paavonen [883] examined 99 women having sex with men with proven chlamydial urethritis. The agent was recovered from the cervix and urethra in 46 women, from the cervix only in 28, and from the urethra only in 25 women. Urethral symptoms were reported by 38 (54%) of those with Chlamydia-positive urethral cultures. These data have resumed the notion that *C. trachomatis* may be the cause of the so-called urethral syndrome. We will note that it is logical to suggest that the cylindrical urethral epithelium, which is rich in glycogen and is well suited to the development of chlamydia, is affected by this infection. Characteristic signs of urethral syndrome are dysuria, urinary burning, and a feeling of uncompleted urination often associated with sterile pyuria. Lower abdominal pain is also registered along with a feeling of pressure in the abdomen, pain during sex, and blood in the urine. Urethral syndrome can also cause discomfort in the vulvar area. It is seen that many of the clinical signs of urethral syndrome are covered with those of chlamydial urethritis. By its nature, patients' complaints are chronic-intermittent. In some cases, isolation of chlamydia is difficult, but in most patients is established dynamics of specific antibodies in MIFT [24].

Shatkin and Mavrov [92], Stamm et al. [884] have observed urethritis with involvement of the paraurethral glands, whereupon purulent exudates are released. According to other data [875], paraurethritis often develops with hyperemia of the mouths of the paraurethral ducts, and mucopurulent contents squeezed out in the form of a drop. Furthermore, the vestibule of the vagina is also involved in the process, which is manifested by the itching and burning of the vulva, and the release of mucus.

Evidence in the literature suggests that urethral syndrome and paraurethritis are very difficult to clinically treat. In many of the partners of women suffering from these diseases have been isolated chlamydiae [884, 885]. The differential diagnosis includes infections with Ureaplasma or Mycoplasma, as well as atrophic urethritis and urethral trauma [886].

11.6.4 Chlamydial Bartholinitis

Davies et al. [887] first identified bartholinitis etiologically associated with *C. trachomatis* by isolation of the agent from 9 women. In fact, only one of the patients had a single chlamydial infection, while the others had gonococcus at the same time. Shatkin and Mavrov [92] also reported chlamydial bartholinitis, as in the majority of cases a mixed infection was also detected. In our practice we had a case of acute chlamydial bartholinitis combined with Proteus. The same patient had chlamydial endocervicitis and cervical erosion [24]. Self etiological involvement of *C. trachomatis* in bartholinitis was announced by Bleker et al. [888] and Garutti et al. [890].

The inflammation of the Bartholin's gland is favored by some of its anatomical and histological features. The glandular capsule with its location, narrow canal, and cylindrical epithelium predispose it to the development of chlamydiae and other microorganisms [24]. In other publications, it is stressed that Bartholin's gland and ducts are lined with columnar epithelium, making them susceptible to infection with *C. trachomatis*. It has been estimated that 30% of the infection of this gland is abscesses initiated by chlamydial infection, although absolute contribution is not known [886–889, 891]. Usually bartholinitis is an acute unilateral inflammatory process. Acute onset with vulvar pain and swelling is noted, which becomes more intense as the abscess expands. The gland is swollen, flushed and painful. In some cases, it has a general malaise. After maturation of abscess, pus runs through the channel or through the resulting fistula. The inflammatory process tends to be chronic, with frequent exacerbations and no tendency to self-healing. In some patients the infection is more moderate and even mild. The treatment consists of complete gland extirpation, while anti-chlamydial and antibacterial treatment is simultaneously performed.

11.6.5 Chlamydial Endometritis

Chlamydia trachomatis affects the uterus of women. Endometritis develops inflammation of the endometrial lining of the uterus. This is a serious disease of the female genital apparatus. The importance of endometritis for infertility in women is undoubtedly proven.

In addition to the endometrium, inflammation may involve the myometrium and, occasionally, the parametrium [892]. Mardh et al. [902] considered endometritis as common component of pelvic inflammatory disease (PID). Cervicitis is also found in a significant proportion of patients with endometritis. It is logical to suppose that the inflammation of the endometrium is the result of an ascending infection. Paavonen et al. [893] conducted cervical examinations for 35 women and in addition made uterine trans-cervical biopsies. As a result, they have established severe endometrial involvement and irregular inter-menstrual uterine bleeding. In 40% of patients, elevated antibody titers against *C. trachomatis* were found. Chlamydial inclusions can be found in the cells of the endometrium and plasma cells during the follicular or ovulatory phases of the menstrual cycle [896, 897]. In a number of publications, endometritis has been reported in infertile women with etiologically proven chlamydial infection [893–895].

The clinical picture of chlamydial endometritis is conditionally subdivided into acute and chronic. In the acute form, the patients have discomfort and low pelvic pain, fever, altered blood count and other disturbances. In acute endometritis, occurring independently without salpingitis, irregular uterine bleeding, central lower abdominal pain and elevated body temperature were observed [898]. In chronic endometritis, symptoms may be limited only to non-constant mucous purulent or urinary tract and inter-menstrual bleeding [24]. Other authors point out that women with chronic endometritis may not experience irregular bleeding, but instead report menorrhagia [899, 900]. In this regard, Jones et al. [901] states that patients with acute, sub-chronic, or chronic endometritis may have menorrhagia or irregular uterine bleeding without having evidence of salpingitis.

11.6.6 Chlamydial Salpingitis

Salpingitis is an infection and inflammation in the fallopian tubes. Chlamydial salpingitis is a disease that affects women's generic functions. The major systemic complication of chlamydial infection in women is acute salpingitis, which may result in long-term consequences such as infertility and/or ectopic pregnancy [903–905]. Convincing data from Scandinavian, British, and American authors suggest that the chlamydial inflammation of the fallopian tubes is the most common acute salpingitis. One of the frequently used methods is the diagnostic laparoscopy that displaces aspirates from tubes. The latter are used to isolate chlamydia [904, 906, 907]. Chlamydial salpingitis has a varied clinical picture and most often starts acutely with fever and pain in the small pelvis [908]. Svensson et al. [910] indicate that the chlamydia-infected patient has generally had abdominal pain for a longer period of time than other salpingitis patients and is less often febrile. In addition, the erithrocyte sedimentation rate is usually higher in comparison to salinititis with other etiology. [909].

It is interesting to note that laparoscopic findings are often more severe than could be expected from the relatively benign clinical course of chlamydial salpingitis [909].

Acute disease is followed by a chronic process. Adhesions, obliterations and tubular cysts are obtained. These pathological changes are a major cause of female infertility. Very common signs are vaginal discharge, irregular menstrual cycle, and inter-menstrual bleeding [235].

11.6.7 Pelvic Inflammatory Disease

Pelvic inflammatory disease (PID) is a concept that covers a number of inflammatory changes in the pelvic organs. This includes the upper part of the female reproductive system – the uterus, fallopian tubes, and ovaries, and inside the pelvis [911, 912]. It is usually caused by a sexually transmitted infections (STIs), like chlamydia or gonorrhea. Paavonen et al. [913] considers the PID as a polymicrobial infection in which *C. trachomatis* plays a significant role.

The defeat of the appendages of the uterus (tubes and ovary) is observed during generalization of the process from the lower divisions. Chlamydia affects the internal surfaces of the tubes, the outer shell of the ovaries, possibly with the transition to the parietal peritoneum [875].

This disease occurs in various forms, but most often begins as acute or sub-acute pelvioperitonitis. Forslin et al. [914] point to the following PID criteria: (A) weight and sensitivity in the lower pelvic region with or without a palpating mass finding; (B) vaginal discharge; (C) menstrual disorders (D) dysuria; (E) dysparrenuria; (F) a temperature higher than 38°C; (F) elevated erythrocyte sedimentation rate; and (G) leukocytosis (10,000 leukocytes). In a number of other publications, the clinical picture of PID is characterized by identical and other signs: cervical motion tenderness; adnexal tenderness; abnormal cervical mucopurulent discharge or cervical friability; pain with intercourse; painful and frequent urination; elevated C-reactive protein; presence of abundant numbers of WBC on saline microscopy of vaginal fluid [915]; Abnormal uterine bleeding, especially during or after intercourse, or between menstrual cycles; difficult urination [919]; and nausea, and vomiting manifested late in the clinical course of the disease [922].

IgA and IgG antibodies to *C. trachomatis* were found in patients with the above complaints [916, 917]. The chronic form of the disease has found such lesions as tube adhesions, cysts obliteration, and ovarian damage. Chlamydiae are isolated from such cysts [918]. The chronic stage of PID is also referred to as chronic genital-induced pelvioperitonitis, which is a serious disease leading in many cases to chronic sterility due to the large adhesions formed in the small pelvis and genital organs [24]. Along with secondary infertility, PID is a hazardous pathological phenomenon for women's health. The most serious complications of the disease, which are associated with *C. trachomatis* infection, are spontaneous abortions, stillbirths, dysmenorrhea, sterility, disseminating infection in other organs, chronic obliterating endometritis, salpingitis and oophoritis, Douglas space inflammation, cysts in the tubes and ovaries, severe adhesions of the peritoneum with the intestine and rectum, ectopic pregnancy, tubo-ovarian abscess, and chronic pelvic pain [24, 919–921].

11.6.8 Fitz-Hugh-Curtis Syndrome (FHCS)

The synonyms of the disease name are Gonococcal Perihepatitis and Perihepatitis Syndrome. The disease is named after Thomas Fitz-Hugh, Jr [926] and Arthur Hale Curtis [927], who first reported this condition in 1934 and 1930 respectively. Originally it was found that the disease was caused by gonococcus. Later it became clear that *C. trachomatis* also causes FHC as prevailing observations of more frequent cases of the syndrome etiologically associated with chlamydiae [933, 934].

The disease is a rare disorder that occurs almost exclusively in women. It is characterized by the inflammation of the membrane lining the stomach (peritoneum) and the tissues surrounding the liver (perihepatitis) [923]. The diaphragm, may also be affected. Although the FHCS is considered to be a stylized nosological unit, its boundaries are not quite clear. Some authors consider the FHCS as a component or consequence of PID, while others treat only the acute perihepatitis. There is also a third opinion, namely that the disease includes both acute and chronic perihepatitis. Although FHCS is primarily associated with PID, there are cases when it comes down to an independent clinical picture [24].

Regardless of the discussions about the nature of this syndrome, its clinical characteristic is clearly outlined. Acute perihepatitis (FHCS) is characterized by a sudden onset of severe soreness in the right upper quadrant of the abdominal area, accompanied by a sense of tension, palpitant sensitivity, and fever. Marked are more headaches, chills, night sweats, vomiting, nausea, malaise. Pain may spread to additional areas including the right shoulder and the inside of the right arm. [24, 923]. FHCS often mimics a biliary crisis (acute cholecystitis) [924, 928] or pancreatitis. It is possible that the infection spreads to the surrounding area, leading to the development of peritoneal adhesive-type processes in the peritoneum located on the front wall of the liver. As a result, Treichi may form encapsulated pus foci between the capsule of the liver and lig. [925].

Diagnostic tests for FHCS over the years have involved various methods – cytology, serology, infectious agent isolation, laparoscopy, imaging, paraclinical studies. Wolner-Hansen et al. [929] used laparoscopy and found *C. trachomatis* in 31 of 52 women (59.6%) with complaints in the upper right abdominal quadrant. In patients with acute hepatic pain and other symptoms characteristic of FHC syndrome, Darougar et al. [930] isolates chlamydia from various sites of the genital system and found high titers of IgG, IgM, and IgA immunoglobulins against *C. trachomatis*. Puolakkainen et al. [931] used a set of methods. High titers of chlamydial IgG were detected in 88–96% of patients by ELISA and 50% by CFT and also 50% by Chlamydia membranes ELISA. Seroconversion was found in 44% of patients The authors recommend this diagnostic approach as convenient for screening for complaints directed to the FHC syndrome, thus avoiding laparoscopy [931]. Paavonen et al. [932] isolate chlamydiae, respectively, in 40% and 78% of different parts of the genital tract in female patients with suspected acute perihepatitis. At the same time, this research group highlights the great diagnostic value of serological research due to the considerable difficulty in trying to isolate chlamydia from the peri-hepatic space.

In recent years, several publications have indicated the usefulness of computed tomography (CT) in the clinical diagnosis of FHC syndrome [928, 935, 936]. Woo et al. [928] suggests instead of abdominal zonography to use abdominal CT to facilitate diagnosis. According to Wang et al. [936], the dynamic enhanced multi-slice computed tomography (MSCT) can accurately display liver capsule lesions and possible pelvic inflammatory diseases related to FHCS, suggesting the infection source, and have a high application value for making early, accurate diagnoses and improved prognosis.

Fitz-Hugh-Curtis syndrome is treated with antibiotics, given by intravenous injection or as medication taken by mouth. If pain continues after treatment with antibiotics, surgery may be done to remove bands of tissue (adhesions) that connect the liver to the abdominal wall and cause pain in the affected individuals [925, 937].

Differential diagnosis should exclude a number of disease states that may mimic FHCS [936]: Ectopic pregnancy; pyelonephritis; cholecystitis; viaral hepatitis; renal colic; and pulmonary embolism.

11.6.9 Chlamydia Trachomatis Placental Inflammation

Chlamydial placental inflammation has been identified by many authors. Pankuch et al. [938] treat *C. trachomatis*, among other agents, as a cause of premature rupture of the amniotic envelopes in the development of chronic chorioamnionitis and placenta. In 64% of patients with early tearing of the amniotic envelopes were detected by MIFT high titers of chlamydial antibodies. Rours et al. [939] examined by PCR placental tissues of 304 patients and found *C. trachomatis* DNA in 76 (25%) of women who had early preterm delivery ($\leq$32 weeks) and was associated with histopathological signs of placental inflammation. The authors consider that the data presented by them to be highly suggestive of *C. trachomatis* being a component of premature delivery through a pathway that involves a trans-membrane or trans-placental route of transmission for chlamydial infection leading to chorioamnionitis [939]. The role of *C. trachomatis* for the occurrence of chorioamnionitis and placentitis and various associated forms of adverse pregnancy outcome are discussed in a number of other publications [24, 940–943].

When considering the above question, useful comparisons and analogies can be made with *C. abortus* in ruminants that cause placentitis leading to frequent consequences such as abortions and related conditions. The same applies to the abortogenic potential of *C. abortus* of sheep and goat origin for women.

11.6.10 Reiter's Syndrome and Reactive Arthritis

In classical course, the Reiter's syndrome (RS) is characterized by a triad of signs – genital, articular, and ocular. In a number of patients it was supplemented by skin-mucosal lesions. A significant part of the cases were manifested by an incomplete clinical picture. Most often it was represented by two signs – urethritis and arthritis, rarely conjunctivitis and arthritis or other combination [20, 58, 944–947].

In one of our studies of 60 men with a proven *C. trachomatis* infection, genital lesions consisted of 96.6%, the joints – 93.3%, the eye manifestations – 66.7% and the mucous membranes – 21.6%. One patient showed five signs, five showed four signs, another 25, three, and the others showed two symptoms, mainly urethritis and arthritis [58].

Urethritis when RS was NGU and PGU, acute or chronic, or a separate complicated with prostatitis. Women had cervicitis and the ratio of the four groups of signs was basically similar.

Table 11.4 Infectious agents associated with reactive arthritis in 108 patients

Type of Microorganism	Positive Cases	
	Number	%
Chlamydia trachomatis	32	30
Yersinia enterocolitica	30	28
Staphylococcus	7	6
Shigella flexneri	4	4
Streptococcus	3	3
Neisseria gonorrhoea	2	2
Chlamydia psittaci (Avian chlamydiosis)	1	1
Salmonella	1	1
Proteus mirabilis	1	1
Unidentified	27	24
Total	108	100

Most patients experienced asymmetric involvement of large joints – ankle (69%), knee (65%), and metatarsophagangal (26%). Often, the joint of the sacrum is also affected. In some cases, we found plantar fasciitis.

In another of our studies [948], 108 patients from the Department of Rheumatology at the Research Institute of Internal Medicine of Sofia with reactive arthritis were subject to comparative etiologic and clinical studies (Table 11.4).

The table shows that *C. trachomatis* accounts for about 30% of the infectious agents, followed by *E. enterocolotica* (26%), while the remaining micro-organisms have a much smaller share.

In Table 11.5 there is some data on the clinical course of reactive arthritis associated with chlamydia and other causative agents. Of the 32 patients affected by *C. trachomatis* infection, four were with monoarthritis, 10 with oligoarthritis and 18 with polyarthritis. 22 cases were acute or sub-acute, and 10 were chronic.

Table 11.5 Comparative clinical data on reactive arthritis etiologically associated with chlamydiae and other microorganisms

Clinical Symptoms	No.	%	Chlamydia n* = 32	Yersinia n = 30	Shigella n = 4	Staphylococci n = 7	Streptococci n = 3	Unidentified n = 27
Monoarthritis	10	9	4	2	–	–	–	4
Oligoarthritis	52	48	10	18	2	2	1	17
Polyarthritis	46	33	18	10	2	5	2	6

*n – Number of patients

Table 11.6 Arthrological signs in separate reactive arthritis subgroups

Signs	PR-ReA		UG-ReA		EC-ReA	
	No.	%	No.	%	No.	%
Joint number	2.6	± 3.8	3.1	± 1.2	3.0	± 1.6
Upper limb arthritis	8	25.8	5	35.7	2	28.6
Lower limb arthritis	29	93.5	14	100	7	100
Sacroiilitis	5	16.1	5	35.7	1	20
Epicondilitis	1	3.2	–	0	1	14.3
Achilitis	8	25.8	8	57.1	2	28.6
Talalgia	8	25.8	10	71.4	–	0
Plantar fasciitis	4	12.9	6	42.8	–	0

We note that the majority of patients are found to have HLA B27 antigen. For example, in three of our targeted trials ranging from 48 to 85 patients with chlamydia-dependent RS, there were HLA B27 on average for men and women between 80.9% and 85.45%. The ocular manifestations of RS were expressed in conjunctivitis, keratitis, iridocyclitis or uveitis. Skin-mucous signs were psoriasiform changes in the skin of the hands and feet of the patients [58, 944, 945].

Subject to the studies of our other research group – Dimov et al. [589, 590] were 67 patients with reactive arthritis aged 17–45 years: 31 after acute respiratory infection (PR-ReA), 14 after urogenital infection (UG-ReA), seven after enterocolitic infections (EC-ReA) 15 after unidentified infections. PR-ReAs were etiologically associated primarily with *C. pneumoniae* and UG-ReAs with *C. trachomatis*. With regard to HLA-antigens, significant association of PR-ReA was found only with B40, whereas UG-ReA and EC-ReA correlated with B27. In terms of the general and extraarticular signs, in PR-ReA was found rarer affection of eyes, mucous membranes, and urogenital tract. This was reflected in the smaller proportion of the RS in this group of reactive arthritis. As regards the changes in the conventional blood and synovial fluid indicators, no statistically significant differences between the separate ReA subgroups were established.

The arthrological signs in the separate ReA subgroup are presented in Table 11.6. The less frequent involvement of the axial and periarticular structures in PR-ReA than in the other subgroups is seen.

Treatment was conducted according to NSAID principles. In some cases, it was necessary to add other drugs: corticosteroids (intra-articular and/or orally in low doses), antibiotics, or sulfasalazine. Note that additional drugs were used more rarely in PR-ReA, than in the UG-ReA and EC-ReA groups.

This fact testifies to the milder and less complicated clinical course of post-respiratory reactive arthritis, etiologically associated with *C. pneumoniae* [589, 590].

11.6.11 Infection with *C. trachomatis* in Newborn Children

In three children, a contingent of the First City Hospital – Sofia, we proved *C. trachomatis* as an etiological agent of conjunctivitis and pneumonia (N.T.P., 14 days old); conjunctivitis and nasopharyngitis (B.V.P, 16 days old); and tracheobronchitis and pneumonia (M.B.B, 2 months and 25 days old). It was about unilateral or bilateral serous conjunctivitis with severe redness and yellowish secretion, small-focal left-sided pneumonia (one case), and bronchopneumonia in the right lung base (one case) [59].

In two severe fatal cases of pneumonia-affected children aged 2 to 3 months (Pirogov Institute for Emergency Care), we proved *C. trachomatis* by DEM of pulmonary necropsy (see Chapter 3). Clinical lethal outcome has been characterized as "viral pneumonia" without conducting etiological studies for possible viral involvement [59].

11.6.12 Use of Immuno-stimulator Urostim in Chronic Recurrent Andrologic Infections with Chlamydial Etiology

Urostim (BulBio – Sofia, Bulgaria) is a polybacterial immunomodulator for the prevention and treatment of inflammatory diseases of the genitourinary system. It contains killed bacterial cells of four bacterial species: *Escherichia coli, Klebsiella pneumoniae, Proteus mirabilis*, and *Entrococcus faecalis* – the most common causes of uroinfections. It is intended for immunoprophylaxis and immunotherapy of acute, recurrent or chronic diseases of the urogenital tract in children and adults, irrespective of the type of the causative agent: cystitis, pyelonephritis, urethritis, prostatitis, asymptomatic bacteriuria, etc., caused by different bacterial species [949].

Our study group (Uzunova et al.) aimed to test this product for chlamydial infections in men [950]. The Urostim was administered orally with one tablet of 50 mg for 2 to 4 consecutive months in 20 patients with chronic prostatitis, chronic exacerbated urethritis, and chronic prostatitis. Along with these patients, another 50 men with the same clinical diagnoses, but with bacterial etiology (*E. coli*, Enterococcus, Proteus, etc.) were treated as well as a control group of 20 patients with chronic androgenic infections treated with antibiotics and chemotherapeutics alone. In all, 20 patients with *C. trachomatis*, combination therapy (Urostim + antibiotics) resulted in a significant reduction of the serologic response to the chlamydial infection and ultimately, negative result. A significant effect of application of the preparation was also found in subjective complaints of patients: burning after urination or sexual contact; Weight in the perineum; Recovery of potency and, last but not least, the impact on their mental state [950].

11.7 Infections Caused by Chlamydia Pneumoniae

11.7.1 Atypical Pneumonia

Our observations on a large number of etiologically proven cases show that four major clinical forms are revealed: moderate, severe, oligo-symptomatic, and asymptomatic. The majority of patients had a clinical picture with subacute onset, pharyngitis, mild-to-moderate fever, unproductive cough, mild leukocytosis, weak chest pain, accelerated ESR. In the severe form requiring hospitalization and intensive treatment, the overall condition was worsened: hypoxemia, difficulty breathing, productive cough, headache, myalgia. The oligosymptomatic cases have been manifested with sub-febrility, rapid tiredness, fatigue and discomfort for prolonged periods. Oligosymptomatic cases have been manifested with sub-febrility, rapid tiredness, fatigue and discomfort for prolonged periods.

11.7.2 Chlamydia Pneumoniae and Asthma

There are studies that have shown that *C. pneumoniae* can be the cause of asthma [951–954]. Other data suggest that *C. pneumoniae* plays a role in the exacerbation of bronchitis or bronchial asthma in children [955] and in adults [956–958].

Based on the literature, we conducted targeted clinical and etiological studies in which we found that *C. pneumoniae* may be the cause of asthma, can exacerbate asthma, and/or cause asthma symptoms to persist. The study covered 20 patients with asthma, of which 5 with newly discovered disease (group 1) and 21 with severe non-atopic asthma – mild persistent and medium-heavy persistent (group 2). Ten healthy, non-asthmatic controls (Table 11.7) were used for the control group [586, 587].

We investigated the titer of specific IgG antibodies against C. *pneumoniae* by indirect solid phase enzyme immunoassay (EIA), and received data on acute or chronic chlamydial infection in 12 patients. We conducted

Table 11.7 Characteristics of patients studied for *Chlamydia pneumoniae*

Indicators	Newly Discovered Asthma	Chronic Asthma	Controls
Number	5	21	10
Age	18–48 years	33–64 years	25–52 years
Sex	3 males + 2 females	8 males + 13 females	6 females + 4 males
Smoking:			
active smokers	3	3	7
former smokers	0	10	2
no smoking	2	8	0
Steroids:	5	21	0
inhaled	2	15	0
systemic	3	6	0

macrolide treatment in 8 patients and in other 4 with fluoroquinolones. Corticosteroids were administered as follows: systemic Urbason 40 mg at 7 and inhaled in 5 patients. The results of therapy and clinical follow-up for a period of 4–6 weeks to 12 months showed that 11 out of 12 patients had asthma remission and 14% spirometry improvement; improvement but still-persisting asthma symptoms was recorded in one patient. All patients after the antibiotic treatment gradually serologically depleted for antibodies against *C. pneumoniae*.

The results presented lead to two conclusions:

- *C. pneumoniae* may possibly trigger asthma and lead to exacerbation and/or persistence of asthma symptoms;
- When the asthma is tightened, the serology test for chlamydia is considered and in case of positive samples, adequate antibiotic therapy is included.

11.8 Clinical Manifestations of Chlamydial Infections in Other Animal Species

Koalas. Chlamydial diseases among Australian koalas are a serious problem because of the threatening range of infection and marked clinical manifestations that in some cases have fatal consequences. This determines the need for scientific research to develop a vaccine against the disease. Etiologically, they are associated with *C. pecorum* and *C. pneumoniae*. Later, by direct sequencing of the nucleic acid fragments, termed "uncultured koala Chlamydiales" (UKC) the presence of non-*C. pecorum*, non-*C. pneumoniae* strains of novel chlamydia-like organisms has been confirmed. Usually UKC has been found as co-infection with *C. pecorum* or *C. pneumoniae* [1084, 1085].

Studies in Australia show that the level of clinical disease is 17%, whereas the clamydial infection status is 83%. It confirms that the infection commonly occurs without visible clinical manifestations [1084]. The clinical spectrum of chlamydial infection in the koalas is too broad. It includes the following symptoms: (a) ocular – unilateral or bilateral keratoconjunctivitis [1086]; (b) Urinary tract infection (urethritis, cystitis, ureteritis and nephritis, hydroureter, hydronephrosis, pyogranulomatous, and chronic interstitial nephritis) [1087–1089]; (c) Reproductive tract infection (Cystic enlargement of the ovarian bursae; salpingitis; hydrosalpingitis; metritis; pyometra; vaginitis; pyovagina [1087, 1090]; and prostatitis [1089]. Affected females are often rendered infertile. Atypical clinical presentations may include throat pain, dysphagia, inflammation of the patellar and carpal tendons, shifting lameness.

Reptilies. Chlamydial infections have been reported in various reptiles including Puff Adders (*Bitis arietans*), Emerald Tree Boas (*Corralus caninus*), Nile Crocodiles (*Crocodylus niloticus*), etc. [1091]. Necrotizing myocarditis, pneumonia, enteritis, and acute hepatitis have been described. Jacobson et al. [1092] report the death of six captive-born puff adders (*Bitis arietans*) housed together. The snakes died within 4 months of acquisition. All snakes occasionally regurgitated the mice within 2 days of feeding. One of these snakes manifests a mild respiratory disease prior to death. At the necropsy, all snakes had exudate within the pericardial sac and two serpents had multifocal white nodules in their fetuses. Histological findings include pneumonia, granulomatous peri- and myocarditis, splenitis, hepatitis, and enteritis with basophilic inclusion bodies of various sizes within the caseated centers of the granulomas. Through direct EM, the observed inclusions are identified as chlamydial with typical ultra-structural features. Although the authors did not isolate and antigenically differentiate the agent, they hypothesized that it is *C. psittaci* [1092]. Soldati et al. [1093] received etiological data by immunohistochemistry with monoclonal antibodies against chlamydial lipopolysaccharide (LPS) and a Chlamydiales order-specific PCR and sequencing for *C. pneumoniae* infection in snakes, chelonians, and lizards with granulomatous inflammation. Chlamydia pneumoniae has been also identified in chameleons and iguanas [1094]. These finds suggest that the reptiles have the potential for zoonotic transmission to humans [1094]. Taylor-Brown et al. [1095] reported a disease caused by Chlamydia pneumoniae in both free-range and captive snakes. There are molecular evidence suggesting that additional novel C. pneumoniae-like strains may also be present [1095]. The infection usually manifests as granulomatous inflammation in inner organs such as spleen, heart, lung, and liver but might also occur in asymptomatic reptiles. This study definitely testifies again for the broad host range for C. pneumoniae. The authors emphasize that reptiles may still contain a significant and largely uncharacterized level of chlamydial genetic diversity that requires further investigation [1095].

Amphibians. Chlamydial infections are found among both wild and captive populations of the anurans, including the African clawed frog (*Xenopus sp*), the giant barred frog (*Mixophyes iteratus*), the Gunther's triangle frog (*Ceratobatrachus guentheri*), and in the European common frog (*Rana temporaria*) [265, 267, 1094, 1102–1104]. Initially as an etiological agent, *C. psittaci* and *C. pneumoniae* were identified. Later, the participation of *C. abortus* and *C. suis* was also revealed. For diagnosis, pathogen isolation was used in CC, immuno-fluorescence, immunohistochemistry, LM, EM, PCR, and histopathology for the detection of intracytoplasmic chlamydial basophilic inclusions in hepatocytes.

Clinical signs include abdominal swelling due to hydrocoeloma, petechiation and sloughing of the skin, skin depigmentation, accumulation of excess fluid in lymphatic sacs, edema, dysequilibrium, and lethargy [1104]. Interstitial pneumonia [265] is also described. In some cases, infected frogs may die peracutely. Blumer et al. [1103] reported subclinical infections by *C. abortus* and *C. suis* in captive African clawed frogs (*Xenopus laevis*) and free-ranging European common frogs in Switzerland. Whitaker [1104] emphasizes that *Chlamydia spp* has also been found in apparently healthy frogs, which raises the question of whether these animals are a reservoir or vector for these infectious organisms. Reed et al. [267] described an epizootic of *C. pneumoniae* infection in a breeding colony of African clawed frogs (*Xenopus tropicalis*), imported into the United States from western Africa, where 90% mortality was recorded. Etiological diagnosis is made by isolation of the agent and by direct detection (EM, LM) in the liver. In this epizootic, however, two frogs have also been found to have cutaneous infection by a chytridiomycete fungus, which suggests that fungal infection may have played the role of a cofactor and contributed to high mortality.

Infected frogs show enlargement of the liver, spleen and kidneys, and the microscopic find is characterized by marked histiocytic or granulomatous inflammation in these organs as well as in the heart and pulmonary interstitium [1102]. Important findings in the liver and spleen are the abovc-mentioned chlamydial intracytoplasmic basophilic inclusions in sinusoidal lining cells of the liver and spleen.

In terms of differential diagnosis, red leg syndrome and ranaviral infections should be considered too similar in a clinical setting, so the etiological diagnosis is crucial.

Treatment is with tetracycline antibiotics. Oral administration of doxycycline (5–10 mg/kg/day) or oxytetracycline (50 mg/kg/day) may be effective against chlamydial infection [1104, 1105].

Current knowledge of epidemiology and prophylaxis of *C. pneumoniae* infection in frogs is insufficient, so people who have contact with this species should be careful to avoid any possible zoonotic risk.

The above and a number of other facts certainly indicate that *C. pneumoniae* is genetically diverse in animals and appears to have crossed the host barrier to humans on some occasions. By analyzing such cases Mitchell et al. [1106] draw attention to the fact that, in addition to infections in humans, *C. pneumoniae* infections have been reported from a range of animals from an equally diverse range of body sites including liver and spleen (frog, iguana, chameleon), respiratory (frog, snake, bandicoot, koala, horse), heart (frog, turtle, snake), conjunctival (koala), and urogenital tract (koala) [265, 1094, 1107].

12

Pathology

12.1 Macroscopic Observations

12.1.1 Domestic Ruminants

In sheep, goats, and cattle with pathology of pregnancy, we observed four major clinical forms that were subject to pathological and histological examinations: abortions, stillbirths, births of non-viable offspring, dying in the next 2 days, and premature births. In these disease states, we found a variable percentage of placental retention (Retentio secundinarum) [966]. Studies on pathological changes in placentas and fetuses in abortions and related conditions in ruminants have always been the focus of researchers in different countries in earlier periods and today [49, 959, 960–968].

In ***chlamydial abortion in sheep***, the aborted fetuses were fresh and well developed, with no signs of maceration and rotting processes. After autopsy, an increased amount of blood transudate was detected in the peritoneal and pleural cavities. The liver had mild to moderate hepatomegaly, and splenomegaly was found in the spleen. Both organs had rounded edges. Mesenteric lymphatic nodes were enlarged. A rare finding was hemorrhagic diathesis in the heart, lungs, and pulmonary pleura. Hyperemia was observed in the kidney and, less frequently, single and pinpoint petechial hemorrhages. In the brain tissue and the meninges are established well distinguishable blood vessels, hyperemia of the meninges and single hemorrhages in the cut surface of the white cerebral matter.

The macroscopic picture described was observed in fresh fetuses thrown out shortly after their death. With longer retention of dead fetuses in the uterus, we found varying degrees of degenerative changes in their parenchymal organs [966].

The most characteristic feature of the disease from a clinical and pathomorphological standpoint is lesions in the placenta. In the vast majority of cases, placentas are hemorrhagic, edematous, thickened, and have a dirty

Figure 12.1 Macroscopic appearance of the placenta in chlamydial abortion in sheep.

red color. The chorion is diffusely thickened and pleated. On its surface, there are white and gray areas of varying size, sometimes merging into wider areas. The necroses are rough, dry and slightly protruding. Cotyledons are hyperemic, edematous, and jagged, with grayish-white deposits (Figure 12.1).

The macroscopic findings described differed in the extent and severity of the damage to the individual affected sheep [49, 333, 965, 966, 984].

12.1.1.1 Chlamydial abortion in goats

In goats with chlamydial abortions, the pathological picture is similar to that of sheep. Placentas and cotyledons are distinguished by macroscopically visible lesions, usually strongly manifested with a full or partial spectrum of the lesions described above. Invariably discolored cotyledons in shades from light brown to dark brown (chocolate) color are notable, while in fetuses, edema of the subcutis.

12.1.1.2 Chlamydial abortion in cows

The pathological anatomical picture in the fetus is characterized by edematous subcutaneous connective tissue, especially in the head area, the presence of pinpoint bleeding on the skin, oral and conjunctival mucosa, upper respiratory tract, esophagus, stomach, pericardium, and thymus. There is also hepatomegaly, lymphadenopathy, and increased quantities of fluid in the abdominal and pleural cavity. Changes in the bovine placenta are similar to those described in small ruminants, but in general they are more moderate.

12.1.1.3 Stillborn lambs, kids, and calves

In stillborn lambs, kids, and calves, we found macroscopically visible changes in the same anatomical locations, but in a variety of combinations and with varying degrees of damage. The same applies to the deliveries of non-viable offspring, where the chlamydial infection is invariably lethal in the next 48 h.

12.1.1.4 Premature births in sheep, goats and cows

Premature births in sheep, goats and cows in late pregnancy were, as a rule, accompanied by pathologic changes in the parenchymatous organs of newborns, which in most cases had a reversible character, and the animals survived with adequate healing actions. Some of the prematurely born animals developed toxic-infectious syndromes and lesions leading to death.

We will emphasize that, as with abortions and other related conditions, pathological changes in placenta and cotyledons were a major and permanent indicator of chlamydial infection both from a pathological and etiological point of view. It is important to note that placental inflammation (placentitis) was detected not only in abortions and similar clinical forms but also in a certain percentage of sheep, goats, and cows that gave birth to normal, fully viable animals. Thus, this clinical and pathological finding was a useful benchmark for etiological detection of the actual extent of lesions in already-infected flocks, or for targeted studies of *C. abortus* in other flocks with unknown status so far.

12.1.1.5 Chlamydial polyarthritis in sheep

Chlamydial polyarthritis in sheep has been macroscopically characterized by joint changes [22, 672]. Tissues in the area around the joints were also affected. The synovial capsules were thickened and the amount of synovial fluid increased. Its color was grayish yellow. The liquid was cloudy and homogeneous in acute cases, and in the advanced cases, it contained fibrin particles and plaques. On the synovial membranes we found tightly adherent fibrin plaque. We have seen more erosions of articular cartilage, joint connections hyperemia, distended tendon sheaths, hyperemia, and edema of the surrounding musculature. These lesions affected single joints (arthritis) or multiple joints (polyarthritis) [22].

12.1.1.6 Chlamydial polyarthritis in cattle

The most frequently and heavily affected large joints are the shoulders, hips, and the atlanto-occipital, tarsal, and carpal joints. They are enlarged and a periarticular edema is observed around them. The synovial capsules are filled with a whitish liquid in large quantities and are subject to pressure fluctuation. Stretching of the tendon sheaths is observed [22, 969, 970]. Findings in other organs could occur in some lambs and calves with polyarthritis – hyperemia, petechiation and edema, swollen lymph nodes, and enlargement of the spleen [979].

12.1.1.7 Chlamydial conjunctivitis in sheep

Chlamydial conjunctivitis in sheep is of a follicular type with a mild, moderate, or severe degree of macroscopic visible changes. Follicular hyperplasia is a characteristic pathological feature of conjunctivitis in sheep etiologically associated with chlamydia [336, 651]. In mild cases, conjunctival mucosal hyperemia was mild or moderate, with or without large areas affected. In advanced cases, we observed strongly manifested hyperemia of the entire conjunctiva, chemosis and numerous, often confluent lymphoid follicles of the lower and the third eyelid. Approximately 10% of cases develop interstitial keratitis with deep vascularization and corneal edema [336].

12.1.1.8 Chlamydial pneumonia in sheep

Dungworth and Cordy [645] and Pavlov [971] classified the chlamydia-induced pneumonia in sheep as an interstitial bronchopneumonia, similar to that described to human chlamydiosis. Since these authors found alveolar epithelialization, the pulmonary lesions were also considered as hyperplastic pneumonia [19]. After autopsy, the lungs are enlarged, with rounded peripheral parts and grayish-white or gray–pink color. Pneumonic changes are located in the ventral areas of the apical parts, and in the later stages of the disease, the dorsal parts are also affected. Bronchial lymph nodes are enlarged [971]. The consolidated zones are slightly elevated and feel solid and lumpy on palpation. The lung lesions have irregular, streaky dark-red bands during the evolving and resolving stages [645, 19]. In chronic (often subclinical) chlamydial infections, macroscopic examination of the respiratory tract reveals only mild lesions or a few foci of atelectasis, predominantly affecting the apical lobes [633]. In our own observations there were highly compacted lesions with dirty yellow and pink color. These tissue sections were slightly raised above the surface of the organ. We found this finding in sheep much more common than in goats affected by chlamydial pneumonia. Depending on the stage of the disease, the appearance of the outer and incisional surfaces of the lungs had a different color shade, from pale gray to pink, in the form of irregular diffuse spots or belts. In almost all cases the regional lymph nodes showed lymphadenitis.

12.1.1.9 Chlamydial respiratory diseases in goats and kids

In macroscopic observations, the lymph nodes were inflamed and hypertrophied (lymphadenitis). We recorded tracheitis, tracheobronchitis and pneumonia. The macroscopic picture of chlamydial pneumonia in goats was similar to that described in sheep. The picture of pulmonary inflammation was characterized by the presence of pneumonic areas of grayish-white, gray-pink or gray-red color in the form of single spots in earlier phases of the disease or as large diffuse zones in advanced cases. Interlobular connective tissue septa were expanded. The lungs were affected unilaterally or bilaterally. The mediastinal lymph nodes in the most cases were enlarged.

In the case of sporadic bovine encephalomyelitis (SBE) hyperemia, edema and increased cerebrospinal fluid are detected macroscopically in the brain [22]. Regular findings are inflammation of the vascular endothelium and mesenchymal tissue, and serositis in the three body cavities [972, 973]. Similar gross lesions can be detected in buffalo chlamydial encephalitis. Hunt et al., 2016 [983] presented clinical and pathological data for *Chlamydia pecorum* encephalitis in calves in New Zealand. Etiological diagnosis was performed by PCR on brain samples. Necropsy findings included fibrinous peritonitis, pleuritis, and pericarditis, with no gross abnormalities visible in the brain or joints. However, histological findings have been obtained, which are discussed below in the section on microscopic observations.

12.1.2 Swine

12.1.2.1 Chlamydial pericarditis

In 12 pigs slaughtered by necessity, the most marked pathological changes we observed were in the heart (Figure 12.2) [47, 334].

After the opening of the pericardial pouch, about 20–30 ml of liquid mixed with fibrin threads (serofibrinous pericarditis) flowed in approximately half of them. The walls of the pericardial envelope were thickened.

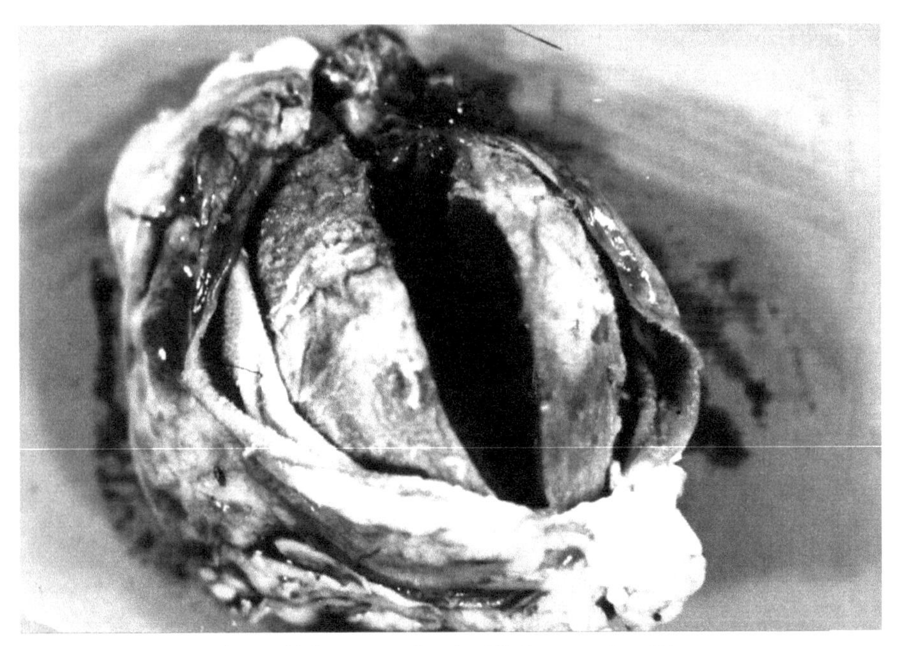

Figure 12.2 Heart of a pig. Fibrinous pericarditis.

The myocardium was hypertrophied. On the epicardium, we found mesh deposits with a grayish-yellow color. In the other cases, we also found an increased amount of pericardial fluid (10–20 ml), but no fibrin deposits were present [47]. Similar macroscopic findings have been identified in two early studies of this disease in pigs [42, 43]. These studies have led to the consideration that a polyserositis appears to be associated with spontaneous chlamydial infection in young pigs [19].

The lungs were without visible macroscopic changes. In six pigs the gastrointestinal mucosa showed a picture of catarrhal inflammation. The liver was light yellow in color [47].

12.1.2.2 Chlamydial vaginitis

Hyperemia, edema, and the presence of follicles (follicular colpitis) was observed in the vagina through colposcopy. The vaginal part of the cervix is inflamed (exo-cervicitis). Bagdonas et al. [335] examined 61 pigs with vaginitis and found that 22.9% of them were etiologically associated with chlamydial infection.

12.1.2.3 Chlamydial epizootic abortion

Macroscopically, subcutaneous hemorrhages were detected in the aborted fetuses in the area of the thymus, nose, pharynx, trachea, mucous membranes of the esophagus, and in the muscle tissue. Exudates were also observed without or with blood impurities in the gastric cavity, thoracic cavity, and pericardial sac. The liver is enlarged, has a brittle texture and small necrotic stains. Spleen and lymph nodes are enlarged. [335]. A number of organs and tissues showed gray areas of 5–10 mm in size. Macroscopic changes in dead non-viable piglets are more pronounced. In the kidneys, the hemorrhages are spotty. Enteritis and meningoencephalitis may also occur [22]. In chlamydia-positive fetal livers, Thoma et al. [974] found mild periportal hepatitis. Käser et al., 2017 [975] conducted experiments with 26-week-old female pigs which were inoculated in standing estrous intra-vaginally and intra-uterinally each with 10^8 inclusion forming units (IFU) of *Chlamydia suis* (Cs) or *Chlamydia trachomatis* (Ct) in sucrose-phosphate-glutamine (SPG) buffer. As a result, uterine horn flushes of 3/5 and 4/5 Cs- and Ct-inoculated pigs were cloudy, indicating an abnormal histotroph with the presence of cells and debris in the upper genital tracts of these pigs. In addition to these findings, the authors established that Cs- and Ct-inoculated pigs also showed significant pathological changes such as congestion throughout the genital tract and low muscle tonus or large vacuoles in the uterine horns. Etiological evidence of successful experimental infections was obtained by qPCR, in which chlamydial DNA was detected in tissue samples of pathological lesions and mucus from Cs- and Ct-inoculated animals indicating that the chlamydia infection was responsible for the pathological changes [975].

12.1.2.4 Chlamydial pneumonia in pigs

Permanent and characteristic macroscopic lesions are found in the lower parts of the cardiac and apical lobules, while the diaphragm lobule is rarely affected. Severe limited compacted inflammatory sections are observed, which are at or below the adjacent normal intact tissue. The color of these pneumonic parts varies from bright red to gray. Sebo-fibrinous inflammation is detected in the pleura and pericardium. Sometimes the peritoneum is also affected [18, 23].

12.1.3 Birds

12.1.3.1 Chlamydiosis (ornithosis) in ducks

A very characteristic finding was the presence of fibrinous inflammatory exudates on the surfaces of the liver, spleen, peritoneum and epicardium.

The air bags were thickened. There were more bleeding and adhesions in the epicardium. The observed changes in the lungs were diffuse inflammation, edema and non-lymphadenitis. Almost all sick ducks had enlarged lungs of greenish color and perihepatitis. About 50% of the ovaries had degenerative changes, atrophy, necroses and bleeding. [53, 579, 580].

12.1.3.2 Chlamydiosis (ornithosis) in hens

The most characteristic were changes in the respiratory tract. Trachea, larynx and air sacs often contained serofibrinous exudate located in the lower third of the organs as well as circular mucosal ulcers. The lungs were hyperemic and edematous. Very pronounced splenomegaly (two to three times enlargement of the spleen) with a "mosaic" of light and dark spots was observed. The liver is slightly enlarged, with rounded edges and yellowish or grayish-brown color. The kidneys are pale and swollen. In some cases, the pericardial sac was filled with gelatin-like material. A common finding was fibrinous pericarditis. [53, 579].

12.1.3.3 Chlamydiosis (ornithosis) in pigeons

The macroscopic picture of pigeons included hyperemia, enlargement and tendency for spleen rupture, hepatomegaly with necrotic foci, enteritis and airsacculitis. In many cases, conjunctivae were inflamed and swollen and the eyelids were covered with crusts [579, 1000].

12.1.3.4 Psittacosis in psittacine birds

A regular finding in acute psittacosis was the enlarged spiced white spleen. The liver was also often enlarged, yellowish and necrotic. In some parrots we observed pericarditis, airsacculitis and enteritis. According to the frequency of detection Mohan [784] rank the major gross lesions in pet parrot birds as follows: splenohepatomegaly, followed by enteritis, sinusitis, airsacculitis, pneumonitis, and pericarditis.

12.1.4 Guinea Pigs

12.1.4.1 Guinea-Pig Inclusion Conjunctivitis (GPIC)

We performed macroscopic observations in natural and experimental chlamydial infection [23, 24]. Permanent morphological features were hemostasis and follicular hyperplasia in the conjunctival mucosa, as well as superficial keratitis. In some cases, the corneal inflammation was more pronounced.

We often found pannus or micro-pannus. Some fibrous particles were found in the front eye of some guinea pigs.

Depending of the infectious dose of *Chlamydia caviae*, the observed ocular macroscopic changes have different intensity. The palpebral and bulbar conjunctivae were evaluated for erythema, edema, and exudation [976]. One the basis of a visual scoring of gross ocular pathology, Filipovich et al., 2017 [976] classified the findings into five categories: (0.5) trace pathologic response, (1) slight erythema or edema of either the palpebral or bulbar conjunctiva, (2) definite erythema or edema of either the palpebral or bulbar conjunctiva, (3) definite erythema or edema of both the palpebral and bulbar conjunctiva, or (4) definite erythema or edema of both the palpebral and bulbar conjunctiva plus the presence of exudate. Possibly there are dose-dependent differences on the immune responses following infection with *C. caviae*. This question should be further investigated in repeated ocular chlamydial infections [976].

12.2 Microscopic Findings

12.2.1 Chlamydial Abortion in Sheep

12.2.1.1 Placentas

In etiologically proven chlamydial infection, necrotic regions containing multiple cell debris of varying sizes are detected histomorphologically in the placenta. Around and under the necrotic areas there were leukocyte deposits with retic nuclei, as well as small hemorrhages (Figure 12.3).

The tissue under the necrosis is strongly thickened and edematous, especially in the perivascular areas, and there are a large number of leukocytes and plasmatic cells with normal or damaged nuclei in the fluid of the edema. The walls of the coronary vessels of the chorion were hyalinized and had a number of leukocytes, and in certain regions they were transformed into broad amorphous unscrupulous fields (Figure 12.4).

Perivascularly, at the base of the villi, leukocyte infiltration and relatively large hemorrhages are detected. There were calcifications between the chorionic villi. Most placentomas were swallowed with erythrocytes. Cotyledons are edematous. Most of their epithelial cells are desquamated. Serofibrinous exudate and calcification of the cristae were detected (Figure 12.5).

Blood vessels are expanded. The adventitial cells of the large vessels were in the process of necrobiosis.

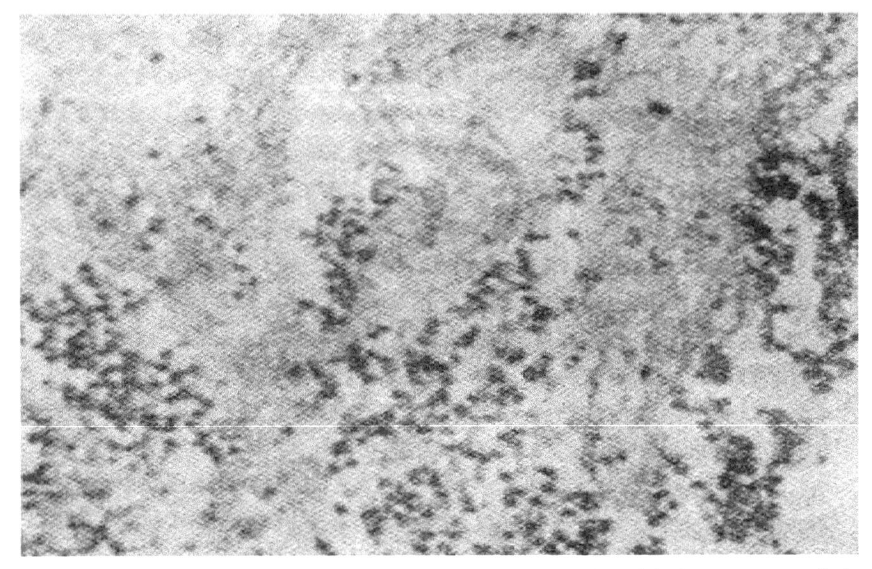

Figure 12.3 Histological section of aborted placenta. Necrosis and leukocyte accumulation is observed. 100× (H. and E. stain).

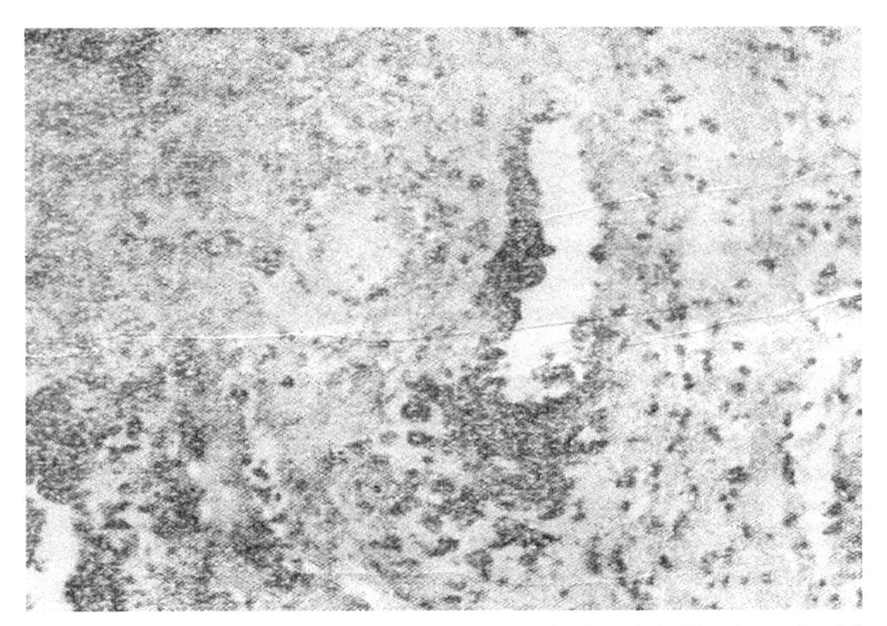

Figure 12.4 Histological section. Placenta – the hyalinization of the blood vessels of the chorion. 50× (H. and E. stain).

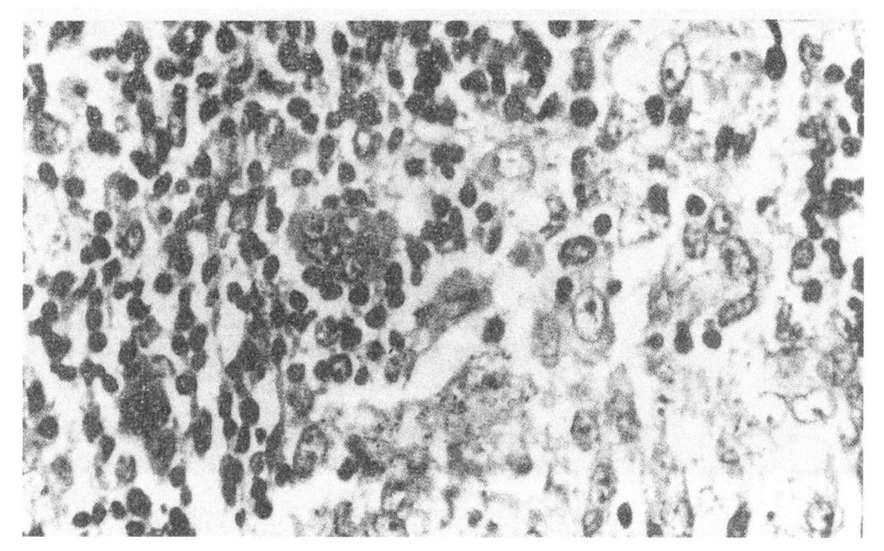

Figure 12.5 Chlamydial abortion in sheep. Histological section. Cotyledon-calcification of the cristae. 200× (H. and E. stain).

Characteristic eosinophilic inclusions of *C. abortus*, colored in red, are observed in the cytoplasm of the epithelial surface layers (Figures 12.6 to 12.8). Especially rich accumulations of chlamydial inclusions are found in the cotyledons.

It is important to note that the intra-cytoplasmic inclusions of the agent are found more frequently, when there is a greater concentration (morphological titer) of the chlamydiae established by DEM [49, 965, 977].

As a model for this correlation, we present a fragment of our work in the study area, including 35 placentas and fetuses (Table 12.1). The table shows that of the 35 studied placentas, 30 (85.72%) are positive for chlamydia by EM. Of these placentas with positive EM findings, 11 strains of chlamydia identified morphologically (EM) and serologically by detection of chlamydial antigen were isolated. The aborted sheep showed seroconversion of chlamydial CF antibodies to a significant percentage. From the placentas, we selected 22 as EM-positive and in histomorphological studies we found that 17 (77.28%) were positive for the presence of *C. abortus* inclusions.

In studies by EM on parenchymal organs from aborted fetuses, chlamydia was found in half of them. The degree of agent isolation from these materials is low, and the indication of eosinophilic chlamydial inclusions by the histological method is problematic (Table 12.1).

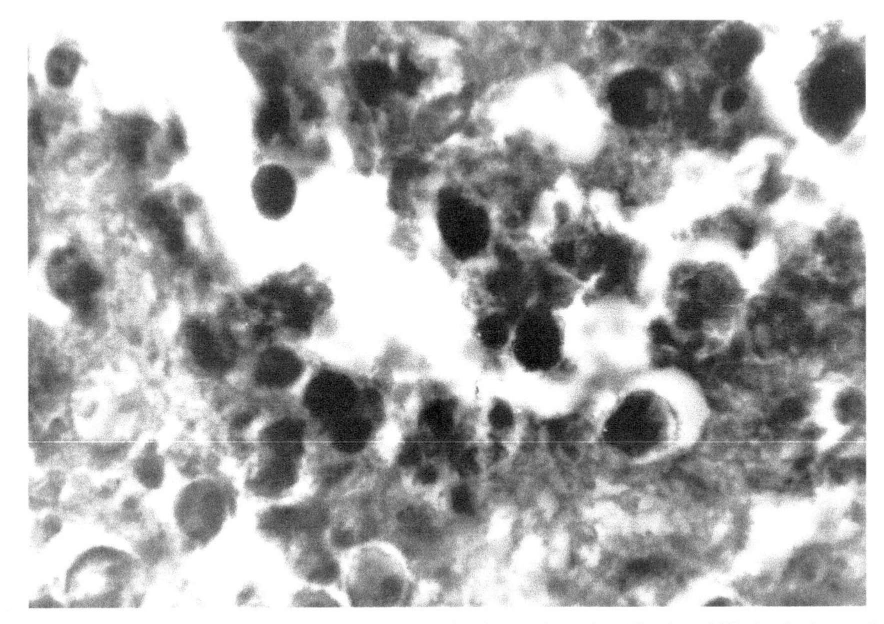

Figure 12.6 Chlamydial abortion in sheep. Histological section. Eosinophilic inclusions of *C. abortus* in the placental epithelial cell cytoplasm. 200× (H. and E. stain).

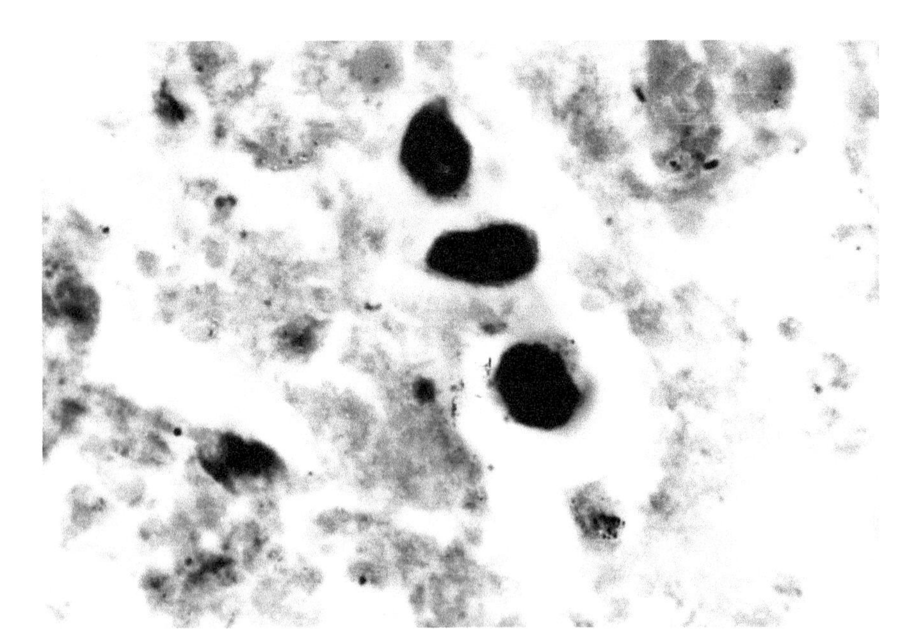

Figure 12.7 Chlamydial abortion in sheep. Histological section. Cotyledon. Intracytoplasmic inclusion of *C. abortus*. 200× (H. and E. stain).

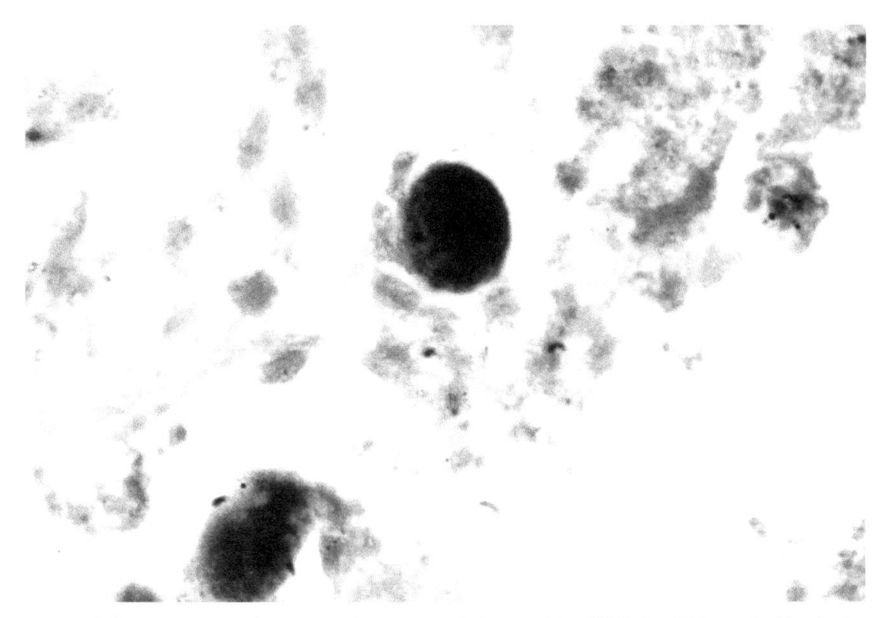

Figure 12.8 Histological section of placenta of sheep with stillbirths. Chlamydial inclusions in the cytoplasm of an epithelial cell. 400× (H. and E. stain).

Table 12.1 Comparative etiological and morphological data on chlamydial miscarriage in sheep

		Etiological Diagnosis			Histopathological Data
Materials	Examined (No.)	EM	CE	CFT	Inclusions of *C. abortus*
Placentas	35	30	11	16	22/17
Fetuses	22	11	4	7	22/0

12.2.1.2 Fetuses

Respiratory tract. The visceral pleura of some fetuses reveal focal thickenings that give it a matte appearance. Pleural capillaries are expanded and filled with erythrocytes. Pulmonary tissue is in the state of atelectasis, but in separate sections the alveolar walls are edematous and infiltrated with lymphocytes, histiocytes, and single leukocytes. Inflammatory infiltration was more pronounced around the bronchi surrounding alveolar tissue and major blood vessels. Endothelial cells are activated and some of them have round nuclei. In the alveoli and desquamated epithelial cells, pigment grains with yellow-green color are deposited.

In the ***liver,*** capillaries are extensively expanded in separate places. There is a diffuse activation of the reticuloendothelium by swelling to the

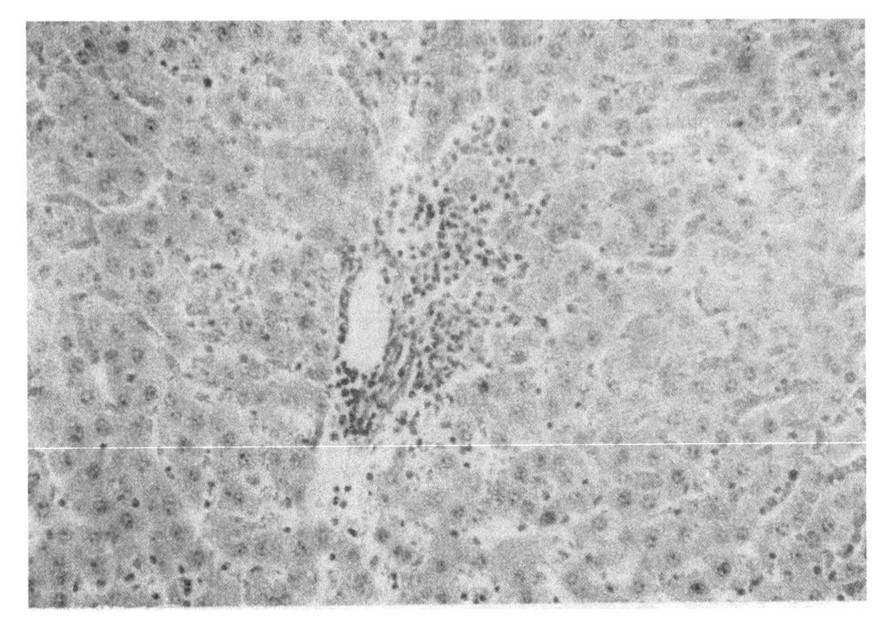

Figure 12.9 Chlamydial abortion in sheep. Liver of aborted fetus. Histological section. Lympho-histiocytic proliferate around an intracellular vein. 100× (H. and E. stain).

capillary lumen and multiplication of the pericytes, thereby forming perivascular and intravascular nodal lymphoid cell and plasmocellular infiltrates. Intralobular nodular connective tissue and histiocytic knots are observed (Figure 12.9).

Interstitial connective tissue proliferates and forms large nodal growths – connective tissue forms with a strong round-vascular lymphocyte infiltration. In the liver, we observed more intracellular necroses, and around them, the growth of single connective tissue cells.

In ***spleens*** observed hemorrhages and in some cases, small strokes. In the centers of some lymph nodes, the cellular elements are watered with a hyaline substance. Among the red pulp we found small globular formations of pink-colored amorphous matter.

In the ***heart*** we observed blood vessel hyperemia, with some areas of the inter-muscular tissue showing small hemorrhages. The myocardium has lymphocyte infiltration.

Kidneys are hyperemic. Some areas of the cortical area have small hemorrhages. Some of the glomeruli are highly atrophied and have desquamated

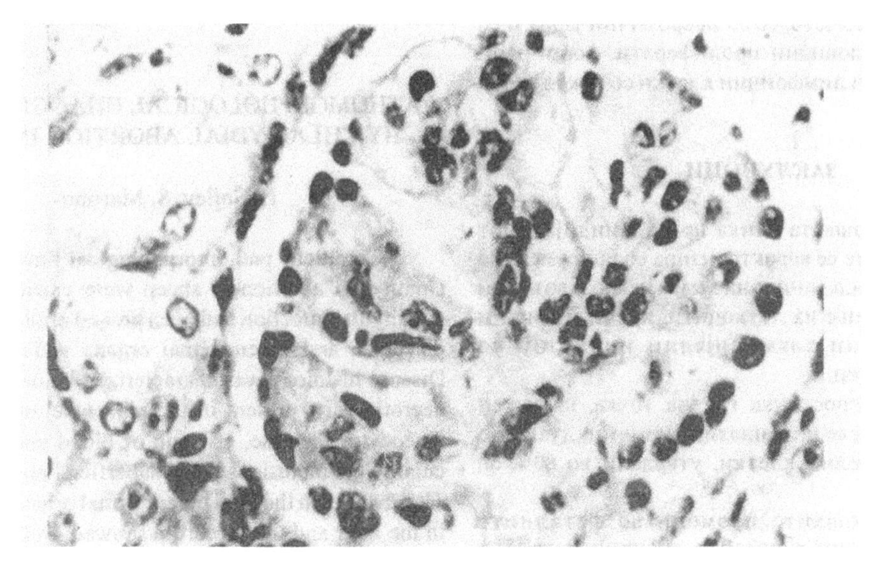

Figure 12.10 Fetal kidney in chlamydial abortion in sheep. Karyorexis, karyolysis, and tubular necrosis. 100× (H. and E. stain).

epithelial cells. The nuclei of some canal cells showed a karyorexis and a karyolysis. Tubular curves are necrotic (Figure 12.10).

Cerebral hyperemia, enlarged blood vessels, and some small perivascular extravasations are found in the ***brain.*** A number of vacuoles (Status spongiosis) are visible around the blood vessels, ganglia, and glia cells. In the white cerebral substance there are perivascular lymphoid cell proliferates and diffuse glial proliferation. Nodular glial proliferates, formed by epithelioid and lymphoid cells as part of the proliferated glial cells at the center of the nodules also show signs of necrobiosis.

Comparison of the etiological evidence of chlamydial infection, predominantly of EM, with histomorphological data indicates that the disease agent is mainly reproduced in the placenta and especially in the cotyledons. At the same time, microscopic changes in a number of organs of the fetus, and especially in the liver and the cerebellum, predominantly indicate a proliferative response to this infection.

Novilla and Jensen [978] observed chlamydial inclusions in the sheep placenta in experimentally induced chlamydial abortion. Our results indicate that the same inclusions are revealed in spontaneous chlamydial abortion in the placenta [49, 965, 977].

Histopathological data correlate well with electron data for inclusion in the placenta. The presence of eosinophilic inclusion in the placenta of sheep with chlamydial abortion may in some cases serve as a diagnostic marker [977].

Our detailed histomorphological picture of chlamydial abortion in sheep has served as a basis for comparison in analogous studies on chlamydial abortions in goats and cows. Essentially, the data obtained are similar to those described for sheep, in the first place to the placenta as a predilection site of infection. In fetal parenchymal organs, the picture is similar with variations in the strength and the expressiveness of the changes, in which the virulence of the strain of the etiological agent obviously plays a role.

In the three types of domestic ruminants, macroscopic and microscopic pictures in placentas and in the fetal organs in stillbirths were virtually identical to those described for abortion.

12.2.2 Chlamydial Polyarthritis in Sheep

Histologically, reactive processes of inflammatory and proliferative type are established. Changes in the synovial membrane are characterized by cell infiltration, edema, focal synovial cell necrosis, and fibrin deposits. Chlamydiae are located extracellularly or in phagocytes where they are apparently destroyed by lysosomal enzymes [980]. According to Shupe and Storz [981], the most common changes involved the joint capsule, synovial membranes, ligaments, tendons and their sheaths, periarticular connective tissue, and muscles. The inflammatory response is consistent with infiltration of plasma cells, lymphocytes, neutrophils, and macrophages. The synovial lining cells are changing. They swell and acquire a hyperplastic appearance. Intra-articularly, the accumulation of fibrin masses containing polymorphonuclear cells, monocytes, and necrotic cellular debris was detected. This leads to fibrotic thickening of the joint capsules and tendon sheaths. Chlamydial inclusions in fibroblasts, monocytes, endothelial, and synovial cells [982] can be found in thin sections of affected joint tissues. Muscles with periarticular position and adjacent to tendinous insertion are affected by infiltration of inflammatory cells and accumulation of edema fluid. Myositis, musculotendon fasciitis, and tendovaginitis develop [981].

12.2.3 Chlamydial Polyarthritis in Calves

The histopathological findings in calves affected by polyarthritis are similar to those described in lambs. They are characterized by peculiar and proliferative changes in the synovial membrane, joint capsule, tendons, ligaments, connective tissue and muscles in this area [22, 981, 989]. In calves, periarticular subcutaneous edema along the tendon sheaths and fluid-filled fluctuating synovial sacs contribute to enlargement of the joints. Most affected joints of lambs or calves contain excessive, grayish-yellow, turbid synovial fluid. Martinov et al. [150, 969] reported isolation of chlamydial strains from synovial fluid of calves affected by polyarthritis from six farms in several districts of Bulgaria. These data confirm earlier isolations of chlamydiae from synovial fluids of polyarthritic calves in the United States [19]. Tendon sheaths of severe cases may be distended and contain a creamy, grayish-yellow exudate. The surrounding muscles are hyperemic and edematous, with petechiae in their associated fascial planes [672]. In chronic cases, erosion of the articular cartilage and fibrotic thickening of the articular capsule and the tendon sheath can be observed, accompanied by severe inflammation and hyperplasia of synovial villi [981, 989].

12.2.4 Chlamydial Encephalitis in Cattle and Buffalos

Histologically, there are inflammatory reactions in the brain, spinal cord, and in the meninges. They consist of vasculitis and perivascular cell infiltration and accumulations of mononuclear cells in brain parenchyma. Sometimes inflammatory processes in the kidneys, the lungs, the pleura and the peritoneum are found [19, 22, 38, 972, 973].

Histopathological examinations of calves with *C. pecorum* encephalitis, cited above [983], have revealed lesions ranging from lymphocytic and histiocytic vasculitis and meningoencephalitis to extensive thrombosis and neutrophilic inflammation. In addition, these cases illustrate that histological lesions in the calves of *C. pecorum* are more variable than previously reported [983]. Other publications [988, 989] describe *C. pecorum*-associated SBE lesions as follows: leptomeningeal hyperemia and edema with exudative meningitis accompanied by lymphohistiocytic perivascular cuffing, predominantly on the ventral surface of the brain stem; a severe, diffuse mononuclear meningo-encephalomyelitis, affecting mainly the medulla and cerebellum, developing as a result of vasculitis and thrombosis, followed by degenerative processes and necrotic foci in the gray matter; Glial cell proliferation [980, 988, 989]. A common histological finding is the fibrinous polyserositis involving the pleural, peritoneal, and pericardial cavities. It develops secondary to vasculitis, which can occur with or without endothelial proliferation in multiple organs [991–993].

12.2.5 Chlamydial Pneumonia in Sheep and Cattle

Histologically in ***lambs*** *is* establishes histiocytic – and in sheep – lympho-histiocytic interstitial pneumonia [971]. The liver shows degenerative changes by activating the reticuloendothelial system. In some cases, lymphoid cell nephritis, myocarditis, and focal encephalitis are also reported. These changes indicate that chlamydial infection causes a wide range of damage to the animal organism [22]. Punke et al. [985] find that polymorphic granulocytes appear in the secondary complication of chlamydial pneumonia with *Pasteurella* and other bacteria and develop a picture of exudative pneumonia.

Initially in the acute stage, an inflammatory reaction is found in terminal bronchioles and adjacent alveoli. This is followed by the proliferation of alveolar cells and the accumulation of macrophages inside alveoli and extensive alveolar epithelialization. Lesions are most prominent adjacent to terminal bronchioles [987]. Bronchopneumonical lesions gradually become confluent, leading to the consolidation of the lung. The development of lesions is accompanied by perivascular and peribronchial accumulation of mixed mononuclear cells, which later develop into characteristic follicles [645]. *C. abortus* bodies are mostly destroyed during the acute phase of inflammation, which then subsides. Resolution can be complete within 3–4 weeks of experimental infection [645, 987].

The microscopic picture of chlamydial pneumonia in ***calves*** is characterized by inflammation of the bronchi and peribronchial tissues, bronchiolitis, and alveolitis. A particularly characteristic finding in the inflamed lung areas is the peribronchial lymphocytic infiltration. Moderately sized cuffs of lymphocytes are also present around small blood vessels at the height of the lesion [645, 987]. The alveolar epithelium is activated. Mononuclear cells and neutrophils are observed in alveolar cells. Chlamydial inclusions can be detected in the cytoplasm of epithelial cells of bronchioles [986]. Biberstein et al. [994] and Reggiardo et al. [995] described the histological findings of chlamydial lung inflammation in sheep and dairy calves as bronchointerstitial and fibrinous, with thickened septa and accumulation of macrophages in the alveoli.

12.2.6 Chlamydial Enteritis in Calves

The pathological changes correspond to the severity of the clinical manifestations. In experimental chlamydial gastroenteritis with pronounced general symptoms, some authors found petechial hemorrhage, epithelial erosions, and ulcerations in the abomasum and in the terminal part of ileum edema [996, 997]. A common finding is the fibrinous intestinal peritonitis of the intestinal tract and the peritoneal cavity. In other cases, only catarrhal enteritis is observed [995]. Post-mortem examination of calves suffering from chlamydial gastroenteritis has revealed hyperplastic ileitis and the agent has been isolated from the walls of the ileum [692].

Histological lesions of chlamydial enteritis have been characterized as acute, necrotizing, and exudative, with neutrophil and monocyte infiltration [999]. Surface epithelium desquamation, dilatation of Lieberkühn glands, and lamina propria infiltration with multiple mononuclear cells and neutrophils have also been reported. Many epithelial cells and lamina propria are highly infected with chlamydia. A suitable method for detecting chlamydial inclusions is the fluorescent-antibody staining [997]. The lumens of crypts of Lieberkühn contain inflammatory cell and sloughed epithelial cells [996]. Histopathological lesions in the intestine and abomasa are rather similar. Foci of granulomatous inflammation have been found in the abomasal and intestinal mucosal, muscular, and serosal layers [996].

12.2.7 Chlamydial Pericarditis in Pigs

Histopathologically, on the surface of the epicardium, we found uneven fibrin deposits formed as a homogeneous compact mass mixed with leukocytes (Figures 12.11 and 12.12). Below them, on the side of the epicardium, there is a growth of granulation tissue. Separate sites in this expanded tissue include fat cells (Figure 12.13) [47].

At the base of the epicardium, edema is detected, as well as infiltration with leukocytes and erythrocytes. Blood vessels in this area were heavily expanded and overloaded with erythrocytes.

In the myocardium we observed neutrophil polymorphonuclear granulocytes and strong dissociation of the surface layer of the muscle fibers (Figure 12.14). In addition, there was a strong edema in the myocardium. Separate muscle fibers had swollen and had grainy cytoplasm and degenerative changes in the nucleus [47].

12.2.8 Chlamydial Conjunctivitis and Keratoconjunctivitis in Pigs

Rogers et al. [716] performed necropsies on seven pigs (2 to 8 weeks old) with mucopurulent conjunctivitis from one farm and on one sow with

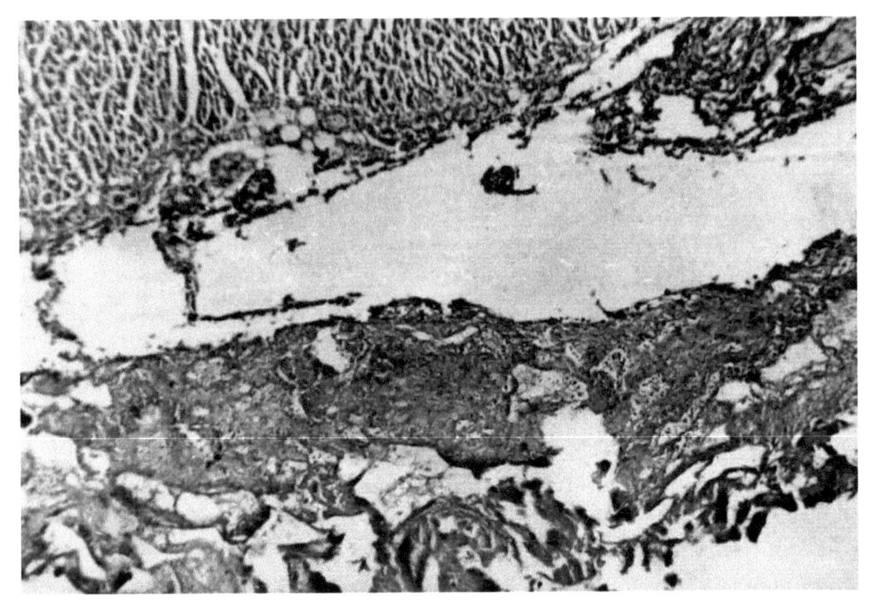

Figure 12.11 Heart of pig. Deposition of fibrin exudate in the epicardium. 40× (H. and E. stain).

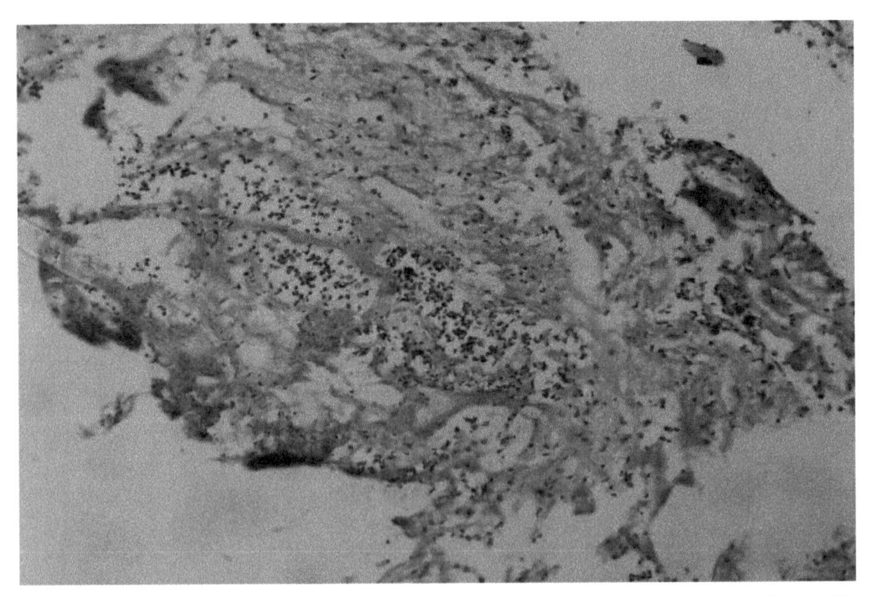

Figure 12.12 Pig heart – fibrin deposits admixed with leukocytes on the epicardium. 250× (H. and E. stain).

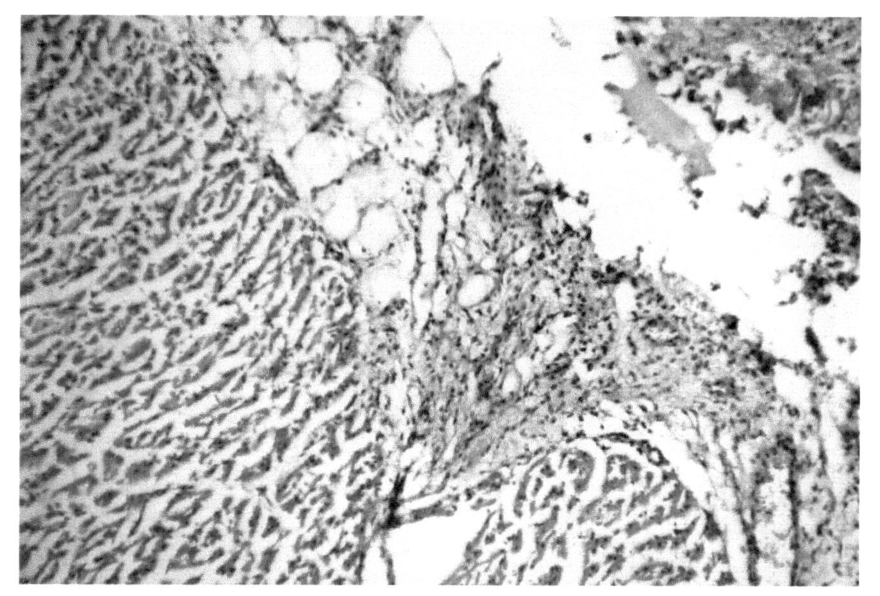

Figure 12.13 Pig Heart. Unorganized (top right) and organized (in the middle) fibrin exudate. Immediately to the myocardium there are fat vacuoles. 500× (H. and E. stain).

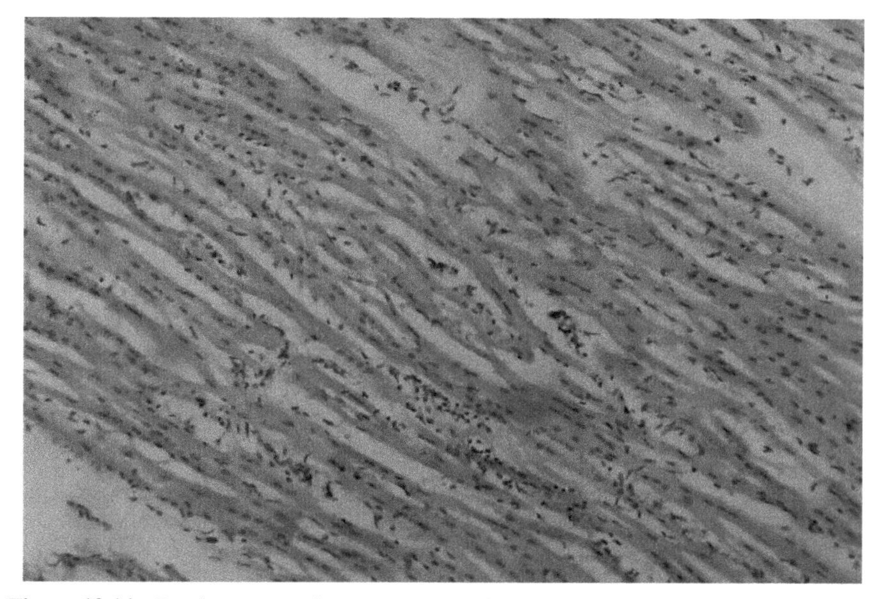

Figure 12.14 Porcine myocardium – presence of neutrophilic leukocytes between muscle fibers. 250× (H. and E. stain).

keratoconjunctivitis from another farm. It was histologically found that the small pigs had lymphoplasmacytic conjunctivitis with mild lymphofollicular hyperplasia. In the sow, a microscopic picture of marked conjunctival lymphofollicular hyperplasia and ulcerative keratitis with neovascularization was found. The authors allow the possibility that the chlamydiae seen ultrastructurally within the intra-cytoplasmic vacuoles containing glycogen in the conjunctival cells of the pigs and sow were *C. trachomatis*. Rogers and Andersen [717] investigated complex gnotobiotic pigs experimentally challenged with *C. trachomatis* strain H7. Bilateral conjunctival lesions have been seen histologically in the three piglets necropsied 7 days post infection (DPI). Palpebral conjunctivae were characterized by mild (one piglet) or moderate (two piglets), focally extensive conjunctivitis. The inflammatory process in the nictitating membranes was mild and multifocal. The authors also found that the cells in the conjunctival propria-submucosa have been predominantly lymphocytes. Also observed were smaller numbers of plasma cells, neutrophils, macrophages, and much less often – eosinophils. A majority of the principal piglets necropsied at 14–28 DPI had histologic lesions of mild conjunctivitis. The authors conclude that chlamydial strain H7 may cause mild or occasionally moderate conjunctivitis in gnotobiotic pigs, but the conjunctival infection is asymptomatic. These results are interesting and useful when trying to assess the actual involvement of chlamydia in the ocular pathology of pigs. In the interpretation of such experiments, the type of strains used and the dose administered should be taken into account, and in our opinion the emphasis should be placed on animals and farms with more massive detection of conjunctivitis and keratoconjunctivitis by conducting studies on spontaneous and experimental infection with the isolated strains.

12.2.9 Chlamydial Pneumonia in Pigs

Histopathologically it is about interstitial intralobular pneumonia.

In experimentally infected pigs with no macroscopically visible lesions, histological investigation revealed alveolar epithelial hyperplasia, alveolar degeneration, edema, serous exudation, and deposition of mononuclear and mucosal cells in the alveolar lumen, intra-alveolar, and peribronchial mycophotic infiltration. The lesions are in the form of islands located in the apical and cardial lobules.

In pigs with macroscopic lesions, intensive zones of bronchopneumonia, lymphocyte infiltrates, bronchial epithelium hyperplasia, and massive deposition of mucosal cells leading to bronchial lumen narrowing have been observed. In both groups of pigs, there is edematous infiltration and hyperplastic inflammatory reaction in the interlobular spaces [18, 696, 1001].

Reinhold et al., 2008 [697] reported an experimentally induced *C. suis* infection in pigs resulting in severe lung function disorders and pulmonary inflammation. The acute pulmonary inflammation in infected pigs has been confirmed by postmortem histology. A prominent dissemination of chlamydial bodies in the lung has been accompanied by an influx of macrophages, granulocytes, and activated T-cells.

12.2.10 Chlamydial Abortion in Pigs

Histopathological examinations of the porcine chlamydial abortion are often hampered by the fact that placentas are not always available. An additional problem is that when there are fetal membranes, they are in most cases autolytic and therefore unsuitable for histopathology and immunohistochemistry. Thoma et al. [974] have found positive fetal liver inflammatory changes (mild periportal hepatitis with the presence of neutrophilic and eosinophilic granulocytes) in chlamydia, which they believe may serve as an indication of a possible chlamydial etiology of abortion.

Okada et al. [1002] described the pathological findings in porcine fetuses, aborted at 95–102 days of gestation. Macroscopically, the fetuses had a free reddish fluid in the thoracic and peritoneal cavities, hepatomegaly, with a reddish-yellow coloring, linear hemorrhaging in the heart and enlargement of the lymph nodes. The histopathological picture is characterized by epi- or endocarditis, non-suppurative choriomeningitis, degenerated and necrotic hepatocytes and angitis. Chlamydial inclusion bodies have been identified in the cytoplasm of infiltrated monocytes, hepatocytes, and trophoblastic cells by immune-histochemical staining. In addition, *C. psittaci* (old nomenclature) has been isolated from 2 fetuses showing severe lesions. With regard to differential diagnosis, serological and microbiological evidence suggesting an abortion in swine due to Aujeszky's disease virus, Japanese encephalitis virus, Getah virus, porcine parvovirus, Escherichia coli and Toxoplasma gondii were excluded. In addition to the aforementioned etiological data for chlamydiosis, chlamydial antibodies have been detected among many sows and fattening pigs on the same farm [1002].

12.2.11 Chlamydiosis in Ducks

In about half of the ducks we found inter-lobar and alveolar pneumonia. Bronchioles and parabronchial spaces were filled with fibrinous exudate (Figure 12.15). In many tissue areas of the stroma and the lung, parenchyma

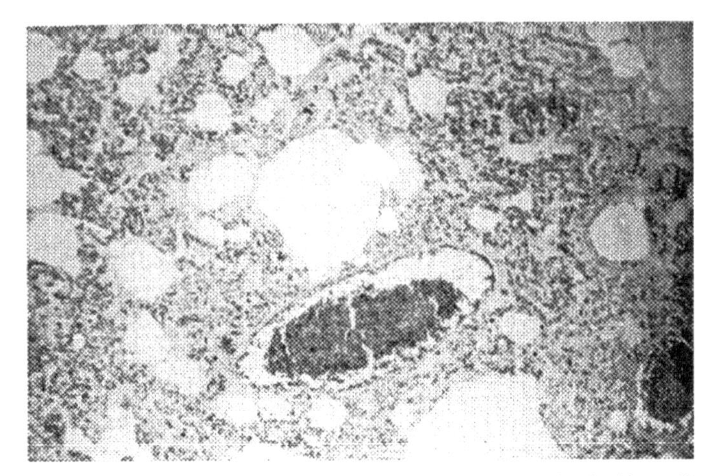

Figure 12.15 Histological section. Lung of a duck with acute fatal chlamydiosis. Interlobular and alveolar pneumonia. Histiocytes and fibrinous exudate in the para-bronchi and bronchioles. 200× (H. and E. stain).

were detected small hemorrhages and cellular necrosis. In 12% of cases, we observed eosinophilic necrotic masses, epithelial and tissue desquamation in the lumen of bronchioles and parabronchi [53, 579].

The air sacs contained a large number of histiocytes and lymphocytes in the membrane propria (Figure 12.16).

The epicardium was infiltrated with macrophages, fibroblasts, lymphocytes, and heterophiles (Figure 12.17). In about 60% of ducks, we recorded a non-purulent myocarditis.

Most often, in 90% of ducks, the liver was affected (hepatitis), where hyperemia and infiltration with lymphocytes and pseudo-eosinophils were observed (Figure 12.18).

Kupffer cells were filled with hemosiderine and cell debris. The body was streaked with necrotic liver cells and less frequently with focal necrosis. The granulomas contained histiocytes, lymphocytes, pseudo-eosinophils and necrotic cells.

A common finding (80% of ducks) was of proliferated reticular cells and necroses in the spleens (Figure 12.19).

In about 40% of cases, the kidneys were affected by degenerative and inflammatory changes (Figure 12.20).

Degenerative lesions, necroses, atrophy, and hemorrhage was observed in about a half of the examined ovaries of the ducks [53, 579].

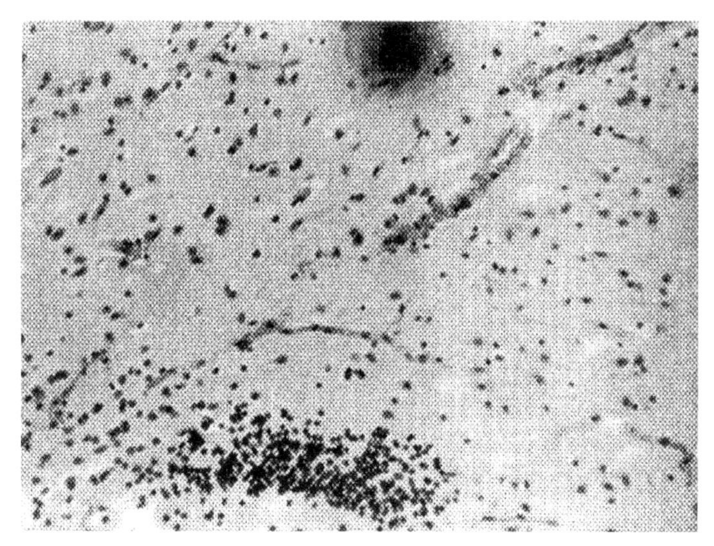

Figure 12.16 Histological section from an air sac of a duck with acute chlamydiosis. A plurality of histiocytes and lymphocytes in the membrana propria. 200× (H. and E. stain).

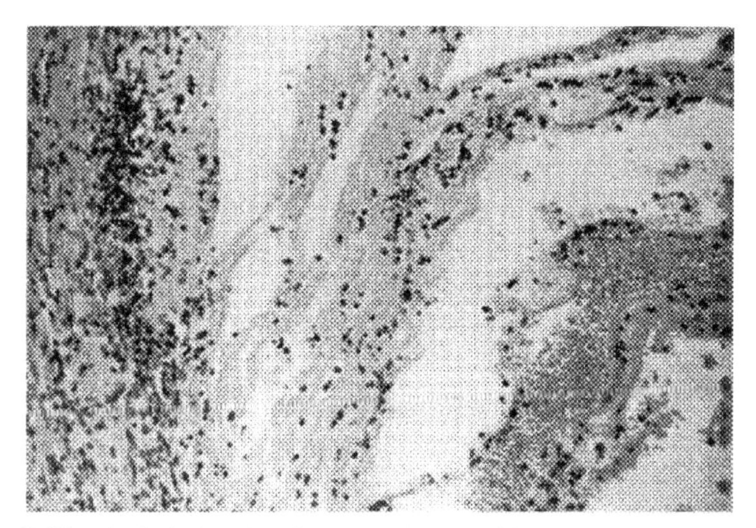

Figure 12.17 Histological section of the heart of a duck with acute chlamydiosis. Infiltration of the epicardium with inflammatory cells. 200× (H. and E. stain).

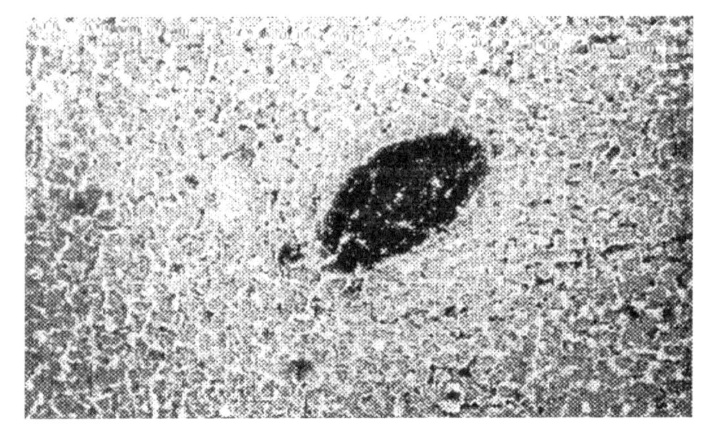

Figure 12.18 Histological section of duck liver affected by acute chlamydiosis. Hepatitis with hyperemia and infiltration of lymphocytes and pseudo-eosinophils. 200× (H. and E. stain).

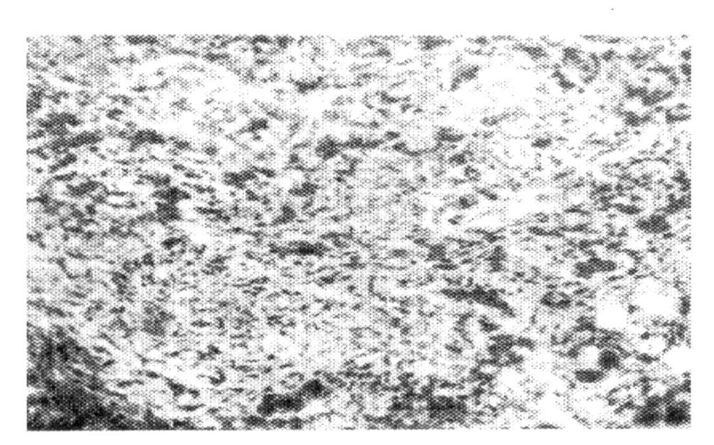

Figure 12.19 Histological section of duck spleen with acute chlamydial infection. Reticular cell proliferation and necrosis. 200× (H. and E. stain).

12.2.12 Chlamydiosis in Hens

The Glisson's zones of the liver had perivascular infiltration of lymphocytes and plasmatic cells and lymphohystiocytic proliferates. Hepatocytes had diffuse hyperemia and parenchymatous degeneration. Some focal necroses was seen in some liver sections.

The spleens of most hens are affected by hyperplasia of the monocyte–macrophage system. Multifocal necrosis in the liver and spleen is associated

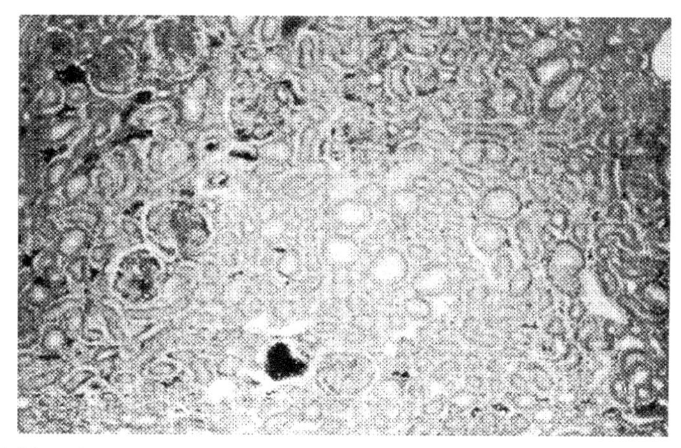

Figure 12.20 Histological section of a duck kidney with acute chlamydial infection. Inflammatory and degenerative changes. 200× (H. and E. stain).

with small, granular, basophilic intra-cytoplasmic bacterial inclusions in multiple cell types; occasional heterophiles; and increased mononuclear cells (macrophages, lymphocytes, and plasma cells) in hepatic sinusoids and splenic sinuses. Necrosis results from direct cell lysis or vascular damage [1003]. These finds are also relevant for other bird species.

Common findings in the kidneys are vacuolar degeneration, hyperemia, and single hemorrhages.

Changes in the lungs were mild to moderate perivascular lympho-histiocyto proliferation, the presence of serous exudate, hyperemia, hemorrhages and interstitial pneumonia in some cases.

In the pancreas we observed degenerative and hemodynamic changes and lymphohystiocytic clusters. The intestinal mucosa in some areas was necrotic, and the submucosa had infiltrates with lymphocytes and plasmatic cells. The lesions in the brains of hens were moderate perivascular edema, thrombosis of the small vessels and fibrinous necrosis of larger vessels [53, 579].

Lesions are usually absent in latently infected birds, even though *C. psittaci* is often being shed [1003].

The typical histopathological findings in various bird species have also been described in a number of other works [734, 998, 1000, 1003, 1004], and other papers printed during the decades after the discovery of avian chlamydiosis.

13

Epidemiology

Chlamydial agents have a wide pathogenicity among the various representatives of the animal kingdom. The infections they cause in farm animals are widespread and economically significant. In human populations, chlamydial agents cause a large number of diseases, which determine their health and social significance. Over the years, various explanations have been sought for the nature of chlamydia and their unprecedented planetary spread: biological, ecological, organizational, and structural factors; virulence, infectivity, and other strain characteristics and their association and relations with the host; disease manifestations, etc. With advancing molecular microbiology and the availability of modern diagnostic tools, the presence of chlamydiae has been frequently noticed in clinically inconspicuous pets and farm animals [633]. The analysis of previous epidemiological studies shows that despite the worldwide dissemination of chlamydioses, the epidemiological features and significance of many of them remain vague and insufficiently studied. This determines the need for continuing research efforts, a realistic assessment of the achievements so far and skillfully combining classical and modern methodology in scientific research on epidemiology of diseases caused by Chlamydia.

13.1 Sero-epizootiology and Nosogeography of Chlamydial Infections in Bulgaria

In Chapter 9, we presented detailed data on the detection and state of foci of chlamydial infections in domestic animals in Bulgaria, based on systemic and multidimensional seprological examinations in the period 1986–2007. Our results, detailed in stages, were obtained from the studies of 39,180 serum samples from different categories of animals with heterogeneous clinical status (Table 13.1). The sero-epizootiological studies had significant territorial coverage, of 420 settlements of 25 districts [643].

Table 13.1 Data on the sero-epidemiology of animal chlamydial diseases in Bulgaria

Animal Species	No. Tested Serologically	No. of Studied Settlements/ Districts	Revealed Outbreaks	General Sero-Positivity (%)	Disclosed Nosological Units
Sheep	10,606	154/20	107	13,3	• Abortions and related conditions • Respiratory disease • Arthritis • Conjunctivitis • Chlamydiosis rams • Latent form
Goats	1,249	45/14	28	12,75	• Abortions and related conditions • Pneumonia • Arthritis • Latent form
Cattle	17,371	118/25	92	4,58	• Abortions and related conditions • Chlamydiosis bulls • Latent form
Buffaloes	31	3/2	1	22,58	• Pneumonia • Abortion • Latent form
Pigs	2,346	30/16	21	7,37	• Pathology of pregnancy • Reproductive disorders • Respiratory • Urogenital • Latent form
Horses	210	1/1	1	0,95	• Latent form
Dogs	393	3/3	3	6,16	• Conjunctivitis • Respiratory • Urogenital
Cats	16	1/1	2	37,5	• Conjunctivitis • Abortion
Ducks	5,181	46/19	27	9,86	• Ornithosis
Geese	505	8/7	8	28,76	• Ornithosis
Hens	1,272	11/8	8	17,3	• Ornithosis
Total tested	39,180	420 settlements	298		

A total of 298 foci of chlamydial infections were revealed, including 228 in ruminants, 21 in pigs and 43 in birds. An entirely new moment in our development was the establishment of the country's earliest etiological data on chlamydial infections in dogs and cats (five outbreaks).

The isolation of the causative agent and its direct detection in various clinical conditions confirm the etiological nature and supplement the epizootiological characteristics of the chlamydioses and their nosogeography, represented by a large number of foci.

The epizootiological situation with regard to chlamydial diseases in the country was studied under radically changed social and political conditions and organizational-structural situation in agriculture and livestock breeding with the respective consequences, influences, and changes in the field of infectious animal pathology.

The nosological structure of etiologically proven chlamydial infections includes several groups of diseases:

- abortions and related conditions of pregnancy pathology;
- respiratory diseases;
- conjunctivitis and keratitis;
- arthritis;
- urogenital diseases;
- reproductive disorders;
- avian chlamydiosis (psittacosis – ornithosis);
- and latent chlamydiosis.

In the wide-ranging studies conducted throughout the country, the detection of new outbreaks of chlamydiosis was achieved through prophylactic or targeted serological screening and virological investigations in clinical diseases, combined with clinical-pathological observations and epidemiological surveys. A permanent element in this activity was the observation of the status, activity and dynamics of already known foci of chlamydial infections in the relevant livestock areas [643].

A very important point in the overall work on the detection and epidemiological characterization of chlamydial animal diseases was the consideration of the real zoonotic significance of some of them (avian chlamydiosis, chlamydial enzootic ovine abortion, and chlamydial keratoconjunctivitis in cattle) and with the suspected but not sufficiently studied zoonotic potential of other chlamydial pathogens.

A second important part of our epidemiologic and environmental studies is the detection of chlamydial diseases in some wild birds and mammals. Etiological studies have revealed chlamydial infection in 13 species, some of which, as far as we know (Japanese and rice finches and pink pelican), are reported for the first time in the literature, and for others (mouflon, heron, Gouldian Finch and zebra finches) – for the first time in Bulgaria. In another group of species with proven zoonotic significance – parrots, pigeons, and canaries – we confirmed the existence of chlamydiosis among these populations. The total number of revealed foci of chlamydial disease in wild species is 12.

From the above it can be concluded that the advanced studies conducted prove the circulation of chlamydia among domestic animals and among some wild animal species in the country.

For a comprehensive study of the problem of chlamydial infections in Bulgaria, we have undertaken extensive serological and virological studies of zoonotic and anthropogenic chlamydial diseases in the human population. The latter designation is conditional to some extent due to the cumulative data on the infection of animals with chlamydial species considered pathogenic to humans only. The results of sero-epidemiological studies of 19,970 selected patients found seropositivity for *Chlamydia psittaci* at 6.67%, for *Chlamydia trachomatis* – 36.48% and for *Chlamydia pneumoniae* – 43.33%. Particular attention is paid to infections in humans with *C. psittaci* in the form of epidemics and epidemic outbreaks acquired from infected waterfowl bred in large massifs for meat and liver [53, 337, 577, 579].

13.2 General Characteristics of Foci of Chlamydial Infections Among Domestic Ruminants

The foci of chlamydial infections are complex dynamic systems composed of a plurality of elements and having a number of features. In the epizootiological studies we took into account the location of the farm – within the locality or outside it, including the summer camps for animals. The size of herds – small, medium, large – and the number of affected animals served for both the study and control of a given chlamydiosis enzootic, as well as for the assessment of the risk of a continuing expansion of the epizootic process involving a larger number of animals. In all cases, we considered the existing farming practices, technological features and the organizational structure of livestock farming – the presence of compact farms with the corresponding capacity or a number of small farms, including single ruminants farmed in the private yards of the owners.

Given the widespread nosogeographic prevalence of chlamydioses in the country and a *de facto* lack of "geographic tropism", we have repeatedly found foci of infection that were close to each other within a region or area. Thus, large areas of massive, often long-lasting foci of chlamydial infections were formed, where the detection of sero-reactors continued for years, and the most frequent clinical manifestations – abortions and similar conditions – were a regular phenomenon. Typical examples in this regard are some areas of the Sofia region (Kostenets, Pchelin, G. Vasilitsa; Troyan and the neighboring villages; Shumen, and others).

We consider the internal structure of the foci of chlamydioses an important epizootiological factor. Some of them – standing herds of cattle, sheep and goats under the previous organization, in the composition of cooperatives and state farms, and currently owned by one person or company – represent a conglomeration of animals located in close daily contact with each other in livestock premises and grazing. In clinical chlamydial disease or massive release of the agent from latent infected animals, the infection spreads directly between them without intermediate hosts or vectors. The human contact of these animals is through their livestock and veterinary staff. The most obvious example in this respect is the fully enclosed cattle-breeding farms for stationary rearing without grazing. The possibilities for chlamydia infection in this type of breeding are mainly related to the introduction of newly acquired infected animals into herds.

Another type of cattle, especially prevalent in recent years, are pooled from individual cows, sheep, and goats in private yards, but daily grazed with other animals of many individual farms, including newly purchased ruminants without clear etiological status. In these herds, contamination opportunities are increased due to the daily collection of animals and their close contact on pastures and roads. Chlamydia-infected ruminants are also in daily close contact with their owners' families, where they live and where births and abortions occur and milking takes place. The proven zoonotic risk associated with chlamydia abortions in sheep and goats and the assumption of a similar epidemiological effect for humans on the same clinical events in cows are part of the characteristics of both herd types.

We distinguish the following categories of animals in the outbreaks of chlamydioses:

- Ruminants with a clinical disease and the most active shedding of chlamydiae;

- Oligo-symptomatic, subclinical, and asymptomatic seropositive animals with varying degrees and duration of agent shedding, but usually less than the previous category;
- Animals suffering from clinically manifested chlamydial disease, but which have turned into latent carriers with excretion of chlamydiae;
- Affected animals – clinically and microbiologically healthy without excretion of chlamydia;
- Contact clinically healthy animals in the incubation period;
- Contact non-infected animals

These categories are in varying proportions. This is determined by the virulence of circulating chlamydial strains, the intensity of the epizootic process, the size and structure of herds and practices for animal husbandry, the systemic implementation of sero-epidemiological surveillance of chlamydioses, the regularity and the effectiveness of the applied control measures [643].

In Figure 13.1, the epidemiological characteristics of the active foci of chlamydioses among domestic ruminants are presented. It can be seen that the foci of chlamydial infection are complex systems with seven categories of animals, a different degree of pathogen shedding with two main options for staging, a rich nosological structure, and a number of predisposing and risk factors.

The epizootiology of individual groups of chlamydial infections in different animal species has specific features that require self-examination.

13.3 Epizootiology of Chlamydial Abortions and Related Conditions of Pathology of Pregnancy

Our observations of a large number of outbreaks of chlamydial abortions, stillbirths and deliveries of weak unviable lambs indicate that the infected sheep is the key unit in the epidemiological chain [23, 49, 54, 333, 576, 643]. Other biological vectors or mechanical carriers have not been established. The lambing period (December to February) is particularly risky for the massive spread of the pathogen from the source of the infection – the aborted sheep, the dead or non-viable lambs dying in the next 48 hours, as well as the normally born but infected animals During this period, *C. abortus* was emitted in large quantities by the amniotic fluid, the placenta, the fetus, and the puerperal uterine and vaginal discharge, leading to the contamination of farms and pastures, thus creating favorable conditions for alimentary

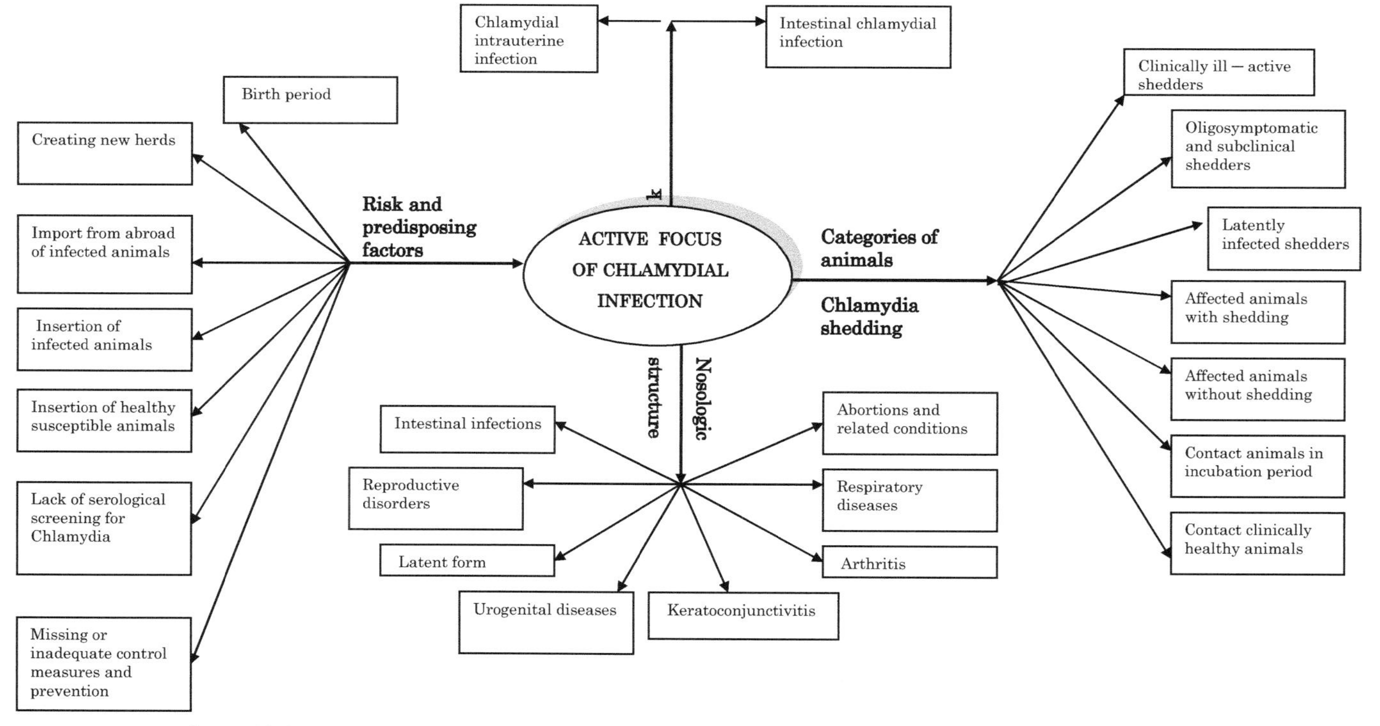

Figure 13.1 Epizootiological characteristics of active foci of chlamydial infection in domestic ruminants.

infection of susceptible contact animals. We also tolerated aerogenic infection by inhalation of infectious aerosols from birth fluids and dust particles contaminated with chlamydiae.

Therefore, sheep infected with *C. abortus* are epizootiologically the most dangerous during the lambing campaign and especially immediately after the abortion or the act of pathological or normal birth. In the case of two-fold insemination during one year, the second lambing period in another season is also risky with regard to the appearance and exaltation of clinical events leading to a wider dissemination of the infection in the herd.

Sheep with abortions and pathological births at the start of the lambing campaign are the primary source of infection for other sheep in an earlier pregnancy. In a number of such cases, we observed a rapid spread of the infection in herds, leading to new abortions in the current and next season.

Animals with chlamydia-induced abortions and similar clinical forms shed the infectious agent not only in the generic routes, but also with the urine, milk and feces for several months. This prolonged shedding is a prerequisite for infecting other sheep long after the end of the lambing season. The shedding of *C. abortus* from the feces due to the inapparent intestinal infection was confirmed by the isolation of the agent from such samples at different times after the abortion (see Chapter 5).

We found persistent chlamydial infection in a number of farms where seropositive reagents were established for years, and *C. abortus* was detected by direct indication methods and by isolation. We believe that this phenomenon is at the heart of the so-called stationarity of abortion and other forms of fetal loss. For the persistence of chlamydial infection, two epizootiological factors play a major role: (a) widespread asymptomatic intestinal infections with chlamydia and their systemic emissivity in the surrounding environment and (b) the maintenance of infection in the outbreak of lambs originating from infected flocks which are infected intrauterine or postpartum and are also inapparent chlamydia carriers and shedders. As the technological process advances, the infected lambs become ewes for the renewal and replenishment of the herd ("Repair animals") or for the formation of whole new flocks. Some of these infected ewes abort and the cycle described above is repeated. Therefore, in herds with stationarity with regard to chlamydia, reproduction and cyclicity of chlamydial infection can be observed [643].

Clinical-epizootiological analyzes, epizootic surveys, and etiological investigations to reveal new foci of the disease have allowed us to trace the epizootiological chain in which we have identified new flocks of infected animals from other known outbreaks or by introducing infected animals into healthy flocks.

In the 107 foci of chlamydial abortion and related conditions in sheep disclosed by us, the range of infected animals varied widely. This allowed us to group the outbreaks into three categories, ranging from >50%, 10–40%, to <10%. We will emphasize that, within the relevant category, the percentages of infected sheep always exceeded those of the animals that lost the fetus. The latter constituted heterogeneous combinations of percentages – from two to 50 on individual farms [572, 643].

The issue of the potential role of rams in the epizootiological chain of chlamydial abortions remains unclear. We have found sero-positive rams in some farms, both clinically healthy and sick with orchitis and orchiepididymitis. Despite the isolation of chlamydia from the semen of the latter, at this stage we have not found conclusive evidence of transmission of the infection by copulation with clinical consequences – abortion or pathological birth. Rather, we assume the role of the rams for the development of the infection by the established mechanism – through the manifestations showing large variations – from single cases to small or larger groups, or real epizootic outbursts. Undoubtedly, the virulence of the circulating chlamydial strain played an important role in the magnitude of clinical manifestations. In the presence of highly virulent chlamydia, there was a clear correlation between the range of seropositive reactions, the isolation of placenta and fetal strains, and the percentage of calf loss. The high seropositivity of *C. abortus* (>50%) in a herd or farm was a serious indicator of the presence of the infection and a prognostic mark of abortions and other clinical forms leading to reduced reproductive growth and reduced calves.

The main source of the infection is the infected animals. During abortion or delivery, the chlamydial agent is released in large quantities with maternity materials and genital discharges in the puerperal period. The latter contaminate the surrounding environment, creating conditions for alimentary and aerogenic (air-drip with aerosols created and airborne powder) contamination of the contact animals.

In herds also establishes sheep that are asymptomatic emitters of chlamydia with their feces. Such animals contribute to the occurrence of alimentary infection. In some cases there may be a venereal transmission of the agent causing minor urogenital infections that do not lead to abortion, [23, 643].

In the case of goats, the season of massive births and respectively abortions and similar disturbances is in March–April [167, 573, 643]. For the conditions of the country, the transition from a centralized cooperative state to a fragmented private stock-breeding was made in the 1990s. This has undoubtedly influenced, in various aspects, the goat-breeding sub-branch. Both before and after the structural reform, traditional goat farming is in the private yard, while the proportion of large and medium-sized goat farms is much smaller. Over the last decade, we have witnessed an increase in the number of goats kept in private yards at the expense of cows as a source of milk. Our observations show that the epidemiological features of chlamydial abortions in goats kept in permanent stationary herds with placement on distinct farms are very similar to those described in sheep grown under the same conditions.

Cultivating a much larger number of goats in individual courtyards with individual grazing for some of them or a common grazing in assemblage flocks with daily gathering and returning habitat generates differences in the clinical and epidemiologic aspect. In the mixed herds of goats, conditions are created for the animals to be infected by oral and inhalation route. In the private yard, the chlamydial abortion or stillbirth leads to localized contamination of the animal breeding site where, with adequate general and special measures against chlamydiosis, the concentration of the infectious agent can be significantly reduced. Conversely, in the case of abortion outbreaks in compact stationary goat herds, massive shedding of the agent leads to strong and disseminated contamination of pens and creates conditions for new infections and re-infections which, despite the auto-immunization of aborted animals, cause clinical illnesses and long-term damage. The activity of such an outbreak of chlamydiosis is maintained by the periodic introduction of new susceptible animals and their transformation into clinically ill or asymptomatic carriers of the pathogen and new shedders of it.

Cows with etiologically proven chlamydial abortions, premature births, stillbirths, and non-viable calves found the essential feature of lack of seasonality in this pathology [436, 643]. The range of clinical abortus bulls and the change in the insemination practices in recent years have raised the question of the possible role of male bovine animals in the epizootiological chain of abortions and other forms. Our targeted observations on a number of occasions indicate that the natural insemination and fertilization of cows from bulls with chlamydiosis do not result in a clinical consequence of abortion or pathological birth. On the other hand, artificial insemination with bull ejaculates with a proven lack of chlamydial infection does not prevent chlamydial abortions. It is important to emphasize, however, that some cows after copulation with chlamydia-infected bulls developed genital inflammation or sterility. Thus, the bulls that are cow contacting obviously play a more widespread epidemiological role, beyond the possibilities of infection by the already-defined model of oral or inhalational infection with chlamydiae shed from their gastrointestinal tract [643].

13.4 Epizootiological Aspects of Chlamydial Respiratory Diseases

In our studies, we found inflamed lower respiratory-tract infections (tracheobronchitis, pneumonia) in three age-groups: adult sheep, adolescent lambs and newborn lambs, some of whom died in the next 2–3 days. For goats, the age structure of the affected was similar [54, 643, 1006].

No predominantly territorial distribution of these diseases was found. They were diagnosed in different regions with their respective landscape, climatic, and ecological features.

In these species of small ruminants, two groups of respiratory chlamydial diseases are distinguished. The first is the self-flowing form. In the second group, respiratory diseases and abortions are simultaneously detected.

The evidence from epizootiological surveys has shown that the occurrence of respiratory chlamydiae is influenced by predisposing factors such as overcrowding, inhalational infection, unbalanced and inadequate nutrition, poor hygiene, transportable conditions, etc.

Diseases were found throughout the year, with the prevalence of cases during the winter season.

The source of infection for adult small ruminants and adolescent lambs and kids are the sick animals that excrete the causative agent from the airways during coughing and the intensified nasopharyngeal secretion. The animals are infected by an inhalational route. In herds with etiologically proven chlamydial infection clinically manifested in abortions, we consider it possible to contaminate sensitive animals with the agent massively released from the genital tract. In these cases, the pathogen apparently finds favorable conditions for reproduction in the lung epithelium. We find this possibility more likely in non-pregnant and adolescent animals. We assume also the mechanism of inhalation infection with chlamydia, excreted by the feces of sheep and goats with latent intestinal chlamydiosis.

The etiological involvement of abortiogenic chlamydial strains in pneumonia in neonates and young lambs is evidenced by the pathomorphological and etiological findings of chlamydia in the lungs of aborted fetuses, stillborn lambs and kids, and non-viable animals born in time. Part of the lambs and kids born and surviving also develop chlamydial pneumonia. These facts clearly indicate the role of the intrauterine mechanism of infection in prenatal pneumonia and in newborns and young small ruminants [54, 643, 1006].

13.5 Epizootiological Aspects of Chlamydial Keratoconjunctivitis and Arthritis

The mechanically prima facie grouping and examination of ocular and joint inflammatory diseases have been adopted due to the fact that the two clinical diseases occur frequently in the same sheep herds and simultaneously in many individual animals [19, 336, 635, 660, 661].

This suggests the existence of a specific antigenic relationship between the chlamydial agents isolated from the conjunctiva and the joints. The fact that some of the cases of sheep arthritis have been found in herds with chlamydial abortions also has epizootic significance. We also registered an autonomous course of each of the two diseases.

Sheep with conjunctivitis release the chlamydial agent with the ocular secretions and serve as the major source of the infection for adjacent animals that are infected after a mechanical fall into their eyes. The close contact between the animals in the herds favors the rapid transmission and spread of the infection. Morbidity varies widely and depends largely on the virulence of the circulating strain.

We have seen multiple cases of chlamydial conjunctivitis in cats. The ocular secretions of the diseased animals were highly infectious, and in colonies of cats the infection was transmitted rapidly through a mechanical contact pathway.

Our epidemiological observations on ovine and caprine arthritis are limited. This is in direct connection with the hypothetical views of the pathogenetic mechanisms leading to joint inflammation. In line with the notion that a generalized chlamydial infection is at the root of the pathogenesis, we assume that the primary source of the infection is the infected animals, especially those with intestinal localization, which systematically release chlamydiae. In all cases, the development of arthritis or polyarthritis implies a chlamydemia in which the pathogen falls into the sensitive joint and causes its inflammation. It also implies possible involvement of certain antigenic structures in the body of the animal similar to the HLA antigens in humans with reactive arthritis, while itself chlamydia has the role of an unlocking, a trigger factor.

We assume that animals already infected with chlamydia are the source of infection for other animals, but the mechanism of transmission of the pathogen remains unclear. In support of this assumption is our observation of the occurrence of mass polyarthritis in fattening lambs soon after they were brought from different farms and their collection in a common complex.

We did not find evidence of transmission between species of the causative agent from cattle to sheep and goats and back due to the existing practices of separately farming these species, nor for participation of birds and rodents in the epidemiological chain.

13.6 Epizootiological Studies of Swine Chlamydiosis

Chlamydial infections in pigs are manifested in different clinical forms. The variety in clinical manifestations is also related to the specific epizootiological features of individual age-groups and nosological units [23, 334, 335, 643].

The diseases we observed were single cases or affecting groups of animals or enzootics. The factors that determine contamination and the form of clinical course are complex: the virulence of the chlamydial strain, climatic conditions, room temperature, overcrowding, hygiene gaps, organization of farming and technological process, and trade and transport of pigs from infected to successful farms and vice versa.

The source of infection is sick pigs. Infection occurs mainly after contact with the infected animals. We have not detected any evidence of transmission of the infection indirectly via biological vectors or mechanical carriers.

In modern terms, keeping the pigs takes place in different types of holdings, including large pig-breeding farms and in small private yards. The level of maintained zoohygienic conditions, nutrition, and the prophylactic-treatment regimen vary considerably, and all of these factors may affect the course of the epizootic process in the positive or negative direction after chlamydial infection. Our observations show that infection with the chlamydial agent is much easier when the source of infection is massive, represented by a large number of animals and, on the other hand, susceptible animals are also numerous [643].

In chlamydial abortions and deliveries of dead and weak unviable offspring, the source of infection are aborted fetuses and stillborn pigs that emit large amounts of chlamydiae with the placenta and amniotic fluid.

Here we will point out that the widespread latent chlamydia infection in pigs is often transformed into a clinically manifested disease in the form of abortion or stillbirth as a result of its activation in the vulnerable period of pregnancy [643].

In farms with separate birth compartments, only the litters of infected mothers are infected intrauterinally or post-natally as a result of the massive shedding of the agent from the generic routes. Pigs from other groups become infected after weaning during mixing to form this age and technology group. Of such herds – a collection of healthy and infected pigs – the infection continues to move on to the fattening commodity farms (groups) by transferring infected animals. Also, infected pigs are also transferred to breeding farms or groups of young female animals. The latter, on the latent carrier of *C. suis* may abort as a result of activation of the infection similar to the above-described mechanism.

The intensity of the epizootic process was greatest in animals that first met with the chlamydial pathogen. Pigs in first pregnancy are more sensitive than those of second and subsequent pregnancies, which were clinically manifested in abortions and stillbirths reaching epizootic proportions in some farms (30–40%).

In pigs where the mothers and pigs are in one room and the births are taking place there, the chlamydial infection often encompassed a large number of pigs at the age of several weeks.

Adolescent pigs (over 3 months) were the most sensitive group in terms of aerosol infection with *C. suis* and development of pneumonia. The source of the infection was the sick pigs emitting the agent with the secretions of the airways, especially in the cough [643].

The epizootiological mechanisms for the occurrence of the pericarditis we have described in adolescent pigs have not been established [47]. This also applies to the other quoted earlier studies [42, 43], as well as to a few unspecified works. We tolerate aerogenically acquired chlamydial infection with subclinical lung involvement and complication of the infectious process by passing it to a new sensitive localization – the heart, manifested clinically and pathomorphologically as pericarditis.

An important clinical and epizootiological feature in pigs is the widespread latent form of chlamydia infection. The infected pigs emit the pathogen from the gastrointestinal tract with feces and systematically contaminate the premises in which they are kept. This creates opportunities for oral and aerosol infections of contact animals that can develop a clinically manifested form of chlamydiosis, chronic infection or latent carrier.

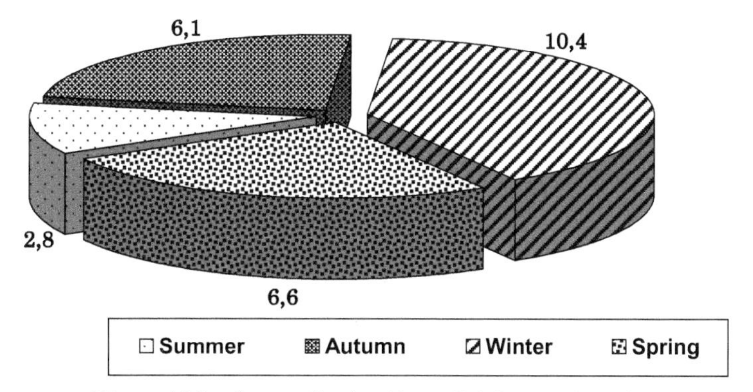

Figure 13.2 Seasonality in chlamydial diseases in swine.

We believe that part of the cases of reproductive disorders and infertility in pigs are basically a chronic *C. suis* infection, and another part is a direct complication of abortions and stillbirths leading to severe genital inflammations.

Our studies show that apart from the chlamydial agent and its virulence, two other factors also play an important role in swine – chlamydial disease: the season and age of the animals. The majority of disease cases were detected in the winter – December and February (10.4%), and most rarely in the summer – in August and June (0% and 2.8%, respectively) [335]. Bortnichuck [708] found that the highest incidence of the disease in Ukraine was in February–April. Spring and autumn figures are approximately the same – 6.6% and 6.1% (Figure 13.2). In terms of age, we recorded the highest incidence in adolescent pigs for fattening (17.6%) and the lowest in new-born piglets (3.4%) [335].

13.7 Epizootiological and Epidemiological Studies of Avian Chlamydiosis

Avian chlamydiosis occurs world-wide. Chlamydiae are known to infect most species of wild birds, domestic poultry, and pet birds [53, 580, 1007–1009] Kaleta and Taday [432] stress that 465 species of 30 orders, including not less than 153 parrot species of the order Psittaciformes are susceptible to infection with chlamydia. Also included are many other species of wild birds – coastal, migratory, pigeons, sparrows, etc. Chlamydiosis can result in serious economic losses in poultry, especially waterfowl (ducks, geese) and turkey. The disease is also a significant zoonosis. Special laboratory handling (biosafety level 3) is recommended because avian chlamydial strains can cause serious illness and possibly death in humans [83].

The reservoir and source of infection are wild, decorative, and domestic birds. The transfer of chlamydia from wild birds to poultry is undoubtedly proven. Thus, infected poultry, in turn, are the source of the infection for other poultry and for wild birds. This creates a closed epizootic circle [21, 580]. The migration of infected wild birds from endemic foci helps to carry the infection to considerable distances. Thus, the chain of the infectious cycle includes natural foci and secondary foci (infected flocks) from which the pathogen can be passed on to the staff of poultry farms, poultry slaughterhouses, and poultry stores as well as to some of the people in the surrounding settlements. The susceptible birds can become infected through inhalation of airborne contaminated material or possibly through ingestion of contaminated feeds. The chlamydia organisms spread over a large number of birds, creating a massive reservoir of infection in the herd. Infected birds shed chlamydia in both the respiratory excretions and in feces, which creates conditions for the transmission of infection from bird to bird. Consideration should be given to the fact that not only the obviously diseased birds but also those with a subclinical course of the infection are the source of infestation of flocks or areas as they periodically shed the chlamydial agent. Reports of persistent chlamydial infections in pigeons, psittacine birds, and other avian species were presented in several publications [566, 1007–1009]. Vertical transmission has also been demonstrated [1008, 1009]. It has been described in ducks, chickens, sea gulls, snow geese, and parakeets [1008]. The real significance of the vertical transmission of the infection in a flock and its potential to introduce chlamydia into biologicals produced in eggs need clarification.

In the previous chapters, Chapters 5 and 9, we presented virological and sero-epizootiological data on chlamydiosis in domestic, wild, and ornamental birds shown in stages over two periods – 1991–1998 and 2000–2007.

The spectrum of affected birds by species covers 13 species in total: four domestic, six wild and three decorative.

The prevalence of avian chlamydosis in Bulgaria is characterized by the presence of 43 foci of the disease among the birds with agricultural use and three among domestic pigeons in over 20 districts and by the finds positive for chlamydia in wild and decorative birds from settlements and natural areas in six districts.

The primary reservoir and source of infection with *C. psittaci* for domestic waterfowl, hens, and chickens from healthy herds free from chlamydia is unknown. Some wild bird species most likely play a role in this regard. For the conditions of the country, freely living pigeons, turtle doves, and sparrows, which are widely spread bird species seeking food in settlements and farm yards, are a very possible primary reservoir and carrier of chlamydial infection. The demonstration of *C. psittaci* infection in pigeons and doves in our current and previous research fueled the suggested assumption of the role of these birds. In an expanded plan, we believe that all wild bird species for which we have received positive etiological data represent a reservoir of chlamydial infection, the transfer of which to poultry is not excluded.

People can also become infected through contact with infected pigeons and decorative birds and songbirds kept in cages. The occurrence of chlamydial infection in humans is favored by active shedding of the pathogen with the nasal secretions of clinically diseased birds and the feces of visibly affected and latent asymptomatic carriers.

The epidemiological significance of pigeons must be emphasized. Already in 1941, Meyer [783] suspected pigeons and barnyard fowls as possible sources of human psittacosis – ornithosis. In 1965, the same scientist summarized the data of 50 authors from 24 countries and estimated that 26.9% of the 16,539 pigeons were seropositive for *C. psittaci* [219]. In the United States and Europe, the infected pigeons can reach 50–90% [580]). In the publication of a team of European researchers, seropositivity was reported from 19.4% to 95.6% among the freely living feral pigeons, *Columbia livia domestica,* in the cities of 11 European countries [582]. The huge increase in urban pigeon populations in Europe and their massive release have a negative impact on the hygiene of the environment and systematically damage the facades of buildings and monuments. Risk factors for the population are exposure to dust contaminated with *C. psittaci* and direct contact with birds in holding and feeding [580, 582, 812].

In the latent chlamydiosis in birds is hidden potential for conversion into a generalized clinical disease, resulting in lower overall resistance of the organism in overcrowding, transportation, sharply lower temperatures, and more.

The period of this extensive study also includes epizootic ornithosis in ducks and geese grown in large massifs for meat and liver extraction that we uncovered in the 1990s. Some of these epidemics have caused ornithosis outbreaks among farm workers and residents in neighboring populations. Below is presented a more detailed description and analysis of these zoonotic diseases.

13.7.1 Epizootlcs of Ornithosis in Ducks – Trustenik (Pl), 1991

In the autumn of 1990, we received the first serological evidence of infection with *C. psittaci* in a large holding growing more than 100,000 ducks. The practice was to breed ducklings for up to two months, then move to forced feeding ("hatch") bases for the purpose of obtaining liver for export. Clinical-epizootiological data showed a significantly increased mortality among ducks and increased cases of flu-like illness among farm staff. There was a great overcrowding of the halls, poor animal health conditions, pollution of the yard with sewage and dead body carcasses. Disease among ducks increased and reached 50%. We serologically examined 130 birds. Of these, 56 (43%) were positive for *C. psittaci*. From the internal parenchymal organs of dead ducks, we isolated strain Trustenik Pl. Between January and early March 1991, an epidemic outbreak of ornithosis broke out among the workers at the poultry farm.

13.7.2 Epizootics of Ornithosis in Ducks – Kostinbrod (SF), 1991

13.7.2.1 Technological and sanitary analysis

The production of ducks is organized in two large poultry farms where hens have been raised for the past 20 years. The owner has handed over the farms to the hiring company in an incorrect condition with many malfunctions. In general, the halls have been brought into technical suitability. According to the technological plan, four batches of one-day ducklings imported from abroad are accommodated in the halls. We found violations in cultivation technology. The supply of special granular mixtures is regular except for three days from July. The feed is excellent but does not contain antibiotics for chlamydiosis prophylaxis. The water regime is insufficient. Ventilation has been compromised due to central power failure. The courtyards in front of the halls are swamped. The population density of ducks is too large.

13.7.3 Analysis of *C. psittaci* Infection and Mortality

Out of a total of 231,762 ducks delivered to Kostinbrod, 53,888 died, representing a mortality rate of 23.2%. The lethality in the different batches is different – the highest (29%) – is in the second batch and the lowest – 18.2% – in the fourth batch. The periods of increased mortality are recorded in detail in days and weeks. These periods include stressful moments such as transportation, accommodation, and adaptation of day-old ducklings, lack of

Table 13.2 Serological evidence of infection with *Chlamydia psittaci* in industrially grown ducks during the epizootic and epidemic of chlamydiosis (ornithosis) in Kostinbrod

Lots of Ducks (No.)	Tested Sera (No.)	Positive (No./%)
1	25	6/24
2	25	8/32
3	25	14/56
4	25	7/28
Total	100	35/35

electricity and ventilation, moving from one room to another, high outdoor temperatures, and lack of fodder on certain days.

We believe that stress factors have led to a disruption of physiological equilibrium in chlamydia-bearing ducks, which is a common occurrence and, as a result, an active chlamydial infection occurs. An indicator of this is the highly susceptible organisms of the caretakers and the erupted epidemic outburst. The disease in people is favored by their performance in an inadequately sanitary state. Laboratory studies confirmed the activation of the latent chlamydia infection (Table 13.2).

The laboratory establishment of the etiological diagnosis was followed by healing measures. The whole herd was treated with chlortetracycline added by the manufacturer to the specialized feed. The treatment was effective. The therapeutic effect confirmed the etiology of the disease and, together with the complex of general measures, led to the control of chlamydial infection, a sharp decrease in mortality and a lack of new disease in humans.

13.7.4 Other Epizootics and Enzootics of Chlamydiosis in Waterfowl

Ornithosis in ducks and geese with a mass character was observed in a number of other settlements. The extent of infection varied over a wide range from 12.85 to 55% (Table 13.3).

In general, the epizootiological features of these diseases were similar to those described in Trustenik and Kostinbrod. We found that for various reasons, the conditions for rearing birds were not always adequate, which has led to the violation of the technological regime. Another common finding of epizootic surveys was insufficient awareness of the high susceptibility of waterfowl to chlamydia contamination and the associated lack of antibiotics in disease prevention schemes in farms. In some of the mentioned outbreaks of chlamydia in watermills, diseases of ornithosis occurred in the service staff.

Table 13.3 Coverage of chlamydial infection during epizootics and enzootics of ornithosis at duck farms in different locations

Settlement (District)	Serologically Tested (No.)	Positive (%)	% Positive in Individual Farm Halls
Lom (Mt)	103	25,24	4 – 46
Shtraklevo (Rs)	265	20	7,27 – 30,72
Yoglav (Lch)	40	55	55
Leshnitsa (Lch)	43	16,27	16,27
Manole (Pd)	80	16,25	10 – 22,5
Manole (Mt)	343	21,28	14 – 33
Dryanovets (Rz)	50	20	20
Balkanski (Rz)	50	40	40
Nozarovo (Rz)	10	40	40
Levski (Pl)	70	12,85	10 – 20
Parvomai (Pd)	15	33,3	33,3

13.7.5 Epidemiological Analysis of the Diseases of Psittacosis–Ornithosis in Humans

The main prerequisite for the occurrence of psittacosis–ornithosis in humans is the presence of foci of chlamydial infection in the birds, which we have described in detail above.

Infected farmed birds, free-living pigeons, pigeons for pleasure and for sale, mongers, and ornamental birds are the main reservoir and source of infection for humans. For human ornithosis, the natural foci of the disease are not essential as people rarely come into direct contact with them.

Vanrompay et al. [1005] studied the zoonotic transmission of *C. psittaci* into 39 breeding centers for parrots and birds, where antimicrobial drugs are often used. In 14.9% of people working in these farms chlamydia of genotypes A or E/B were found [1005]. According to Smith et al., in the United States for the period 1988–2003, CDC – Atlanta reported 935 human cases of psittacosis but some positive cases may be missed due to a misdiagnosis and failure to undertake etiological studies for chlamydiae [230]. Morbidity is increasing in industrialized countries and this is related to imports of exotic birds. The owners of such birds, the staff of pet shops, zoos and veterinary specialists are at risk of infection.

For people working in the poultry and poultry-processing industry, chlamydiosis (ornithosis) has the character of an occupational disease. In addition to sporadic diseases, we have seen epidemic outbreaks of ornithosis in humans directly related to epizootic diseases and chlamydial enzootic diseases in ducks and geese grown industrially [53, 577, 579–581]. In the cases of ornithosis that we observed in persons professionally involved in poultry and veterinary services, the link between the epizootiological and the epidemiological chain of the disease was clearly visible. The main mechanism for human infestation in ornithosis is respiratory (airborne) in direct contact with diseased or dead birds or by indirect delivery of the agent by fluff, dust, nasal secretions, and excrements contaminated with it. In rare cases, respiratory transmission of human-to-human infection is possible.

Similar is the mechanism of contamination in case of sporadic cases of psittacosis–ornithosis in owners of parrot and other ornamental birds, as well as racing pigeons.

The large epizootics among the waterfowl birds (Trustenik, Kostinbrod, Shtruklevo, Razgrad, etc.) and the associated epidemics or group diseases have clearly established the link between the intensity of ornithological epizootic diseases and human diseases.

During the two epidemic outbreaks in Trustenik and Kostinbrod in January–August 1991, 45 clinically ill persons were hospitalized, of which 14 were in Trustenik and 31 in Kostinbrod. These 31 patients from Kostinbrod were hospitalized for a period of 35 days (end of June–2 August, 1991). Diagnosis is serological in the majority of patients [803].

The following results were obtained from the research of contact persons among the bird breeders in Trustenik and the families of the patients: six (17,64%) positive for *C. psittaci* of the 34 tested (I sample) and six (23.07%) positive of the 26 tested (sample II) as a diagnostic increase in titer was established in three people. In the case of the epidemic in Kostinbrod, 12 (9.67%) of the 124 investigated serum samples from contact persons at the place of work and residence were positive. All these cases of positive outcomes without clinical manifestations of the disease are indicative of the development of ornithosis also in the form of subclinical infections.

The age-based distribution of patients (Table 13.4) shows that the most affected were age-groups 40–49, 30–39, and 50–59 (35.55%, 26.67%, and 22.22% respectively), which corresponds to professional employment with this type of production activity mainly by persons of active age [803].

Clinically, the disease is manifested by a toxic infection syndrome, clinical and radiological evidence of bronchopneumonia and pneumonia, and in some cases only with signs of the upper respiratory tract without evidence of lung involvement. Two of the cases in Trustenik were fatal, with a mortality rate of 14.28% for this epidemic explosion. The primary and anti-rebound course with tetracycline preparations showed the necessary therapeutic effect.

Table 13.4 Distribution of people with ornithosis in two epidemics by age

	Age Groups (years). No./% of Total Number of Patients												
	15–19		20–29		30–39		40–49		50–59		60–69		All
Epidemic location	No.	%	No.	%	No.	%	No.	%	No.	%	No.	%	Ages (No.)
Trustenik	1	7,14	2	14,28	3	21,4	4	28,57	3	21,43	1	7,14	14
Kostinbrod	1	3,22	1	3,22	9	29	12	38,71	7	22,58	1	3,22	31

The Kostinbrod outbreak covered a significant part of the service staff. It was mainly in moderate form without fatalities. There were also single severe cases. Etiological treatment was made possible on the basis of early evidence of the first case of chlamydial infection. Tetracycline was applied in the two-week course, which led to timely etiological mastery and successful conclusion of these diseases.

It should be noted that the implemented complex of anti-epizootic and anti-epidemic measures led to a sharp reduction of the morbidity in the described epidemics. Veterinary control included remedial actions in relation to bird diseases on a scheme developed by us including serological and virological research; elimination of violations of health requirements at existing facilities for waterfowl, foreclosure measures, and others. Anti-epidemic measures included epidemiological surveillance of farm staff and population in affected and adjacent settlements; monitoring of contact persons for a maximum incubation period (14 days) and urgent prophylactics with tetracycline; searching and hospitalization of the suspected patients; control of work requirements in this production (protective working clothes), ongoing cleaning and disinfection, decontamination of dead birds [580, 643, 803].

In the studies at Shtraklevo (RS) in 1993/1994, we revealed chlamydiosis of ducks and geese from a large farm of the same company. Due to the increased morbidity among farm staff, febrile conditions with unclear etiology, flu-like conditions and pneumonia, serological tests were performed and 16 seropositive individuals were found. The purposeful etiological treatment of the sick people was effective. The applied set of technological, sanitary, and veterinary-sanitary measures against zoonosis led to the cessation of active chlamydial infection among ducks and geese and to the prevention of new diseases in humans [643].

Systemic epidemiological surveillance is required – serological screening for the detection of infected farms, isolation and quarantine of affected farms. Imports of ornamental birds should be limited. When such birds are introduced, it is necessary to quarantine them for 45 days and carry out chemoprophylaxis and chemotherapy under certain schemes. Disinfections are a constant element of complex anti-epizootic measures. Eradication of avian chlamydiosis is not possible due to the large number of reservoirs and vectors of the infection, including many free-living birds, and a lack of reliable vaccines.

The described epidemic outbreaks of ornithosis in humans up to the mid-1990s were followed by a follow-up period still ongoing. This period was marked by mostly sporadic cases with diverse sources of infection – waterfowl, pigeons, parrots, or of unknown origin. Significant factors for the mitigated situation regarding the epidemic outbreak of ornithosis in humans are the improved technological and sanitary regime, the targeted anti-chlamydial prophylaxis of herds and the improved personal prophylaxis of the staff. However, the discovery of new outbreaks of chlamydial infection in the extensive sero-epizootiological research of birds, especially waterfowl and the serological status of the period 2005–2007, requires systematic surveillance in view of the potential risks to the population associated with it.

14

Vaccines for Chlamydia-induced Diseases

14.1 Introduction

The issue of vaccines and immunoprophylaxis of chlamydial diseases in animals and humans has long been the focus of the scientific community working in this field. This is largely due to the fact that obtaining effective vaccines for the diverse host and clinical manifestations of chlamydia-induced diseases has been a complex and difficult problem with controversial results. The reasons for this are ambiguous and should be sought both in the unique nature and properties of the chlamydial agents and in the characteristics and specificities of the hosts that determine their specific immune response and protection.

All this determines the continued relevance of the issue of chlamydial vaccines and the need for new ideas and quests in this direction.

14.2 Avian Chlamydiosis

There are no commercial vaccines available for chlamydiosis in poultry. In fact, antibiotics are the only current means of control. Attempts to produce a vaccine have met with limited success, and most have been based on bacterins produced by formalin inactivation of concentrated suspensions of chlamydiae [83]. According to Smith et al. [230]; and Beeckman and Vanrompay [440] the immunity involves cell-mediated immune responses. At this stage, however, vaccine manufacture has not been directed toward reactions of this type [83]. Analysis of the literature shows that interest in vaccines aimed at inducing a cell-mediated immune response goes back to the 1970s when Page [448] reported that such a vaccine protected 90% of turkeys against challenge. At the end of the nineties, Vanrompay et al. [1010, 1011] found that vaccination of turkeys with a *Chlamydia psittaci* DNA vaccine was successful in generating both humoral and cell-mediated responses, resulting in a significant level of protection in turkeys.

Andersen et al. [1009] and Andersen and Vanrompay 2000 [1012] expressed economic and organizational considerations for a similar type of vaccine, namely that the expense and the practicalities of vaccinating large numbers of birds for the prevention of sporadic epidemics may not be justified.

14.3 Chlamydial Abortion in Sheep

The creation of immunoprophylactic agents against chlamydial miscarriage of sheep is an important part of scientific research on this economically significant and zoonotic disease. The medical prophylaxis of these abortions requires the oral administration of chlamydia suppressants throughout the pregnancy, which is economically irrational [1013]. A number of scientists have shown a positive assessment of EAE-immune prophylaxis. Wachendorfer et al. [1013] states that vaccination is a specific control measure and the only effective means of reducing chlamydial abortions and related conditions in sheep. Such an opinion is also held by Foggie [1014].

The OIE, summarizing the worldwide experience of the problem, indicates that inactivated and live vaccines are available that have been reported to prevent abortion and reduce excretion of infective organisms. They assist in the control of the disease but will not eradicate it. In addition, it emphasizes the importance of identifying infected flocks with the help of serological screening during the period after parturition as control measures can then be applied in these flocks [216]. Both types of vaccines are available commercially. They are administered intramuscularly or subcutaneously at least 4 weeks prior to breeding to aid in prevention of abortion.

For a fuller understanding of the problem, we find it helpful to make a brief retrospective of OEA vaccines. Attempts for their development date back to the 1950s, when a team of colleagues in Scotland headed by McEwen received formalized vaccines from chlamydia-infected fetal membranes or crude yolk sacs of chicken embryos [1015–1019]. Later, a number of authors prepared the vaccines of placentas, parenchymal organs of infected fetuses and in particular from chicken embryos, and for inactivation using formalin [1020–1026]. In other vaccines, the chlamydiae were inactivated with silver nitrate [18, 1026–1028]. Vaccines were without adjuvant [18, 1020–1028] or with three different adjuvants: mineral oil [18, 1015, 1019, 1022, 1024, 1025], aluminum hydroxide [1026, 1027], or saponin [1023, 1029].

Schoop et al. [1030] and Nevjestic et al. [1031, 1032] reported live vaccines without adjuvant. Mitscherlich [1033–1035] and Yilmaz [1036–1038] offered live vaccines with aluminum hydroxide as an adjuvant. The commercial live, attenuated vaccine is a chemically induced temperature-sensitive mutant strain of the organism that grows at 35°C but not at 39.5°C, the body temperature of sheep [1060].

Attempts were made to produce associated vaccines: Chlamydia and Vibriosis [1039] and Chlamydia and Salmonellosis [1040].

The data on the effect of the above-mentioned vaccines are contradictory. Some authors reported positive results from the use of their vaccines [1013, 1019, 1029, 1035]. Other researchers, however, reported the failure of immunization experiments with inactivated or live laboratory and commercial vaccines [1025, 1030, 1038, 1041, and others]. However, Linklater and Dyson [1042] conclude that this vaccine appears to have lost its prophylactic properties due to biological changes in vaccine strains and possible other causes. As is known, the Scottish commercial vaccine against EAE, initially containing the A22 strain and having subsequently added strain S26/3, was successfully used for 20 years but was withdrawn in 1992 due to loss of efficacy. The failure of the vaccine was most likely due to production problems, since these strains continue to provide good levels of protection in experimental challenge model systems [1059]. Other publications highlighted shortcomings such as high reactogenicity [18, 1013, 1041] and low or relatively low immunogenicity [18, 1025, 1030, 1038, 1041]. The general characteristics of these vaccines indicate that they are native, unpurified and un-concentrated.

There are attempts to develop tissue-culture-grown inactivated vaccines against OEA, with the effect of their experimental testing being variable. Different adjuvants and doses have been tested. For example, the findings of Jones et al. [1058] suggested the efficacy of inactivated EAE vaccines to be antigen-dependent, and showed that good protection against heterologous challenge was achieved when 16 μg of chlamydial protein (containing approximately 30% MOMP) was combined with the ISCOM matrix adjuvant and administered twice or more.

Regardless of varying success, failures, and difficulties in the preparation and use of vaccines against EAE, most experts suggest the perspective of immuno-prophylaxis in this disease. In the absence of other possibilities for more effective control, reports of a significant reduction of chlamydiae for at least two to three lambs after vaccination clearly show the usefulness of the vaccines.

Initial immunization experiments for sheep immunization with live vaccines were accompanied by optimistic opinions about their effectiveness. For example, Mitscherlich [1033] expressed the view that live vaccines against EAE had a better immunizing power than precipitated vaccines. More recently, there are also optimistic views on the issue, even reaching the claim that the live vaccine could be an aid to eradication of disease [1044]. However, Schoop et al. [1030], and later, other authors pointed out that in a number of herds, injection even with live vaccines did not bring the expected success. Some disadvantages of live vaccines are the cause of these unsatisfactory results, – changes in vaccine strains, insufficient chlamydial concentration in individual doses of the vaccine [1038], limited durability of the non-lyophilized vaccine, and antibiotic susceptibility [83, 1013]. Inactivated vaccines are safe for administration during pregnancy, whereas live vaccines cannot be used in pregnant animals [83]. In our opinion, not least in the negative aspects of live vaccines is the possibility of creating new outbreaks and the likelihood of reversing the pathogenic properties of the chlamydiae. This view is confirmed by the recent publication of Wheelhouse et al. [1043], who found that the live vaccine strain 1B has been detected in the placentas of vaccinated animals that have aborted as a result of OEA, suggesting a possible role for the vaccine in causing the disease. Despite the controversies, doubts, and shortcomings highlighted, Stuen and Longbottom [1045] believe that the use of live vaccines remains the most effective method of protecting against the disease.

Several years earlier, Longbottom & Livingstone [1059] expressed the view that a multi-component recombinant vaccine against *Chlamydia abortus* remains a future goal of chlamydial vaccine research. Attempts with DNA vaccines in large animals have not so far yielded satisfactory results. Babiuk et al. [1061] indicate that the cause of the lower efficacy of DNA vaccines in animals is unknown, but is likely to be a consequence of transfection efficiency. The OIE [216] emphasizes that no biotechnology-based vaccines are currently in use for EAE.

Our own experience includes the preparation and testing of an experimental series of inactivated, purified and concentrated vaccine against chlamydial abortion in sheep [1045–1052].

We used four strains of *C. abortus*: GV, Pk, and Kt, isolated by us upon EAE [23, 166] and the D strain isolated by Ognianov [37]. The first three strains were used on the level of the 6th–15th passage in the YS of 7-day-old CE, while strain D was passed over 150 times. The infectious titer of the strains was equal to LD_{50}log 5 × $10^{4-6.5}$/0.3 ml for CE. Two series of vaccines were obtained, designated as PM-1 and PM-2. The initial material – the YS of an infected CE – was collected in a sterile vessel and three times frozen and unfrozen using liquid oxygen, which we consider a useful new aspect in the preparation of inactivated vaccines. For the purpose of obtaining a series of vaccines, we took about 300 infected YS. A volume of 350 ml corresponded to YS with a weight of 250 g. The material was homogenized in an automatic homogenizer. The last unfreezing was followed by inactivation. After the addition of an equal volume of formol solution we obtained 50% suspension, and in this manner the inactivation of the chlamydiae was achieved with 0.25% formol solution. Inactivation with formol continued for 48 h. For the purification and concentration of the vaccine we used differential centrifugation and super-centrifugation. The homogenized suspensions were centrifuged at 1500 g for 30 min. The supernatants obtained were centrifuged at 20,000 g for 90 min. The sediments after the second centrifugation were re-suspended in sterile saline solution (SS) to a concentration of 1:100 and were homogenized at 7,000 r.p.m. for 10 min. The inactivated chlamydial suspension was deposited with sterile 10% aluminum hydroxide at pH 6.8–7.0 added in small proportions. The next step was homogenization at 7,000 r.p.m. for 10 min. The vaccine was kept in a liquid state at 4°C. It was shaken once a day during the first day after obtaining it. The two series of vaccines differ in that PM-1 is concentrated 100 times, while PM-2 is concentrated 10 times.

Here we will comment on the choice of a method to purify our vaccine. In general, chlamydial infections in animals have not been administered concentrated and purified vaccines, especially in routine immunization practice. Human medicine has long been making attempts to prepare purified vaccines against trachoma. In these attempts to purify the propagated chlamydiae in the YS of CE, extraction with ether and centrifugation was used [1053]; Trypsin digestion of infected YS and high-speed centrifugation [1054]; differential high- and low-speed centrifugation [1055]; Sephadex column chromatography [1056, 1057]; gradient centrifugation in a 0–40% sucrose gradient [19]; and fluorocarbon treatment [1056]. Storz [19] discusses the advantages and disadvantages of these purification methods. We have developed a method for the preparation of purified and concentrated chlamydial suspensions characterized by the use of differential centrifugation and super-centrifugation, gel-filtration on Sepharose, and ultracentrifugation on a sucrose pad [250, 23]. It has a multipurpose application [23, 99, 100, 250, 251]. The combined use of the three basic procedures allowed us to obtain fully purified chlamydiae. For the purposes of this work, we used the first step of our method – differential centrifugation and super-centrifugation. At EM control we found that, compared to native, unpurified, and un-concentrated chlamydial suspensions, the suspensions in the vaccine had a good degree of purity and high agent concentration. As regards the selection of guinea pigs as an experimental model for laboratory testing, we will note that in previous experiments to produce labeled gamma globulins, we obtained high-titer guinea-pig sera against concentrated, purified and deposited chlamydia, with a large repeatability of results. Thus, guinea pigs proved to be very suitable for the purposes of this work.

The sterility tests carried out on six nutrient media showed no bacterial growth after inoculation with the supernatant after centrifugation at 1500 g for 30 min, with the aggregate suspension after the re-suspension of the sediments from the centrifugation at 20,000 g for 90 min and with the ready vaccine. The aggregate suspension from the coarse sediment after 1500 g/30 min. was also sterile.

The EM control included observations of ultrathin sections of the initial YS and of negatively stained preparations from the concentrated and purified suspensions before the addition of aluminum hydroxide. Only YS with a rich accumulation of chlamydial elementary bodies (EBs) and initial bodies (IBs) were included in the vaccine. The EM observations of the aggregate suspension showed that the chlamydiae are efficiently purified. Their concentration was 10^9 and there were no bacteria. During the comparative EM investigations of the coarse sediment obtained after centrifugation at 1500 g/30 min, we observed large quantities of coarse debris masses and large membranous structures. The chlamydiae were in very low concentrations. There were no bacteria.

The vaccine was tested for innocuousness on guinea pigs weighing 400 g which were injected deeply s.c. with 1 ml of the preparation. The animals were observed over a period of one month. During this period, the animals showed no deviations in their growth, development and health condition. Consequently, the vaccine is harmless for guinea pigs.

The experiments for immunogenicity were carried out on guinea pigs and sheep. In the sera of the animals investigated in advance, there were no chlamydial CF-antibodies. We tested two immunization schemes. In the first pattern, we gave the animals a single injection i.m. in the rear leg with 0.2 ml of vaccine. The second diagram included twofold muscular injections on the 1st and 30th days with the same dose. The bleeding of the animals was carried out in diagram I on the 30th day after the single injection, and in diagram II, this was done on the 45th day. The results are shown in Table 14.1.

Table 14.1 Chlamydial CF – antibodies in the blood sera of guinea pigs after immunization with vaccines PM-1 and PM-2

Vaccine Immunization	Group of Guinea Pigs, No.	Term of Bleeding Days	Immunity Positive No./%	Immunity Negative No./%	Cf-Titer (Reciprocal Values) <4	4	8	16	32	64	128	256	512
PM-1													
Single	15	30	14/93.3	1/6.67	1	–	1	4	6	2	1	–	–
Double	17	45	17/100	0/0	–	–	4	–	2	2	1	1	7
PM-2													
Single	5	30	5/100	0/0	–	–	–	1	1	2	1	–	–
Double	13	45	13/100	0/0	–	–	–	–	–	–	3	5	5

Table 14.1 shows that after single immunization of guinea pigs, chlamydial CF-antibodies with titers from 1:8 to 1:128 or an average of 1:32–1:64 (30th day) are accumulated in their sera. In the groups of twice-immunized animals (1st and 30th days), there is a predominance of high titers of 1:128 to 1: 512 in the bleeding on the 45th day (15 days after the second injection). These data testify to a high immunogenic effect of vaccines PM-1 and PM-2 for guinea pigs in the immunization diagrams used.

After keeping the PM-1 in a liquid state at 4°C for 5 months, we carried out a new test with guinea pigs to assess the preservation of the immunogenic properties. The results obtained showed that the vaccine preserves its immunogenic properties after being kept in the conditions and period referred above. There were no local reactions or general disturbances in the experiments with guinea pigs.

The immunogenicity experiments in sheep were carried out with two flocks. In the first flock (200 animals), made up of the sheep of two or three lambings, we made deep s.c. injections with the vaccine PM-2 at a dose of 0.5 ml. On the 45th day we revaccinated 159 sheep at a dose of 1 ml. The second flock consisted of 200 sheep one to two years old and they were injected with vaccine prepared under the same technological conditions from the coarse sediment obtained after centrifugation at 1500 g. Bleeding was carried out prior to the vaccination, on the 45th, 111th and 260th days.

The testing of the PM-2 vaccine for immunogenicity in sheep was carried out in a flock where there were 8% abortions during the previous years. We carried out studies during which we established the participation of the chlamydia in the etio-pathogenesis of the abortions by direct EM of placentas, isolation of strain, and CFT of the aborted sheep. The parallel investigations for other infectious pathogens proved negative. In the serological examination of the entire flock one week prior to vaccination, chlamydial CF-antibodies with low titers of 1:4–1:8 were found only in 4 out of 200 sheep. No antibodies were established in the second flock made up of 200 ewes. Vaccination was performed in a period when two-thirds of the animals had been artificially inseminated during the last 10 days, while the remaining part were to be inseminated during the next 7 days.

Table 14.2 shows that after a single immunization with 0.5 ml vaccine, antibodies are discovered in the sera of groups of sheep selected at random on the 45th and 111th days, at the rate of 91.42% and 96.43%, respectively, with titers from 1:8 to 1:512. On the 45th day, 56.62% of the sheep possess titers of 1:8–1:16, and 43.38% possess titers of 1:32–1:512; and on the 111th day 40.74% of the sheep have titers 1:8–1:16, while 59.26% have 1:32–1:256. These data testify to a very good immune response by the sheep after a single immunization with vaccine PM-2.

The same table shows that on the 45th day 19 (86.37%) of 22 singly vaccinated sheep are seropositive, 11 (57.89%) have titers 1:8–1:16 and 8 (42.11%) have titers of 1:32–1:128. After repeated immunization on the 45th day with 1 ml vaccine, 100% of the sheep are seropositive on the 111th day, 3 of them (13.63%) having titers of 1:16, and the remaining 19 sheep (86.37%) having titers of 1:64–1:256. In these animals there is an increase of the titers of over 4 times on the 111th day (16 sheep) or two times (6 sheep). At the same time it becomes clear from Table 14.2 that 79 (49.69%) of 159 twice-immunized sheep are seropositive on the 260th day. Predominant are the titers of 1:4–1:8 (83.55%), while in 16.45% of the sheep they are 1:32–1:512.

No chlamydial CF-antibodies were established upon the investigation of 200 sheep, injected with vaccine from the coarse sediments serving as a kind of technological control. This is an indication that the product has no immunogenic effect, while on the other hand it confirms the

Table 14.2 Chlamydial CF-antibodies in the blood sera of sheep after immunization with vaccine PM-2

Group of Sheep (No.)	Term of Bleeding (days)	Immunity Positive No./%	Immunity Negative No./%	CF Titer (Reciprocal Values) <4	4	8	16	32	64	128	256	512
		Single Immunization										
70	45	64/91.42	6/8.58	6	–	34	2	11	8	8	–	1
28	111	27/96.43	1/3.57	1	–	5	6	0	7	8	1	–
		Double Immunization										
22	45	19/86.37	3/13.63	3	–	10	1	4	2	2	–	–
22	111	22/100	0/0	–	–	–	3	–	7	8	4	–
159	260	79/49.69	80/50.31	80	23	24	19	3	5	2	1	2

correct methodological orientation of the technology for obtaining purified vaccine.

There were no local or general reactions in the vaccinated sheep, and consequently the vaccine is harmless and non-reactogenic. There were no abortions, stillbirths, or feeble lambs in the vaccinated flock. These results point out to the epizootiological effectiveness of the concentrated and purified vaccine.

Based on the positive experience of the preparation and testing of PM-1 and PM-2 vaccines, we have developed a modified method for the production and testing of PM-3 vaccine, presented in detail below.

14.3.1 Method for the Preparation of PM-3 Vaccine Against Chlamydial Miscarriage in Sheep

14.3.1.1 Chlamydial strains

We used reference strains A-22 and S26/3 of *C. abortus* isolated in Scotland from aborted sheep placentas. They had an infective titer of $4 \times 10^{7.5}LD_{50}/0.2$ ml for CE. The strains were delivered to us by Moredun Research Institute, Edinburgh (courtesy of Dr. I. D. Aitken and Dr. G. E. Jones).

14.3.1.2 Preparation of suspensions

300 YS of strains A-22 and S26/3 were mixed with 800 ml of PBS. Homogenizing was done in a mechanical homogenizer at 300 r.p.m. for 2 min, followed by 6000 r.p.m. for 5 min. These suspensions were 30%.

14.3.1.3 Electron microscopic examination of suspensions

By negative staining, it was found that the starting suspensions were well homogenized but contained cell fragments and debris. They consisted of chlamydial EB with a concentration (morphological titer) of $10^{7.5}$. There were still the same-sized protein and lipid macromolecular particles (Figure 14.1).

14.3.1.4 Bacteriological control of suspensions

The starting suspensions were sterile in normal broth, conventional agar, blood agar, phenol-rot, thioglycolate medium, and soya-casein medium. Samples in the broth were visually examined. The remaining samples were examined using a phase-contrast microscope in which no bacteria were detected.

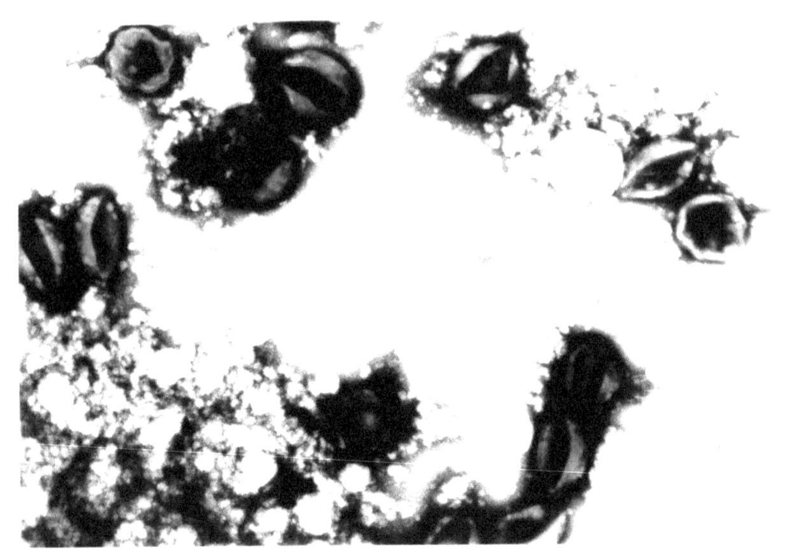

Figure 14.1 Chlamydia abortus. Vaccine strain S26/3 – starting suspension comprising a plurality of elementary bodies, cell fragments, and debris. Transmission electron microscopy. Negative staining. Magnification 20,000×.

14.3.1.5 Centrifuging the suspension at 1500 g for 40 min

Each of the two pooled suspensions of both strains is placed into sterile banks. Each bank contained 100 ml of the corresponding pooled suspension. This is followed by centrifugation at 1500 g corresponding to 7000 rpm. The resulting supernatants are separated. The resulting precipitates are homogenized by thorough stirring with a Pasteur pipette, then mixed to give a volume of 50 ml. Sterile PBS washes the residue remaining on the walls of the banks and was added to the above (50 ml) and this solution was brought to 100 ml. The supernatant liquids from each pool were collected in a total vessel to give a total of 150 ml.

14.3.1.6 EM-control of the aggregate sediment and the pooled supernatant after centrifugation at 1500 g

In the aggregate sediment from the two pools, by negative staining a number of chlamydial EBs were revealed at a concentration of 10^9 (Figure 14.2).

In the aggregate supernatant from the two pools were observed chlamydial bodies at a concentration of 10^7. There were a number of EBs and distinct IBs (Figure 14.3).

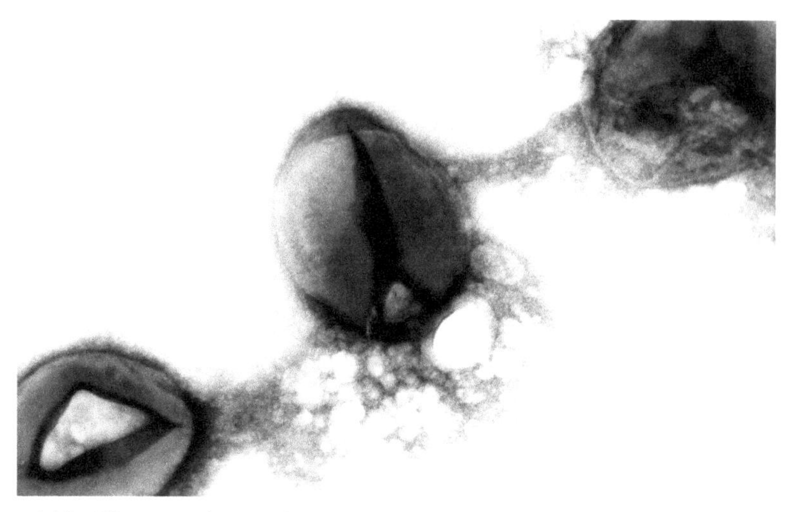

Figure 14.2 Electron microscopic control in the process of obtaining the vaccine PM-3. Negative staining of the suspension of the aggregate pellet after centrifugation at 1500 g. Elementary corpuscles of *Chlamydia abortus* strain S26/3. Magnification 80,000×.

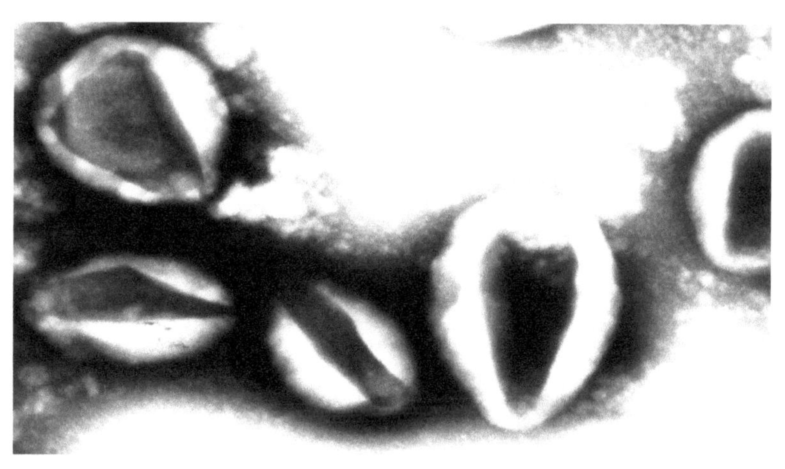

Figure 14.3 Electron microscopic control in the process of obtaining the vaccine PM-3. Negative staining of suspension of pooled supernatant after centrifugation at 1500 g. Elementary bodies of *Chlamydia abortus* strain A22. Magnification 60,000×.

14.3.1.7 Inactivation of the chlamydiae

It was preceded by the separation of material for serological, molecular-biological, and Product IDEM studies. Inactivation EM studies. Inactivation was performed with 2/1000 formalin solution. The above aggregate precipitate after 1500 g and pooled supernatant after the same centrifugation were separately treated with formalin. Inactivation with formalin was continued for 48 h at 4°C.

14.3.1.8 Determination of the antigenic content in the aggregate sediment and the pooled supernatant after centrifugation at 1500 g

CFT micro-method. This study was performed before inactivation with formalin. Results: CF-titer of the aggregate precipitate – 1:128; CF-titer of the pooled supernatant liquid – also 1:128 (Figure 14.4).

■ Antigen 1 - supernatant after 1500 g

Chlamydial Ag1 antigen

8 16 32 64 128 256 512 51 2

	8	16	32	64	128	256 / 512 51 2
8	4+	4+	4+	4+	4+	2+
16	4+	4+	4+	4+	4+	2+
32	4+	4+	4+	4+	4+	2+
64	4+	4+	4+	4+	4+	1+
128	2+	2+	2+	1+	1+	–
256	–	–	–	–	–	–
512	–	–	–	–	–	–
AgC	–	–	–	–	–	–

S+

■ Antigen 2 - sediment after 1500 g

Chlamydial Ag2 antigen

8 16 32 64 128 256 512

	8	16	32	64	128	256 / 512
8	4+	4+	4+	4+	4+	2+
16	4+	4+	4+	4+	4+	2+
32	4+	4+	4+	4+	4+	1+
64	4+	4+	3+	3+	3+	1+
128	–	–	–	–	–	–
256	–	–	–	–	–	–
512	–	–	–	–	–	–
AgC	–	–	–	–	–	–

S+

Figure 14.4 CFT micro-method. Determination of chlamydia antigen content in *Chlamydia abortus* suspensions after centrifugation at 1500 g (prior to formalin inactivation).

Table 14.3 ELISA. Determination of the antigen content of *Chlamydia abortus* vaccinal strains after the centrifugation at 1500 g

Test Material	Dilution	Chlamydial Antigen Content (IFU/ml)
Pooled supernatant fluid after 1500g	1:50	438
	1:100	181
	1:200	88
	1:400	59
Pooled precipitate after 1500g	1:50	424
	1:100	206
	1:200	115
	1:400	81

ELISA. These results are shown in Table 14.3. We investigated the same samples indicated in the CFT micro-method. Results: pooled supernatant – at 1:50 dilution of the suspension, the concentration of chlamydial antigen is 438 IFU/ml/; at a dilution of 1:100, concentration is 181 IFU/ml; at 1:200, concentration is 88 IFU/ml; at 1:400, concentration is 59 IFU/ml; aggregate precipitate – at dilution of the precipitate of 1:50, the concentration of chlamydial antigen is 424 IFU/ml; at a dilution of 1:100, concentration is 206 IFU/ml; at 1:200, concentration is 115 IFU/ml; at 1:400, concentration is 81 IFU/ml.

The above results show that both serological tests detect large amounts of chlamydial antigen at this stage.

14.3.1.9 Determination of antigenic content in inactivated chlamydial suspensions

The study was performed after 48 h of inactivation with formalin and prior to super-centrifugation.

CFT micro-method. CF-titer of the inactivated supernatant fluid from the two pools – 1:64. CF-titer of the inactivated pooled precipitate from the two pools after centrifugation at 1,500 g – 1:128 (Figure 14.5).

ELISA. A pooled sample of the above materials (supernatant + sediment from the two pools) was examined in four dilutions. At a 1:50 dilution, the chlamydial antigen concentration is 162 IFU/ml; at 1:100, it is 108 IFU/ml; at 1:200, it is 70 IFU/ml; at 1:400, it is 57 IFU/ml (Table 14.4).

It can be seen from the table that the chlamydial antigen content is retained in the aggregate sediment from the two pools (CFT), but is reduced on average by two times in the pooled supernatant fluid (CFT) and in the aggregate sample of the supernatant and sediment (ELISA).

■ **Antigen 3 - the supernatant after inactivation**

Chlamydial Ag3 antigen

	8	16	32	64	128	256
8	4+	4+	4+	3+	–	–
16	4+	4+	4+	2+	1+	–
32	4+	4+	4+	2+	–	–
64	4+	3+	3+	2+	–	–
128	2+	2+	2+	2+	–	–
256	1+	1+	1+	–	–	–
512	–	–	–	–	–	–
AgC	–	–	–	–	–	–

S+

■ Antigen 4 - sediment after inactivation

Chlamydial Ag 4 antigen

	8	16	32	64	128	256
8	4+	4+	4+	4+	4+	2+
16	4+	4+	4+	4+	3+	1+
32	4+	4+	4+	4+	3+	1+
64	4+	4+	4+	3+	2+	1+
128	2+	–	–	–	–	–
256	–	–	–	–	–	–
512	–	–	–	–	–	–
AgC	–	–	–	–	–	–

S+

Figure 14.5 CFT micro-method: Determination of chlamydia antigen content in Formalin-inactivated suspensions of *Chlamydia abortus*.

Table 14.4 ELISA. Determination of the antigen content in the inactivated chlamydial suspensions

Test Material	Dilution	Chlamydial Antigen Content (IFU/ml)
Cumulative sample of inactivated supernatants and sludge from the two pools	1:50	162
	1:100	108
	1:200	70
	1:400	57

At 48 h after the addition of formalin, practically no sediments were found in both suspensions (supernatant pooled after 1500 g and pooled sediment after 1500 g). Both suspensions were uniform and homogeneous.

Super-centrifugation was performed at 20,000 g for 2 h. Suspensions of the supernatant and sediment were in separate centrifuge tubes (eight counts) containing 31.25 ml. After super-centrifugation, three large sediments were obtained. They had the size of a man's watch. Five small sediments were also obtained. With a sterile scalpel, the lipids were removed from the surface. The total amount of collected lipids filled about 1/3 of the volume of the average petri dish. The removal and collection of supernatants in a common vessel followed. The latter had a gray–black color. Under EM, chlamydiae were not detected in them. There were large quantities of debris and rubbish. The supernatant liquid was retained for molecular-biological research.

Re-suspension of the above sediments followed. To the big ones we added 10 ml of PBS, and to the small ones – 5 ml. PBS was added in small portions for good re-suspension. The EM-analysis of sediments showed the presence of an EB concentration of 10^7. The suspensions were of good purity (Figures 14.6 and 14.7).

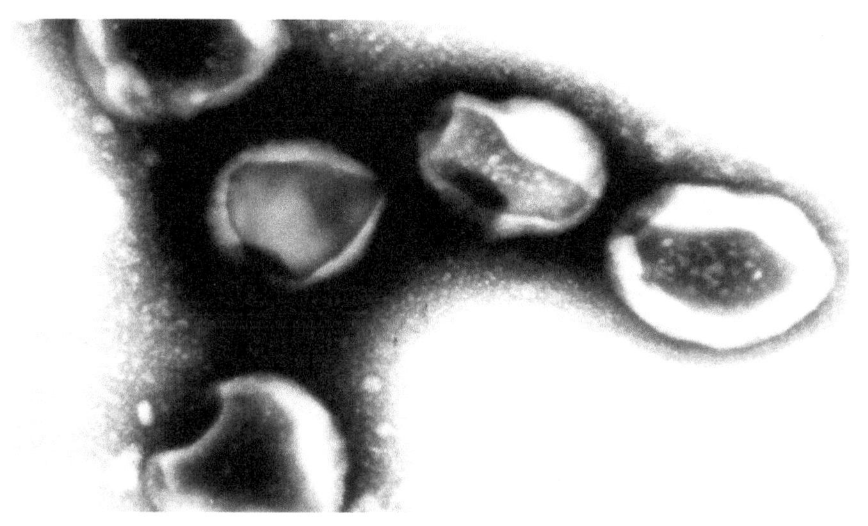

Figure 14.6 Electron microscopic control in the preparation of a vaccine PM-3. Elementary bodies of *Chlamydia abortus* in negatively stained preparation after super-centrifugation at 20,000 g. Magnification 50,000×.

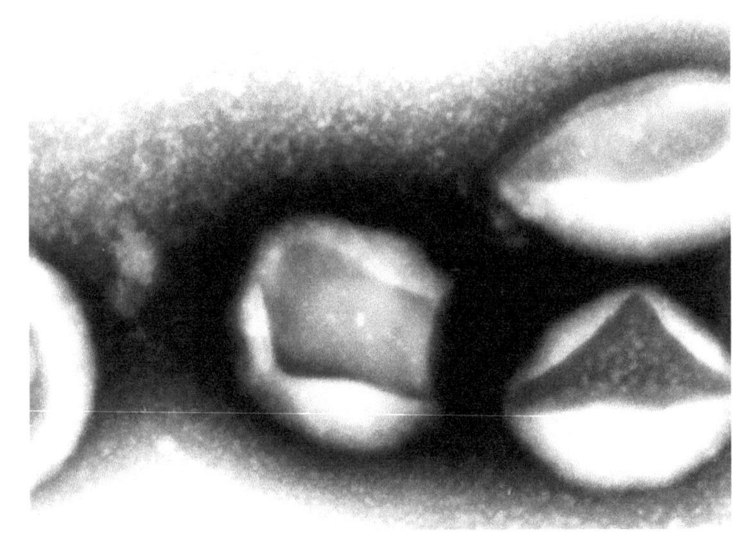

Figure 14.7 Electron microscopic control in the preparation of the vaccine PM-3. Elementary bodies of *Chlamydia abortus* in a negatively stained preparation after super-centrifugation at 20,000 g. Magnification 40,000×.

14.3.1.10 Treatment of the suspensions by ultrasound

It was once for 3 s. Following the collection of suspensions of various glass in common court. Separation of samples for EM, ELISA, and SDS-PAGE and for testing of infectivity on cell cultures.

14.3.1.11 Infectivity control

The suspension obtained over the entire cycle, including sonication was examined for residual chlamydial infectivity on EBT cell culture. The EM observations of the mono-layer and culture medium at 72 h after inoculation did not detect chlamydia multiplication in the cell culture; therefore, the vaccine suspension had no residual infectivity.

14.3.1.12 Determination of antigenic content in the ultrasound treated suspension

Using an ELISA, the accumulation of an enormous amount of chlamydial antigen was found: in sample I (dilution of 1:50) – 5885 IFU/ml; in sample II (1:100) – 387 IFU/ml; in sample III (1:200) – 337 IFU/ml (Table 14.5).

Table 14.5 ELISA. Determination of antigenic content in the ultrasound-treated suspension

Test Material	Dilution	Chlamydial Antigen Content (IFU/ml)
Final suspension after	1:50	5885
treatment with ultrasound	1:100	387
	1:200	337

14.3.1.13 Polyacrylamide gel electrophoresis and immunoblotting

Two identical gels were coated with Sigma molecular weight plus two other markers: 130 kDa beta-galactosidase and 93 kDa phosphorylase; with samples of different stages in the vaccine preparation: starting suspensions from the two strains; pooled supernatant from the two strains after 1500 g before inactivation; precipitate after 1500 g treated with formalin; sludge after super-centrifugation (20,000 g) of sediment of 1500 g, sounded; sludge after super-centrifugation (20,000 g) of sediment of 1500 g, sounded; final suspension – a mixture of sludge in Tracks 6 and 7; final supernatant after 20,000 g and control – purified EBs from strain S 26/3. The gels are 3–12.5% polyacrylamide; a mini-protein apparatus (Biorad) is used. Electrophoresis was performed at 200 V according to the manufacturer's instructions and at 93 mA to 50 mA at the end of the procedure (40 min). One gel was stained with Coomassie Brilliant Blue solution for 1 h. Treatment of the gel with acetic acid 7% (V/V) and 23% (V/V) ethanol in distilled water with a change of paper to adsorb the excess paint to background enrichment. The following is a photograph of the gel (Figure 14.8).

The other identical gel was transferred to a nitrocellulose membrane to perform western blotting (Figure 14.9).

The results of SDS-PAGE and immunoblotting showed that there are high levels of egg albumin, glycolipids and glycol-polysaccharides in the chlamydial suspensions, but the main outer membrane protein (MOMP) can be seen in strain S26/3. In the pooled supernatant of the two strains after 1500 g before inactivation, MOMP appears, and there are still large amounts of ovalbumin and glycolipids. The sediment seems clear after 1500 g of formalin-treated MOMP, and the amount of ovalbumin and glycolipids is less than the pre-existing gels. The sediment treated with ultrasound after 20,000 g of the sediment of 1500 g MOMP seems clear and the decrease in ovalbumin and glycolipids continues. The precipitate treated with ultrasound is clear after centrifugation at 20,000 g of supernatant from 1500 g MOMP, but significantly less than in the above gel; Ovalbumin is more than in the

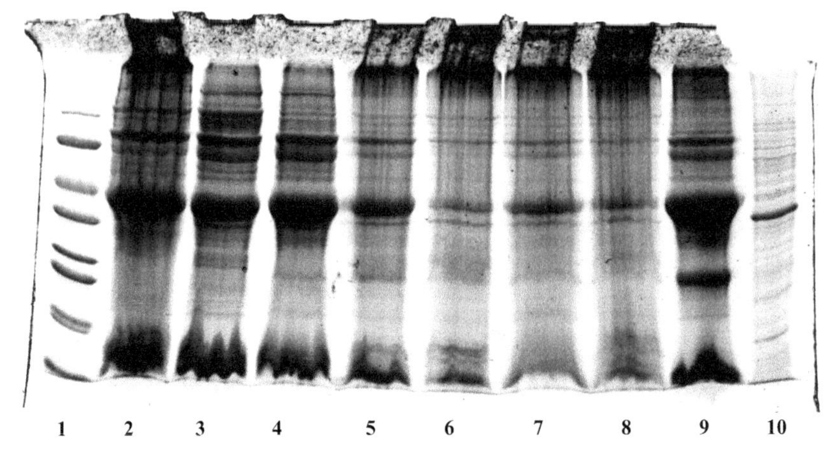

Figure 14.8 Molecular biological control in the process of obtaining the vaccine PM-3. Polyacrylamide-gel electrophoresis (PAGE). **Line 1 –** Sigma standards for molecular weights + β-galactosidase of 130 kDa and phosphorylation of 93 kDa; **Lane 2** – Starting suspension of *Chlamydia abortus* strain A22; **Lane 3** – Starting suspension of *Chlamydia abortus* strain S26/3; **Lane 4** – Cumulative supernatant fluid from the two strains after centrifugation at 1500 g and pre-inactivation with formaldehyde; **Lane 5 –** Sediment after centrifugation at 1500 g treated with formaldehyde; **Lane 6** – Sediment after super centrifugation at 20,000 g of sediment after 1500 g treated with ultrasound; **Lane 7 –** Sediment after super-centrifugation at 20,000 g of supernatant after 1500 g, treated with ultrasound; **Lane 8** – Final Vaccine Suspension – a mixture of sludge on Lines 6 and 7 after 20 000 g; **Line 9** – Ultimate supernatant after 20,000 g containing large amounts of ballast substances and MOMP missing; **Line 10** – Control-purified elementary bodies of *Chlamydia abortus* strain S26/3.

preceding gel. In the final suspension, a mixture of sediments after super-centrifugation of the sediment and the supernatant of 1500 g (shown in gels No. 6 and No. 7 samples), MOMP is also visible; comparison with the starting suspensions and those from the initial stages shows that a significant degree of purification has been achieved. Conversely, large amounts of ballast substances – egg albumin, lipids, polysaccharides, but lacking MOMs – are found in the gray-black final supernatant fluid after super-centrifugation (20,000 g). Immunoblotting results showed that the vaccine obtained by the method described contained MOMP.

14.3.1.14 Deposition with aluminum hydroxide

The treatment with ultrasound of the chlamydial suspension is followed by deposition with a 10% aluminum hydroxide. Thus, the final product was obtained – the ready-made vaccine. By its concentration of chlamydial

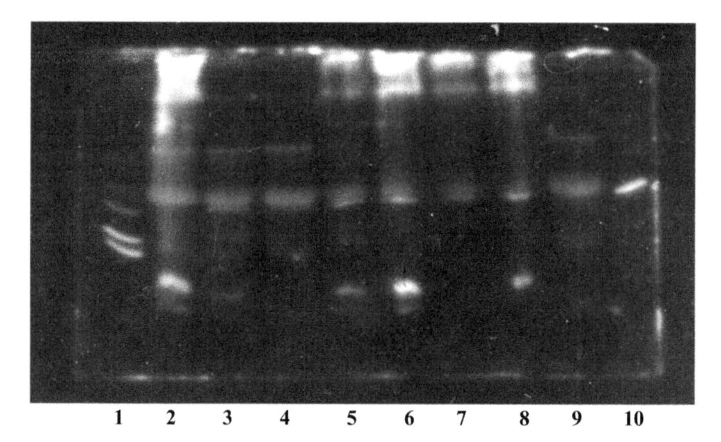

Figure 14.9 Molecular biological control in the process of obtaining the vaccine PM-3. Western blotting. Lanes 1–10 of the gel contained the same suspensions as described in Figure 14.8. The main outer membrane protein (MOMP) was observed in the middle of lines 5, 6, 7 and 8.

elementary bodies – 10^7, the vaccine is a significantly concentrated product which, as noted above, has a good degree of purity.

14.3.1.15 Sterility controls

In addition to the starting chlamydial suspensions, bacterial growth was lacking in the nutrient media inoculated with the formalin-inactivated suspension and separately with the ready-to-use vaccine.

14.3.1.16 Innocuousness control

The six guinea pigs injected with 1 ml of vaccine in the 1-month observation showed no deviation from their normal health status, growth, and development. This shows that the vaccine is harmless for guinea pigs.

Immunogenicity

- *For guinea pigs.* The single immunization of guinea pigs resulted in the accumulation of CF-antibodies with titers ranging from 1:16 to 1:128. Titers 1:32 to 1:64 predominate. The geometric mean titer (GMT) is 45 (30th day).

Double immunization on days 1 and 30 lead to the accumulation of the 45th day of CF antibodies from 1:128 to 1:512, and the GMT is 256.

The graphical representation of the results of the two immunization schedules of the laboratory animals is shown in Figure 14.10. The data obtained testify to a very good immunogenic effect of the vaccine for guinea pigs.

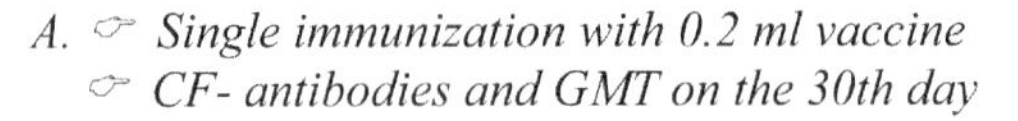

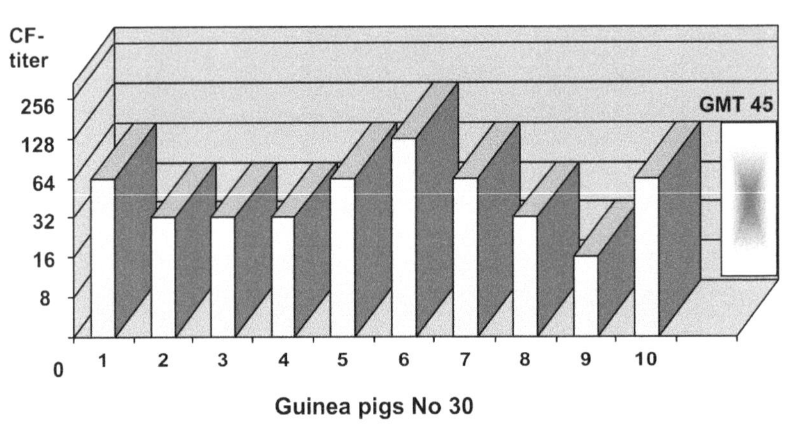

B. ☞ Two-fold immunization (1st, 30th day)
☞ CF- antibodies and GMT on the 60th day

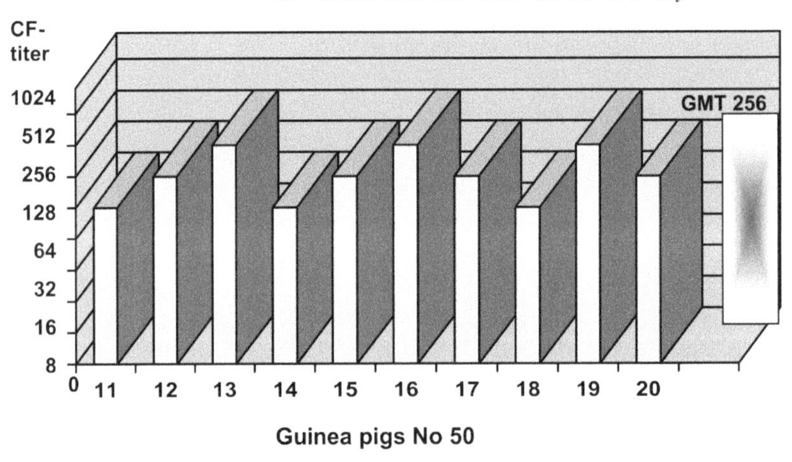

Figure 14.10 Testing of vaccine PM-3 immunogenicity of model guinea pigs.

Table 14.6 Chlamydial CF-antibodies in the blood sera of sheep after immunization with vaccine PM-3

Duration of Study (Months)	One-fold Immunized (No.)	CFT/+/ No./%	GMT	Two-Fold Immunized (No.)	CFT/+/ No./%	GMT	GMT for the Two Subgroups
1.5	14	14/100	17.1	13	13/100	18.4	17.8
2.5	14	14/100	64	13	13/100	60	62
3	14	14/100	112	13	13/100	94	104
3.5	14	14/100	120	13	13/100	94	104
4	14	14/100	69	13	13/100	74	69
7	14	12/85,7	13,9	13	11/84,6	9,2	11,3
10	14	11/78,5	4,6	13	8/61,5	3,2	4

- *For sheep* Table 14.6 provides immunogenicity data for the vaccine PM-3 based on the detection of chlamydial CF- antibodies in the 10-month period.

From the table, it is seen that up to 4 months after vaccination, 100% of the animals have CF-antibodies; at the 7th month, 85% of the sheep are seropositive and at the 10th month, 70%.

Antibody titers 1.5 months after immunization ranged from 1:4 to 1:64, dominating those from 1:16 to 1:32 (77.8%). The 1-ml immunized patients had a lower GCT (17.1%) than those immunized with a 0.75 ml dose (18.4%).

Antibody titers 2.5 months after single immunization ranged from 1:32 to 1:128 as the 1:64–1:128 (92.86%) values prevailed. Over the same period, those vaccinated twice with 0.75 ml showed no higher titers; they range from 1:16 to 1:128 (most commonly 1:64–1:128 (76.93%). The GMT for both subgroups are 64 and 60, respectively.

On the 98th day, the trend of absolute titers and GMT increases. The titers range from 1:64 to 1:256, and those of 1:128–1:256 make up 78.57% in the one-fold immunized and 53.84% in the two-fold immunized. The GMT is 112 and 94, respectively.

At 3.5 months, absolute titers do not exceed 1:128, but they account for 92.9%; the remaining reactions were 1:64 (7.1%). Within this period, the maximum GMT of 120 (once vaccinated) is established. The GMT is kept close to 94.

The 4th month is the beginning of the reverse trend – to reduce individual titers and GMT. From the table, it is seen that GMT in single-vaccinated patients is 69, whereas in the case of two-fold vaccinated patients, it is 74. Within this period, the GMT also had significant values.

At 7 months, titers range from 1:4 to 1:64, with 1:32 to 1:64 being 66.7% of the positive responses. The sero-positivity in both subgroups is similar – about 85% and GMT 13.9 and 9.2, respectively.

The last titration at 10 months indicates that chlamydial CF antibodies are detected in a significant percentage of animals – 78.58 and 61.54, although absolute titers decrease (1:4–1:64 but predominate 1:8). The MGTs are also lower – 4.6; 3.2 – average 4.

Therefore, immunization with the vaccine PM-3 results in the accumulation of chlamydial CF antibodies with progressively increasing absolute and mean geometric titers up to 3.5 months post-immunization, to titer retention and GMT at 4 months with significant values that are still relatively lower than the maximum registered titers. This was followed by a gradual reduction in titers up to the 10th month when about two-thirds of the animals were still seropositive. The data presented show a very good immune response in sheep after immunization with the vaccine PM-3 (Table 14.6).

Quite the opposite is the picture of Table 14.7, reflecting the immunogenicity test of a sheep control group immunized with the Cooper vaccine. The table shows that the maximum rate of seropositivity does not exceed 72.73. Geometric mean titers are low. At 1.5 months after immunization, GMT is 6.1; one month later – 5.3; on the third month – 3.2, a value that lasts for one month and a negligible GMT (<1) on the 7th and 10th month.

The GMT values shown in Table 14.7 reflect the low individual titers in the immunized sheep. At the first titration (1.5 months) titers are from 1:4 to 1:32, but 10 animals are seronegative. At 2.5 months, a similar trend was observed, with GMT showing a slight decrease.

Between the 3rd and 4th month, the titers varied between 1:4 and 1:16, and the GMT was maintained at 3.2. It will be noted that with one exception, the sheep that were sero-negative at first titration remained negative until the end of the experiment. On the 7th month, 87.85% of the sheep were

Table 14.7 Chlamydial CF antibodies in the blood serum of sheep of control group immunized once with an inactivated vaccine "Coopers"

Duration of Study (Months)	No. of Sheep Tested	CFT/+/No./%	CFT/–/No./%	GMT
1.5	33	23/69.70	10/30.30	6.1
2.5	33	24/72.73	9/27.27	5.3
3	33	22/66.67	11/33.33	3.2
3.5	33	22/66.67	11/33.33	3.2
4	33	22/66.67	11/33.33	3.2
7	33	5/15.15	28/87.85	<1
10	33	2/6.06	31/93.94	<1

seropositive and the five positive animals had a low titer of 1:4. At 10 months, the sheep negative for chlamydial antibodies were 31 (93.94%) and the positive were only two with a titer of 1:4 (Table 14.7).

The above titrations of serum from sheep immunized with the vaccines PM-3 and Coopers were accompanied by a study of control groups of unvaccinated ewes. Chlamydial CF antibodies have not been established.

The differences in GMT values in immunizations with both vaccines are presented graphically in Figure 14.11.

Using ELISA, we tested serum samples from sheep immunized with PM-3 and Cooper's, taken at different times (Table 14.8). The table also shows CF-titers of the respective samples. It can be seen that titers in ELISA

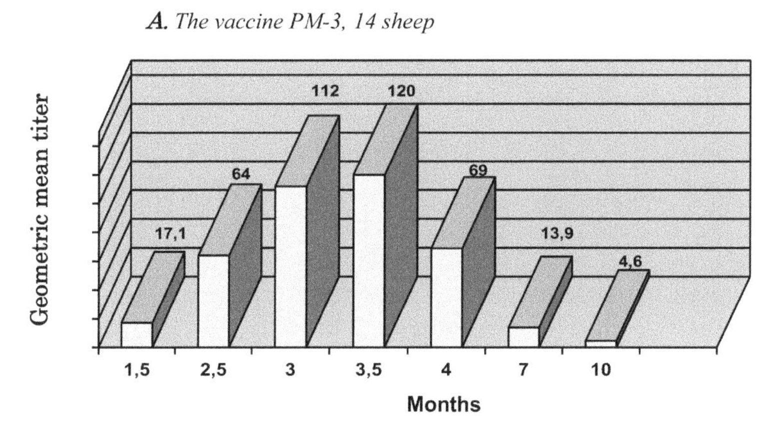

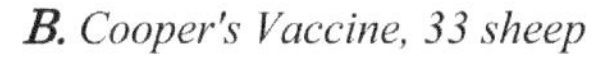

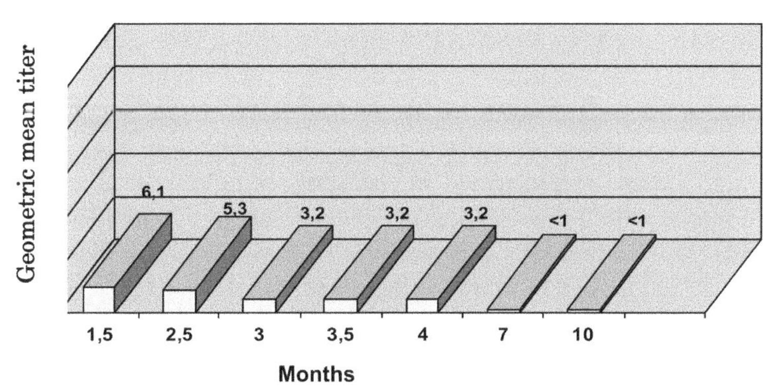

Figure 14.11 Comparative data for GMT of CF antibodies in blood sera of sheep immunized one-fold with the vaccine PM-3 and Cooper's.

Table 14.8 Comparative serological tests by ELISA and CFT of blood sera from sheep immunized with PM-3 and Cooper's vaccines

Sheep Ear No.	Vaccine	Time after Immunization in Months	Immunization Scheme	Titer in CFT	Titer in ELISA
80	PM-3	1.5	Once with 1 ml	1:32	1:256
96	PM-3	1.5	Once with 0.75 ml	1:32	1:512
114	Cooper's	1.5	Once with 1 ml	0	1:16
84	PM-3	2.5	Once with 1 ml	1:128	1:1024
91	PM-3	2.5	Once with 1 ml	1:128	1:1024
99	PM-3	2.5	Twice × 0.75 ml	1:128	1:1024
114	Cooper's	2.5	Once with 1 ml	0	0
85	PM-3	3	Once with 1 ml	1:128	1:1024
95	PM-3	3	Once with 0.75 ml	1:256	1:1024
94	PM-3	3	Twice × 0.75 ml	1:256	1:1024
86	PM-3	3	Once with 1 ml	1:128	1:512
107	Cooper's	3	Once with 1 ml	1:8	0
105	PM-3	3	Twice × 0.75 ml	1:128	1:256
86	PM-3	4	Once with 1 ml	1:128	1:1024
96	PM-3	4	Twice × 0.75 ml	1:128	1:512
118	Cooper's	4	Once with 1 ml	1:8	>1:16<1:32
82	PM-3	10	Once with 1 ml	1:8	1:128
90	PM-3	10	Once with 1 ml	1:4	1:256
95	PM-3	10	Once with 0.75 ml	1:8	1:512
106	PM-3	10	Twice × 0.75 ml	1:4	1:256
104	PM-3	10	Once with 0.75 ml	1:4	1:128
107	Cooper's	10	Once with 1 ml	0	0

range from 1:128 to 1:1024 for PM-3 vaccine and 0 to >1:16 <1:32 for Cooper's. As a rule, the ELISA titers were from 4 to 64 times higher than CF-titers. In PM-3 to 1.5 month ELISA-titers were 1:256 to 1:512; at 2.5 months were 1:1024; at 3 months were 1:512–1024; at 4 months, from 1:256 to 1:1024; and at 10 months, from 1:128 to 1:512 (Table 14.8).

Two of the tested sheep immunized with Cooper's had no CF antibodies; no ELISA antibodies were detected in the same samples. Another serum sample (Cooper's) with a low CF-titer of 1:8 was negative in ELISA. In the fourth sample, the CF-titer was zero and the ELISA-titer, 1:16. One of the samples with a CF-titer of 1: 8 showed ELISA-titers >1:16 <1:32.

The investigations carried out have shown that there is a correlation between the results obtained by using both serological tests.

Table 14.9 presents the results of a serum neutralization reaction carried out with blood sera from PM-3 vaccinated sheep obtained in seven different periods up to the 6th month, including post-immunization.

Table 14.9 Serum neutralization reaction with blood serum of sheep immunized with PM-3 vaccine

Ewe (No.)	% Reduction of Infectivity Produced by Sera Obtained after Vaccination in the Below-Mentioned Deadlines								Outcome of Pregnancy	Presence of Chlamydia in the Placenta
Days after Vaccination	0	28	56	84	97	112	140	180		
1	6	98	100	45	46	70	52	67	Birth	–
2	0	33	50	7	12	32	44	73	Birth	+
3	0	100	100	1	57	19	23	74	Birth	+
4	0	94	100	0	15	68	48	82	Birth	–
5	60	96	76	56	57	37	54	99	Birth	–
6	0	100	100	32	63	56	29	100	Birth	–
7	0	untested.	100	0	27	39	25	untested.	Birth	–
8	0	100	85	51	51	0	19	83	Birth	–
9	0	58	89	3	13	56	58	100	Birth	–
10	0	42	97	39	untested.	57	67	100	Birth	–
Average:	6,6	80,1	89,7	18,8	37,9	43,4	41,9	77,8	Birth	

The table shows that the mean percentages of the reduction in infectivity produced by the sera increased progressively until the second month (56th day). Later, between the second and the third month (84th day), these values, respectively, decreased the level of immunity; however, then – at the beginning of the 4th month (97th day) they rose and stabilized at a higher level. On the 6th month, they reached the level of the 28th day. The results obtained show that the PM-3 vaccine produces a very good immune response to the serum neutralization test.

14.3.1.17 Challenge inoculation

It was conducted on 40 pregnant ewes, conditionally divided into 5 groups: I – 10 ewes vaccinated once with PM-3; II – 10 ewes vaccinated twice with PM-3; III – 10 ewes vaccinated once with Cooper's; IV – 5 controls, contact, unvaccinated; and V – 5 controls, also contact, unvaccinated. In the first four groups the challenge inoculation was performed by s.c. injection of 2.5 ml of live chlamydiae at a concentration of $10^6 LD_{50}$, and in the fifth group by 2.5 ml of the suspension administered intravenously. The challenge inoculation was carried out at an average of 85 days of gestation, followed by daily clinical observations for 1 month and clinical monitoring by the end of the experiment.

The single- and double-immunized ewes with PM-3 after challenge inoculation showed no marked increase in body temperature or disorders in general condition. There were no abortions, premature births, stillbirths, or deliveries of unviable lambs.

Three of the Cooper's vaccine-immunized ewes showed an increase in body temperature of 0.5 to 1.0°C, which lasted for one week. These animals were negative for chlamydial antibodies in the previous series of serological tests. Two of the ewes aborted 40–45 days after the challenge inoculation in the last month of pregnancy. The chlamydia etiology of the abortion has been demonstrated by DEM of placentas at which high concentrations (10^8) of chlamydial EBs, CBs and RBs were discovered.

In the groups of unvaccinated but challenged controls, we noted an increase in body temperature by 1°C and malaise (decreased appetite, stinginess) in 4 out of 10 animals – 3 i.v. injected and 1 s.c. injected. Subsequently, one of the ewes that were i.v. injected aborted 35 days post-challenge, and a second ewe, also i.v. injected, gave birth to a feeble non-viable lamb that died after 2 days. In both cases, chlamydial infection was demonstrated by placental DEM and serologically by CFT: titers 1:32–1:64 on the 30th day after abortion and on the 45th day after birth, respectively.

In 17 of the 20 ewes immunized with PM-3, the challenge inoculation did not affect the dynamics of chlamydial antibodies previously induced by the adsorbed vaccine. Obviously, the non-adsorbed antigen, as represented by living chlamydiae, is practically non-immunogenic, and therefore the vast majority of animals did not show antibody growth. The lack of a pronounced immunogenic effect and the lack of clinical manifestations following the challenge of living chlamydiae demonstrate a solid immunity and protection of the animals after single or double immunization with the PM-3 vaccine.

In the group of challenged ewes (immunized with Cooper's) immunogenic effect of the challenging un-adsorbed antigen was also not established. Here, however, we find this feature that 3 of the ewes before the challenge inoculation were with a zero titer. These three animals, apparently devoid of vaccine protection, showed clinical manifestations after challenge, including abortion. These results testify to insufficient immunity and protection of the ewes following a single immunization with Cooper's vaccine.

The above findings for the very weak immunogenic effect of non-adsorbed chlamydial antigen were confirmed by observations of the unvaccinated control group. Here again, 80% of the inoculated animals did not develop antibodies. Only 2 animals after intravenous infection were found to have low titers (1: 4). The clinical signs that occurred in 4 cases are indicative of contamination of these animals with complete lack of specific protection in the body of unvaccinated ewes.

14.3.1.18 Other clinical observations

Sheep immunized with a PM-3 vaccine showed no local or systemic reactions and therefore the vaccine is safe and non-reactogenic. All vaccinated sheep gave birth normally. There were no miscarriages and stillbirths in the whole group.

Sheep immunized with a Cooper's vaccine showed the presence of local inflammatory reactions at the site of injection. The magnitude of these events was from hazelnut or egg. These reactions are fully consistent with the information the manufacturer provides about the use of the vaccine associated with such effects. However, there were no general reactions in the vaccinated animals.

14.3.1.19 Storage of the vaccine

The vaccine should be stored in a liquid state at 4°C in a refrigerator, with daily shaking in the first week of its receipt.

14.4 Vaccines for Cattle

There is no commercial vaccine against chlamydial miscarriage in cows. This is due, on the one hand, to the inconclusive results of testing in experimental inactivated and live vaccines and, on the other hand, to underestimating the problem and insufficient research into the development of effective vaccines.

In the 1960s, McKercher et al. [1062, 1063] tested inactivated and live vaccines against epizootic bovine abortion in the United States. The inactivated vaccine was prepared according to the scheme of McEwan et al. [1018] for a vaccine against EAE. In the live vaccines, either the chlamydial agent of feline pneumonitis or the pathogen of bovine abortion is used. The strains are passed in mouse lungs over 120 transfers before the final propagation in the YS of CE. The positive effect of these vaccines consists in reducing abortions, but immunized cows harbor and excrete chlamydia [694]. In addition, on challenge of the immunity of the vaccinated cows with virulent chlamydial agent of bovine abortion, febrile response and abortions is registered in all vaccinated groups of cows. Later, McKercher et al. [1064] reported that vaccination against bovine abortion caused by chlamydia which, in earlier trials, failed to protect cattle against challenge by the pathogen administered either by s.c. or i.m. injection, protected against intradermal challenge. The authors discuss this finding in the light of the apparent epidemiological pattern of the disease, in particular regarding the possible nature of the immune response and the natural mode of transmission of the causative agent.

Unlike the above unsatisfactory results in chlamydial abortion in cattle, De Graves et al. [1065] reported that the experimental use of a *C. abortus* vaccine provided evidence of immunoprotection against *C. abortus*-induced suppression of bovine fertility. Kaltenboeck et al. [1066] interpreted these results, considering that they give reason to consider the bovine Chlamydophila infection as a pervasive, low-level infection of cattle rather than a rare, severe disease. Such infections proceed without apparent disease or with only subtle expressions of disease, but potentially have a large impact on bovine herd health and fertility [1066].

Biesenkamp-Uhe et al. [639] conducted therapeutic *C. abortus* and *C. pecorum* vaccination, as a result of which, a transient reduction of bovine mastitis associated with chlamydial infection was achieved. The experimental vaccine was prepared from inactivated *C. abortus* and *C. pecorum* elementary bodies and administered twice subcutaneously. The protective effect peaked at 11 weeks after vaccination and lasted for a total of 14 weeks. As a result of the vaccination, a significant reduction in the number of milk somatic cells has been achieved, thus reducing bovine mastitis and increasing antibody levels against Chlamydia. This vaccine, however, did not eliminate shedding of *C. abortus* in milk as detected by PCR [639]. Regardless of the temporary effect of the vaccine, this type of therapeutic immunization is useful for reduction of damage of bovine mastitis, a herd disease of critical importance for the dairy industry.

14.5 Vaccines Against Chlamydial Infections in Cats

Attempts for active prophylaxis of chlamydial conjunctivitis and respiratory diseases in cats have been conducted with inactivated and live vaccines. Wills et al. [1068] described experimentally induced chlamydial conjunctivitis in specific-pathogen-free cats exhibiting marked clinical signs. Subcutaneous vaccination with live feline pneumonitis *C. psittaci* (old classification) 4 weeks before ocular challenge significantly reduced the severity of the conjunctivitis. However, there was no effect on the shedding of organisms from the eye or on the transmission of infection to the gastrointestinal and genital tracts [1068].

In an early McKercher study, several different preparations of vaccines were tested which contained either inactivated or live organisms propagated in yolk sacs, allantoic fluid, or mouse lungs. The vaccine was administered subcutaneously. The challenge inoculation was made intra-nasally with chlamydiae propagated in the YS of CE 5 weeks later. Better results were obtained with the live vaccine as the majority of vaccinated kittens were clinically protected [1068].

Mitzel and Strating carried out vaccination against feline pneumonitis with a modified live CE-origin commercial vaccine administered intramuscularly. Thirty days later, an aerosol challenge was performed with a YS-grown chlamydial agent of feline pneumonitis. As a result, a reduction in clinical manifestations in vaccinated cats was achieved, unlike unvaccinated controls. This positive result, however, contrasts with the fact of excretion of the agent and its isolation from individual vaccinated animals on the 27th day following the aerosol challenge [1069].

Shewen et al. communicate not very satisfactory results from the vaccination of cats with of a live and four inactivated vaccines [1070]. The Cello study ends with completely negative results – a lack of protective effect of vaccination [1071].

The generally accepted conclusion about the efficacy of these vaccines is that they do not provide complete protection from infection and shedding of chlamydia after challenge, but they reduce and minimize replication of the pathogen and disease severity and duration [767, 769, 1072, 1074]. The vaccines have the potential to cause serious side effects. Transient fever, lethargy, anorexia, and lameness have been described after vaccination in a certain percentage of cats [769, 1073]. The feline chlamydia vaccine is considered a non-core vaccine, meaning it is an optional vaccine that cats may benefit from based on their risk of exposure to the disease [769, 1073]. Risk situations for chlamydial cat infection are their introduction and residence in catteries with laboratory-proven endemic chlamydiosis, participations in cat shows, exhibitions, stud cats, etc. Cats who go outside, live with other cats, or visit grooming or boarding facilities are at greater risk of exposure than cats who stay indoors and have limited contact with other cats [767, 769, 1073, 1074].

When the condition of the individual animal and the risks of chlamydial infection are assessed, a decision may be made to vaccinate it. In these cases the instructions of the manufacturer and of the authorized professional organization are observed. For example, according to the recommendations of the American Feline Practitioners Vaccination Guidelines [1072], the feline chlamydiosis vaccine is recommended for cats according to the following schedule: Kittens should receive an initial dose as early as 9 weeks of age; and the second dose is administered 3–4 weeks later. Adult cats (older than 16 weeks) should receive an initial dose, followed by a booster 3–4 weeks later [1072]. Tasker points out that limited information is available about the duration of immunity [1075]. There is some evidence that previously infected cats can become vulnerable to re-infection after a year or more. Annual boosters are recommended for cats that are at continued risk of exposure to infection.

14.6 Attempts to Develop Vaccines Against C. trachomatis-induced Diseases in Humans

The widespread prevalence of human infections caused by *C. trachomatis* in the world and their great health and social significance has long been a cause for a search for opportunities for active prophylaxis. Vaccine development is key to preventing or eliminating the severe reproductive consequences of untreated infections such as PID and tubal factor infertility. There can be no doubt that vaccines against *C. trachomatis* are of benefit to health and likely to have a significant impact on healthcare costs [1076]. A number of attempts have been made in this regard, but there is no vaccine currently available. Elwell et al. [1110] broadened the scope of this conclusion by including *Chlamydia* spp. as important causes of diseases in humans, for which no effective vaccine exists. Apparently, research over the last thirty years has been unproductive [1077]. The main efforts over the years have been directed toward identifying and testing vaccine candidates and appropriate delivery systems.

Hafner and Timms [1078] indicate that virtually all efforts toward vaccine development have focused on the use of recombinant proteins either singly or occasionally as mixtures, but the results relating to the levels of "protection" obtained have usually been relatively modest. The most commonly used model for the evaluation of the experimental vaccines was the *C. muridarum*-mouse model. For the same purpose, the guinea-pig model was successfully used – Andrew et al. [1079]; Wali et al. [1080]; Neuendorf et al. [1081].

It is noteworthy that a number of new works on experimental vaccines use the nasal method of administration considered to be more effective than traditional old-fashioned methods of s.c. or i.m. administration [1077, 1079, 1080, 1082]. According to U.S. scientists from Harvard Medical School, their vaccine is administered at a mucosal site, such as in the nose or under the tongue, using nanoparticles that are small enough to travel to the lymph nodes, producing an immune response that can be spread to other parts of the body [1082].

Encouraging results are reported by Canadian researchers Bulir et al. In their study, they show that a novel chlamydial antigen known as BD584 is a promising potential vaccine candidate for *Chlamydia trachomatis*. This fusion protein antigen (BD584) consists of three T3SS proteins from *C. trachomatis* (CopB, CopD, and CT584). Intranasal immunization with BD584 elicited serum neutralizing antibodies in mice that inhibited *C. trachomatis* infection *in vitro*. On intra-vaginally infecting the immunized mice with *Chlamydia muridarum*, they found the antigen reduced chlamydial shedding from the vagina at the peak of infection by 95% and cleared the infection sooner than control mice.The immunization with the BD584 antigen also decreased the rate of hydrosalpinx, another *C. trachomatis* condition which involves the fallopian tubes by 87.5% compared to control mice [1077].

We will note that there is currently no vaccine to protect against *Chlamydia pneumoniae.*

Bibliography

[1] Halberstädter, L., and von Prowazek, S. (1907). Über Zelleinschlüsse parasitärer Natur beim Trachom. *Arb. Gesundheitsamt Berlin* 26, 44–47.

[2] Morange, A. (1895). *De la psittacose, ou infection spéciale déterminée par des perruches*. Thesis, Paris.

[3] Levinthal, W. (1930). Die Ätiologie der Psittakosis. *Klin. Wschr.* 9, 654–659.

[4] Coles, A. C. (1930). Microorganisms in psittacosis. *Lancet* 1, 1011–1012.

[5] Lillie, R. D. (1930). Psittacosis: rickettsia – like inclusions in man and experimental animals. *Public Health Rep.* 45, 773–778.

[6] Bedson, S. P., Western, S. L., and Simpson, S. L. (1930). Observation on the etiology of psittacosis. *Lancet* 1, 235–236.

[7] Bedson, S. P., and Bland, J. O. W. (1932). Morphological study of the psittacosis virus, with a description of the developmental cycle. *Brit. J. Exp. Path.* 31, 461–466.

[8] Bedson, S. P., and Bland, J. O. W. (1934). The developmental forms of psittacosis virus. *Brit. J. Exp. Path.* 15, 243–254.

[9] Yanamura, H. Y., and Meyer, K. F. (1941). Studies on the virus of psittacosis cultivated in vitro. *J. Inf. Dis.* 68, 1–15.

[10] Haagen, E., and Maurer, G. (1938). Ueber eine auf den Menschen übcrtragbarc Viruskrankhcit bci Sturmvögcln und ihrc Bczichung zur Psitakose. *Zbl. Bakt. (I Orig.)* 143, 81–93.

[11] Gönnert, R. (1941). Die Bronchopneumonie, eine neue Viruskrankheit der maus. *Zbl. Bact. (Orig)* 148, 161–174.

[12] Greig, J. R. (1936). Enzootic abortion in ewes; a preliminary note. *Vet. Rec.* 48, 1225–1227.

[13] Stamp, J. T., McEwen, A. D., Watt, J. D., and Nisbet, D. J. (1950). Enzootic abortion in ewes. I. Transmission of the disease. *Vet. Rec.* 62, 251–254.

[14] Tang, F. F., Chang, H. L., Huang, Y. T., and Wang, K. C. (1957). Studies on the etiology of trachoma with special reference to isolation of the virus in chick embryo. *Clin. Med. J. (Peking)* 75, 429–446.
[15] Jones, B. R., Collier, L. H., and Smith, C. H. (1957). Isolation of virus from inclusion blenorrhea. *Lancet* 1, 902–905.
[16] Gordon, F. B., and Quan, A. L. (1965). Isolation of the trachoma agent in cell cultures. *Proc. Soc. Exp. Biol. Med.* 118, 354–359.
[17] Grayston, J. T., Kuo, C. C., Wang, S. P., and Altman, J. (1986). *Chlamydia psittaci* strain TWAR, isolated in acute respiratory tract infections. *N. Engl. J. Med.* 315, 161–168.
[18] Semerdjiev, B., and Ognianov, D. (1969). *Neorickettsioses in Domestic Animals*. Sofia: Zemizdat, 252.
[19] Storz, J. (1971). Chlamydia and Chlamydia-induced diseases. Springfield, IL: Charles C. Thomas Publishers, 1971, 358.
[20] Schachter, J., and Dawson, C. R. (1978). Human chlamydial infections. Littleton, MA: PSG Publishing Company, Inc., 273.
[21] Terskih, I. I. (1979). *Ornithosis and Other Chlamydial Infections*. Moscow: Medicine, 223.
[22] Martinov, S., and Popov, G. (1980). *Chlamydia and Chlamydial Infections. Natl. Agro-Industrial Union – Center for Scientific, Technical and Economic Information*, Sofia, 88.
[23] Martinov, S. P. (1983). *Studies of Some Biological, Morphological and Immunological Properties of the Chlamydiae*. Ph.D. dissertation, Central Res. Vet. Med. Inst., Sofia, 318.
[24] Popov, G, Martinov, S., and Dimitrov, K. (1989). *Sexually Transmitted Chlamydial Infections*. Sofia: Medicine and Phisical Culture Publisher, 166.
[25] Mardh, P.-A., Puolakkainen, M., and Paavonen, J. (1990). *Chlamydia*. New York, NY: Plenum Medical Book Company.
[26] Kang, J.-B, Martinov, S, and Jun, M.-H. (1996). "Recent trends in diagnosis and control of chlamydial infections," in *Current Approaches and Visions for Veterinary Science and Technology*, Institute of Veterinary Science, Chungnam National University, Daejeon, 20–73.
[27] Aitken, I. D. (2003). "Ovine enzootic abortion," in *Rickettsial and Chlamydial diseases of Domestic Animals*, eds Z. Voldehivwet, and M. Ristic (Oxford: Pergamon Press, Ltd.), 349–360.
[28] Longbottom, D., and Rocchi, M. (eds). (2006). "Diagnosis, pathogenesis and control of animal chlamydioses," in *Proceedings of the 4th*

Annual Workshop of COST Action 855, (Edinburgh: Moredun Res. Institute), 196.

[29] Niemczuk, K., Sachse, K., and Sprague, L. D. (eds). (2007). "Pathogenesis, epidemiology and zoonotic importance of animal chlamydioses," in *Proceedings of the 5th Annual Workshop of COST Action 855* (Pulawy: Natl. Vet. Res. Inst.), 94.

[30] Kuyumdjiev, I. (1956). Psittacosis, Reports of Dept. of Biol. and Med. Sciences. *Bulg. Acad. Sci.* 1:87.

[31] Kuyumdjiev, I. (1959). Reports of Dept. of Biol. and Med. Sciences, *Bulg. Acad. Sci.* 10, 75–86, (by Semerdjiev, B., Ognianov, D., 1969. Neorickettsioses in domestic animals. Zemizdat, Sofia, 252).

[32] Vachev, B., Savov, D., and Genov, I. (1956). *"Psittacosis in parrots,"* in *Proceedings of the Res. Institutes at Ministry of Agriculture of Bulgaria, VII*, 43–46.

[33] Mincheva, N., Genov, I. (1961). *Ornithosis in Chickens.* Agricultural Thought, 1.

[34] Genov, I., and Savov, D. (1962). Scientific articles of Vet. Inst. *Infect. Parasit. Dis.*, Sofia [by Semerdjiev, B., Ognianov, D. (1969). Neorickettsioses in domestic animals. Sofia: Zemizdat, 252].

[35] Nikolov, Z. V. (1965). Epidemiological study of ornithosis. *Epidemiol. Microbiol. Infect. Dis.* 2, 68–80.

[36] Nikolov, Z. V. (1965). Contemporary medicine, 16, 537–543 [by Semerdjiev, B., and Ognianov, D. (1969). Neorickettsioses in domestic animals. Sofia: Zemizdat, 252].

[37] Ognianov, D. (1968). *Studies on the viral enzootic abortion in sheep.* Ph.D. dissertation, VIIPD, Sofia, 174.

[38] Ognianov, D., Panova, M., and Arnaudov, H. (1974). A case of sporadic encephalomyelitis in buffalo calves. *Vet. Sci.* 1, 3–10.

[39] Semerdjiev, B., Ognianov, D., and Makaveeva-Simova, E. (1966). Cases of viral abortion in goats. *Vet. Sci. III*, 2, 115–121.

[40] Semerdjiev, B., Ognianov, D., and Makaveeva–Simova, E. (1964). Isolierung von pneumoniekranken Kälbern. *Zbl. Bact. (Orig)*, 192, 12–16.

[41] Martinov, S., and Popov, G. (1979). Chlamydiae from avian and mammalian origin and relevance to human health. *Vet. Sbirka* 12, 13–16.

[42] Guenov, I. (1961). Etudes sur la péricardite fibrineuse des porcelets due au virus de l'ornithose. *Bull. Office Int. Epizoot.* 55, 1465–1473.

[43] Natscheff, B., Gabraschanski, P, Ognjanoff, D., Djuroff, D., and Gentscheff, P. (1965). *Beitrag zur Frage der Ornithose beim Schwein und ihren Bethandlung*. Berlin: München Tieraerztl. Woschr. 79, 368–369.

[44] Martinov, S., Panova, M., and Popov, G. (1980). New enzootics of chlamydial abortion in sheep. *Vet. Sbirka* 12, 12–15.

[45] Martinov, S. P., and Popov, G. V. (1980). Nachweis von Chlamydien bei enkrankten personen. Z. der Humboldt-University zu Berlin. *Math.- Nat, R., XXIX*, 1, 11–17.

[46] Martinov, S. P., and Popov, G. V. (1982). Inherent trends in the pathogenicity of various chlamydial strains for chicken embryos upon infection in the yolk sac. *Compt. Rend. Acad. Bulg. Sci.* 35, 1017–1020.

[47] Martinov, S., Schoilev, H., and Popov, G. (1985). Study on the chlamydial infection in pericarditis in pigs. *Vet. Sci. XXII* 6, 20–26.

[48] Martinov, S., and Popov, G. (1988). "*Chlamydia psittaci* Infection in Eastern Europe," in *Proceedings of the Eur. Soc. Chlamydia Research*, Bologna, 69.

[49] Martinov, S. P., Popov, G. V., and Schoilev, H. (1988). Ätiologische und pathomorphologische parallelen bein Chlamydien-abort der schafe. *DTW* 95, 3, 114–116.

[50] Martinov, S. P., and Popov, G. V. (1992). Electronmicroscopic diagnosis of chlamydial infections. *Symposium of W.A.V.M.I*, Davis, CA, 394.

[51] Martinov, S. P. (1996). "A vaccine against chlamydial abortion in sheep," in *Proceedings of the The Korea - Bulgaria Int. Seminar on Q-fever and Chlamydiosis*, Anyang, 5.

[52] Martinov, S. P., Tcherneva, E., and Jersek, B. (1996). Repetitive extragenic palindromic sequence polymorphism based polymerase chain reaction (REP-PCR) is a useful tool in diagnosis of *Chlamydia psittaci* infection in sheep. *Infect. Diseas. Obstetr. Gynecol. (USA)*, 4, 180–183.

[53] Martinov, S. P. (2006). "Avian chlamydiosis in Bulgaria and zoonotic imlications," in *Diagnosis, Pathogenesis and Control of Animal Chlamydioses,* eds D. Longbottom and M. Rocchi (Scotland: Meigle Color Printers), 159–169.

[54] Martinov, S. P. (2009). Etiological and clinical studies of chlamydioses in domestic ruminants. *Vet Medicine. XIII* 1–2, 26–34.

[55] Martinov, S. P., Popov, G. V., and Ognyanov, D. K. (1981). Etiological study of orchiepididymitis. *Compt. Rend. Acad. Bulg. Sci.* 34, 863–866.

[56] Martinov, S. P., Popov, G., and Dimitrov, K. (1983). Serological studies of the chlamydial infection in nongonococcal urethritis and cervicitis. *Dermatol. Venerol.* 22, 30–37.

[57] Martinov, S. P., Popov, G. V., Dimitrov, K., and Bonev, A. (1984). Studies on the chlamydial infection in postgonococcal uretrites in men. *Compt. Trend. Acad. Bulg. Sci.* 37, 831–834.

[58] Martinov, S. P., Popov, G. V., and Dimitrov, K. D, (1987). Untersunhungen über die Chlamydien infection beim Reitersyndrome (RS). *H+G Z. Hautkr. (Germany)*, 62, 216–219.

[59] Martinov, S., Popov, G., Gavrilov, T., and Zarcheva, V. (1987). Chlamydia trachomatis inclusion conjunctivitis and pneumonia in neonates. *Pediatrics XXVI* 6, 49–55.

[60] Martinov, S. P., Popov, G. V., Dimitrov, K., Zlatkov, N. B., and Bonev, A. (1988). "*Chlamydia trachomatis* infections in Bulgaria," in *Proceedings of the Eur. Soc. Chlamydia Research*, Bologna, 21.

[61] Bedson, S. P. (1958). The Harben Lectures. The psittacosis-lymphogranuloma group of infective agents. Lecture I: The history and the characters of these agents. *J. Roy. Inst. Pub. Health. Hyg.* 22, 67–78.

[62] Moulder, J. W. (1966). The relation of the psittacosis group (chlamydiae) to bacteria and viruses. *Ann. Rev. Microbiol.* 20, 107–130.

[63] Moulder, J. W. (1968). The life and death of the psittacosis virus. *Hosp. Pract.* 3, 35–45.

[64] Weiss, E. (1967). Comparative metabolism of Chlamydia with special reference to catabolic activities. *Am. J. Ophthal.* 63, 1098/72–1101/75.

[65] Weiss, E. (1968). Comparative metabolism of rickettsiae and other host dependent bacteria. *Zbl. Bact. (Orig)* 206, 292–298.

[66] Storz, J., and Page, L. A. (1971). Taxonomy of the Chlamydiae: Reasons for classifying organisms of the genus *Chlamydia*, family *Chlamydiaceae*, in a separate order, *Chlamydiales* ord. nov. *Int. J. Sys. Bacteriol.* 21, 332–334.

[67] Friis, R. R. (1972). Interaction of L cells and *Chlamydia psittaci*: Entry of the parasite and host responses to its development. *J. Bacteriol.* 110, 706–721.

[68] Giroud, P. (1969). Des agents de la psittacose a ceux du tracome. Agents bedsoniens ou neorickettsiens. *La Presse Med.* 77, 475–479.

[69] Page, L. A. (1966). Revision of the family *Chlamydiaceae* Rake (*Rickettsiales*): Unification of the the psittacosis-lymphogranuloma venereum-trachoma group of organisms in the genus *Chlamydia* Jones, Rake and Stearns, 1945. *Int. J. Sys. Bacteriol.* 16, 223–252.

[70] Page, L. A. (1971). "Order II. Chlamydiales Storz and Page, Part 18. The Rickettsias, pp 914–928," in *Bergey's Manual of Determinative Bacteriology*, 8th Edn, eds R. E. Buchaman and N. E. Gibbons (Baltimore: The Willians and Wilkins Co).

[71] Gordon, F. B., and Quan, A. L. (1965). Occurrence of glycogen in inclusions of the psittacosis-lymphogranuloma venereum-trachoma agents. *J. Infect. Dis.* 115, 186–196.

[72] Lin, H.-S, and Moulder, J. W. (1966). Patterns of response to sulfadiazine, D-cycloserine and D-alanine in members of psittacosis group. *J. Infect. Dis.* 116, 372–376.

[73] Herring, A. J. (1993). Typing *Chlamydia psittaci* - a review of methods and recent fidings. *Br. Vet.* 149, 455–475.

[74] Storz, J., and Kaltenboeck, B. (1993). "The Chlamydiales," in *Rickettsial and Chlamydial Diseases of Domestic Animals*, eds Z. Woldehiwet and M. Ristic (Oxford: Pergamon Press), 363–393.

[75] Everett, K. D., and Anderson, A. A. (1999). Identification of nine species of the *Chlamydiaceae* using PCR-RFLP. *Int. J. Syst. Bacteriol.* 49 (Pt 2), 803–813.

[76] Everett, K. D., Bush, R. M., and Anderson, A. A. (1999). Emended description of of the order Chlamydiales, proposal of *Parachlamydiaceae fam.* nov. and *Simkaniaceae fam.* nov., each containing one monotypic genus, revised taxonomy of the family *Chlamydiaceae*, including a new genus and five new species, and standarts for the identification of organisms. *Int. J. Syst. Bacteriol.* 49 (Pt 2), 415–440.

[77] Bush, R. M., and Everett, K. D. (2001). Molecular evolution of the *Chlamydiaceae. Int. J. Syst. Evol. Microbiol.* 51, 203–220.

[78] Schachter, J., Stephens, R. S., Timms, P., Kuo, C., Bavoil, P. M., Birkelund, S., et al. (2001). Radical changes to taxonomy are not necessary yet. *Int. J. Syst. Evol. Microbiol.* 51, 249.

[79] Stephens, R. S., Myers, G., Eppinger, M., and Bavoil, P. M. (2009). Divergence without difference: phylogenetics and taxonomy resolved. *FEMS Immunol. Med. Microbiol.* 55, 115–119.

[80] Greub, G. (2010). Int. Committee on Systematics of Prokaryotes – Subcommittee on the taxonomy of the *Chlamydiaceae*. Minutes of the inaugural dosed meeting, 21 March 2009, Little Rock, AR. *Int. J. Syst. Evol. Microbiol.* 60, 2691–2693.

[81] Greub, G. (2010). Int. Committee on Systematics of Prokaryotes - Subcommittee on the taxonomy of the *Chlamydiaceae*. Minutes of the inaugural dosed meeting, 21 June 2010, Hof. Bei Salzburg, Austria. *Int. J. Syst. Evol. Microbiol.* 60, 2694–2694.

[82] Kuo, C.-C., Stephens, R. S., Bavoil, P. M., Kaltenboeck, B. (2015). "Chlamydia," in *Bergy's Manual of Systematics of Archaea and Bacteria* (Berlin: Springer), 1–28.

[83] OIE Terrestrial Manual (2017). *Chapter 3.1 Avian Chlamydiosis*, 401–414.

[84] Sachse, K., Laroucau, K., Riege, K., Wehner, S., Dilcher, M., Creasy, H. H., et al. (2014). Evidence for the existence of two new members of the family *Chlamydiaceae* and proposal of *Chlamydia avium* sp. nov. and *Chlamydia galinacea* sp. nov. *Syst. Appl. Microbiol.* 37, 79–88.

[85] Sachse, K., Bavoil, P. M., Kaltenboeck, B., Stephens, R. S., Kuo, C. C., Rosselló-Móra, R., et al. (2015). Emendation of the family *Chlamydiaceae*: proposal of a single genus, *Chlamydia*, to include all currently recognized species. *Syst. Appl. Microbiol.* 38, 99–103.

[86] Lindner, K. (1909). Uebertragungsversuche von gonokokkenfreien Blennorrhoea neonatorum auf Affen. *Wien. Klin. Wochenschr.* 22, 1555.

[87] Schmeichler, L. (1909). Ueber Chlamydozoenbefunde bei nicht-gonorrhoischer Blennorrhöe der Neugeborenen. *Klin. Wochenschr.* 46, 2057–2058.

[88] Stargardt, K. (1909). Über Epithelzellveränderungen beim Trachom und andern Konjunctivalerkrankungen. *Graefe's Arch. Ophtalmol.* 69, 525–542.

[89] Heymann, B. (1910). Ueber die Fundorte der Powazek'schen Körperchen. *Berl Klin. Wochenschr.* 47, 663–666.

[90] Lindner, K. (1910). Zur Ätiologie der gonokokkenfreien Urethritis. *Wien. Klin. Wochenschr.* 23, 283–284.

[91] Carlson, J. R., Marshall, R. L., and Giligan, P. H. (1981). Abstr. *AMASM* 1, 292.

[92] Shatkin, A. A., and Mavrov, I. I. (1983). *Urogenital chlamydiosis.* Kiev, Health.

[93] Crocker, T. T., and Williams, R. C. (1955). Electron microscopic morphology of frozen-dried particles of meningo-pneumonitis. *Proc. Soc. Exp. Biol. Med.* 88, 378–380.

[94] Plas, P. (2015). *The Aurion Immunogold Silver Staining Workshop.* University of Maryland, Baltimore, EM Core ImagingEM Core Imaging Faculty [Accessed: October, 2016].

[95] Murray, P. R., Rosenthal, M. A., and Pfaller, M. A. (2006). *Microbiologica Medica*, 5th Edition. Madrid: Elsevier, 523–531.

[96] Murray, P. R., Rosenthal, M. A., and Pfaller, M. A. (2015). *Chapter 35. Chlamydia and Chlamydophila, pp. 353–354. Medical Microbiology.* London: Elsevier Health Sciences, 836.

[97] Popov, G, Martinov, S., and Dimitrov, K. (1989). Sexually Transmitted Chlamydial Infections. Sofia: Medicine and Phisical Culture Publisher, p. 166.

[98] Matsumoto, A. M., Manire, G. P. (1970). Electron microscopic observations on the fine structure of cell walls of *Chlamydia psittaci. J. Bact.* 104, 803–818.

[99] Popov, G., and Martinov, S. (1980). Morphogenesis of Chlamydia. *Vet. Sci. XVII* 3, 60–69.

[100] Popov, G. V., and Martinov, S. P. (1980). Morphologie und Morphologenese einiger Chlamydienstamme. Wiss. Z. der Humboldt-Universitat zu Berlin. *Math.-Nat. R. XXIX* 1, 11–17.

[101] Matsumoto, A. M. (1979). Recent progress of Electron microscopy in microbiology and its development in future: from a study of the obligate intracellular parasites, Chlamydia organisms. *J. Electron Micros.* 28 (Suppl), 57–64.

[102] Ward, M. E. (1983). Chlamydial classification, development and structure. *Brit. Med. Bull.* 39, 109–115.

[103] Matsumoto, A., and Manire, G. (1970). Electron microscopic observations on the effect of penicillin on the morphology of *Chlamydia psittaci. J. Bact.* 101, 278–285.

[104] Wyrick, P. B. (2010). *Chlamydia trachomatis* persistence in vitro: An Overview. *J. Infect. Dis.* 202 (Suppl. 2), 588–595.

[105] Martinov, S., Dimitrov, K., and Popov, G. (1988). Studies on cervicitis, caused by *Chlamydia trachomatis. Dermatol. I Venerol. XXVII* 2, 12–19.

[106] Santalucia, M., and Farinaty, A. (2015). Chlamydia trachomatis infections and their impact in the adolescent population. *DST–J. Bras. Doencas. Sex. Transm.* 27, 112–115.

[107] Cheema, M. A., Schumacher, H. R., and Hudson, A. P. (1991). RNA – directed molecular hybridization screening evidence for innaparent chlamydial infection. *Am. J. Med. Sci.* 302, 261–268.

[108] Popov, G., and Martinov, S. (1982). Morphologic and morphogenetic characteristics of Chlamydia infection in diseased animals. Vet. Sci. XIX, 6, 15–23.

[109] Popov, G., and Martinov, S. (1984). Contemporary state of the problem Chlamydia and Chlamydial diseases. *Nature (Sofia)*, 3, 31–35.

[110] Pospischil, A., Borel, N., Chowdhury, E. H., and Guscetti, F. (2009). Aberant chlamydial developmental forms in the gastrointestinal tract of pigs spontaneously and experimentally infected with *Chlamydia suis*. *Vet. Microbiol.* 135, 147–156.

[111] Tanami, Y., and Tamada, Y. (1973). Miniature cell formation in Chlamydia psittaci. *J. Bacteriol.* 114, 408–412.

[112] Phillips, D. M. (1984). Ultrasructure of *Chlamydia trachomatis* infection in the mouse oviduct. *J. Ultrastruct. Res.* 88, 244–256.

[113] Gregory, W. Gardner, W., Byrne, C. T., and Moulder, J. W. (1979). Arrays of hemispheric surface projections on *Chlamydia psittaci* and *Chlamydia trachomatis* observed by scanning electron microscopy. *J. Bacteriol.* 138, 241–244.

[114] Matsumoto, A. (1973). Fine structures of cell envelops of Chlamydia organisms as revealed by freese - etching and negative staining technics. *J. Bacteriol.* 116, 1355–1363.

[115] Nichols, B. A., Setzer, P. Y., Pang, R., and Dawson, C. R. (1985). New view of the surface projections of Chlamydia trachomatis. *J. Bacteriol.* 164, 344–349.

[116] Matsumoto, A., Ikegami, M., Uehira, K., Ohmori, K., and Tanaka, Y. (1979). The morphology of *Chlamydia trachomatis* plasmid - free strains. Symposium on Chlamydia infections. *Japan Soc. for Chlamydia Research*, 15–18, Life Science Medica Co., Tokyo.

[117] Higashi, N. (1965). Electron microscopic studies on the mode of reproduction of trachoma virus (psittacosis virus) in cell cultures. *Exp. Mol. Path.* 4, 24–39.

[118] Anderson, D. R., Hopps, H. E., Barile, M. F., and Bernheim, B. C. (1965). Comparision of the ultrastructure of several rickettsiae, ornithosis virus and *Mycoplasma* in tissue cultures. *J. Bact.* 90, 1387–1404.

[119] Cultip, R. C. (1970). Electromikroscopie von Zellkulturen, die mit Polyarthritis der Lämer verursachenden Chlamydien in fiziert wurden. *Inf. Immun.* 1, 499–502.

[120] Friis, R. R. (1972). Interaction of L cells and *Chlamydia psittaci*. Entry of the parasite and host responses to its development. *J. Bacteriol.* 110, 706–721.

[121] Costerton, J. W., Poffenroth, L., Wilt, J. C., and Kordova, N. (1976). Veränderungen in der Ultrastructur von aus der Allantoisflüssigkeit von Hühnerembryonen gewonnenen *Chlamydia psittaci* 6BC. *Can. J. Microbiol.* 22, 9–15.

[122] Eb, F., Orfila, J., and Devauchelle, G. (1973). The polymorphism of *Chlamydia trachomatis* in cell cultures. *Ann. Microbiol. (Paris)*, 124B, 41–71.

[123] Todd, W., Doughri, A., and Storz, J. (1976). Ultrastrukturelle Veränderungen in Wirtszellorganellen im Verlauf des chlamydialen Entwicklungszyklus. *Zbl. Bact. Hyg. I. Abt. Orig. A* 236, 359–376.

[124] Litwin, J. Officer, J., Brown, E., and Moulder, J. W. (1961). A comparative study of the growth cycles of different members of the psittacosis group in different host cells. *J. Infect. Dis.* 109, 251–279.

[125] Martinov, S. P. (1983). *Studies of Some Biological, Morphological and Immunological Properties of the Chlamydiae. Ph.D.* dissertation, Central Res. Vet. Med. Inst., Sofia, 318.

[126] Podolyan, V. Y., Miljutin, V. N., Gudima, O. S., and Lukina, R. N. (1964). Morphogenesis of ornithosis virus. *Vop. Virus* 9, 208–212.

[127] Schachter, J. (1978). Chlamydial infections. *N. Engl. J. Med.* 298, 428–435.

[128] Corsaro, D., Valassina, M., and Venditti, D. (2003). Increasing diversity within Chlamydiae. *Crit. Rev. Microbiol.* 29, 37–78.

[129] Becerra, V. M. (1969). *Conditions for the Interaction of Chlamydial (Psittacosis) Agents with Cultured Cells.* M. S. thesis, Colorado State University, Fort Collins.

[130] Avakian, A. A., Bikovski, A. F. (1970). *Atlas of Anatomy and Ontogenesis of the Viruses in Humans and Animals.* Moscow: Medicine.

[131] Stokes, G. V. (1973). Formation and destruction of internal membranes in L cells infected with *Chlamydia psittaci*. *Infect. Immun.* 7, 173–177.

[132] Kordova, N. (1978). Chlamydiae, rickettsiae and their cell wall defective variants. *Can. J. Microbiol.* 24, 339–352.

[133] Kramer, M. J., and Gordon, F. B. (1971). Urtrastructural analysis of the effect of penicillin and chlortetracycline on the development of a genital tract *Chlamydia. Infect. Immun.* 3, 333–341.
[134] Tamura, A. (1967). Isolation of ribosome particles from meningopneumonitis organisms. *J. Bact.* 93, 2008–2016.
[135] Tamura, A., and Iwanaga (1965). RNA Synthesis in cells infected with meningo- pneumonitis virus. *J. Mol. Biol.* 11, 97–108.
[136] Becker, Y., and Yall, A. (1974). Deoxyribonucleic and acid-dependant ribonucleic acid polymerase activity in purified *Chlamydia trachomatis* initial bodies. *Isr. J. Med. Sci.* 10, 777–781.
[137] Tanaka, A. (1976). Detection of DNA polymerase activity in *Chlamydia psittaci. Jpn. J. Exp. Med.* 46, 181–185.
[138] Storey, C., Lusher, M., Yates, P., and Richmond, S. (1993). Evidence of *Chlamydia pneumoniae* of non-human origin. *J. Gen. Microbiol.* 139, 2621–2626.
[139] Callegos-Avila, A., Rodriges, J. A., Ortega, M., and Jaramillo-Rangel, M. (2010). "Infection and phagocytosis: analysis of semen with transmission electron microscopy." in *Microscopy: Science, Technology, Application and Education,* eds A. Mendes-Vilas and L. Dias (Badajoz: Formatex Research Center), 85–91.
[140] Lepinay, A., Orfila, J., Anteunls, A., Boutry, J. M., Orme-Rossell, L., and Robineaux, R. (1970). Etude en microscopie electronique du development et de la morphologie de l'agent de l'ornithose dans les macrophages de souris. *Ann. Inst. Pasteur*, 119, 222–231.
[141] Popov, G. V, Martinov, S. P., and Dimitrov. K. D. (1988). Morphology and morphogenesis of *C. trachomatis* in the sexually transmitted urogenital infections. *Proc. Eur. Soc. Chlamydia Res*, 103.
[142] Abdel Rahman, Y. M., and Belland, R. (2005). The chlamydial developmental cycle. *FEMS Microbiol. Rev.* 29, 949–959.
[143] Popov, G. V, and Martinov, S. P. (1982). Electron microscopic diagnostics of *Chlamydiae* in animals, birds and man. *Compt. Trend. Acad. Bulg. Sci.* 35, 561–564
[144] Popov, G., and Martinov, S. (1982). Studies on the electron microscope diagnostics of *Chlamydia. Vet. Sci. XIX* 4, 3–12.
[145] Popov, G. V, and Martinov, S. P. (1985). "Electronmicroscopic diagnosis of the Chlamydial abortion in ewes," in *Proceedings of the International Meeting Edinburgh, Moredun Inst* [Accessed: June 23–26, 1985].

[146] Popov, G. V, Martinov, S. P., and Dimitrov, K. (1985). Direct EM-diagnostics of the Chlamydial urethrites and cervicites in man. *Compt. Rend. Acad. Bulg. Sci.* 38, 133–136.
[147] Popov, G. V, and Martinov, S. P. (1986). "Electronmicroscopic diagnosis of the chlamydial abortion in sheep," in *Proceedings of the 4th Inter. Symposium of Vet. Lab. Diagnosticians*, Amsterdam, 737–740.
[148] Popov, G. V, Martinov, S. P., and Dimitrov. K. D. (1988). "Methods for direct electronmicroscopic diagnostics of the Chlamydia sexually transmitted infections," in *Proceedings of the Eur. Soc. Chlamydia Research*, Bologna, 241.
[149] Martinov, S. P., Popov, G. V. (1992). "Electronmicroscopic diagnosis of Chlamydial infections," in *Proceedings of the Symposium of W.A.V.M.I*, Davis, CA, 394.
[150] Martinov, S. P., Popov, G. V., and Bahchenvandzhieva, M. P. (1980). Isolation of *Chlamydia psittaci* upon polyarthritis in calves. *Compt. Rend. Acad. Bulg. Sci.* 33, 1425–1428.
[151] Martinov, S., Buchvarova, Y, Ognyanov, D., and Panova. M. (1978). "On the etiology of the infectious kerotoconjuctivitis in cattle," in *Proceedings of the 7th Int. Congress of Infect. and Parasit. Diseases*, Varna, 549–552.
[152] Popov, G. V., Martinov, S. P., Hristov, L., Mateva, M. H., and Pavlov, N. (1980). Isolation of Chlamydia from Rodents, reservoirs of HFRS in natural foci of this infection. *Compt. Rend. Acad. Bulg. Sci.* 33, 1697–1700.
[153] Manor, E., and Sarov, I. (1986). Fate of *Chlamydia trachomatis* in human monocytes and monocyte-derived macrophages. *Infect. Immun.* 54, 90–95.
[154] Zvillich, M., and Sarov, I. (1985). Interaction between human polymorphonuclear leucocytes and *Chlamydia trachomatis* elementary bodies: electron microscopy and chemiluminiscent response. *Gen. Microbiol.* 131, 2627–2635.
[155] Moulder, J. W. (1991). Interaction of chlamydiae and host cells in vitro. *Microbiol. Rev.* 55, 143–190.
[156] Register, K. B., Morgan, P. A., and Wyrick, B. (1986). Interaction between *Chlamydia* spp. and human polymorphonuclear leucocytes in vitro. *Infect. Immun.* 52, 664–670.
[157] Buendia, A. J., de Oka, R. M., Navarro, J. A., Sanchez, J., Cuello, F., and Salinas, J. (1999). Role of polymorphonuclear neutrophils

in a murine model of *Chlamydia psittaci* - induced abortion. *Infect. Immun.* 67, 2110–2116.
[158] Young, E. C., Klebanoff, S. J., and Kuo, C. C. (1982). Toxic effect of human polymorphonuclear leucocytes on *Chlamydia trachomatis*. *Infect. Immun.* 37, 422–426.
[159] Ojcius, D. M., Souque, P., Perfettini, J. L, and Dautry-Warsat, A. (1998). Apoptosis of epithelial cells and macrophages due to infection with the obligate intracellular pathogen *Chlamydia psittaci*. *J. Immunol.* 161, 4220–4226.
[160] Gibellini, D., Panaya, R., and Rumpianesi, F. (1998). Induction of apoptosis by *Chlamydia psittaci* and *Chlamydia trachomatis* infection in tissue culture cells. *Zblt. für Bact.* 288, 35–43.
[161] Mital, J. (2013). Role for chlamydial inclusion membrane proteins in inclusion membrane structure and biogenesis. *Plos One*, 8:e63426.
[162] Gauliard, E., Quellette, S. P., Rueden, K. J., and Ladani, D. (2015). Characterization of interactions between inclusion membrane proteins from *Chlamydia trachomatis*. *Front. Cell Infect. Microbiol.* 5:13.
[163] Agaisse, H., and Derre, I. (2014). Expression of the effector protein IncD in *Chlamydia trachomatis* mediates recruitment of the endoplasmatic reticulum-resident protein VAPB to the inclusion membrane. *Infect. Immun.* 82, 2037–2047.
[164] Moore, E. R., and Quellette, S. P. (2014). Reconceptualizing the chlamydial inclusion as a pathogen - specified parasitic organelle: an expanded role for Inc proteins. *Front. Cell Infect. Microbiol.* 4:157.
[165] Martinov, S. P., Popov, G. V., and Dimitrov, K. (1985). Isolation of *Chlamydia trachomatis* in yolk sacs of chicken embryous from patients with urogenital infections. *Compt. rend. Acad. Bulg. Sci.* 38, 1257–1260.
[166] Martinov, S., and Popov, G. (1982). Diagnostic methods upon chlamydial abortion in sheep. *Vet. Sci. XIX* 6, 29–38.
[167] Martinov, S. (2009). Etiological investigations on the *Chlamydophila abortus* in goats. *Vet. Med. XIII* 3–4, 11–18.
[168] Stephens, R. S., Kalman, S., Lammel, C., Fan, J., Marathe, R., Aravind, L., et al. (1998). Genome sequence of an obligate intracellular pathogen of humans: *Chlamydia trachomatis*. *Science* 282, 754–759.

[169] Kalman, S., Mitchell, W., Mazare, R., Lammel, C., Fan, J., Hyman, R. W., et al. (1999). Comparative genome of *Chlamydia pneumoniae*. *Nature Genet.* 21, 385–389.
[170] Read, I. D., Brunhann, C., Shen, C., Gill, S. R., Heidelberg, J. F., White, O., et al. (2000). Genome sequences of *Chlamydia trachomatis* - MoPn and *Chlamydia pneumoniae* AR 39. *Nucl. Acids Res.* 28, 1397–1406.
[171] Cristiansen, G., Vandahl, B., and Birkelund, S. (2005). Cell and molecular biology of *C. pneumoniae*, pp. 24–40," in *Infection and Disease*, eds H. Friedman, Y. Yamamoto, and Bendinnel (Dordrecht: Kluwer Academic Publishers).
[172] Dalevi, D. A., Eriksen, N., Eriksson, K., and Anderson, S. (2002). Measuring genomic divergence in bacteria: a case study of using Chlamydian data. *J. Mol. Evol.* 55, 24–36.
[173] Boman, J., and Hammerschlag, M. R. (2002). *Chlamydia pneumoniae* and atherosclerosis: Critical diagnostic methods and relevance to treatment studies. *Clin. Microbiol. Rev.* 15, 1–20.
[174] Smieja, M. Mahoni, J., Petrich, A., Boman, J., and Chernesky, M. (2002). Association of circulating *Chlamydia pneumoniae* DNA with cardiovascular disease: a systematic review. *BMC Infect. Diseas.* 2:21.
[175] Knudsen, K., Madsen, A. F., Mygind, P., Cristiansen, G., and Birkelund, S. (1999). Identification of two novel genes, encoding 97 - to 99 kilodalton outer membrane proteins of *Chlamydia pneumoniae*. *Infect. Immun.* 67, 375–383.
[176] Binet, R., Fernandes, E. R., Fisher, D. J., and Maurell, A. T. (2011). Identification and characterization of *C. trachomatis* L2S - Adenosyl-methionin transporter. *mBio* 2, 10:200051-11.
[177] Stephens, R. S., Sanches-Pescador, E. A., Wagar, C., Inouge, C., and Urdea, M. S. (1987). Diversity of Chlamydia trachomatis major outer membrane protein genes. *J. Bacteriol.* 169, 3879–3885.
[178] Zhang, H. Y. (1994). The sequence of the groES and groEL genes from mouse pneumonitis agent of *C. trachomatis*. *Gene* 141, 143–144.
[179] Henderson, B., (ed.). (2013). *Moonlighting Cell Stress Proteins in Microbial Infections*. Dordrecht: Springer Science and Business Media, 406 p.
[180] Roberts, M. C. (2005). Update on acquired tetracycline resistance genes. *FEMS Microbiol. Lett.* 245, 195–203.

[181] Flores, R., Luo, J., Chen, D., Sturgeon, G., Shivshankar, P., Zhong, Y., et al. (2007). Characterization of the hypothetical protein Cpn 1027, a newly identified inclusion membrane protein unique to *C. pneumoniae*. *Microbiology* 153, 777–786.

[182] Thompson, N. R., Yeats, C., Bell, K., Holden, M. T., Bentley, S. D., Livingstone, M., et al. (2005). The *Chlamydophila abortus* genome sequence reveals an array of variable proteins that contribute to interspecies variation. *Genome Res.* 15, 629–640.

[183] Reed, T. D., Meyers, G. S., Bruncham, R. S., and Nelson, W. C. (2003). Sequence of *Chlamydophila caviae* (*Chlamydia psittaci* GPIC): examing the role of niche-specific genes in the evaluation of the *Chlamydiaceae*. *Nucleic acids Res.* 31, 2134–2147.

[184] Unknown author. (2004). Chlamydial cell biology. Regulation of the chlamydial development cycle. *MEW Cell Biology.* Available at: Chlamydiae.com/twiki/bin/view/

[185] Wichian, D., and Hatch, T. P. (1993). Identification of an early-stage gene of *Chlamydia trachomatis*. *J. Bacteriol.* 179, 7233–7242.

[186] Plaunt, M. R., and Hatch, T. P. (1988). Protein synthesis early in the developmental cycle of *Chlamydia psittaci*. *Infect. Immun.* 56, 3021–3025.

[187] Kaul, R. A., Hoang, A., Yau, P., Bradbury, E. M., Wenman, W. M. (1997). *The chlamydial EUO genes encodes a histone H-1 specific protease*. *J. Bacteriol.* 179, 5928–5934.

[188] Kaul, R. A., and Wenman, W. M. (1997). Eucariotik-like histones in *Chlamydia*. *Front. Biosci.* 3:d300-5.

[189] Zhang, L., Douglas, A. L, and Hatch, T. P. (1997). Characterization of *Chlamydia psittaci* DNA binding protein (EUO) synthesizied during the early and middle phases of the developmental cycle. *Infect. Immun.* 6, 113–1176.

[190] Zhang, L., Howe, M., and Hatch, T. P. (2000). Characterization of in vitro DNA binding sites of the protein EUO of *Chlamydia psittaci*. *Infect. Immun.* 6, 1337–1349.

[191] Nickolson, T. L., Olinger, L., Chong, K., Schoolnik, G., and Stephens, R. S. (2003). Global stage-specific gene regulation during the developmental cycle of *Chlamydia trachomatis*. *J. Bacteriol.* 185, 3179–3189.

[192] Belland, R. J, Zhong, G., Crane, D. D., Hogan, D., Sturdevant, D., Sharma, J., et al. (2003). Genomic transcriptional profiling of the

developmental cycle of *Chlamydia trachomatis*. *Proc. Natl. Acad. Sci. U.S.A.* 100, 8478–8483.

[193] Hatch, T. P., Miceli, M., and Sublett, J. E. (1986). Synthesis of the disulfide-bonded outer membrane proteins during the developmental cycle of *Chlamydia psittaci* and *Chlamydia trachomatis*. *J. Bacteriol.* 165, 379–385.

[194] Hackstadt, T. (1999). "Cell biology," in *Chlamydia Intracellular Biology, Pathogenesis and Immunity*, ed. R. S. Stephens (Washington, DC: ASM Press), 29–67.

[195] Rockey, D. D., Heinzen, R. A., and Hackstadt, T. (1995). Cloning and characterization of *Chlamydia psittaci* gene coding for a protein localized in the inclusion membrane of infected cells. *Mol. Microbiol.* 15, 617–626.

[196] Everett, K. D., and Hatch, T. P. (1995). Architecture of the cell envelope of *Chlamydia psittaci* 6BC. *J. Bacteriol.* 177, 877–882.

[197] Sardinia, L. M., Segal, E., and Ganem, D. (1988). Developmental regulation of the cystein-rich outer membrane proteins of murine *Chlamydia trachomatis*. *J. Gen. Microbiol.* 134, 997–1004.

[198] Hanson, B. R., and Tan, M. (2015). Transcriptional regulation of *Chlamydia* heat shock stress response in an intracellular infection. *Mol. Microbiol.* 97, 1158–1167.

[199] Rosario, C. J., and Tan, M. (2016). Regulation of *Chlamydia*-Gene expression by tandem promoters with different temporall patterns. *J. Bacteriol.* 198, 363–369.

[200] Hafner, L. M., Collet, T. A., and Hickey, D. K. (2014). "Immune regulation of *Chlamydia trachomatis* infections of the female genital tract," in *Immune Response Activation*, eds G. H. T. Duc (London: InTech), pp. 177–225.

[201] Orillard, E., and Tan, M. (2016). Functional analysis of three topoisomerases that regulate DNA supercoilling levels in *Chlamydia*. *Mol. Microbiol.* 99, 484–496.

[202] Greber, U. F. (1998). Virus assembly and disassembley the adenovirus cystein protease as a trigger factor. *Rev. Med. Virol.* 8, 213–222.

[203] Maurer, A. P., Mehlitz, A., Mollenkopf, H. L., and Meyer, T. F. (2007). Gene expression profiles of *Chlamydophila pneumoniae* during the developmental cycle and iron depletion-mediated persistence. *PLoS Pathol.* 3:e83.

[204] Albrecht, M., Sharna, C. M., Ditrich, M. T., Müler, T., Reinhardt, R., Vogel, J., et al. (2011). The transcriptional landscape of *Chlamydia pneumoniae*. *Genome Biol.* 12:R98.

[205] Domman, D., and Horn, M. (2015). Following the Footsteps of *Chlamydiae* Gene Regulation. *Mol. Biol. Evol.* 32, 3035–3046.

[206] Hackstadt, T., Baehr, W., and Ying, Y. (1991). Chlamydia trachomatis developmentally regulated protein is homologous to eucariotic histone H1. *Proc. Natl. Acad. Sci. U.S.A.* 88, 3937–3941.

[207] Perara, E. M., Ganem, D., and Engel, J. N. (1992). A developmentally regulated chlamydial gene with apparent homology to eucariotic histone H1. *Proc. Natl. Acad. Sci. U.S.A.* 89, 2125–2129.

[208] Hatch, T. P. (1999). "Cell biology," in *Chlamydia, Intracellular Biology, Pathogenesis and Immunity*, ed. R. S. Stephens (Washington, DC: Am. Soc. for Microbiology), 101–138.

[209] Shirari, M., Hirakawa, H., Kimoto, M., Tabuchi, M., Kishi, F., Ouchi, K., et al. (2000). Comparison of whole genome sequences of *Chlamydia pneumoniae* J138 from Japan, and CWLO29 from USA. *Nucleic Acids Res.* 28, 2311–2314.

[210] Golub, O. J. (1948). A single-dilution method for the estimation of LD_{50} titers of the psittacosis-LGV group of viruses. *J. Immunol.* 59, 71–83.

[211] Crocker, T. T., and Benett, B. M. (1955). The slope assay for measurement of lethal potency of meningo-pneumonitis virus in the chick embryo. *J. Immunol.* 75, 239–248.

[212] Dougherty, R. M., McCloskey, R. V, and Steward, R. B. (1960). Analysis of the single dilution method for titration of psittacosis virus. *J. Bact.* 79, 899–903.

[213] Reeve, P., and Taverne, J. (1967). The significance of strain differences in the behavior of TRIC agents in the chick embryo. *Am. J. Ophthal.* 63, 1162/136–1166/140.

[214] Andersen, A. A., and Vanrompay, D. (2003). "Avian Chlamydiosis (Psittacosis, Ornithosis)," in *Diseases of Poultry*, 11th Edn, ed. Y. M. Saif (Ames, IA: Iowa State University Press), 863–879.

[215] Andersen, A. A., and Vanrompay, D. (2005). "Chlamydiosis," in *A Laboratory Manual for the Isolation and Identification of Avian Pathogens*, Fifth Edition, eds D. E. Swayne, J. R. Glisson, M. W. Jackwood, International Book Distributing Co.

[216] OIE (2016). "Enzootic abortion of ewes (Ovine chlamydiosis)," in *OIE Terestrial Manual*, Chapter 2.7.7, 1008–1016.

[217] Rake, G., McKee, C. M., and Shafter, M. F. (1940). Agent of lymphogranuloma venereum in the yolk sac of the developing chick embryo. *Proc. Soc. Exp. Biol. Med.* 43, 332–335.

[218] Parker, H. D. (1940). A virus of ovine abortion - isolation from sheep in the United States and characterization of the agent. *Am. J. Vet. Res.* 21, 243–250.

[219] Meyer, K. F. (1965). "Ornithosis," in *Diseases of Poultry*, 5th Edition, eds H. E. Biester and L. H. Schwarte (Ames, IA: Iowa State University Press), 675–770.

[220] Meyer, K. F. (1967). The host spectrum of psittacosis-lymphogranu loma venereum (PL) agents. *Am. J. Ophthal.* 63, 1225/199/1246/220.

[221] Meyer, K. F., Eddi, B., and Schachter, J. (1969). "Psittacosis-lymphogranuloma venereum agents," in *Diagnostic Procedures for Viral and Rickettsial Infections*, 4th Edn, eds F. H. Lennette and N. J. Schmidt (New York, NY: Am. Public Health Assoc., Inc.), 869–903.

[222] Steward, R. B. (1962). Growth and lethal effects in the psittacosis-agent in chicken embryos of different ages. *J. Bact.* 83, 423–428.

[223] Schachter, J., Storz, J., Tarizzo, M., and Bogel, K. (1973). *Chlamydiae as agents of human and animal dus eases. Bull. WHO* 49, 443–449.

[224] Heddema, E. R., Van Hanem, E. J., Vandenbroucke-Grauls, C. M., and Pannekoek, Y. (2006). Genotyping of *Chlamydophila psittaci* in human samples. *Emerg. Infect. Dis.* 12, 1989–1990.

[225] Vanrompay, D., Harkinezhad, T., Walle, M. V., Beckman, D., Droogenbroeck, C., Verminen, K., et al. (2007). *Chlamydophila psittaci* transmission from birds to humans. *Emerg. Infect. Dis.* 13, 1108–1110.

[226] Gaede, W., Reckling, F., Dresenkamp, B., Kenklies, S., Schubert, E., Noack, U., et al. (2008). *Chlamydophila psittaci* infections in humans during an outbreak of psittacosis from poultry in Germany. *Zoonoses and Public Health* 55, 154–188.

[227] Sachse, K., Laroucau, K., Hotzel, H., Schubert, E., Enrich, R., Slickers, P. (2008). Genotyping of *Chlamydophila psittaci* using a new DNA microarray assay based on sequence analysis of ompA genes. *BMC Microbiol.* 8:63.

[228] Bevan, B. J., and Gullen, G. A. (1978). Isolation of *Chlamydia psittaci* from avian sourses using growths in cell culture. *Avian Pathol.* 7, 203–211.

[229] Spencer, W. N., and Johnson, F. W. A. (1983). Simple transport medium for the isolation of *Chlamydia psittaci* from clinical material. *Vet. Rec.* 113, 535–536.

[230] Smith, K. A., Bradley, K. K., Stobieski, M. G., and Tengelsen, L. A. (2008). Compendium of measures to control *Chlamydophila psittaci* (formerly *Chlamydia psittaci*) infection among humans (psittacosis) and pet birds. *J. Am. Vet. Med. Assoc.* 226, 532–539.

[231] Dickx, V., Geens, T., Deschliyffeleer, T., Tyberghien, L., Harkinezdat, T., Beeckman, D. S., et al. (2010). *Chlamydophila psittaci* zoonotic risk assessment in a chicken and turkey slaughter house. *J. Clin. Microbiol.* 48, 3244–3250.

[232] OIE Terrestrial Manual (2015). Chapter 1.1.4. Standart for managing biological risk in the veterinary laboratory and animal facilities. *OIE Terrestrial Manual*, 1–16.

[233] Marriot, M. E., and Storz, J. (1966). The behavior of psittacosis agent of diverse origin and extreme antigenic structure in the chicken chorioallantois. *Zbl. Bact. (Orig)* 200, 304–323.

[234] Wentworth, B. B., and Alexander, E. R. (1974). Isolation of *Chlamydia trachomatis* by use of 5-iodo-2 deoxyuridine-treated cells. *Appl. Microbiol.* 27, 912–916.

[235] Oriel, J. D., and Ridgway, G. L. (1982). *Genital Infections by Chlamydia trachomatis*. London: Edward Arnold.

[236] Mohammed, N. R., and Hillary, I. B. (1985). Improved method for isolation and growth of *Chlamydia trachomatis* in McCoy cells treated with cycloheximide using polyethylene glycol. *J. Clin. Pathol.* 38/9, 1052–1054.

[237] Ripa, K. T. (1974). Cultivation of *Chlamydia trachomatis* in cycloheximide – treated McCoy cells. *J. Clin. Microbiol.* 6, 328–331.

[238] Maass, M., and Haris, U. (1995). Evaluation of culture condition used for isolation of *Chlamydia pneumoniae*. *Clin. Microbiol. Infect. Dis.* 103, 141–148.

[239] Becker, J., and Asher, Y. (1972). Synthesis of trachoma agentin emetine – treated cells. *J. Bact.* 109, 966–970.

[240] Bushell, A. C., and Hobson, D. (1978). Effect of cortical on the growth of *Chlamydia trachomatis* in McCoy cells. *Infect. Immun.* 21, 946–963.

[241] Sreward, R. B. (1960). Effect of cortisone on the growth of psittacosis virus in cultures of L cells. *J. Bact.* 80, 25–29.

[242] Sompolinsky, D., and Richmond, S. (1974). Growth of *Chlamydia trachomatis* in McCoy cells treated with cytochalasin B. *Appl. Microbiol.* 28, 912–914.
[243] Croy, W. B., Kuo, C.-C., and Wang, S.-P. (1975). Comparative susceptibility of eleven mammalian cell lines in infection with trachoma orhanisms. *J. Clin. Microbiol.* 1, 434–439.
[244] Kuo, C.-C., Wang, S.-P., Wentworth, B. B., and Grayston, T. J. (1972). Primary isolation of TRIC organisms in HeLa 229 cells treated with DEAE Dextran. *J. Infect. Dis.* 125, 665–668.
[245] Yong, D., and Paul, N. R. (1986). Micro direct inoculation method for the isolation and identification of *Chlamydia trachomatis*. *J. Clin. Microbiol.* 23, 536–538.
[246] Murthy, A., Aralarandam, B., and Zhong, G. (2012). "Chlamydia vaccine, progress and challenges," in *Intracelular Pathogens I. Chlamydiales* eds M. Tan and P. Bavoil (Washington, DC: ASM Press), 311–333.
[247] Allan, I., Hatch, T. P., and Pearce, J. H. (1985). Influence of cystein deprivation of chlamydial differentiation from reproductive to infective life-cycle forms. *J. Gen. Microbiol.* 131, 3171–3177.
[248] Allan, I., and Pearce, J. H. (1983). Amino acid requirements of strains of *C. trachomatis* and *C. psittaci* growing in McCoy cells. Relationship with with clinical syndrome and host origin. *J. Gen. Microbiol.* 129, 2001–2007.
[249] De Graves, T., Kim, T. Y., Lee, J. B., Schlapp, T., Hehnen, H.-R., and Kaltenboeck, B. (2004). Reinfection with *Chlamydophila abortus* by uterine and indirect cohort routes reduces fertility in cattle preexposed to *Chlamydophila*. *Infect. Immun.* 72, 2538–2545.
[250] Popov, G., Martinov, S., and Ignatov, G. (1981). A method for the production of concentrated and purified chlamydial suspensions. *Vet. Sci. XVIII* 1, 84–92.
[251] Popov, G., Martinov, S., and Dimitrova, Z. (1981). The production of labeled gammaglobulin for the detection of Chlamydia. *Vet. Sci. XVIII* 6, 3–11.
[252] Vanrompay, D., Ducatelle, R., and Haeselbrohck, F. (1992). Diagnosis of avian chlamydiosis; specifity of the modified Gmenez staining on smears and comparison of the sensitivity of isolstion in eggs and three different cellcultures. *J. Vet. Med (B)* 39, 105–112.
[253] Van Wetere, A. J. (2010). *Overview of Avian Chlamydiosis (Psittacosis, Ornithosis, Parrot fever)*. MSD Veterinary Manual.

[254] Green, W., Xian, Y., Huang, Y., McClarty, and Zhong, G. (2004). Chlamydia-infected cells continue to undergo mitosis and resist induction of apoptosis. *Infect. Immun.* 72, 451–460.

[255] Prozialeck, W. C., Fay, M. J., Lamar, P. C., Pearson, C. A., Sigar, I., and Ramsey, K. H. (2002). *Chlamydia trachomatis* disrupts N-Cadherin-dependant Cell–Cell Junctions and sequesters β–Catenin in Human Cervical epithelial cells. *Infect. Immun.* 70, 2605–2613.

[256] Smith, T. F. (1977). Comparative recoveries of Chlamydia from urethral speciments using glass vials and plastic microtiter plates. *J. Clin. Pathol.* 67, 496–498.

[257] Kumar, S. (2015). *Essentials of Microbiology. Section 3: Systematic Bacteriology*. New Delhi: J. P. Medical Ltd.

[258] Gaydos, C. A., Summersquill, J. T., Ramirez, N. N., and Quiun, T. C. (1996). Replication of *Chlamydia pneumoniae* in vitro in human macrophages, endothelial cells, and aortic artery smooth muscle cells. *Infect. Immun.* 64, 1614–1620.

[259] Vielma, S. A., Krings, G, and Lopes-Virella, M. F. (2003). *Chlamydophila pneumoniae* induces ICAM-1 expressioin in human aortic endothelial cells via protein kinase C-dependant activation of nuclear factor-Kb. *Circ. Res.* 92, 1130–1137.

[260] Krull, M., Maass, M., Suttorb, N., and Rupp, J. (2005). *Chlamydophila pneumoniae.* mechanism of target cell infection and activation. *Thromb. Haemost.* 94, 319–326.

[261] Uruma, T. (2005). *Chlamydia pneumoniae* growth inhibition in human monocytic THP-1 cell and human epithelial HEp-2 cells. *J. Med. Microbiol.* 54 (Pt 12), 1143–1149.

[262] Pospischil, L., and Canderle, J. (2004). *Chlamydia (Chlamydophila) pneumoniae* in animals: a review. *Vet. Med. - Czech* 49, 129–134.

[263] Di Francesco, A., Donati, M., Mattioli, L., Naldi, M., Salvatore, D., Poglayen, G., et al. (2006). *Chlamydophila pneumoniae* in horses, a seroepidemiological survey in Italy. *New Microbiologica* 29, 303–305.

[264] Glassick, T., Giffard, P., and Timms, P. (1996). Outer membrane protein 2 gene sequenses indicate that *Chlamydia pecorum* and *Chlamydia pneumoniae* cause infections in koala. *Syst. Appl. Microbiol.* 19, 457–464.

[265] Berger, L., Volp, K., Mathews, S., Speare, R., and Timms, P. (1999). *Chlamydia pneumoniae* in a free-ranging giant frog (*Myxophyes iteratus*) from Australia. *Clin. Microbiol.* 37, 2378–2380.

[266] Kumar, S., Kutlin, A., Robin, P., Kohlhoff, S., Bodetti, T., Timms, P., and Hammerschlag, M. R. (2007). Isolation and antimicrobal susceptibilities of chlamydial isolates from western barred bandicoots. *J. Clin. Microbiol.* 45, 392–394.

[267] Reed, K. D., Ruth, G. R., Meyer, J. A., and Shukla, S. K. (2000). *Chlamydia pneumoniae* in a breeding colony of African clawed frog (*Xenopus tropicalis*). *Emerg. Infect. Dis.* 6, 196–199.

[268] Wehr, J., and Ilchman, G. (1980). Experimentalle Untersuchungen zur immunoprophylaxe de respiratorischen Chlamydien infection des kalbes. Wiss. Z. der Humboldt-Universitat zu Berlin, *Math.-Nat. R. XXIX* 1, 113–122.

[269] Del Rio, L., Ortega, A., Buendia, A. J., Caro, M. R., et al. (2006). “Role of B cells in the protection induced by several vaccines against *Chlamydophila abortus* infection in a mouse model, pp. 61–62,” in *Diagnosis, Pathogenesis and Control of Animal Chlamydioses,* eds D. Longbottom and M. Rocchi (Tweedbank: Meigle Color Printers, Ltd.) in *Proceedings of the 4th Annual Workshop of COST Action 855* (Edinburgh: Moredun Res. Institute), 196.

[270] Buendia, A. J., Navarro, J. A., Caro, M. R., Martinez, C. M., et al. (2007). “Intragastric model of infection in pregnant mice: comparative study in two mouse strains with different susceptibility, p. 13,” in *Pathogenesis, Epidemiology and Zoonotic Importance of Animal Chlamydioses,* eds K. Niemczuk, K. Sachse, L. D. Sprague, in *Proceedings of the 5th Annual Workshop of COST Action 855* (Pulawy: Natl. Vet. Res. Inst.), 94.

[271] De Kruif, M. D., Gorp, C. M. E, Keller, T. T., Ossewaarde, J. M., and Gate, H. (2005). *Chlamydia pneumoniae* infections in mouse models: relevance for atherosclerosis research. *Cardiovascular Res.* 62, 317–327.

[272] O’Meara, C. P., Andrew, D. W., and Beagly, K. W. (2014). The mouse model of Chlamydia genital tract infection: a review of infection, disease, immunity and vaccine development. *Curr. Mol. Med.* 14, 396–421.

[273] Spar, Q. (ed.). (2012). *Project 16. TLR2 and iNOS in Chlamydia muridarum Dissemination in a mouse model of Chlamydia-induced Reactive arthritis*. The University of Queensland, Diamantina Institute, Brisbane, 60.

[274] Patel, M., Lin, S.-A., Mellisa, A., Boddicker, A., et al. (2015). Quantitative in vivo detection of *Chlamydia muridarum* associated inflammation in a mouse model using optical imaging. *Mediators Inflamm.* 2015:264897.

[275] Kuo, C. C., and Chen, W. J. (1980). A mouse model for *Chlamydia trachomatis* pneumonitis. *J. Infect. Dis.* 141, 655–657.

[276] Harrison, H. R. (1982). *Chlamydia trachomatis*. Pneumonitis in the C576l/Ks Mouse: Pathologic and immunologic features. *J. Lab. Clin. Med.* 100, 953–962.

[277] Beale, A. S., and Upshone, P. A. (1994). Characteristics of murine model of genital infection with *Chlamydia trachomatis* and effect on therapy with tetracyclines, amoxicillin-clavulanic acid, or azithromycin. *Antimicrob. Agents Chemother.* 38, 1937–1943.

[278] Swenson, C. E., Donegan, E., and Schachter, J. (1983). *Chlamydia trachomatis* – induced salpingitis in mice. *J. Infect. Dis.* 148, 1101–1107.

[279] Rank, R. G. (2007). "ChlamydiaL diseases," in *The Mouse in Biomedical Research,* eds J. Fox, S. Barthold, C. Newcomer, A. Smith, F. Quimbly, and M. Davisson (New York, NY: Academic Press).

[280] Darville, T., Andrews, C. W., Sikes, J. D., Fraley, P. G., and Rank, R. (2001). Early local cytokine profiles in strains of mice with different outcomes from genital tract infection. *Infect. Immun.* 69, 3556–3561.

[281] Montes, O. R., Buendia, A. J., Del Rio, L., Sanches, J., Salinas, J., and Navarro, J. A. (2000). Polymorphonuclear neutrophils are necessary for the recruitment of CD8+ T cells in the liver of a pregnant mouse model of *Chlamydophila abortus* (*Chlamydia psittaci* Serotype 1) infection. *Infect. Immun.* 68, 1745–1751.

[282] Rodolakis, A., Bernard, F., and Lantier, F. (1989). Mouse models for evaluation of virulence of *Chlamydia psittaci* isolated from ruminants. *Res. Vet. Sci.* 41, 34–39.

[283] Li, D., Borovkov, F., Vaglenov, A., Wang, C., Kim, T., Cad, D., and Kaltenboeck, B. (2006). Mouse model of respiratory *Chlamydia pneumoniae* infection for a genomic screen of subunit vaccine candidates. *Vaccine* 24, 2917–2926.

[284] Donati, M., Di Paolo, M., Favaroni, A., Aldini, R., et al. (2015). A mouse model for *Chlamydia suis* genital infection. *Pathog. Dis.* 73, 1–3.

[285] Jupelli, M., Shimada, K., Chiba, N., Siepenkin, A., Alsabeh, P., et al. (2013). *Chlamydia pneumoniae* infection in mice induces chronic lung inflammation, BALT formation, and fibrosis. *PLoS one* 8:e77447.

[286] Cabbage, S., Leronimakis, N., Preusch, M., Lee, A., Ricks, J., Janebodin, K., et al. (2014). *Chlamydia pneumoniae* infection of lungs and macrophages indirectly stimulates the phenotypic conversion of smooth muscle cells and mesenterial stem cells: potential roles in vascular calcification and fibrosis. *Pathogen. Dis.* 72, 61–69.

[287] Kapourchali, F. R., Surendiram, G., Chen, L, Uitz, E., Bahadoni, B., and Moghadasian. (2014). Animal models of atherosclerosis. *World J. Clin. Case.* 2, 126–132.

[288] Little, G. S., Hammond, C. J., MacIntyre, A., Balin, B. J., and Appelt, D. M. (2004). *Chlamydia pneumoniae* induces Alzheimer – like Amyloid plaques in brains of BALB/c mice. *Neurobiol. Aging* 25, 419–429.

[289] Murray, E. S. (1964). Guinea pigs inclusion conjunctivitis virus. I. Isolation and identification as a member of the psittacosis-lymphogranuloma –trachoma group. *J. Infect. Dis.* 114, 1–12.

[290] Mount, D. T., Bigazzi, P. E., and Barron, A. L. (1972). Infection of genital tract and transmission of ocular infection to newborns by the agent of guinea pig inclusion conjunctivitis. *Infect. Immun.* 5, 921–926.

[291] Mount, D. T., Bigazzi, P. E., and Barron, A. L. (1973). Experimental genital infection of male guinea pigs with the agent of guinea pigs inclusion conjunctivitis. *Infect. Immun.* 8, 925–930.

[292] Alexander, A. J. (ed.). (1979). *Animal Models for Research on Contraception and Fertility*. Hagers town, MD: Harper and Row.

[293] De Clercq, E., Kalmar, I., and Vanrompay, D. (2013). Animal models for studying female genital tract infection with *Chlamydia trachomatis*. *Infect. Immun.* 81, 3060–3067.

[294] Sisk, D. B. (1976). "The biology of guinea pigs," in *Physiology*, eds J. E. Wagner and P. T. Manning (New York, NY: Academic Press Inc.), 69–93.

[295] Rank, R. G., White, H. J. Hough, Jr. A. J., Pasley, N., and Barron, A. L. (1982). Effect of estradiol on chlamydial genital infection of female guinea pigs. *Infect. Immun.* 38, 699–705.

[296] Agrawal, T., Vats, V., Wallace, P. K., Salhan, S., and Mital, (2013). A. Role of cervical dendritic cell subsets, co-stimulatory molecules, cytokine secretions profile and beta-estradiol in development of sequalae to *Chlamydia trachomatis* infection. *Repr. Biol. Endocrionol.* 6:46.

[297] Rank, R. G., Hough, Jr. A. J., Jacobs, R. F., Cohen, C., and Barron, A. L. (1985). Chlamydial pneumonitis in newborn guinea pigs. *Infect. Immun.* 48, 153–158.

[298] Rank, R. G., White, H. J., and Barron, A. L (1979). Humoral immunity in the resolution of genital infection of female guinea pigs. *Infect. Immun.* 26, 573–579.

[299] Rank, R. G., Soderberg, L. S., Sanders, M. M., and Batteiger, B. E. (1989). Role of cell-mediated immunity in the resolution of secondary chlamydial genital infection in guinea pigs infected with the agent of guinea pigs inclusion conjunctivitis. *Infect. Immun.* 57, 706–710.

[300] Rank, R. G., Batteiger, B. E., and Soderberg, L. S. (1989). Susceptibility to re- infection after a primary chlamydial genital infection. *Infect. Immun.* 56, 2243–2249.

[301] Patton, D. L., Halbert, S. A., and Wang, D. A. (1982). Experimental salpingitis in rabbits provoked by Chlamydia. *Fertil. Steril.* 35, 691–700.

[302] Fong, I. W., Chin., B., Vitra, E., Fong, M. W., Jang, D., and Mahony, J. (1997). Rabbit model for *Chlamydia pneumoniae* infection. *J. Clin. Microbiol.* 35, 48–52.

[303] Muhlestein, J. B., Anderson, J. A., Hammond, E. H., Zhao, L., Trehan, S., Schwobe, E. R., et al. (1998). Infection with *C. pneumoniae* accelerates the development of atherosclerosis and treatment with azithromycin prevents it in a rabbit model. *Circulation* 97, 633–636.

[304] Iversen, J. O., Spalatin, J., Fraser, C. E., and Hanson, R. P. (1974). Ocular involvement with *Chlamydia psittaci* (strain M56) in rabbits inoculated intravenously. *Can. J. Comp. Med.* 38, 298–302.

[305] Kane, J. L., Woodland, R. M., Elder, M.-G., and Darougar, S. (1985). Chlamydial pelvic infection in cats: a model for the study of human pelvic inflammatory disease. *Genitourin. Med.* 61, 311–318.

[306] Vanrompay, D., Hoang, T. Q., De Vos, L., Verminnen, K., Hackinezdad, T., Chiers, Morre, S. A., et al. (2006). Specific pathogen-free

pigs as an animal model for study *Chlamydia trachomatis* genital infection. *Infect. Immun.* 73, 8317–8321.

[307] Bray, N., Dubchak, I., and Pachter, L. (2003). AVID: a global aligument program. *Genome Res.* 13, 97–102.

[308] Tuggle, C. K., Green, J., Fitzsimmons, C., Woods, R., Prather, R. S., et al. (2003). EST-based gene discovery in pig: virtually expression patterns and comparative mapping to human. *Mamm. Genome* 14, 565–579.

[309] Patton, D. L., Sweeney, V. C., Rabe, L. K., and Ailler, S. L (1996). The vaginal microflora of pig-tailed macaques and the effect of chlorhexidine and benzal- konium on this ecosystem. *Sex. Trans. Dis.* 23, 489–493.

[310] Patton, D. L., Sweeney, V. C., and Paul, K. J. (2009). A summary of preclinical topical microbicide rectal safety and efficacy evaluations in a pigtailed macaque model. *Sex. Trans. Dis.* 36, 350–356.

[311] Miyan, I., Ramsey, V. H., and Patton, D. L. (2010). Duration of untreated chlamydial genital infection and factors associated with clearance: review of animal studies. *J. Infect. Dis.* 201 (Suppl. 2), S96–S103.

[312] Ripa, K. T., Moller, B. P., Mardh, P.-A, Freundt, E. A., and Melsen, F. (1979). Experimental acute salpingitis in grivet monkeys provoked by *Chlamydia trachomatis*. *Acta Pathol. Microbiol. Scand.* 87b, 65–70.

[313] Johnson, A. P., Hetherington, C. M., Osborn, M. F., Thomas, B. J., and Taylor-Robinson, D. (1980). Experimental infection of the marmoset genital tract with *Chlamydia trachomatis*. *Br. J. Exp. Pathol.* 61, 291–295.

[314] Lindner, K. (1909). Uebertragungsversuche von gonokokkenfreien Blennorrhoea neonatorum auf Affen. *Wien. Klin. Wochenschr.* 22:1555.

[315] Darougar, S., Jones, B. R., and Kinnison, J. R. (1971). "Chlamydial isolates from the rectum in association with chlamydial infection of the eye or genital tract. I. Laboratory aspects," in *Trachoma and Related Disorders*, ed. R. L. Nickols (Amsterdam: Excerpta Medica), 501–506.

[316] Taylor-Robinson, D., Purcell, R. H., Landon, W. T., Thomas, B. J., and Evaans, R. T. (1981). Mophological, serological and histopathological features of experimental *Chlamydia trachomatis* urethritis in chimpazees. *Br. J. Vener. Dis.* 57, 36–40.

[317] Thygeson, P., and Mengert, W. F. (1936). The virus of inclusion conjunctivitis. Further observations. *Arch. Ophthalmol.* 15, 377–410.

[318] Braley, A. E. (1939). Relation between the virus of trachoma and the virus of inclusion blenorrhea. *Arch. Ophthalmol.* 22, 393–398.

[319] Moller, N. F., and Mardh, P.-A. (1980). Experimental salpingitis in grivet monkeys by *C. trachomatis. Acta Pathol. Microbiol. Scand.* B88, 107–114.

[320] Jacobs, N. F. Arum, E. S., and Krauss, S. J. (1978). Experimental infection of the chimpanzee urethra and pharynx with *C. trachomatis. Sex. Trans. Dis.* 5, 132–136.

[321] Patton, D. L., Halberg, S. A., Kuo, C. C., Wang, S. P., and Holmes, K. K. (1983). Host response to primary *Chlamydia trachomatis* infection in the fallopian tube of pig-tailed monkeys. *Fertil. Steril.* 40, 829–840.

[322] Patton, D. L., Kuo, C. C., Wang, S. P., Brenner, R. M., Stemfield, M. D., Morse, S. A., et al. (1987). Chlamydia infection of subcutaneous fimbrial transplants in cynomolgus and rhesus monkeys. *J. Infect. Dis.* 155, 229–235.

[323] Patton, D. L., Wolner-Hanssen, P., Cosgrove, S. J., and Holmes, K. K. (1990). The effects of *Chlamydia trachomatis* on the female reproductive tract of the *Macaca nemestrina* after a single tubal challenge following repeated cervical inoculations. *Obstet, Gynecol.* 76, 643–650.

[324] Bell, J. D., Bergin, I. L., Schmidt, K. Zochowska, M. K., Aronoff, D. M., and Patton, D. L. (2011). Nonhuman primate models used to study pelvic inflammatory disease caused by *Chlamydia trachomatis. Infect. Dis. Obstet. Gynecol.* 2011:675360.

[325] Papp, J. P., Schachter, J., Gaydos, C. A., and Van der Pol. (2014). *Recommendations for the laboratory-based detection of Chlamydia trachomatis and Neisseria gonorrhoe – 2014.* Recommendations and Reports, 63 (RR02), 1–19.

[326] OIE Terrestrial Manual (2012). *Enzootic abortion in ewes (Ovine chlamydiosis; Infection with Chlamydophila abortus). Chapter 2.7.6,* 1008–1016.

[327] Borel, N., Thoma, R., Spaeni, P., Welilenmann, R. et al. (2006). Chlamydia-related abortions in cattle from Graubunden, Switzerland. *Vet. Pathol.* 43, 702–706.

[328] DeGraves, F., Gao, D., Hehnen, H.-R, Schlapp, T., and Kaltenboeck, B. (2003). Quantative detection of *Chlamydia psittaci* and *Chlamydia*

pecorum by high-sensitivity real time PCR reveals high prevalence of vaginal infection in cattle. *J. Clin. Microbiol.* 41, 1726–1729.

[329] Jee, J., DeGraves, F. J., Kim, T., and Kaltenboeck, B. (2004). High prevalence of natural *Chlamydophila* species infection in calves. *J. Clin. Microbiol.* 42, 5664–5672.

[330] Pantchev, A., Sting, R., Bauerfeind, R., Tyczka, J., and Sachse, K. (2009). New real time PCR tests for species-specific detection of *Chlamydophila psittaci* and *Chlamydophila abortus* from tissue samples. *Vet. J.* 181, 145–150.

[331] Watson, E. A., Templeton, A., Rusell, I., and Paavonen, J. (2002). The accuracy and efficacy of screening tests for *Chlamydia trachomatis*: a systematic review. *Med. Microbiol.* 51, 1021–1031.

[332] Black, C. M., Marrazzo, J., Johnson, R. E., Hook, E. W. III, Jones, R. B., Green, T. A., et al. (2002). Head-to-head multicenter comparison of DNA probe and nucleic acid amplification for *Chlamydia trachomatis* in women with use of an improved reference standard. *J. Clin. Microbiol.* 40, 3757–3763.

[333] Martinov, S., Schoilev, C., and Popov, G. (1989). Chlamydien abort dei schafen in der Volkrepublik Bulgarien. *Mh. Vet. Med.* 44, 361–400.

[334] Martinov, S. P. (1994). "Chlamydial infection in pigs with different clinical conditions," in *Proceedings of the 13th Int. Pig Vet. Soc. Congress*, Bangkok, 230.

[335] Bagdonas, J., Mauricas, M. Gerulis, G., Masilionis, K., and Martinov, S. (2004). Incidence of pig chlamydiosis in Lithuania revealed by different techniques. *Biotechnol. Biotechnol. Eq.* 18, 166–176.

[336] Ognianov, D., Genchev, G., and Panova, M. (1978). Chlamydial keratoconjunctivitis in cattle. *Vet. Sci.* 3, 12–16.

[337] Martinov, S. P., Schoilev, H., and Kazachka, D. (1996). "Studies on the *Chlamydia psittaci* infection in dicks and humans," in *Proceedings of the 7th Int. Congress for Infect. Diseases*, Hong Kong, 243.

[338] Fraser, C. E., and Berman, D. T. (1965). Type-specific antigens of the psittacosis-lymphogranuloma venereum group of organisms. *J. Bact.* 89, 949–947.

[339] Byrne, G. I. (2010). *Chlamydia trachomatis* strains and virulence: rethinking links to infection prevalence and disease severity. *J. Infect. Dis.* 201 (Suppl. 2), S126–S133.

[340] Vora, G. L., and Stuart, E. S. (2003). A role for the glycolipid exoantigen (GLXA) in chlamydial infectivity. *Current Microbiol.* 46, 217–223.

[341] Neff, L., Dahar, S., Muzzin, P., Spenato, U., Gulaar, F., Gabey, C., and Bas, S. (2007). *J. Bacteriol.* 189, 4739–4748.

[342] Henderson, I. R., and Nataro, J. P. (2001). Virulence functions of autotransporter proteins. *Infect. Immun.* 69, 1231–1243.

[343] Vretou, E., Katsiki, E., Psarrou, E., Vougas, K., and Tsangaris, G. (2006). "Identification of Inc 766: a member of TMH / Inc protein family in *Chlamydophila abortus*. Diagnosis, Pathogenesis and Control of Animal Chlamydiosis," in *Proceedings of the 4th Ann. Workshop COST Action 855* (Edinburgh: Moredun Res. Inst.), 51–52.

[344] Page, L. A. (1967). Comparison of "Pathotypes" among chlamydial psittacosis strains recovered from diseased birds and mammals. *Bull. Wildlife Dis. Ass.* 3, 166–175.

[345] Iversen, J. O., Spalatin, J., Fraser, C. E. O., and Hanson, R. P. (1974). Ocular involvement with *Chlamydia psittaci*/strain M56 in rabbits, inoculated intravenously. *Can. J. Comp. Med.* 38, 298–302.

[346] Iversen, J. O., Spalatin, J., and Hanson, R. P. (1976). Experimental chlamydiosis in wild and domestic lagomorph. *J. Wildl. Dis.* 12, 215–220.

[347] Page, L. A. (1978). "Avian chlamydiosis (Ornithosis)," in *Diseases of Poultry*, 7th Edn, ed. M. S. Hofstad (Ames, IA: Iowa State University Press), 337–366.

[348] Hearn, H. J., and Hearing, A. S. (1971). Derivation of a Borg strain of *C. psittaci*, attenuated for mice. *Infect. Immun.* 3, 504–505.

[349] Rake, J., and Jones, H. P. (1940). Studies on lymphogranuloma venereum. II. The association of specific toxin with the agent of the lymphogranuloma-psittacosis group. *J. Exp. Med.* 79, 463–485.

[350] Manire, G. P., and Meyer, K. F. (1950). The toxins of the psittacosis-lymphogranuloma group of agents. I. The toxicity of various members of the PLV group. *J. Infect. Dis.* 86, 226–232.

[351] Cristoffersen, G., and Manire, G. P. (1969). The toxicity of meningopneumonitis organisms (*Chlamydia psittaci*) at different stages of development. *J. Immunol.* 103, 1085–1088.

[352] Kordova, N., Wilt, J. C., and Poffenroth, L. (1973). Lysosomes and the "toxicity" of Rickettsiales. V. In vivo relationship of peritoneal phagocytes and egg-attenuated *C. psittaci* 6BC. *Can. J. Microbiol.* 19, 1417–1423.

[353] Kordova, N., Wilt, J. C., and Martu, C. (1975). Lysosomes and the "toxicity" of Rickettsiales. VI. In vivo response of mouse peritoneal phagocytes to L-cell grown *C. psittaci* 6BC strain. *Can. J. Microbiol.* 21, 323–331.

[354] Moulder, J. W., Hatch, T. P., Byrne, G. I., and Kellogg, K. R. (1976). Immediate toxicity of high multiplicities of *Chlamydia psittaci* for mouse fibroblasts (L cells). *Infect. Immun.* 14, 277–289.

[355] Wyrick, P. B., and Brownridge, E. A. (1978). Growth of *Chlamydia psittaci* in macrophages. *Infect. Immun.* 19, 1054–1060.

[356] Wyrick, P. B., Brownridge, E. A., and Ivins, B. E. (1978). Interactions of *Chlamydia psittaci* with mouse peritoneal macrophages. *Infect. Immun.* 19, 1061–1067.

[357] Belland, R. J., Scidmore, M. A., Crane, D. D., Hogan, D. M., Whitmire, W., McClarty, G., et al. (2001). *Chlamydia trachomatis* cytotoxicity-associated with complete and partial cytotoxin genes. *PNAS* 98, 13984–13989.

[358] Hosseinzadeh, S., Pacey, A. A., Eley, A. (2001). *Chlamydia trachomatis*-induced death of human spermatozoa is caused primarily by lipopolysaccharide. *J. Med. Microbiol.* 52, 193–200.

[359] Carlson, J. H., Hughes, S., Hogan, D., Cieplak, G., Sturdevant, D. E., McClarty, G., et al. (2004). Polymorphism in the *Chlamydia trachomatis* cytotoxin locus associated with ocular and genital isolates. *Infect. Immun.* 72, 7063–7072.

[360] Lyons, J. M., Ito Jr. J. I., Pena, A. S., and Morre, S. A. (2005). Differences in growth characteristics and elementary body associated cytotoxicity- between *Chlamydia trachomatis* oculogenital serovars D and H and *Chlamydia muridarum*. *J. Clin. Pathol.* 58, 397–401.

[361] Rajaram, K., Giebel, K. A., Toh, E., Hu, S., Newman, J. N., Morrison, G., et al. (2015). Mutational analysis of the *Chlamydia muridarum* Plasticity zone. *Infect. Immun.* 83, 2870–2881.

[362] Fehiner-Gardiner, C., Boshick, C., Carlson, J. H., Hughes, S., Belland, R. J., Caldwell, H. D., et al. (2002). Molecular basis defining human *Chlamydia trachomatis* tissue tropism. A possible role of Tryptophan synthase. *J. Biol. Chemistr.* 277, 26893–26903.

[363] McClarty, G., Caldwell, H. D., Nelson, D. E. (2007). Chlamydial interferon gamma immune evasion influences infection tropism. *Curr. Opt. Microbiol.* 10, 47–51.

[364] Newman, J. (2015). *Characterization of putative cytotoxin genes of Chlamydia muridarum*. Thesis, 62.

[365] Bard, J., and Levitt, D. (1986). *Chlamydia trachomatis* (L2 serovar) binds to distinct subpopulation of human peripheral blood leucocytes. *Clin. Immunol. Immunopathol.* 38, 150–160.

[366] Hackstadt, T., and Caldwell, H. D. (1985). Effect of proteolytic cleavage of surface-exposed proteins on infectivity of *Chlamydia trachomatis*. *Infect. Immun.* 48, 546–571.

[367] Yang, Y.-S., Kuo, C. C., and Chen, W. J. (1988). Reactivation of *Chlamydia trachomatis* lung infection in mice by cortisone. *Infect. Immun.* 39, 655–658.

[368] Hanna, L., Dawson, C. Briones, O., Thygeson, P., and Javetz, E. (1968). Latency in human infections with TRIC agents. *J. Immunol.* 101, 43–50.

[369] Clark, V. L., and Bavoil, P. M. (eds). (1994). *Bacterial Pathogenesis. Part A: Identification and Regulation of Virulence Factors*. San Diego: Academic Press, 83–92.

[370] Allen, J. E., Cerrone, M. C., Beatty, P. R., and Stephens, R. S. (1990). Cystein- rich outer membrane proteins of *Chlamydia trachomatis* display compensatory sequence changes between biovariants. *Mol. Microbiol.* 4, 1543–1550.

[371] Caldwell, H. D., Kromhout, J., and Schachter, J. (1981). Purification and partial characterization of the major outer membrane protein of *Chlamydia trachomatis*. *Infect. Immun.* 31, 1161–1176.

[372] Bavoil, P. M., and Hsia, R.-C. (1998). Type III secretion in Chlamydia: a case of déjà vu? *Mol. Microbiol.* 28, 860–862.

[373] Winstanley, G., and Hart, C. A. (2001). Type III secretion systems and pathogenicity islands. *J. Med. Microbiol.* 50, 116–126.

[374] Hermann, M. (2004). *Identification and Characterization of Type III Secreted Chlamydial Pathogenicity Factors and Their Impact on the Infected Host Cell*. Ph. D. dissertation, University of Konstanz, Konstanz, 163.

[375] Hefty, P. S., and Stephens, R. S. (2007). Chlamydial type III secretory system is encoded on ten operons preceed by sigma 70-like promoter elements. *J. Bacteriol.* 189, 198–206.

[376] Horn, M., Coliingro, A., Schmitz-Esser, S., et al. (2004). Illuminating the evolutionary history of chlamydiae. *Science* 304:728.

[377] Voigh, A., Schölf, G., and Saluz, H. P. (2012). The *Chlamydia psittaci* genome: a comparative analysis of intracellular pathogen. *PLoS One* 7:e35097.

[378] Miyeiri, I., Laxton, J. D., Wang, X., Obert, G. A., et al. (2011). *Chlamydia psittaci* genetic variants differ in virulence by modulation of host immunity. *J. Infect. Dis.* 204, 654–663.

[379] Carlson, J. H., Whitmire, W. M., Crane, D. D., Wirke, L., Virtaneva, K. Sturdevant, D. E., et al. (2008). The *Chlamydia trachomatis* plasmid is transcriptional regulator of chromosomal genes and avirulence factor. *Infect. Immun.* 76, 2273–2283.

[380] Hillemannn, M. K., and Nigg, C. (1948). Studies of lymphogranuloma venereum complement-fixing antigens. IV. Fractionation with organic solvents of antigens of the psittacosis-lymphogranuloma group. *J. Immunol.* 59, 349–364.

[381] Richmond, S. J., and Stirling, P. (1981). Localization of chlamydial group antigen in McCoy cells monolayers infected with *Chlamydia trachomatis* and *Chlamydia psittaci*. *Infect. Immun.* 34, 561–570.

[382] Martinov, S., and Popov, G. (1984). Our experience in obtaining and use of group-specific chlamydial antigens. *Vet. Sci. XXI* M4, 15–25.

[383] Graystone, J. T., and Wang, S.-P. (1975). New knowledge of chlamydiae and the disease they cause. *J. Infect. Dis.* 132, 87–105.

[384] Reeve, R., and Taverne, J. (1962). Some properties of the complement-fixing antigens of the agents of trachoma and inclusion blennorhoea and the relstionship of the antigens to the developmental cycle. *J. Gen. Microbiol.* 27, 501–508.

[385] Witkins, N. G., Caldwell, H. D., and Hackstadt, T., J. (1987). Chlamydial hemagglutinin identified as lipopolysaccharide. *J. Bacteriol.* 189, 3826–3828.

[386] Dhir, S. P., Hakomori, H., Kenny, G. E., and Graystone, J. T. (1972). Immunochemical studies of chlamydial group antigen (presence of 2-keto-3-deoxycarbohydrate as immunodominant group. *J. Immunol.* 109, 116–122.

[387] Donati, M., Storni, E., Mazzeo, C., Di Francesko, A., Marangoni, A., et al. (2006). "Role of *Chlamydophila psittaci* protein, pgp3 as a marker of infection in cats, infected with *Chlamydophila felis*," in *Diagnosis, Pathogenesis and Control of Animal Chlamydioses*, eds D. Longbottom and M. Rocchi, 115–116, in *Proceedings of the 4th Annual Workshop of COST Action 855* (Edinburgh: Moredun Res. Institute), 196.

[388] Monnickendam, M. A. (2012). "Molecular biology of chlamydiae. Chlamydial antigens," in *Molecular and Cell Biology of Sexually*

Transmitted Diseases, eds D. J. Wright, L. S. Archard (Dordrecht: Springer Science and Business Media), 29–30.

[389] Caldwell, H. D., and Hitchcock, P. J. (1984). Monoclonal antibody against a genus-specific antigen of *Chlamydia* species: location of the epitope on chlamydial lipopolysaccharide. *Infect. Immun.* 44, 306–314.

[390] Bedson, S. P., Barwell, C. F., King, E. J., and Bishop, L. W. J. (1949). The laboratory diagnosis of lymphogranuloma venereum. *J. Clin. Pathol.* 2, 241–249.

[391] Caldwell, H. D., Kuo, C. C., and Kenny, G. E. (1975). Antigenic analysis of chlamydiae by two-dimensional immunoelectrophoresis. I. Antigenic heterogenicity between *C. trachomatis* and *C. psittaci. J. Immunol.* 115, 963–968.

[392] Caldwell, H. D., and Kuo, C. C. (1977). Purification of a *Chlamydia trachomatis*-specific antigen by immunoadsorbtion with monospecific antibody. *J. Immunol.* 118, 437–471.

[393] Caldwell, H. D., Kuo, C. C., and Kenny, G. E. (1975). Antigenic analysis of chlamydiae by two-dimensional immunoelectrophoresis. II. A trachoma-LGV-specific antigen. *J. Immunol.* 115, 969–975.

[394] Martinov, S., Panova, M., and Ognyanov, D. (1978). "Type-specific differentiation among some chlamydial strains," in *Proceedings of the 7th Int. Congress of Infect. and Parasit. Diseases*, Varna, 553–556.

[395] Mueller-Loennies, S., Heine, H., Brade, L., Zamyatina, A., Kosma, P., Mackenze, R., et al. (2007). "Chlamydisl lipopolysaccharides and antibodies against it," in *Pathogenesis, Epidemiology and Zoonotic Importance of Animal Chlamydioses*, eds K. Niemczuk, K. Sachse, and L. D. Sprague, in *Proceedings of the 5th Annual Workshop of COST Action 855* (Pulawy: Natl. Vet. Res. Institute), 94.

[396] Kuo, C. C., Wang, S.-P., Graystone, J. T., and Alexander, E. R. (1974). TRIC, Type K, a new immunological type of *Chlamydia trachomatis*. *J. Immunol.* 113, 591–596.

[397] Wang, S.-P., Graystone, J. T., and Gale, J. L. (1973). Three new immunological types of trachoma-inclusion conjunctivitis- lymphogranuloma venereum organisms. *J. Immunol.* 110, 873–879.

[398] Wang, S.-P., and Graystone, J. T. (1973). Immunologic relationsheep between genital TRIC, LGV and reated organisms in a new microtiter indirect immunofluorescence test. *Am. J. Ophthalmol.* 70, 367–374.

[399] Stuard, E. S., Troidle, K. M., and MacDonald, A. (1994). Chlamydial glicolipid antigen: extracellular accumulation, biological activity, and antibody recognition. *Clin. Microbiol.* 28, 85–90.

[400] Zdrodowska-Stefanow, I., Ostaszewska-Puchalska, I., and Pucilo, K. (2003). The immunology of *Chlamydia trachomatis*. *Arch. Immunol. Therap. Exp.* 51, 289–294.

[401] MacDonald, B. A. (1985). Antigens of *Chlamydia trachomatis*. *Rev. Infect. Dis.* 7, 731–736.

[402] Hourihan, J. T., Rota, T. R., and MacDonald, B. A. (1980). Isolation and purification of a type-specific antigens from *Chlamydia trachomatis* propagated in cell culture utilizing molecular shift and chromatography. *J. Immunol.* 124, 2399–2404.

[403] Jude, R. C., and Caldwell, H. D. (1985). Identification and isolation of surface-exposed portions of the MOMP of *C. trachomatis* by two-dimensional peptide mapping and high-performance liquid chromatography. *J. Liquid Chromatogr.* 8, 1559–1571.

[404] Perez-Martinez, J. A., and Storz, J. (1985). Antigenic diversity of *Chlamydia psittaci* of mammalian origin determined by microimmunofluorescence. *Infect. Immun.* 50, 905–910.

[405] Andersen, A. A. (1991). Serotyping of *Chlamydia psittaci* isolates using serovar-specific monoclonal antibodies with the microimmunofluorescence test. *J. Clin. Microbiol.* 29, 707–711.

[406] Villegas, E., Sorlozano, A., and Gutieez, J. (2010). Serological diagnosis of *Chlamydia pneumoniae* infection: limitations and perspectives. *J. Med. Microbiol.* 59, 1267–1274.

[407] Montigiani, S., Falugi, F., Scarselli, M., and Finko, O. (2002). Genomic approach for analysis of surface proteins of *C. pneumoniae*. *Infect. Immun.* 70, 368–379.

[408] Mukhopadhyay, S., Miller, R. D., Sullivan, E. D., Theodoropoulos, C., Matheus, S. A., Timms, P., et al. (2006). Protein expression profiles of *Chlamydia pneumoniae* in models of persistence versus those of heat shock stress response. *Infect. Immun.* 74, 3853–3863.

[409] Heleström, S. (1962). In Handbuch der Haut-und Geschlechtskranheiten. Erg. Work, VI/2, Part B, (Jadassonhn, J. ed.). Berlin: Springer, 426.

[410] Kiraly, K. (2013). "Chapter 78. Immunoallergologic aspect of *Lymphogranuloma venereum*," in *Allergic Responses to Infectious Agents*, eds E. Rajka and S. Korossy (Dordrecht: Springer Science and Business Media), 364, 209–216.

[411] Gerlach, H. (1994). "Chapter 34. Chlamydia, pp. 984–996," in *Avian Medicine: Principles and Applications,* eds B. W. Ritche, G. J. Harrison, and L. R. Harrison (Florida: Wingers Publishers Inc.).

[412] Samplaski, M. K., Domes, T., and Jarvy, K. A. (2014). Chlamydial infection and its role in male infertility. *Adv. Androl.* 2014:307950.

[413] Schramek, S., Kasar, J., and Sadecky, E. (1980). Serological cross-reaction of lipid A components of lipopolysaccharides isolated from *Chlamydia psittaci* and *Coxiella burnetii*. *Acta Virol.* 24:224.

[414] Urata, K., Narahara, H., Tanaka, Y., Egashira, T., Takayama, F., and Miyakawa, I. (2001). Effect of endotoxin-induced reactive oxygen species on sperm motility. *Fertility and Sterility*, 76, 163–166.

[415] Allan, I., and Pearce, J. H. (1979). Host modification of chlamydiae: presence an egg antigen on the surface of chlamydiae grown in the chick embryo. *J. Gen. Microbiol.* 112, 61–66.

[416] Newhall, W. J., Batteiger, R. B., and Jones, R. B. (1982). Analysis of the human serological response to proteins of *Chlamydia trachomatis*. *Infect. Immun.* 38, 1181–1189.

[417] Brade, L., Nurminen, M., Makela, P. H., and Brade, H. (1985). Antigenic properties of *Chlamydia trachomatis* lipopolysaccharide. *Infect. Immun.* 48, 569–572.

[418] Essig, A., Simnacher, U., Susa, M., and Marrie, R. (1999). Analysis of humoral immune response to *Chlamydia pneumoniae* by immunoblotting and immunoprecipitation. *Clin. Diagn. Lab. Immunol.* 6, 819–825.

[419] Hunter, J. B., Suresh, M. R., Keshvarz, E., Venman, W. M., and Micetich, R. G. (1986). Purification of lectins from Artocarpus altilis and Ficus deltoidea by gel filtration fast protein liquid chromatography. Biochem. *Archives* 2, 99–106.

[420] Schachter, J., and Caldwell, H. D. (1980). Chlamydiae. *Ann. Rev. Microbiol.* 34, 285–309.

[421] Caldwell, H. D, and Schachter, J. (1982). Antigenic analysis of the major outer membrane protein of *Chlamydia* spp. *Infect. Immun.* 35, 1024–1031.

[422] Caldwell, H. D, and Judd, R. C. (1982). Structural analysis of chlamydial outer membrane protein. *Infect. Immun.* 38, 960–968.

[423] Cevenini, R., Rumpianesi, F., Sambri, V., and La Placa, M. (1986). Antigenic specifity of serological response of Chlamydia. trachomatis. urethritis by immunoblotting. *J. Clin. Pathol.* 39, 325–327.

[424] Dhindra, P. N., Agarwal, L. P., Mahajan, M., Adlakha, S. C., and Baxi, K. K. (1981). Chlamydial group antigen. *Zbl. Vet. Med. Reie B* 28, 336–340.

[425] Schachter, J., Banks, J., Sugg, N., Sung, M., Storz, J., and Meyer, K. F. (1974). Serotyping of *Chlamydia*. I. Isolates of ovione origin. *Infect. Immun.* 9, 92–94.

[426] Treharne, J. D., Darougar, S., and Jones, B. R. (1977). Modification of the microimmunofluorescence test to provide a routine serodiagnostic test for chlamydial infections. *J. Clin. Pathol.* 38, 510–517.

[427] Matikainen, M. T., and Terho, P. (1983). Immunochemical analysis of antigenic determinants of *Chlamydia trachomatis* by monoclonal antibodies. *J. Gen. Microbiol.* 129, 2343–2350.

[428] Batteiger, B., Newhall, W. J. V., and Jones, R. B. (1982). The use of Tween 20 as a blocking agent in the immunological detection of proteins transferred to nitrocellulose membrane. *J. Immunol. Method.* 55, 297–307.

[429] Ohtani, A., Kubo, M., Shimoda, H., Ohya, K., Iribe, T., and Ohishi, T. (2015). Genetic and antigenic analysis of *Chlamydia pecorum* strains isolated from calves with diarrhea. *J. Vet. Med. Sci.* 77, 777–782.

[430] Longbottom, D., Aitchison, K. D., Imre, L. H., Livingstone, M., and Inglis, L. H. (2006). "Proteomic analysis of chlamydial outer membrane protein fractions by liquid chromatography combined with electrospray tandem mass spectroscopy," in *Diagnosis, Pathogenesis and Control of Animal Chlamydioses,* eds D. Longbottom and M. Rocchi (Tweedbank: Meigle Color Printers), 45–46.

[431] Lin, X., Afraine, M., Clemmer, D. E., Zhong, G., and Nelson, D. E. (2010). Identification of *Chlamydia trachomatis* outer membrane complex proteins by differential proteomics. *J. Bacteriol.* 192, 2852–2860.

[432] Kaleta, E. F., and Taday, E. M. (2006). Avian host range of *Chlamydophila* spp. based on isolation, antigen detecton and serology. *Avan. Pathol.* 23, 435–461.

[433] Strauss, J., and Sery, V. (1964). Ornithose in der CSSR. Epidemiologiach – virologische aspecte. *Arch. Exp. Vet. Med.* 18, 61–75.

[434] Ilinskii, Y. A. (1974). *Ornithosis. Clinic, Diagnosis, Treatment.* Moscow: Medicine.

[435] Rodolakis, A., and Souriau, A. (1980). Clinical evaluation of immunity following experimental and natural infection of ewes with *Chlamydia psittaci* (var. ovis). *Ann. Rech. Vet.* 111, 215–223.

[436] Martinov, S. (1984). Studies of Chlamydia infection in cattle in herds with abortions. *Vet. Sci. XXI*, 10, 81–88.

[437] Taylor-Robinson, D., and Thomas, B. J. (1980). The role of *Chlamydia trachomatis* in genital tract and associated diseases. *J. Clin. Pathol.* 33, 205–233.

[438] Ford, D. K., and McCandlish, L. (1971). Isolation of genital TRIC agents in nongonococcal urethritis and Reiter's disease by an irradiated cell culture method. *Br. J. Ven. Dis.* 47, 196–197.

[439] Janeway, C. A., Travers, P., Walport, M., and Shlomchic, M. J. (2001). *The Immune System in Health and Disease. Immunobiology,* 5th Edn, New York Garland, Science.

[440] Beeckman, D. C., and Vanrompay, D. (2010). Biology and intracellular pathogenesis of high and low virulent *Chlamydophila psittaci* strains in chicken macrophages. *Vet. Microbiol.* 141, 352–353.

[441] Rank, R. G., Whittum-Hudson, J. A. (2010). Protective immunity to Chlamydial. genital infection: evidence from animal studies. *J. Infect. Dis.* 15 (Suppl. 2), S168–S177.

[442] Unknown author. (2012). "Chapter 5. Vaccines, development of a *Chlamydia trachomatis* T cell vaccine," *Chlamydial infections: New Insights for the Health Care Professionals*, 65.

[443] Stevens, M. (2015). *Bacterial Ghosts Modulation of Innate Immunity Immune Responses during Chlamydia Infection*. Ph.D. dissertation, Clark Atlanta University, Atlanta, 85.

[444] Picard, M. D., Bodner, J.-L., Bodner, Gierah, T. M., et al. (2015). Resolution of *Chlamydia trachomatis* infection is associated with a distinct T cell response profile. *Clin. Vaccine Immunol.* 22, 1206–1218.

[445] Koroleva, E. A., Koberts, N. V., Shcherbinin, D. N., Zigangirova, N., Shmarov, M., et al. (2017). Chlamydial type III secretion system needle protein induces protective immunity against *Chlamydia muridarum* intravaginal infection. *Bio Med Res. Int.* 2017:3865 802, 14.

[446] Kiyazimova, A. A., and Smorodintsev, A. A. (1968). Vopr. Virus. 13, 1968, 466–470 (by Terskih, I. I. (1979). Ornithosis and other chlamydial infections. Moscow: Medicine, 223.

[447] Ahmad, A., Dawson, C. R., Yeneda, C., Togni, B., and Schachter, J. (1977). Resistance to re-infection with chlamydial agent (Guinea pig inclusion conjunctivitis agent). *Invest. Ophthalmol. Vis. Sci.* 16, 549–552.

[448] Page, L. A. (1978). Stimulation of cell-mediated immunity of chlamydiosis in turkeys by inoculation of chlamydial bacterin. *Am. J. Vet. Res.* 39, 473–480.

[449] McKercher, D. G., Wada, E. M., Ault, S. K., and Theis, J. H. (1980). Preliminary studies on transmission of *Chlamydia* to cattle by thicks (*Ornithodorus coriaceus*). *Am. J. Vet. Res.* 41, 922–924.

[450] Brunham, R. C., Martin, D. H., Kuo, C. C., Wang, S. P., Stevens, C. E., Hubbard, T., et al. (1981). Cellular immune response during uncomplicated genital infections with *C. trachomatis* in humans. *Infect. Immun.* 34, 98–104.

[451] Wakefield, D., and Penny, R. (1983). Cell-mediated immune response to *Chlamydia* in anterior uveitis: role of HLA B27. *Clin. Exp. Immunol.* 51, 191–196.

[452] Senyk, G. M., Sharp, D., Stites, P., et al. (1980). *Med. Microbiol. Immunol.* 168, 91–102.

[453] Levitt, D., Danen, R., and Bard, J. (1986). Both species of *Chlamydia* and two biovars of *Chlamydia trachomatis* stimulate mouse B lymphocytes. *J. Immunol.* 136, 4229–4242.

[454] Levitt, D., Newcomb, W., and Beem, M. O. (1983). Excessive number and activity of peripheral blood B cells in infants with *C. trachomatis* pneumonia. *Clin. Immunol. Immunopathol.* 29, 424–432.

[455] Bard, J., and Levitt, D. (1984). *Chlamydia trachomatis* stimulates human peripheral blood B lymphocytes to proliferate and secrete polyclonal immunoglobulins in vitro. *Infect. Immun.* 43, 84–92.

[456] Heggie, A. D., Wyrick, P. B., Chase. P. A., and Sorensen, R. U. (1986). Cell-mediated immune response to *Chlamydia trachomatis* in mothers and infants. *Proc. Soc. Exp. Biol. Med.* 181, 586–595.

[457] Tuffrey, M., Falder, P., and Taylor-Robinson, D. (1985). Effect of *Chlamydia trachomatis* infection of the murine genital tract of adoptive transfer of congenic immune cells or specific antibody. *Br. J. Exp. Path.* 66, 427–433.

[458] Ovigstad, E., and Hirschberg, H. (1984). Lack of cell-mediated cytotoxicity towards *Chlamydia trachomatis* infected target cells in humans. *Acta Pathol. Microbiol. Immunol. Scand.* 92, 153–159.

[459] Hammerschlag, M. R., Suntharalingam, K., and Fikrig, S. (1985). The effect of *Chlamydia trachomatis* on immunol-dependent chemiluminiscence of human polymorphonuclear leucocytes-requirements. *J. Infect. Dis.* 151, 1045–1051.

[460] Onsrud, M., and Ovigstad, E. (1984). Natural killer cell activity after gynecological infections with *Chlamydia. Acta Obst. Gynecol. Scand.* 63, 613–615.

[461] Robinsonn, P., Wakefield, D., Graham, D. M., et al. (1985). The chemiluminescence response of normal human leucocytes to *Chlamydia trachomatis. Diagn. Immunol.* 3, 119–125.

[462] Huang, H.-S, Buxton, D., Burrels, C., Anderson, I., and Miller, H. R. P. (1991). Immune response of the ovine lymph node to *Chlamydia psittaci.* A cellular study of popliteal lymph. *J. Comp. Pathol.* 105, 191–202.

[463] McCafferty, M. C. (1994). The development of proliferative responses of ovine peripheral blood mononuclear cells to *Chlamydia psittaci* during pregnancy. *Vet. Immunol. Immunopathol.* 41, 173–180.

[464] Brown, J., Howie, S. F., and Entrican, G. (2001). A role for tryptophan in immune control of chlamydial abortion in sheep. *Vet. Immunol. Immunopathol.* 82, 107–119.

[465] Graham, S. P., Jones, G. E., McLean, M., Livingstone, M., and Entrican, G. (1995). Recombinant ovine interferon gamma inhibits the multiplication of *Chlamydia psittaci* in ovine cells. *J. Comp. Pathol.* 112, 185–195.

[466] Rocchi, M. S., Wattegedera, S., Meridiani, I., and Entrican, G. (2009). Protective immunity to *Chlamydophila abortus* infection and control of ovine enzootic abortion (OAE). *Vet. Microbiol.* 135, 112–121.

[467] Hein, W. R., and Griebel, P. J. (2003). A road less travelled: large animal models in immunological research. *Nat. Rev. Immunol.* 3, 79–84.

[468] Del Rio, L., Barbera-Cremades, M., Navarro, J. A., Buendia, A. J., Cuello, F., Ortega, N., et al. (2013). IFN expression in placenta is associated with resistance to *Chlamydia abortus* after intragastric infection. *Microb. Pathog.* 56, 1–7.

[469] Poston, T., Qu, Y., Russell, A., Girardi, J., Nagarajan, U., and Darville, T. (2016). Towards of understanding of protective cellular immunity. *J. Immunol.* 196 (1 Suppl.), 66.3.

[470] Käser, T., Pasternak, J. A., Rieder, M., Hamonic, G., Lai, K., Delgado-Ortega, M., et al. (2016). "Towards the understanding of porcine cellular immune response to *Chlamydia trachomatis* and *suis* infections," in *Proceedings of the IAD-2016 – Symp. French Network of Domestic Animal Immunology*, Ploufragan, 34.

[471] Ovigstad, E., Digranes, S., and Thorsby, E. (1983). Antigen-specific proliferative human T-lymphocyte clones with specifity for *Chlamydia trachomatis*. *Scand. J. Immonol.* 18, 291–297.

[472] Cevenini, R., Costa, S., and Rumpianesi, F. (1981). Cytological and histopathological abnormalities of the cervix in genital *C. trachomatis* infections. *Br. J. Vener. Dis.* 57, 334–337.

[473] Mitao, M., Reumann, W., Winkler, B. B., Richard, R. M., Fujima, A., and Crum, C. P. (1984). Chlamydial cervicitis and cervical intraepithelial neoplasia: an immunohistochemical analysis. *Gynecol. Oncol.* 19, 90–97.

[474] Markey, B. (2010). "Immunopathogenetic studies and the development of novel diagnostic methodologies with application in the targeted surveillance of reproductive loss in sheep due to *Toxoplasma gondii* and *Chlamydophila abortus,*" DAFF Project Ref. No. RSF06 394, 2006–2010, School of Agriculture, Food science and Vet. Medicine, University College, Dublin.

[475] Casadevall, A., and Pirotski, L.-A. (2006). A reappraisal of humoral immunity based on mechanisms of antibody–mediated protection against intracellular pathogens. *Adv. Immunol.* 91, 1–44.

[476] Li, L.-X., and McSorley, S. J. (2015). A re-evaluation of the role of B-cells in protective immunity to *Chlamydia* infection. In: The lung and the Heart. *Immunol. Lett.* 164, 88–93.

[477] Richmond, S. J., and Gaul, E. O. (1977). "Single-antigen indirect immunofluorescence test for chlamydial antibodies," in *Non-gonococcal Urethritis and Related Infections*, eds D. Hobson and K. K. Holmes (Washington, DC: Am. Soc. Microbiol), 259.

[478] Cremer, N. E., Devlin, V. L., Riggs, J. T., and Hagens, S. J. (1984). Anomalous antibody responses in viral infections: specific stimulation of polyclonal activation? *J. Clin. Microbiol.* 20, 468–472.

[479] Newer, A., Lang, K. N., Tiller, F. W., Kiesd, L., and Witkin, S. S. (1997). Humoral immune response to membrane, component of *C. trachomatis* and expression of human 60 kDa heat shock protein in follicular fluid of in-vitro fertilization patients. *Hum. Reprod.* 12, 925–929.

[480] Domeika, M., Domeika, K., Paavonen, J., Mardh, P.-A., and Witkin, S. S. (1998). Humoral immune response to conserved epitopes of *Chlamydia trachomatis* and human 60 kDa heat shock protein in women with pelvic inflammatory disease. *J. Infect. Dis.* 177, 114–119.

[481] Osser, S., and Persson, K. (1984). Postabortal pelvic infection associated with *Chlamydia trachomatis* and the influence of humoral immunity. *Am. J. Obstet. Gynecol.* 150, 599–703.

[482] Skaug, K., Otnaess, A. B., Orstaik, and Jerve, F. (1982). Chlamydial secretory IgA antibodies in human milk. *Acta Pathol. Microbiol. Immunol. Scand.* 90, 21–26.

[483] Schachter, J., Gles, L. D., Ray, R. M., and Hesse, F. E. (1983). Is there immunity to chlamydial infection of the human genital tract. *Sex. Transm. Dis.* 10, 123–125.

[484] Johnston, A. R., Osborn, M. F., Thomas, B. J., Hetherington, C. M., and Taylor-Robinson, D. (1983). Immunity to re-infection of the genital tract of marmosets with *Chlamydia trachomatis*. *Br. J. Exp. Pathol.* 62, 606–613.

[485] Williams, D. M., Schachter, J., Grubbs, B., and Samaya, C. V. (1982). The role of antibody in host defense against the agent of mouse pneumonitis. *J. Infect. Dis.* 145, 200–205.

[486] Barnes, R. C., Suchland, R. J., Wang, S. P., Kuo, C. C., and Stamm, W. E. (1985). Detection of multiple serovars of *Chlamydia trachomatis* in genital infections. *J. Infect. Dis.* 152, 985–989.

[487] Brunham, R. C. (2013). Immunity to *Chlamydia trachomatis*. *J. Infect. Dis.* 207, 1796–1796.

[488] Farris, C. M., and Morrison, R. P. (2011). Vaccination against chlamydial genital infection utilizing the murine *C. muridarum* model. *Infect. Immun.* 79, 986–996.

[489] Moore, T., Ekworomadu, C. O., Eko, F. O., MacMillan, L., Ramey, K., Ananaba, G. A., et al. (2003). Fc receptor-mediated antibody regulation of T cell immunity against intracellular pathgens. *J. Infect. Dis.* 188, 617–624.

[490] Farris, C. M., Morrison, S. G., and Morrison, R. P. (2010). CD^{4+} T cells and antibody are required for optimal major outer membrane protein vaccine-induced immunity to *Chlamydia muridarum* genital infection. *Infect. Immun.* 78, 4374–4383.

[491] Su, H., and Caldwell, H. D. (1991). In vitro neutralization of *Chlamydia trachomatis* by monovalent Fab antibody specific to the major outer membrane protein. *Infect. Immun.* 59, 2843–2845.

[492] Wyrick, R. B., and Brownridge, E. A. (1978). Growh of *Chlamydia psittaci* in macrophages. *Infect. Immun.* 19, 1050–1060.

[493] Armitage, C. W. (2012). *The Role of Immunoglobulins and Their Transporters in Urogenital Chlamydial Infections*. Ph.D. dissertation,

Inst. of Health and Biomed. Innoviations, School of Biomed. Science, Queensland University of Technology, Brisbane.

[494] Redgrove, K. A., and McLaughlin, E. A. (2014). The role of immune response to *Chlamydia trachomatis* infection of the male genital tract: a double edged sword. *Front. Immunol.* 5:534.

[495] Bartolini, E., Ianni, E., Frigimelica, E., Pehacca, R., Galli, G., Berlanda, F., et al. (2013). Recombinant outer membrane vesicles carring *Chlamydia muridarum* HtrA induce antibodies that neutralize chlamydial infection in vitro. *J. Extracell. Vesicles* 17, 315–316.

[496] Moore, T., Atanaba, G. A., Bolier, J., Bowers, S., Belay, T., Eko, F. O., et al. (2002). Fc receptor regulation of protective immunity against *C. trachomatis*. *Immunology* 105, 213–221.

[497] Igietseme, J. M., Eko, F. O., He, Q., and Black, C. M. (2004). Antibody regulation of T cell immunity; implication for vaccine strategis against intracellular pathogens. *Expert. Rev. Vaccines* 3, 23–24.

[498] Hafner, L. M., Volp, K., Mathews, S., and Timms, P. (2001). Peptide immunization of guinea pigs against *Chlamydia psittaci* (GPIC agent) infection induces good vaginal secretion antibody response, in vitro neutralizatioin and partial protection against the challenge. *Immunol. Cell Biol.* 79, 245–250.

[499] De Clercq, E., Devriendt, B., Yin, L., Chiers, K., Cox, E., and Vanrompay, D. (2014). The immune response against *Chlamydia suis* genital tract infection partially protects against re-infection. *Vet Res.* 445:95.

[500] Khan, S. A., Polkinghome, A., Waugh, C., Hanger, J., Loader, J., Beagley, K., et al. (2016). Humoral immune response in coalas (*Phascolarctos cinereus*) either naturally infected with *Chlamydia pecorum* or following administration of a recombinant chlamydial major outer membrane protein vaccine. *Vaccine* 34, 775–782.

[501] Hagemann, J., Simnacher, U, Longbottom, D., Livingston, M., Maile, J., Soutschek, E., et al. (2016). Analysis of humoral immune response against surface and virulence-associated *Chlamydia abortus* proteins in Ovine and human abortion using newly developed line immunoassay. *J. Clin. Microbiol.* 54, 1983–1890.

[502] Fuentes, V., Lefebver, J., Lema, F., Bissau, E., and Orfila, J. (1985). Establishment of hybridomas secreting moinoclonal antibody to *Chlamydia psittaci*. *Immunol. Lett.* 10, 325–326.

[503] Clark, R. B., Nachamkin, I., Schatzki, R. F., and Dalton, H. P. (1982). Localization of distinct surface antigens on *Chlamydia trachomatis*

HAR-13. Immune electron microscopy with moinoclonal antibodies. *Infect. Immun.* 38, 1273–1278.
[504] Stephens, R. S., Tam, M. R., Kuo, C. C., and Nowinski, R. C. (1982). Monoclonal antibodies to *Chlamydia trachomatis*: antibody specifities and antigen characterization. *J. Immunol.* 128, 1083–1089.
[505] Gillespie, S. H. (2014). *Medical Microbiology Illustrated*, 1st Kindle Edn. Oxford: Butterworth-Heinemann, 286.
[506] Alexander, I., Paul, I. D., and Caul, E. O. (1984). Evaluation of a genus reactive moinoclonal antibody in rapid identification of *Chlamydia trachomatis* by immunofluorescence. *Genitourin. Med.* 61, 252–254.
[507] Dimitrov, K., Martinov, S., and Popov, G. (1986). "Diagnosis of chlamydial cervicitis by monoclonal antibodies," in *Proceedings of the 4th Congress of the Bulgarian Dermatologists with Intl. Participation*, Varna, 165.
[508] Forbes, B. A., Bartholoma, N., McMillan, J., Roefaro, M., Weiner, L., and Welvch, L. (1986). Evaluation of a moinoclonal antibody test io detect Chlamydia in cervical and urethral speciments. *Clin. Microbiol.* 23, 1136–1137.
[509] Shafer, M. A., Vaugan, E., and Lipkin, E. S., Moscicki, B. A., and Schachter, J. (1986). Evaluation of fluorescein-conjugated moinoclonal antibodies for detection of *Chlamydia trachomatis* endocervical infections in adolescent girls. *J. Pediatr.* 108, 779–783.
[510] Chang, F. Y., and Yu, V. L. (1999). "Acute pneumonia," in *Clinical Infectious Diseases: a Practical Approach,* ed. R. K. Rot (Oxford: Oxford University Press), 1013.
[511] Unknown author (2016). *Basic medical key. Chlamydia spp.* Antigens.
[512] Kumar, A., and Mital, A. (2006). Production and characterization of monoclonal antibodies to *Chlamydia trachomatis. Hybridoma (Larchmt)*, 25, 293–299.
[513] Zhang, J., Ding, T., Chen, Z., Fang, H., Li, H., Lu, H., et al. (2015). Preparation and evaluation of monoclonal antibodies against chlamydial protease-like activity factor to detect *Chlamydia pneumoniae* antigen in early pediatric pneumonia. *Eur. J. Clin. Microbiol. Infect. Dis.* 34, 1319–1326.
[514] ViroStat. (2017). *Chlamydia pneumoniae* monoclonal antibodies. Available at: www.virostatinc.com.

[515] Morrison, S., and Morrison, R. P. (2006). "Humoral immunity to *Chlamydia muridarum* genital tract infection," in *Proceedings of the 11th Intl. Symp. On Human Chlamydial Infection*, Niagara-on-the Lake, ON.

[516] Pal, S., and De la Masa, L. M. (2013). Mechanism of T cell mediated protection in newborn mice against a Chlamydia infection. *Microbes Infect.* 15, 807–814.

[517] Vats, A. A., and Mital, A. (2013). Study of time kinetics of monoclonal antibodies to *Chlamydia trachomatis* to define an optimum titer. *Sri Ramachandra J. Med.* 68, 1–6.

[518] Essig, A., and Longbottom, D. (2015). *Chlamydia abortus*: New aspects of infectious abortion in sheep and potential risk for pregnant women. *Current Clin. Microbiol. Rep.* 2, 22–34.

[519] Van Veen, L. M., Keessen, E. C., and Van Knapen, F. (2012). Control of zoonotic risk for pregnant animal handlers in zoos. *Dept. Vet. Pub. Health, IPAS*, 1–9.

[520] Aitken, I. (2008). "Chlamydia abortion," in *Diseases of Sheep*, ed. I. Aitken (Hoboken: John Wiley and Sons), 624.

[521] Ortega, M., Caso, R., Galligo, M. C., Murcia-Belmonte, A., Álvarez, D., del Río, L., et al. (2015). Isolation of *Chlamydia abortus* from a laboratory worker diagnosed with atypical pneumonia. *Ir. Vet. J.* 69:8.

[522] Bedson, S. P. (1935). The use of complement fixation test in diagnosis of human psittacosis. *Lancet* 2, 1277–1280.

[523] Fraser, C. E. O. (1969). "Analytical serology of the *Chlamydiaceae*," in *Analytical Serology of Microorganisms*, ed. J. B. G. Kwapinski (New York, NY: Wiley), 257–329.

[524] Volkert, M., and Christensen, P. M. (1955). Two ornithosis complement fixing antigens from infected yolk sacs. I. The phosphatide antigen, the virus antigen, preparation and methods. *Acta Path. Microbiol. Scand.* 37, 211–219.

[525] Chervonsky, V. M., and Popova, O. M. (1959). Antigen for CFT in Ornithosis-Psittacosis. *Voproci Virusologii* 1, 68–71.

[526] OIE Terrestrial Manual (2016). *Chapter 2.7.6. Enzootic Abortion in Ewes*, 1008–1016.

[527] Serion. (2012). *Complement Fixation Test (CFT)*. Inst. Virion/Serion GmbH, Germany.

[528] Zatulovski, B. G., Popovich, G. G., Bortnichuk, B., and Carapata, A. P. (1978). *Materials for the etiological role of chlamydiae in animals*

in human infections pathology, Reports 7th Intl. Congress Infect. Parasit. Dis. I, Varna, 545–548.

[529] Niemczuk, K. (2007). "Seroprevalence of antibodies anti-*Chlamydia* spp. in pigs from Poland. The comparison of serological methods. pp. 86–88," in *Pathogenesis, Epidemiology and Zoonotic Importance of Animal Chlamydioses*, in *Proceedings of the 5th Annual Workshop of COST Action 855,* eds K. Niemczuk, K. Sachse, L. D. Sprague (Pulawy: Natl. Vet. Res. Inst.), 94.

[530] Rypula, A., Kumala, K., Ploneczka-Janeczko, K., and Karuga-Kuzniewska, E. (2016). Chlamydia prevalence in Polish pig herds. *Epidemiol. Infect.* 144, 2578–2586.

[531] Wilson, M. R., and Plummer, P. (1966). A survey of pig sera for the presence of antibodies to the psittacosis-lymphogranuloma venereum group of organisms. *J. Comp. Pathol.* 76, 427–433.

[532] Szymanska-Czerwinska, A., Niemczuk, K., and Wojuk, A. (2011). Prevalence of *Chlamydia suis* in population of swine in Poland and comparison of complement fixation test and PCR used in the diagnosis of chlamydiosis. *Bull. Vet. Inst. Pulawy* 55, 381–383.

[533] McCauley, L. M. E., Lancaster, M. J., Young, P., and Butler, K. L., Ainsworth, C. G. V. (2007). Comparison of ELISA assays for *Chlamydophila abortus* antibodies in ovine sera. *Aust. Vet. J.* 85, 325–328.

[534] Wilson, K., Livingstone, M., and Longbottom, D. (2009). Comparative evaluation of eight serological assays for diagnosing *Chlamydophila abortus* infection in sheep. *Vet. Microbiol.* 135, 38–45.

[535] Stamp, J. T., Watt, J. A., and Cockburn, R. B. (1959). Enzootic abortion in ewes: complement fixation test. *J. Comp. Pathol.* 62, 93–101.

[536] Sachse, K., Vretou, E. Livingstone, M., Borel, N., and Pospischil, A. (2009). Recent developments in t5he laboratory diagnosis of chlamydial infections. *Vet. Microbiol.* 135, 2–21.

[537] Brabetz, W., Lindner, B., and Brade, H. (2000). Comparative analysis of secondary gene products of 3-deoxy-D-manno-oct-2 ulosonic acid transferases from *Chlamydiaceae* in *Escherichia coli* K-12. *Eur. J. Biochem.* 267, 5458–5465.

[538] Brade, L., Rozalski, P., Cosma, P., and Brade, H. (2000). A monoclonal antibody recognizing the 3-deoxy-D-manno-oct-2 ulosonic

acid (Kdo) trisaccharide alpha Kdo (2 → 4) alpha Kdo *Chlamydophila psittaci* 6BC lipopolysaccharide. *J. Endotoxin Res.* 6, 361–368.

[539] Andersen, A. A., and Vanrompay, D. (2009). "Chapter 24, Section II: Bacterial diseases, avian chlamydiosis," in *Diseases of Poultry*, 12th Edn, ed. Y. M. Saif.

[540] Evans, A. S., and Brachman, P. S. (eds). (2013). "Chapter 10: Chlamydial infections," in *Bacterial Infections of Humans. Epidemilogy and Control* (Berlin: Spinger), 702.

[541] Mabey, D., and Peeling, R. W. (2002). Lymphogranuloma venereum. *Sex. Trans. Infect.* 78, 90–94.

[542] De Ory, F., Guisasola, M. E., Coccola, F., Teilez, A., and Echevarria, J. M. (2004). Evaluation of an automated complement fixation test (Seramat) for diagnosis of acute respiratory infections, caused by viruses and atypical bacteria. *Clin. Microbiol. Infect.* 10, 220–223.

[543] Magnino, S., Giovannini, S., Paoli, C., Ardenghi, P., and Sambri, V. (2005). Evaluation of an automated complement fixation test (Seramat) for detection of chlamydial antibodies in sheep and goat sera. *Vet. Res. Commun.* 29 (Suppl. 1), 157–161.

[544] Cevenini, R., Donati, M., and Rumpianesi, F. (1981). Elementary bodies as single antigen in a micro-ELISA test for *Chlamydia trachomatis*. *Microbiologica* 4, 347–351.

[545] Evans, R. T., and Robinson, T. (1982). Development and evaluation of an enzyme-linked immunosorbent assay (ELISA) using group antigen to detect antibodies to *Chlamydia trachomatis*. *J. Clin. Pathol.* 35, 1122–1128.

[546] Caldwell, H. D., and Schachter, J. (1983). Immunoassay for detecting *Chlamydia trachomatis* major outer membrane protein. *J. Clin. Microbiol.* 18, 539–545.

[547] Jones, M. F., Smith, T. F., Houglum, A. J., and Hermann, J. E. (1984). Detection of *Chlamydia trachomatis* in genital speciments by the Chlamydiazyme test. *J. Clin. Microbiol.* 20, 465–467.

[548] Mumtaz, G., Mellars, B. J., Ridgway, G. L., and Oriel, J. D. (1985). Enzyme immunoassay for the detection of *Chlamydia trachomatis* antigen in urethral and endocervical swabs. *J. Clin. Pathol.* 38, 740–742.

[549] Matikainen, M. T., and Lentonen, O. P. (1984). Relations between avidity and specifity of monoclonal anti-Chlamydial antibodies

in culture supernatant and ascetic fluids determined by enzyme immunoassay. *J. Immunol. Methods* 72, 341–348.

[550] Matikainen, M. T. (1984). Effect of pH on reactivity of monoclonal antibodies to *Chlamydia*. *J. Immunol. Methods* 75, 211–216.

[551] Griffits, P. C., Plater, J. M., Horigan, M. W., Rose, M. P., Venables, C., and Dauson, M. (1996). Serological diagnosis of ovine enzootic abortions by comparative inclusion immunofluorescence assay, recombinant lipopolysaccharide enzyme-linked immunosorbent assay, and complement fixation test. *J. Clin. Microbiol.* 34, 1512–1518.

[552] Brade, L., Brunnemann, H., Erust, M., Fu, Y, Holst, O., Cosma, P., et al. (1994). Occurance of antibodies againt chlamydial lipopolysaccharide in human sera as measured by ELISA using an artificial glucoconjugate antigen. *FEMS Immunol. Med. Microbiol.* 8, 27–41.

[553] Vretou, E., Radouani, F., Psarrou, E., Kritikos, I., Xylouri, E., and Mangana, O. (2007). Evaluation of two commercial assays for the detection of *Chlamydophila abortus* antibodies. *Vet. Microbiol.* 123, 153–161.

[554] Jones, G. E., Low, J. C., Machell, J., and Armstong, K. (1997). Comparison of five tests for the detection of antibodies against chlamydial (enzootic) abortion in ewes. *Vet. Rec.* 141, 164–168.

[555] Vretou, E., Psarrou, E., Kaisar, M., Vlisidou, I., Salti Monte, V., and Longbottom, D. (2001). Identification of protective epitops by sequencing of the major outer membrane protein gene of a variant strain of *Chlamydia psittaci* serotype 1 (*Chlamydophila abortus*). *Infect. Immun.* 69, 607–612.

[556] Buendia, A. J., Cuello, F., Del Rio, l., Gallego, M. C., Caro, M. R., and Salinas, J. (2001). First evaluation of a new commercially available ELISA based on recombinant antigen for diagnosing *Chlamydophila abortus* (*Chlamydia psittaci* serotype 1) infection. *Vet. Microbiol.* 78, 229–239.

[557] Moore, S. A., Munk, C., Persson, K., Kruger-Kiaer, S., Van Diik, R., Meijer, C. J., et al. (2002). Comparison of three commercially available Peptide-based immunoglobulin IgG and IgA assays to microimmunofluorescence assay for detection of *Chlamydia trachomatis* antibodies. *J. Clin. Microbiol.* 40, 584–587.

[558] Menon, S., Stanfield, S. H., Logan, B., Hocking, J. S., Timms, P., Rommbauts, L., et al. (2016). Development and evaluation of a multiantigen peptide ELISA for the diagnosis of *C. trachomatis* related infertility in women. *J. Med. Microbiol.* 65, 915–922.

[559] Salti-Montesanto, V., Tsoli, E., Papavassilou, P., Psarrou, E., Markey, B. K., Jones, G. E., et al. (1997). Diagnosis of ovine enzootic abortion, using a competitive ELISA based on monoclonal antibodies against variable segment 1 and 2 of the major outer membrane protein of *Chlamydia psittaci* serotype 1. *Am. J. Vet. Res.* 58, 228–235.

[560] Longbottom, D., Psarrou, E., Livingstone, M., and Vretou, E. (2001). Diagnosis of ovine enzootic abortion, using an indirect ELISA (rOMP91B ELISA) based on recombinant protein fragment of the polymorphic outer membrane protein POMP91B of *Chlamydophila abortus. FEMS Microbiol. Lett.* 157–161.

[561] Longbottom, D., Fairley, S., Chapman, S., Psarrou, E., Vretou, E., and Livingstone, M. (2002). Serological diagnosis of ovine enzootic abortion by enzyme-linked immunosorbent assay using a recombinant protein fragment of the polymorphic outer membrane protein POMP90 of *Chlamydophila abortus. J. Clin. Microbiol.* 40, 4235–4243.

[562] Stephens, R. S., Kuo, C. C., and Tam, M. R. (1982). Sensitivity of immunofluorescence with monoclonal antibodies for detection of *Chlamydia trachomatis* in cell cultures. *J. Clin. Microbiol.* 16, 4–7.

[563] Stam, R. S., Tam, M., Koester, M., and Cles, L. (1983). Detection of *Chlamydia trachomatis* incklusions in McCoy cell cultures with Fluorescein-conjugated monoclonal antibodies. *J. Clin. Microbiol.* 17, 666–668.

[564] Eb, F., and Orfila, J. (1982). Serotypig of *Chlamydia psittaci* by the microimmunofluorescence test: isolates of ovine origin. *Infect. Immun.* 37, 1289–1291.

[565] Eb, F., Orfila, J., Milon, A., and Geral, M. F. (1986). Epidemiogical significance of the immunofluorescence typing of *Chlamydia psittaci. Ann. Inst. Pasteur Microbiol.* 137B, 77–93.

[566] Andersen, A. A. (1991). Serotypig of *Chlamydia psittaci* using serovsr-specific monoclonal antibodies with the microimmunofluorescence test. *J. Clin. Microbiol.* 29, 707–711.

[567] Andersen, A. A. (2005). Serotypig of US isolates of *Chlamydophila psittaci* from domestic and wild birds. *J. Vet. Diagn. Invest.* 17, 479–482.

[568] Newhall, W. J., Terho, P., Wilde, C. E., Batteiger, B. E., and Jones, R. B. (1986). Serovar determination of *Chlamydia trachomatis* isolates by using type-specific monoclonal antibodies. *J. Clin. Microbiol.* 23, 333–338.

[569] Verminen, K., Duquenne, B., De Keukeleire, Duim, B. M, Pannekoek, L. Braeckman, L., et al. (2008). Evaluation of a *Chlamydophila psittaci* diagnostic platform for zoonotic risk assessment. *J. Clin. Microbiol.* 46, 281–285.

[570] Di Francesco, A., Baldelli, R., Cevenini, R., Magnino, S., Pignanelli, S., Salvatore, S., et al. (2006). Seroprevalence to chlamydiae in pigs in Italy. *Vet Rec.* 159, 849–850.

[571] Borel, N., Sachse, K., Rassbach, A., Brucner, L., Vretou, E., Psarrou, E., et al. (2006). Ovine enzootic abortion (OAE) antibody response in vaccinated sheep compared to naturally infected sheep. *Vet Res. Commun.* (Suppl. 1), 151–156.

[572] Martinov, S. (1983). Serological study on chlamydial abortion in sheep. *Vet Sci. XX* 8, 8–16.

[573] Martinov, S. (1984). Study on the chlamydial infection in mass abortion in goats. *Vet. Sci. XXI* 6, 33–40.

[574] Reinhold, P., Sachse, K., and Kaltenboeck, B. (2011). *Chlamydiaceae* in cattle: commensals, trigger organisms, or pathgens? *Vet. J.* 189, 257–267.

[575] Martinov, S., Popov, G., and Panova, M. (1980). "Studies on chlamydial miscarriages in domestic animals in recent years," in *Proceedings of the 5th Congress Bulg. Microbiologists*, *I*, 359–398.

[576] Popov, G. V., and Martinov, S. P. (1985). "New data on the problem chlamydial abortion in sheep," in *Proceedings of the 6th Congress of Microbiology*, Varna, 597–601.

[577] Martinov, S. P., and Popov, G. V. (1992). "Recent outbreaks of ornithosis in dicks and humans in Bulgaria," in *Proceedings of the Eur. Soc. Chlamydia Research*, Stockholm (abs).

[578] Magnino, S., Haag-Wackernagel, D., Geigenfeind, I., Dovc, A., Prukner-Radovcic, E., Donati, M., et al. (2006). "Report on the Cost 855 meeting held in Pavia on urban pigeons and their implications on public health," in *Diagnosis, Pathogenesis and Control of Animal Chlamydioses*, eds D. Longlottom and M. Rocchi (Tweedbank: Meigle Colour Printers Ltd.), 10–11.

[579] Martinov, S. P., Schoilev, H., Kazachka, D., and Runevska, P. (1997). Studies on the *Chlamydia psittaci* infection in dicks and humans. *Biotechnol. Biotechnol. Eq.* 1, 74–80.

[580] Martinov, S. (2012). "Psittacosis in birds, pp. 325–329," in *Zoonoses in Humans and Animals,* eds S. Martinov, and R. Komitova (Sofia: Medicine and Physical Education Publishers), 530.

[581] Martinov, S. P., Kazachka, D., and Schoilev, H. (1998). "Study on psittacosis-ornithosis in birds and humans," in *Proceedings of the Natl. Conference on Zoonotic Diseases with International Participation*, Stara Zagora, 10–11.

[582] Magnino, S. Haag-Wackernagel, D., Geigenfeind, I., Helmecke, S. Dovc, A., Prukner-Radovcic, E. et al. (2009). Chlamydial infections in feral pigeons in Europe: Review of data and focus on public health implications. *Vet. Microbiol.* 135, 54–67.

[583] Ferreira, V. L., Diaz, R. A., and Raso, T. F. (2016). Screening of feral pigeons (*Columba livia*) for pathogens of veterinary and medical importance. *Rev. Bras. Cienc. Avic.* 18:4.

[584] Haag-Wackernagel, D., and Moch, H. (2004). Health hazards posed by feral pigeons. *J. Infection* 48, 307–313.

[585] Dickx, V., Beeckman, D. S. A., Dossche, L., Tavernier, P., and Vanrompay, D. (2010). *Chlamydia psittaci* in homing and feral pigeons and zoonotic transmission. *J. Med. Microbiol.* 59, 1348–1353.

[586] Youroukova, V., Ivanov, S., and Martinov, S. (2001). *Chlamydia pneumoniae* and its potential role in asthma. *Pneumol. Phthisiatr.* 37, 15–19.

[587] Youroukova, V., Ivanov, S., and Martinov, S. (2003). *Chlamydia pneumoniae* and asthma. *Eur. Resp. J.* 22:421.

[588] Dimov, D., Zhutev, I., Raichev, R., Kassanski, V., Christov, M., Martinova, F., et al. (1998). "Characteristics of reactive arthritis (ReA) in the Bulgarian military men," in *Proceedings of the 3rd Congress of Balkan Military Medical Committee*, Athens, 116 (abs).

[589] Dimov, D., Zhutev, I., Raichev, R., Martinova, F., Christov, M., and Martinov, S. (2001). "Investigations on postrespiratory reactive arthritis," in *Proceddings of the 3rd Central European Congress of Rheumatology*, Bratislava, 11.

[590] Dimov, D., Zhutev, I., Raichev, R., Martinova, F., Christov, M., and Martinov, S. (2001). Postrespiratory reactive arthritis: etiology, clinical and laboratory characteristics and early prognosis. *Therapevtic Archive* 73, 65–68.

[591] Kaltenboeck, B., Schmeers, N., and Schneide, P. (1997). Evidence of numerous omp 1 alleles of porcine *Chlamydia trachomatis* and novel chlamydial species obtained by PCR. *J. Clin. Microbiol.* 35, 1835–1841.

[592] Joshida, H., Kishi, Y., Shige, S., and Hagiwara, T. (1998). Differentiation of *Chlamydia* species by combined use of polymerase chain

reaction and restriction endonuclease analysis. *Microbiol. Immunol.* 42, 411–414.

[593] Everett, D. A., Hornune, L. J., and Andersen, A. A. (1999). Rapid detection of *Chlamydiaceae* and other families in the order Chlamydiales: three PCR tests. *J. Clin. Microbiol.* 37, 575–580.

[594] Messmer, T. O., Skelton, S. K., Daugharty, J. F., and Fields, B. S. (1997). Application of a nested multiplex PCR to psittacosis outbreak. *J. Clin. Microbiol.* 35, 2043–2046.

[595] Sachse, K., and Hotzel, H. (2003). Detection and differentiation of chlamydiae by nested PCR. *Microbiol. Methods Mol. Biol.* 216, 123–136.

[596] Van Look, M., Verminen, K. T. O., Messmer, T. O., Volckaert, G., Gooddeeris, B. M., and Vanrompay, D. (2005). Use of a nested PCR-enzyme immunoassay with an internal control to detect *Chlamydophila psittaci* in turkey. *BMC Infect. Dis.* 5:76.

[597] Madico, G., Quinn, T. C., Boman, J., and Gaydos, C. A. (2000). Touchdown enzynme release-PCR, for detection and identification of *C. trachomatis*, *C. pneumoniae* and *C. psittaci* using 16S and 23S spacer rRNA genes. *J. Clin. Microbiol.* 38, 1085–1093.

[598] Hartley, J. C., Kage, S., Stevenson, S., Bennett, J., and Ridgway, G. (2001). PCR detection and molecular identification of *Chlamydiaceae* species. *J. Clin. Microbiol.* 39, 3072–3079.

[599] Laroucau, K., Trichereau, A., Vortimore, F., and Mahe, A. M. (2007). A pmp genes based PCR as a valuable tool for the diagnosis of avian chlamydiosis. *Vet. Microbiol.* 121, 150–157.

[600] De Graves, A., Gao, D., and Kaltenboeck, B. (2003). High-sensitivity quantative PCR-platform. *Biotechniques* 34, 106–115.

[601] Alfalter, P., Reischl, U., and Hammerschlag, M. R. (2005). In-house nucleic acid amplification assays in research: how much quality control is needed before one can rely upon the results? *J. Clin. Microbiol.* 43, 5835–5841

[602] Livak, K. J., Flood, S. J., Marmaro, J., Ginsti, W., and Deetz, K. (1995). Oligonucleotides with fluorescent dyes at opposite ends provide a quenched probe system useful for detecting PCR products and nucleic acid hybridization. *PCR Methods Appl.* 4, 357–362.

[603] Heid, C. A., Stevens, J., Livak, K. J., and Williams, P. M. (1996). Real-time quantative PCR. *Genome Res.* 6, 968–994.

[604] Wittwer, C. A., Hermann, M. G., Moss, A. A., and Rasmussen, R. P. (1996). Continuos fluorescence monitoring of rapid cycle DNA amplification. *Biotechniques* 22, 130–131, 134–138.

[605] Geens, T., Dewitte, A., Boon, N., and Vanrompay, D. (2005). Development of a *Chlamydophila psittaci* species-specific and genotype-specific real-time PCR. *Vet. Res*. 36, 1–11.

[606] Dorak, M. T. (ed.). (2006). *Real-time PCR (Advanced Method Series).* Oxford: Taylor and Francis.

[607] Enricht, R., Slickers, P., Goellner, S., Hotzel, H., and Sachse, K. (2006). Optimized DNA microarray assay allows detection and genotyping of single PCR amplifiable target copies. *Moll. Cell. Probes* 20, 60–63.

[608] Menard, A., Clerc, M., Subtil, A., Megcaud, F., Bebear, C., and De Barbeyrac, B. (2006). Development of a real-time PCR for the detection of *Chlamydia psittaci*. *J. Med. Microbiol*. 55, 471–473.

[609] Jalal, H., Stephen, H., Curran, M. D., Burton, J., Bradley, M., and Carne, C. (2006). Development and validation of a Rotor-Gene Real-time PCR assay, for detection, identification and quatification of *Chlamydia trachomatis* in a Single reaction. *J. Clin. Microbiol*. 44, 206–213.

[610] Butcher, R., Houghton, J., Derrick, T., Ramadhaui, A., Herrera, B., Last, A. P. et al. (2017). Reduced-cost *Chlamydia trachomatis*- specific multiplex real-time PCR diagnostic assay evaluated for ocular swabs and use by trachoma research programmes. *J. Microbiol. Methods* 139, 95–102.

[611] De Puysseleyr, K., De Puysseleyr, L., Geldhof, J., Cox, E., and Vanrompay, D. (2014). Development and validation of a Real-time PCR for *Chlamydia suis* diagnosis in swine and humans. *PLoS ONE* 9:e96704.

[612] Hardick, J., Maldeis, N., Theodore, M., Wood, B. J., Yang, S., Lin, S., et al. (2004). Real-time PCR for *C. pneumoniae* utilizing the Roche Light cycler and a 16S rRNA gene target. *J. Mol. Diuagn*. 6, 132–136.

[613] Cuervo, A., Petrucca, A., and Cassone, A. (2003). Identification and quantification of *Chlamydia pneumoniae* in human atherosclerotic plaques by Light-Cycler real-time PCR. *Moll. Cell. Probes* 17, 107–111.

[614] Tcherneva, E., and Martinov, S. P. (1996). "Repetitive element sequence based polymerase chain reaction is a useful tool for typing *Chlamydia psittaci* strains," in *Proceedings of the Eur. Soc. For differentiation of Research*, 3, 49, Uppsala.

[615] Martinov, S. P., Bonovska, M., Alexandrov, E., and Tcherneva, E. (2006). "Detection and identification of rickettsiae by polymerase chain reaction," in *Proceedings of the 7th Natl. Congress Med, Geography with intl. participation*, Sofia, 336–347.

[616] Martinov, S. P., Bonovska, M., and Tcherneva, E. (2008). Polymerase chain reactions for proof and differentiation of coxiellas and rickettsiae. *Vet. Sci. XII* 3–4, 5–13.

[617] Tcherneva, E., Rijpens, N., Naydensky, C, and Hermann, M. F. (1996). Repetitive element sequence based polymerase chain reaction for typing of *Brucella* strains. *Vet. Microbiol.* 51, 169–178.

[618] Tcherneva, E., Ljutzkanov, M., and Ivanov, P. (1999). Typing of *Brucella* species by application of different PCR techniques. *Bulg. J. Vet. Med.* 2, 17–32.

[619] Pollard, D. R., Tyler, S. D., Nug, C. W., and Rosee, K. (1989). A polymerase chain reaction (PCR) protocol for the specific detection of *Chlamydia* spp. *Moll. Cell. Probes* 3, 383–389.

[620] Martinov, S, and Petrova, Z. (2011). "Use of real-time polymerase chain reaction for detection and differentiation of *Chlamydophila psittaci* and *Chlamydophila abortus*," in *Proceedings of the Sci. Reports 8th Natl. Congress Med, Geography with Intl. Participation*, eds N. Dimov, S. Martinov, D. Philipov, Z. Spassova, Sofia, 197–202.

[621] Wilson, W. J., Strout, C. L., DeSantis, T. Z., Stilwell, J. L., Carrano, A. V., and Andersen, G. L. (2002). Sequence-specific identification of 18 pathogenic microorganisms using microarray technology. *Moll. Cell. Probes* 16, 119–127.

[622] Borucki, M. K., Krug, M. J., Muraoka, W. T., and Call, D. R. (2003). Discrimination among *Listeria monocytogenes* isolates using a mixed genome DNA microarray. *Vet. Microbiol.* 92, 351–362.

[623] Monecke, S., Leube, I., and Enricht, R. (2003). Simple and robust array-based methods for parallel detection of resistance genes of *Staphylococcus aureus*. *Genome Lett.* 2, 106–118.

[624] Call, D. R., Brockman, F. J., and Chandler, D. P. (2001). Detection and genotyping *Escherichia coli*, 0157:H7 using multiplex PCR and nucleic acid microarray. *Intern. J. Food Microbiol.* 67, 71–80.

[625] Troesch, A., Nguen, H., Miyada, C. G., Desvarenne, S., Gingeras, T. R., Kaplan, P. M., et al. (1999). *Mycobacterium* species identification and rifampin resistance testing with high density DNA probe assay. *J. Clin. Microbiol.* 37, 49–55.
[626] Chizhikov, V., Rasooly, A., Chumakov, K., and Levy, D. D. (2001). Microarray analysis of microbial virulence factors. *Appl. Environ. Microbiol.* 67, 3258–3263.
[627] Antony, R. M., Brown, T. J., and French, G. L. (2000). Rapid diagnosis of bacteremia by universal amplification of 23S ribosomal DNA followed by hybridization to an oligonucleotide array. *J. Clin. Microbiol.* 38, 781–788.
[628] Sachse, K., Hotzel, H., Slickers, P., Ellinger, T., and Enricht, R. (2005). DNA microarray-based detection and identification of *Chlamydia* and *Chlamydophila* spp. *Moll. Cell. Probes* 19, 41–50.
[629] Borel, N, Kempf, E., Hotzel, H., Schubert, E., Torgerson, P., Slickers, P., et al. (2008). Direct identification of chlamydiae from clinical samples using a DNA microarray assay-a validation study. *Moll. Cell. Probes* 22, 55–64.
[630] Enricht, R., Slickers, P., Coellner, S., Hotzel, H., and Sachse, K. (2006). Optimized DNA microarray assay allows detection and genotyping of single PCR amplification target copies. *Moll. Cell. Probes* 20, 60–63.
[631] Sachse, K., Laroucau, K., Vorimore, F., Magnino, S. Feige, J., Muller, W., et al. (2005). DNA microarray-based genotyping of *Chlamydophila psittaci* strains from culture and clinical samples. *Vet. Microbiol.* 135, 22–30.
[632] Reinhold, P., Jaeger, J., Berndt, A., Bachmann, R., Schubert, E., Melzer, M., et al. (2008). Impact of latent infections with *Chlamydophila species* in young cattle. *Vet. J.* 175, 202–211.
[633] Reinhold, P. (2006). *Overview of Chlamydiosis*. MSD Veterinary Manual, 1–4.
[634] York, C. J., and Baker, J. A. (1956). *Miyagawanella bovis* infection in calves. *Ann. N. Y. Acad. Sci.* 66, 210–214.
[635] Walker, E., Leu, J., Timms, P., and Polkinghome. (2015). *Chlamydia pecorum* infections in sheep and cattle: A common and under-recognized infectious disease with significant impact on animal health. *Vet. J.* 206, 252–260.

[636] Storz, J., and Thornley, W. R. (1966). Serologische und aetiologische Studien über die intestinale Psittacose-lymphogranuloma-Infection der Schafe. *Zbl. Vet. Med.* 13, 14–24.

[637] De Graves, F. J., Cao, D., Hehnen, H.-R, Schlapp, T., and Kaltenboeck, B. (2003). Quantative detection of *Chlamydia psittaci* and *Chlamydia pecorum* by high-sensitivity real time PCR reveals high prevalence of vaginal infection in cattle. *J. Clin. Microbiol.* 41, 1726–1729.

[638] Wehrend, A., Failing, K., Hauser, B., Jafer, C., and Bostedt, H. (2005). Production, reproductive and metabolic factors associated with chlamydial serropositivity and reproductive tract antigens in dairy herds with fertility disorders. *Theriogenology* 63, 923–930.

[639] Biesenkamp-Uhe, C., Li, Y., Hehnen, H.-R., Sachse, K., and Kaltenboeck, B. (2007). Therapeutic *Chlamydophila abortus* and *C. pecorum* vaccinations transitely reduce bovine mastitis associated with *Chlamydophila* infection. *Infect. Immun.* 75, 870–877.

[640] Poudel, A., Elsasser, T. H., Shamsur Rahman, K. H., Chowdhury, E. U., and Kaltenboeck, B. (2012). Asymptomatic endemic *Chlamydia pecorum* infections reduce growth rates in calves by up to 48 percent. *PLoS ONE* 7:e44961.

[641] Kemmerling, K., Muller, U, Mieleuz, M., and Sauerwein, H. (2009). *Chlamydophila* species in dairy farms. Polymerase chain reaction prevalence, disease association, and risk factors identified in a cross-sectional Study in Western Germany. *J. Dairy Sci.* 92, 4347–4354.

[642] Aitken, I. D., Longbottom, D. (2008). "Chlamydial abortion," in *Diseases of Sheep*, 4th Edn, ed. I. D. Aitken (Oxford: Blackwell Scientific Ltd), 105–112.

[643] Martinov, S. P. (2009). Seroepizootiology and nosogeography of the chlamydial infections in Bulgaria," in *Proceedings of the Symposium "Nature and Health,"* Sofia, Union of Scientists in Bulgaria, 117–134.

[644] McCowan, B., Moulton, J. E., and Shultz, G. (1957). Pneumonia in California lambs. *J. Am. Vet. Med. Ass.* 131, 318–323.

[645] Dungworth, D. L., and Cordy, D. R. (1962). The pathogenesis of ovine pneumonia. II. Isolation of virus from faeces: comparison of pneumonia caused by faecal, enzootic abortion and pneumonitis viruses. *J. Comp. Pathol. Therap.* 72, 71–79.

[646] Ognianov, D. (1963). Rep. Inst. Virololy, 2, 25–30. [by Semerdjiev, B., Ognianov, D. (1969). Neorickettsioses in domestic animals. Sofia: Zemizdat, 252].

[647] Giroud, P., Roger, F., and Dumas, N. (1956). Résultats concernants l'avortement de la femme dû à un agent du groupe de la psitacose. C. R5. *Acad. Sci. (Paris)*, 242, 697–698.
[648] Pospischil, A. Thoma, R., Hilbe, M., Grest, P., and Geblerts, F. O. (2002). Abortion in woman caused by caprine *Chlamydophila abortus* (*C. psittaci* serovar 1). *Swiss Med. WKLY*, 232, 64–66.
[649] Martinov, S. (2009). Etiological investigation of the *Chlamydophila abortus* infection in goats. *Vet. Med. XIII* 3–4, 11–18.
[650] Cello, R. M. (1967). Ocular infections in animals with PLT (Bedsonia) group agents. *Amer. J. Ophthalmol.* 63, 1270/244–1273/247.
[651] Storz, J., Pierson, J., and Marriott, M. E. (1967). Isolation of psittacosis agents from follicular conjunctivitis of sheep. *Proc. Soc. Exp. Biol. Med.* 125, 875–860.
[652] Ognianov, D. Karadzhov, Y, and Martinov, S. (1979). "Infectious kerato-conjunctivitis in commercially reared domestic animals (A review)," in *Proceedings of the Center for Scientific, Technical and Economic Information at Ministry of Agriculture*, Sofia, 71.
[653] Schoenian, S. (2008). *Contagious keratoconjunctivitis (Pinkeye). Maryland Small Ruminant Page.*
[654] Gupta, S., Chalota, R, Bhardwaj, B, Malik, P., Verma, S., and Sharma, M. (2015). Identification of chlamydiae and mycoplasma species in ruminants with ocular infections. *Lett. Appl. Microbiol.* 60, 135–139.
[655] Mendlowski, B., and Segre, D. (1960). Polyarthritis in sheep. I. Description of the disease and experimenral transmission. *Amer. J. Vet. Res.* 21, 68–73.
[656] Pierson, J. (1968). Polyarthritis in Colorado feedlot lambs. *J. Amer. Vet. Med. Ass.* 150, 1487–1492.
[657] Page, L. A., and Cultip, R. C. (1968). Chlamydial polyarthritis in Iowa lambs. *Iowa Vet.* 39, 10–11; 14–18.
[658] Constable, P. D., Hinchliff, K. W., Done, S. H., and Greenberg, W. (2016). "Infectious diseases of musculoskeletal system: *Chlamydial polyarthritis*," in *Vet. Medicine–E-Book: A Textbook of the Diseases in Cattle, Horses, Sheep, Pigs and Goats* (London: Elsevier Health Sciences), 2278.
[659] Walker, E., Moore, C., Shearer, P., Jelocnic, M., Bommana, S., Timms, P., et al. (2016). Clinical, diagnostic and pathologic features of presumptive cases of *Chlamydia pecorum*-associated arthritis in Australian sheep flocks. *BMC Vet. Res.* 12:193.

[660] Jelocnic, M., Frentiu, F. D., Timms, P., and Polkinghome, A. (2013). Multilocus sequence analysis provides insights into molecular epidemiology of *Chlamydia pecorum* in Australian sheep, cattle and koalas. *J. Clin. Microbiol.* 51, 2625–2632.

[661] Polkinghome, A., Borel, N., Becker, A., Lu, Z. H., Brugnera, E., Pospischil, A., et al. (2009). Molecular evidence for chlamydial infections in the eyes of sheep. *Vet. Microbiol.* 135, 142–146.

[662] Watt, B. (2011). *Chlamydial Infection in Sheep*. Flock and Herd-case notes.

[663] Goat Kingdom Index. (2016). *Other Causes of Arthritis*. Available at: http://goat kingdom tripodcom/arthritis

[664] Nietfield, J. C. (2001). Chlamydial infections in small ruminants. *Vet. Clinics N. Am. Food Animal Pract.* 17, 301–314.

[665] Gupta, P. P., Singh, B., and Dhingra, P. N. (1976). Chlamydial pneumonia in a buffalo calf. *Zoon. Publ. Health* 23, 779–781.

[666] Dhingra, P. N., Agarwal, L. P., Mahgjan, V. M., and Adalakha, S. C. (1980). Isolation, of Chlamydia from pneumonic lungs of buffalos, cattle and sheep. *Zentralbl. Veterinarmed. B*, 27, 680–682.

[667] Rowe, L. W., Hedger, R. S., and Smale, C. (1978). Isolation, of *Chlamydia psittaci*-like agent from a free-living African buffalo (*Sincerus caffer*). *Vet. Rec.* 103, 13–14.

[668] Magnino, S., Galiero, G., Paliadino, M., Nigo, P. G., Bazzocchi, C., De Guilli, L., et al. (2000). "An outbreak of chlamydial encephalomyelitis in water buffalo calves," in *Proceedings of the 4th Meeting of Eur. Soc. Chlamydia Research*, Helsinki, 273.

[669] Greco, G., Corrente, M., Buonavoglia, D., Campanile, G., Di Palo, R., Martella, V., et al. (2008). Epizootic abortion related to infections by *Chlamydophila abortus* and *Chlamydophila pecorum* in in water buffalo (*Bubalus bubalis*). *Theriogenology* 69, 1061–1069.

[670] Storz, J., Smart, R. A., and Shupe, J. L. (1964). "*Virusbededingte polyarthritis*, bei kalbern," in *Proceedings of the 3rd Int. Meet. on Diseases of Cattle, Copenhagen. Nord Veterinaermed*, 16, 109–115.

[671] Köble, O., and Psota, A. (1968). Miyagawanellen–Isolierungen bei Polyarthritis, Pneumoniae, Encephalomyelitis und interstitieller Herdnephritis (Fleckniere) der Kälber, Wien Tieraerztl. *Monat.* 55, 443–445.

[672] McIlwraith, C. W. (2016). *Chlamydial Polyarthritis-Serositis in Large Animals*. MSD Veterinary Manual. Hunterdon, NJ: Merck Sharp and Dohme Corp.

[673] Xiao, D., Jiang, H., Huang, Y., Wang, Z., Zheng, J. Wang, Y., et al. (2006). Study on newborn calf chlamydial arthritis. *China Dairy Cattle* 34:37.

[674] McNutt, S. H. (1940). A preliminary report on an unfectious encephalomyelitis in cattle. *Vet. Med.* 35, 228–231.

[675] Konrad, J., and Bohac, J. (1959). Encephalomyelitis in cattle. *Vet Gas.* 8, 228–238.

[676] Lear, A. S., and Callan, R. J. (2016). *Overview of Sporadic Bovine Encephalomyelitis.* MSD Veterinary Manual, 1–2. Hunterdon, NJ: Merck Sharp and Dohme Corp.

[677] Hunt, H., Orbell, G., Buckle, K. N., Ha, H. J., Lawrence, K. E., Fairley, R. A., et al. (2016). First report and histologic features of *Chlamydia pecorum* encephalitis in calves in New Zealand. *N. Z. Vet. J.* 64, 364–368.

[678] Matumoto, M., Omori, T., Morimoto, T., Harada, K., Inaba, Y., and Ishii, S. (1955). Studies on the disease of cattle caused by a psittacosis-lymphogranuloma group virus. VIII. Sites for the virus to leave the infected host. *Jap. J. Exp. Med.* 25, 223–245.

[679] Mecklinger, S., Wehr, J., Horsch, F., and Seffner, H. (1980). Experimental Chlamydia mastitis. Wiss. Zeitsch. Humboldt-University zu Berlin. *Math.-Nat. R. XXIX* 1, 75–80.

[680] Wehner, C., Wehr, J. Teichmann, G., Mecklinger, S., and Zimmerhackel, W. (1980). Investigations on the participation of *Chlamydia* in a Mastitis Event. Wiss. Zeitsch. Humboldt-University zu Berlin. *Math.-Nat. R. XXIX* 1, 71–73.

[681] Kaltenboeck, B. (2015). *Undenstanding and mitigating thr economic impact of bovine Chlamydia spp. infections.* A project. (2012–2015, Auburn University, College of Vet. Medicine, Alabama).

[682] Ahluwalia, S. K. (2009). *Influence of Natural Chlamydia spp. Infection on the Health of the Ruminant Mammary Gland.* Ph.D. dissertation, Auburn University, Auburn, AL.

[683] Blanco, L. (1969). Isolierung eines Erregers der Psittacose - Lymphogranuloma venereum Group (PLT) bei boviner Mastitis. *Revista Del Patronado De Biologia Animal* 179–186.

[684] York, C. J., and Baker, J. A. (1951). A new member of the psittacosis - lymphogranuloma group of viruses that causes infection in calves. *J. Exp. Med.* 93, 587–604.

[685] Omori, T., Morimoto, T., Harada, K., Inaba, Y., Ishii, S., and Matumoto, M. (1957). Miyagawanella: Psittacosis - Lymphogranuloma

group of viruses. I. Excretion of goat pneumonia virus in feces. *Jap. J. Exp. Med.* 27, 131–143.

[686] Kawakami, Y., Kaji, T., Sugimura, K., Omori, T., and Matumoto, M. (1958). Miyagawanella: Psittacosis - Lymphogranuloma group of viruses. V. Isolation of a virus from feces of naturally infected sheep. *Jap. J. Exp. Med.* 28, 51–58.

[687] Popovici, V. A. (1964). A study of respiratory and digestive Miyagawanella infections in cattle," in *Proceedings of the 3rd Int. Meet. Dis. Cattle, Copenhagen*, 1, 70.

[688] Storz, J., and Thornley, W. R. (1966). Serologische und aetiologische Studien über die intestinale Psittacose - Lymphogranuloma-Infektion der aschafe. *Zbl. Veterinaermed.* 13, 14–24.

[689] Storz, J., Collier, J. R., and Altera, K. P. (1968). Pathogenetic studies on the intestinal psittacosis (*Chlamydia*) in newborn calves," in *Proceedings of the 5th Int. Meet. Dis. Cattle*, Opatja, 83–85.

[690] Eugster, A. K. (1970). *Pathogenetic Studies on Intestinal Chlamydial Infection in Calves*. Ph.D. thesis, Colorado State University, Fort Collibs.

[691] Ehret, W. J., Schutte, A. P. Pienaar, J. G., and Henton, M. M. (1975). Chlamydiosis in a beef herd. *J. S. Afr. Vet. Assoc.* 46, 171–179.

[692] Ronsholt, L. (1978). *Chlamydia psittaci* infection in Danish cattle. *Acta Pathol. Microbiol. Scand. 86B,* 291–297.

[693] Li, J., Cuo, W., Kaltenboeck, B., Sachse, K., Yang, Y., Lu, G., et al. (2016). *Chlamydia pecorum* is the endemic intestinal species, while *C. gallinacea*, *C. psittaci* and *C. pneumoniae* associate with sporadic systemic infections. *Vet. Microbiol.* 193, 93–98.

[694] Storz, J., and Krauss, H. (1985). "Chlamydial infections and diseases in animals," in *Handbook of Bacterial Infections in Animals,* eds H. Blobel and I. Scheisser (Jena: Gustav Fischer Verlag) 447–453, 337–531.

[695] Rank, R. G., and Veruva, L. (2014). Hidden in plain sights: chlamydial gastrointestinal infection and its relevance to persistence in human genital infection. (Andrews-Polymenis, H. L., ed.). *Infect. Immun.* 82, 1362–1371.

[696] Sorodoc, C., Surdan, C., and Sarateanu, D. (1961). Investigation on the identification of the virus of enzootic pneumonia in swine. *Stud. Cerv. Infra-microbiol.* 12, 355–364.

[697] Reinhold, P., Kirschvik, N., Theegarten, D., and Berndt, A. (2008). An experimentally induced *Chlamydia suis* infection in pigs results

in severe lung function disorders and pulmonary inflammation. *Vet. Res.* 39, 35–54.

[698] Hoelzle, L. E., Steinhausen, G., and Wittenbrink, M. M. (2000). PCR-based detrection of chlamydial infection in swine and subsequent PCR-coupled genotypoing of chlamydial omp1-gene amplicons by DNA-hybridization, RFLP-analysis and nucleotide sequence analysis. *Epidemiol. Infect.* 125, 427–439.

[699] Eggeman, G., Wendt, M., Hoelzle, L. E., Jager, C., Weiss, R., and Failing, K. (2000). Prevalence of Chlamydia infections in breeding sows and their importance in reproductive failure. *Dtsch Tierarztl. Wochenschr.* 107, 3–10.

[700] Camenisch, U., Lu, Z. H., Vaughan, L., Corboz, L., Cimmermann, D. R., Wittenbrink, M. M., et al. (2004). Diagnostic investigation into the role of chlamydiae in cases of increased rates of returns to oestrus in pigs. *Vet. Rec.* 155, 593–596.

[701] Vanrompay, D., Geens, T., Desplanques, A., Hoang, I. Q., De Vos, L., Van Loock, M., et al. (2004). Immunoblotting, ELISA and culture evidence for *Chlamydiaceae* in sows on 258 Belgian farms. *Vet. Microbiol.* 99, 59–66.

[702] Longbottom, D. (2004). Chlamydial infections in domestic ruminants and swine: new nomenclature and new knowledge. *Vet. J.* 168, 9–11.

[703] Schautteet, K., and Vanrompay, D. (2011). *Chlamydiaceae* infections in pigs. *Vet. Res.* 42, 29–10, 1186–1297.

[704] Surdan, C., Anthansiu, P., Sorodoc, C., Copetovici, D., Strulovici, D., Enach, A., et al. (1965). Various epizootiologic and anatomo-clinical aspects of pararickettsial infection in pigs. II. Sow enzootic abortion and orchovaginalitis in boars. *Rev. Roum. Inframicrobial.* 2:165.

[705] Bohac, J., and Menzik, J. (1965). "Bedsonia group agents as a cause of enzootic abortion in pigs," in *Proceedings of the Conference on Diseases of Pigs*. Coll. Vet. Med, Brno.

[706] Popovici, V., Hiastru, F., Draghici, D., Berbinschi, C., and Dorobantu. (1972). Isolation of Chlamydia from pigs with various disorders. *Lucr. Inst. Cerc. Vet. Bioprep. Pasteur, B*: 19–28.

[707] Shcherban, G. P., Firsova, G. D., and Voskresenskaya. (1978). Chlamydiosis of pigs. *Veterinaria* 8, 55–58.

[708] Bortnichuck, V. A. (1991). *Chlamydiosis of Pigs*. Kiev: Urozhai, 5–192.

[709] Schiller, I., Koesters, R., Weilermann, R., Thoma, R., Kaltenboeck, B., Heiz, P., et al. (1997). Mixed infections with porcine *Chlamydia trachomatis/pecorum* and infections with ruminant *Chlamydia psittaci* serovar 1 associated with abortions in swine. *Vet. Microbiol.* 58, 251–260.

[710] Kölbl, O. (1969). Untersuchungen über das Vorkommen von Miyagawanellen beim Scwein. *Wien. Tieräerztl. Monat.* 56, 332–335.

[711] Kölbl, O., Burtscher, H., and Hebenstreit, J. (1970). Polyarthritis bei Schachtsweinen. Mikrobiologische, histologische und fleischhygienische. Untersuchngen und Aspekte. *Wien. Tieräerztl. Monat.* 57, 355–361.

[712] Wittenbrink, M. M. (1991). Detection of antibodies against Chlamydia in swine by an immunofluorescent test and an enzyme immunoassay. *Berl. Munch. Tierarztl. Wochenschr.* 104, 270–275.

[713] Willingam, D. A., and Beamer, P. D. (1955). Isolation of transmissible agent from pericarditis in swine. *J. Am. Vet. Med. Assoc.* 126, 118–122.

[714] Yaeger, M. J. (2012). "Disorders of Pigs," in Kirkbride's Diagnosis of Abortion and Neonatal Loss in Animals, Fourth Edn, ed. B. L. Njaa (Oxford: Wiley-Blackwell). doi: 10.1002/9781119949053.ch4

[715] Pavlov, P., Milanov, M., and Tschilev, D. (1963). Recerchers sur la rickettsiose kerato-conjunctivale du pore en Bulgarien. *Ann. Inst. Pasreur* 105, 450–454.

[716] Rogers, D. G., Anderson, A. A., Hogg, A., Nielsen, D. L., and Huebert, M. A. (1993). Conjunctivitis and kerato-conjunctivitis associated with chlamydiae in swine. *J. Am. Vet. Med. Assoc.* 203, 1321–1323.

[717] Rogers, D. G., and Anderson, A. A. (1999). Conjunctivitis caused by *C. trachomatis*-like organism in gnotobiotic pigs. *J. Vet. Diagn. Invest.* 11, 341–343.

[718] Becker, A., Lutz-Wohlgroth, L., Brugnera, E., Lu, S. H., Zimmerman, P. D., Grimm, F., et al. (2007). Intensively kept pigs pre-disposed to Chlamydial associated conjunctivitis. *J. Vet. Med. A. Physiol. Pathol. Clin. Med.* 54, 307–313.

[719] Englund, C. H., Segerstad., C. H., Arulund, F., Westergren, E., and Jacobson, M. (2011). The occurance of *Chlamydia* spp. in pigs with and without clinical disease. *BMC Vet. Res.* 8:9.

[720] Schautteer, K., Beeckman, D. S., Delava, P., and Vanrompay, D. (2010). Possible pathogenic interplay between *Chlamydia suis*,

Chlamydophila abortus and PCV-2 on a pig production farm. *Vet. Rec.* 166, 329–333.

[721] Schautteer, K. (2010). *Epidemiological research on Chlamydiaceae in pigs and evaluation of a Chlamydia trachomatis DNA vaccine.* Gent University, Gent.

[722] Pollmann, M., Nordhoff, M., Pospischil, A., Tedin, K., and Wieler, L. H. (2005). Effect of a probiotic strain of *Enterococcus faecium* on the rate of a natural Chlamydia infection in swine. *Infect. Immun.* 73, 4346–4353.

[723] Hunter, A. R., and McMartin, D. A. (1984). Experimental chlamydial pneumonia in pigs. *Comp. Immun. Microbiol. Infect. Dis.* 71, 19–24.

[724] Bortnichuk, V. A. (1992). *Chlamydiosis in Swine. (Etiology, Diagnostics, Epizootiology and Control Measures).* Ph.D. dissertation, Vet. Institute, Kiev, p. 43.

[725] Sarma, D. K., Tamuli, M. K., Kahman, T., Baro, B. R., Deka, B. C., and Raikonwar, C. K. (1983). Isolations of Chlamydia from a pig with lessions of the urethra and prostate gland. *Vet. Rec.* 112:525.

[726] Kauffold, F., Melzer, F., Henning, K., Schulze, K., Leiding, and Sachse, K. (2006). Prevalence of chlamydiae in boars and semen used for artificial insemination. *Theriogenology* 65, 1750–1758.

[727] Pospischil, A., and Wood, R. I. (1987). Intestinal Chlamydia in pigs. *Vet. Pathol.* 24, 568–570.

[728] Zahn, I., Szeredi, L., Schiller, I., Straumann-Kunz, U., Buergi, E., Guscetti, F., et al. (1995). Immunohistochemical determination of *Chlamydia psittaci/pecorum* and *C. trachomatis* in the piglet gut. *Zentralblt. Veterinarmed.* B 42, 266–276.

[729] Szeredi, L., Schiller, I., Sydler, T., Guscetti, F., Heinen, E., Corboz, L., et al. (1996). *Vet. Pathol.* 33, 369–374.

[730] Nietfield, J. C., Leslie-Steen, P., Zeman, H., and Nelson, D. (1997). Prevalence of intestinal chlamydial infection in pigs in the Midwest, as determined by immunoperoxidase staining. *Am. J. Vet. Res.* 58, 260–264.

[731] Schiller, I., Koesters, R., Weilenmann, R., Kaltenboeck, B., and Pospischil, A. (1997). Polymerase chain reaction (PCR) detection of porcine *Chlamydia trachomatis* and ruminant *Chlamydia psittaci* serovar 1 DNA in formalin-fixed intestinal speciments from swine. *Zentralblt. Veterinarmed.* B 44, 185–191.

[732] Guscetti, F., Schiller, I., Sydler, T., Heinen, E., and Pospischil, A. (2009). Experimental enteric unfection of gnotobiotic piglets with *Chlamydia suis* strain S45. *Vet. Microbiol.* 135, 157–168.

[733] Schiller, I., Sydler, T., Corboz, L., and Pospischil, A. (1998). Experimental *Chlamydia psittaci* serotype 1 enteric infection in gnotobiotic piglets: histopathological, immunohistochemical and microbiological findings. *Vet. Microbiol.* 62, 251–263.

[734] Rogers, D. G., and Anderson, A. A. (2000). Intestinal lesions caused by a strain of *Chlamydia suis* in weaning pigs infected at 21 days of age. *J. Vet. Diagn. Invest.* 12, 233–239.

[735] Hoffman, K., Schott, F., Donati, M., Di Francesco, A., Hassig, M., Wanninger, S., et al. (2015). Prevalence of chlamydial infection to fattening pigs and their influencing factors. *PLoS ONE* 10:e0143576.

[736] Unknown author. (2014). *Chlamydial Diseases*. S. Diego, CA: Mol. Diagn. Services.

[737] Sellon, D. C., and Maureen, T. L. (eds). (2007). "Other infections causing abortions," in *Equine Infectious Diseases* (London: Elsevier Health Sciences), 653.

[738] Henning, K., Sachse, K., and Sting, R. (2000). Demonstration of Chlamydia from an equine abortion. *Dtsch. Tierarztl. Wochenschr.* 107, 49–52.

[739] Szeredi, L., Hotzel, H., and Sachse, K. (2005). High prevalence of chlamydial (*Chlamydophila psittaci*) infection in fetal membranes of aborted equine fetuses. *Vet. Res. Commun.* 29, 37–49.

[740] Wills, J. M., Watson, C., Lusher, M., Wood, D., and Richmond, S. (1990). Characterization of *Chlamydia psittaci* solated from horse. *Vet. Microbiol.* 24, 11–19.

[741] Kneg, N. R., Ludwig, W., Whitman, W., Hedlund, B. P., Paster, B. J., Staley, J. T., et al. (eds). (2011). "Family *Chlamydiaceae*," in *Bergey's Manual of Systematic Bacteriology* (New York, NY: Springer-Verlag), 4, 949.

[742] McChesney, S. L., England, J. J., and McChesney, A. E. (1982). *Chlamydia psittaci* induced pneumonia in a horse. *Cornell Veterinarian* 72, 92–97.

[743] Moorthy, A. R., and Spradbrow, P. P. (1978). *Chlamydia psittaci* infections in horses with respiratory disease. *Equine Vet. J.* 10, 38–42.

[744] Burell, M. H., Chambers, W. S. K., and Kewly, D. R. (1986). Isolation of *Chlamydia psittaci* from respiratory tract and conjunctivae of thorough-bred horses. *Vet. Rec.* 119, 302–303.

[745] Di Francesko, A., Donati, M., Mattioli, L., Naidi, M., Salvatore, D., Polglayen, G., et al. (2006). *Chlamydophila pneumoniae* in horses: a seroepidemiological survey in Italy. *New Microbiol.* 29, 303–305.

[746] McChesney, A. E., Becerra, V., and England, J. J. (1974). Chlamydial polyarthritis in a foal. *J. Am. Vet. Med. Assoc.* 165, 259–261.

[747] Robinson, N. E., Derksen, F. E., Otszewski, M. A., and Büchner. (1996). The pathogenesis of chronic obstructive pulmonary disease in horses. *Brit. Vet.* 152, 283–306.

[748] Theegarten, D., Sachse. K., Mentrup, B., Fey, K., Hotzel, H., and Anhenn, O. (2008). *Chlamydophila* spp. infection in horses with recurrent airways obstruction: similarities to human chronic obstructive disease. *Resp. Res.* 29:14.

[749] Giroud, P., Groulade, P., Roger, F., and Dortois, N. (1954). Reaction positives vis-á-vis d'un antigene du groupe de la psittacose chez le chien, au cours de divers syndromes infectieux. *Bull. Acad. Vet. France.* 27, 309–311.

[750] Groulade, P., Roger, F., and Dortois, N. (1954). Contribution á l'etude d'un syndrome infectieux duy chien repondant, serologiquement á une souche de *Rickettsia psittaci*. *Rev. Path. Comp.* 54, 1426–1432.

[751] Maierhofer, C. A., and Storz, J. (1969). Clinical and serological response of dogs to infection with chlamydial agent of ovine polyarthritis. *Am. J. Vet. Res.* 30, 1961–1966.

[752] Fraser, G., Norwall, J., Withers, A. R., and Gregor, W. W. (1969). A case history of psittacosis in the dog. *Vet. Rec.* 85, 54–58.

[753] Contini, A. (1956). Su di un caso della malattia di Giroud e Groulade del cane. *Atti Soc. Ital. Sci. Vet.* 10, 736–740.

[754] Popovici, V. (1966). Contribution to the epizootiology of virus abortion in ewes. *Inst. Cerv. Vet. Bioprep. Pasteur* 1, 29–39.

[755] Terzin, A. L., and Miskov, D. (1965). Cross infection with Bedsonia agents in different categories of people and animals. *J. Hyg. Epidemiol. Microbiol. Immunol.* 9, 336–345.

[756] Scott, J. G. (1968). TRIC infections (bedsoniae). *Med. Proc. Mediese Bydraes*, 14, 410–414.

[757] Voigt, A., Dietz, O., and Schmidt, V. (1966). Klinische und experimentelle Untersuchungen zur Ätiologie der Keratitis superficial chronica (Überreiter). *Arch. Exp. Vet. Med.* 20, 259–274.

[758] Arizmendi, F., Grimes, J. E., and Relford, R. L. (1993). Isolation of *Chlamydia psittaci* from pleural effusion in a dog. *J. Vet. Diagn. Invest.* 4, 440–463.

[759] Saco, T., Takanashi, T., Takehana, K., Nakade, T., Umemura, T., and Tanagama, T. (2002). Chlamydial infection in canine atherosclerotic lesions. *Atherosclerosis* 162, 253–259.

[760] Lambrechts, N., Picard, J., and Tustin, R. C. (1999). Chlamydia-induced septic polyarthritis in a dog. *J. South Afr. Vet. Ver.* 70, 40–42.

[761] Cheng, T., Zou, S.-S, Wang, X.-Q., Liang, Y., and Lin, R. Q. (2015). Prevalence of antibodies to *Chlamydiaceae* in pet dogs in Shenzhen Guandonge Province, China. *Acta Vet. Brno* 84, 13–17.

[762] Fukushi, H., Ogawa, H., Minamoto, M. M., Yagami, K., Tamura, H., Shimakura, S., et al. (1985). Seroepidemiological surveillance of *Chlamydia psittaci* in cats and dogs in Japan. *Vet. Rec.* 117, 503–504.

[763] Gresham, A. C., Dixon, C. E., and Bevan, B. J. (1996). Domiciliary outbreak of psittacosis in dogs: potential for zoonotic infection. *Vet. Rec.* 138, 622–623.

[764] PET MD. (2017). *Upper Respiratory Infection (Chlamydia) in cats. Cat Conditions.* A Pet 360 Media Neowork Property.

[765] Cello, R. M. (1967). Ocular infections in animals with PLT (Bedsonia) group agents. *Am. J. Ophthalmol.* 63, 1270–1274.

[766] MSD Animal Health (2017). *Feline Chlamydophila.* Dublin: MSD Animal Health.

[767] Sykes, J. E. (2016). *Overview of Chlamydial Conjunctivitis.* MSD Veterinary Manual.

[768] Kahu, S. (ed.). (2007). *Chlamydial pneumonia (Feline Chlamydiosis; Pneumonitis). The Merck/Merial Manual for Pet Health.* New York, NY: Simon and Schuster, 215.

[769] Sykes, J. E. (2013). *Chapter 33. Chlamydial infections. Canine and Feline Infectious Diseases (E-book).* London: Elsevier Health Sciences, 928.

[770] Dickie, C. W., and Sniff, E. S. (1980). Chlamydial infection associated with peritonitis in a cat. *J. Am. Vet. Med. Assoc.* 176, 1256–1259.

[771] Hargis, A. M., Prieur, D. J., and Gailard, F. T. (1983). Chlamydial infection of the gastric mucosa in twelve cats. *Vet. Pathol.* 20, 170–178.

[772] Sparkes, A. H., Caney, S. M., Sturgess, C. P., and Gruffydd-Jones, T. J. (1999). The clinical efficacy of for the treatment of feline ocular chlamydiosis. *J. Feline Med. Surg.* 1, 31–35.

[773] Donati, M., Piva, S., Di Francesco, A., Mazzeo, C., Pietra, M., Cevenini, R., et al. (2005). Feline ocular chlamydiosis: Clinical

and microbiological effect on topical and systemic therapy. *New Microbiol.* 28, 369–372.

[774] Gruffydd-Jones, T., Addie, D., Belak, S., Boucraut-Baralon, C., Egberink, H., Frymus, T., et al. (2009). *Chlamydophila felis* infection. ABCD guidelines on prevention and management. *J. Feline Med. Surg.* 11, 605–609.

[775] Dean, R., Harley, R., Helps, C., et al. (2005). Use of quantitative real-time PCR to monitor the response of *Chlamydophila felis* infection to doxycycline treatment. *J. Clin. Microbiol.* 43, 1858–1864.

[776] Sturgess, C. P., Gruffydd-Jones, T., Harbour, D. A. et al. (2001). Controlled study of the efficacy of clavulanic acid-potenciated amoxicillin in the treatment of *C. psittaci* in cats. *Vet. Rec.* 149, 73–76.

[777] Hartmann, A. D., Helms, C. P., and Lappin, M. R. (2008). Efficacy of pradofloxacin in cats with feline upper respiratory diasease due to *Chlamydophila felis* or *Mycoplasma* infections. *J. Vet. Intern. Med.* 22, 44–52.

[778] Owen, W. M., Sturges, C. P., Harbour, D. A., Egan, K., Gruffydd-Jones, T. J., et al. (2003). Efficacy of azithromycin for the treatment of feline chlamydophilosis. *J. Feline Med. Surg.* 5, 305–311.

[779] Kubasek, M., and Strauss, J. (1956). "Epidemic and epizootic appearance of ornithosis in one district, (abs)," in *Proceedings of the Congress on Anthropozoonoses*, Prague, 1.

[780] Strauss, J., and Sery, V. (1964). Ornithose in der CSSR. Epidermiologisch-virologische Aspekte. *Arch. Exp. Vet. Med.* 18, 61–75.

[781] Schmitt, U., Urbaneck, D., Pehl, K. H., and Benndorf, J. (1964). Untersuchungen über die Ornithose bei Enten. I. Mitteilung, Befundl nach intratrachealer Aplikation des Erregers bei erwachsenen Enten. *Arch. Exp. Vet. Med.* 18, 33–60.

[782] Page, L. A. (1978). "Avian chjlamydiosis," in *Diseases of Poultry*, 7th Edn, ed. M. C. Hofstad (Ames, IA: Iowa State University Press).

[783] Meyer, K. F. (1941). Pigeons and barn yard fowls as possible source of human psittacosis or ornithosis. *Schweiz Med. Wschr.* 71, 1377–1379.

[784] Mohan, R. (1984). Epidemiologic and laboratory observations of *Chlamydia psittaci* infection in pet birds. *J. Am. Vet. Med. Assoc.* 184, 1372–1374.

[785] Weyer, F. (1964). Weitere Beobachtungen im Rahmen von diagnostischen Tierversuchen bei Ornithose-Psittacose mit Bemerkungen über

die Entwicklung der Ornithose-situation in Deutscland während der letzten Jahre. *Zbl. Bakt. (Orig)* 193, 147–178.

[786] Dustin, L. P. (1985). Clinical conditions in a pet canary practice. *J. Assoc. Avian Vet.* 53–70.

[787] Unknown author. (2016). *Finch and Canary Diseases*. Omlet. Available at: www.omlet.co.uk.

[788] Soucek, G. (2008). *Gouldian Finches: Everything about Purchase, Housing, Nutrition, Health Care and Breeding*. Hauppauge, NY: Barron Educational Series, 95.

[789] National Assoc. State Publ. Health Veterinarians (NASPHV). (2006). *Compendium of measures to control Chlamydophila psittaci infection among humans and pet birds.* Available at: http://www.naasphv.org/83416/index.html.

[790] Donelly, T. M. (2016). *Guinea Pigs*. MSD Veterinary Manual.

[791] Rank, R. G., White, H. J., Soloff, B. L., and Barron, A. L. (1981). Cystitis associated with chlamydial infection of the genital tract of male guinea pigs. *Sex. Transm. Dis.* 8, 203–210.

[792] Balsamo, G., Maxted, A. M., Midla, J. W., Murphy, J. M., Elding, T. M., Fish, P. H., et al. (2017). Compendium of measures to control *Chlamydophila psittaci* infection among humans (psittacosis) and pet birds (Avian Chlamydiosis). *J. Avian Med. Surger.* 31, 262–282.

[793] Gaede, W., Reckling, K.-F., Dresen-Kamp, B., Kenklies, S., Schubert, E, Noack, U., et al. (2008). *Chlamydophila psittaci* infection in humans during an outbreak of psittacosis) from poultry in Germany. *Zoonoses Public Health* 55, 184–188.

[794] CDC (2016). *Pneumonia. Psittacosis. CDC 24/7 Saving Lives, Helping People*. Atlanta: CDC.

[795] Kovacova, E., Majian, J., Botek, R., Bokor, T., Blaskovicová, H., Solavová, M., et al. (2007). A fatal case of psittacosis in Slovakia, January 2006. *Euro Surveill*, 12:E070802.

[796] Beeckman, D. S. A., and Vanrompay, D. (2009). Zoonotic *Chlamydophila psittaci* infections from a clinical perspective. *Clin. Microbiol. Infect.* 16, 11–17.

[797] Stewardson, A. J., and Grayston, M. L. (2010). Psittacosis. *Inf. Dis. Clin. North. Am.* 24, 7–25.

[798] Schinkel, A. F. L., Bax, J. J., van der Wall, V., Jonkers, G. J. (2000). Endocardiographic follow-up of *C. psittaci* myocarditis. *Chest* 117, 1203–1205.

[799] Lanham, J. G., and Doyle, D. V. (1984). Reactive arthritis following psittacosis. *Br. J. Rheumatol.* 23, 225–226.

[800] Birkhead, J. S., and Apostolov, K. (1974). Endocarditis caused by a psittacosis agent. *Br. Heart J.* 36, 728–731.

[801] Carr-Locke, D. L., and Mair, H. J. (1976). Neurological presentation of psittacosis as part of psittacosis. *Br. Med. J.* 2, 853–854.

[802] Walder, G., Meusburger, H., Hotzel, H., Ochme, A., Neunteufel, W., Dierich, M. P., et al. (2003). *Chlamydophila abortus* pelvic inflammatory disease. *Emerg. Infect. Dis.* 9, 1642–1644.

[803] Georgiev, P., Martinov, S., Borisova, M., Novkirishki, S., Runevska, P., Teoharova, M., et al. (1992). "Epidemic spread of *Chlamydia psittaci* in the country in 1991 – epidemiological and epizootiological studies," in *Proceedings of the 7th Congress on Infect. and Parasit. Diseases with Intl. Participation, Velingrad*, 38–43.

[804] Enright, J. B., and Sadler, W. W. (1954)0. Presence in human sera of complement fixing antibodies to virus of sporadic bovione encephalomyelitis. *Proc. Soc. Exp. Biol. Med.* 85, 466–468.

[805] Barwell, C. F. (1955). Laboratory infection of man with virus of enzootic abortion of ewes. *Lancet* 2, 1369–1371.

[806] Giroud, D., Roger, F., and Dumas (1956). Résultats consernant l'avortement de la femme dû à un agent du groupe de la p[sittacose. *C. R. Acad. Sci. (Paris)* 260, 4874–4876.

[807] Meyer, K. F. (1965). "Psittacosis-lymphogranuloma venereum agents," in *Viral and Rickettsial Infections of Man*, 4th Edn, eds F. L. Horsfall and J. Tamm (Philadelphia: Lippincott), 1006–1041.

[808] Prat, J. (1955). Contribution à l'étude de étiologie de la bronchopneumonia contagieuse des Bovides. *Rev. Med. Vet.* 106, 668–675.

[809] Fiocre, B. (1959). Les broncho-pneumonies à néo-rickettsies des bovins. Contagiosité à l'homme. *Rev. Med. Vet.* 135, 199–209.

[810] Sarateanu, D., Sorodoc, G., Surdan, C., and Fuhrer-Anagnoste, B. (1961). Pneumonia in calves due to Pararickettsia and the possibility of human contamination. *Stud. Cerret. Inframicrtobiol.* 12, 353–362.

[811] Martinov, S. P. (1983). *Studies of Some Biological, Morphological and Immunological Properties of the Chlamydiae.* Ph.D. dissertation, Central Res. Vet. Med. Inst., Sofia, 318.

[812] Martinov, S., and Popov, G. (1979). Chlamydiae of avian and mammalian origin and their significance for the human health. *Vet. Sbirka, Sofia XII* 13–16.

[813] McKinley. A. W., White, N., Buxton, D., Inglis, J. M., Johnson, F. W. A., Kurtz, J. B., et al. (1985). Severe *Chlamydia psittaci* sepsis in pregnancy. *QJM* 57, 689–696.

[814] Johnson, F. W., Matheson, A., Williams, B. A., and Laing, H. (1985). Abortion due to infection with *Chlamydia psittaci* in a sheep farmer's wife. *Br. Med. J.* 290, 592–594.

[815] Buxton, D. (1986). Potential danger to pregnant women of *Chlamydia psittaci* from sheep. *Vet. Rec.* 118, 510–511.

[816] Wong, S. Y., Gray, E. S., Buxton, D., and Johnson, F. W. A. (1985). Acute placentitis and spontaneous abortion caused by *Chlamydia psittaci* of sheep origin: a histological and ultrastructural study. *J. Clin. Pathol.* 38, 707–711.

[817] Longbottom, D., and Coutler, L. J. (2003). Animal chlamydioses and zoonotic implications. *J. Comp. Pathol.* 128, 217–244.

[818] Walder, G., Hotzel, H., Brezinka, C., Gritsch, W., Tauber, R., Würzner, R., et al. (2005). An unusual cause of sepsis during pregnancy: recognizing infection with *Chlamydophila abortus*. *Obstet. Gynecol.* 106, 1215–1217.

[819] Jansen, M. L., Van de Wetering, K., and Arabin, B. (2006). Sepsis due to gestational psittacosis: a multidisciplinary approach within a perinatological center–review of reported cases. *Int. J. Fert. Women's Med.* 51, 17–20.

[820] Meijer, A., Brandenburg, A., Vries, J. J., Beentjes, J., Roholi, P., and Dercksen, D. (2004). *Chlamydophila abortus* infection in a pregnant woman, associated with indirect contact with infected goats. *Eur. J. Clin. Microbiol. Infect. Dis.* 23, 487–490.

[821] De Puysseleyr, K., De Puysseleyr, L., Dhondt, H., Geens, T., Braeckman, L., Morre, S. A., et al. (2014). Evaluation of the presence and zoonotic transmission of *Chlamydia suis* in a pig slautherhouse. *BMC. Infect. Dis.* 14:560.

[822] Dugan, J., Rockey, D. D., Vores, L., and Andersen, A. A. (2004). Tetracycline resistance in *Chlamydia suis* mediated by genomic islands inserted into the chlamydial inv-like gene. *Antimicrob. Agents Ch.* 48, 3989–3995.

[823] De Puysseleyr, K., De Puysseleyr, L., Braeckman, L., and Morre, A. (2015). Assessment of *Chlamydia suis* infection in Pig Farmers. *Transboundry Emer. Diseas.* 64, 826–833.

[824] Rohde, G., Straube, E., Essig, A., Reinhold, P., and Sachse, K. (2010). Chlamydial Zoonoses, Dtsch. *Arztebl. Int.* 107, 174–180.

[825] Sprague, L. D., Schubert, E., Hotzel, H., Scharf, S., and Sachse, K. (2009). The detection of *Chlamydophila psittaci* genotype C infection in dogs. *Vet. J.* 181, 274–279.
[826] Fraser, G., Norwall, J., Withers, A. R., and Gregor, W. W. (1969). A case history of psittacosis in a dog. *Vet. Res.* 85, 54–58.
[827] Eliot, D. L., Tolle, S. W., Goldberg, L., and Miller, J. B. (1985). Pet-associated illness. *N. Engl. J. Med.* 313, 985–995.
[828] Werth, D., Schmeer, N., Miller, H. P., Karo, M., and Krauss, H. (1989). Demonstration of antibodies against *Chlamydia psittaci* and *Coxiella burnetii* in dogs and cats: comparison of the enzyme immunoassay, immunoperoxidase techniques, complement fixation test and agar gel precipitation test. *Zentralb. Veterinarmed. B* 34, 165–176.
[829] Ostler, H. B., Schachter, J., and Dowson, C. R. (1969). Acute follicular conjunctivitis of epizootic origin. *Arch. Ophthalmol.* 82, 587–591.
[830] Ostler, H. B., and Schachter, J. (1972). Bedsonia/Chlamydia of epizootic origin. *Rev. Int. Trachoma* 49:37.
[831] Griffiths, P. D., Lechter, R. I., and Treharne, J. D. (1978). Unusual chlamydial infection in a human renal allograft recipient. *Br. Med. J.* 2, 1264–1265.
[832] Regan, R. J., Dathan, J. R. E., and Treharne, J. D. (1979). Infective endocarditis with glomerunephritis associated with chlamydia (*Chlamydia psittaci*) infection. *Br. Heart J.* 42, 349–352.
[833] Cotton, M. M., and Patridge, M. R. (1998). Infection. with feline *Chlamydia psittaci*. *Thorax* 53, 75–76.
[834] Vlahovich, K., Dovc, A., and Lasta, P. (2006). Zoonotic aspects of animal chlamydioses – a review. *Vet. Arhiv.* 76 (Suppl), 259–274.
[835] Hartley, J. C., Stevenson, S., Robinson, A. J., Underwood, J. D., Carder, C., Cartledge, J., et al. (2001). Conjunctivitis due to *Chlamydophila felis* (*Chlamydia psittaci* feline pneumonitis agent) acquired from a cat: a case report with molecular characterization of isolates from the patient and cat. *J. Infect.* 43, 7–11.
[836] Browning, G. F. (2004). Is *Chlamydophila felis* a significant zoonotic pathogen? *Aust. Vet. J.* 82, 695–696.
[837] Tasker, S. (2016). *Chlamydia felis*. Chlamydia guidelines. *Eur. Adv. Board Cat Diseases* [Accessed: December 2016].
[838] Lietman, T., Brooks, D., Monkada, J., Schachter, J., Dowson, C. R., and Dean, D. (1998). Cronic follicular conjunctivitis associated with

Chlamydia psittaci or *Chlamydia trachomatis*. *Clin. Infect. Dis.* 26, 1335–1340.

[839] Schachter, J., Ostler, H. B., and Meyer, K. F. (1969). Human infection with the agent of feline pneumonitis. *Lancet* 31, 1063–1065.

[840] Lutz-Wohlgroth, L., Becker, A., Brugnera, E., Huat, Z. L., Zimmerman, D., Grimm, F., et al. (2006). Chlamydiales in guinea-pigs and their zoonotic potential. *J. Vet. Med. A. Physol. Pathol. Clin. Med.* 53, 185–193.

[841] Unknown author. (2017). *Zoonotic Chlamydiae maintained in Mammals. Chlamydiosis*. Ames, IA: CFSPH–Iowa State University.

[842] Dimitrov, K., Martinov, S., and Popov. G (1984). Klinisch äetiologische befunde bei Chlamydienurethritiden bei männern. *H+G Zeitschrift für Hautkrankhneiten (Germany)* 59, 1229–1236.

[843] Dimitrov, K., Martinov, S., and Popov, G. (1984). Clinico-etiological parallels of *Chlamydia urethritis* in men. *Dermatol. Venerol XXIII* 2, 9–18.

[844] Dimitrov, K. D. (1985). *Studies on the Most Common Genital Chlamydial Infections (Clinic, Diagnosis, Treatment)*. Ph.D. disertation, Medical Academy, Sofia, 164.

[845] Illin, I. I. (1983). Non-gonococcal urethritis in men. Moscow: Medicine, 255.

[846] Siboulet, A., Bohlot, J. M., Catalan, F., and Henry-Suchet, J. (1982). *Bull. Mem. Soc. Med. Paris* 10, 103–113.

[847] Martinov, S., Popov, G., Dimitrov, K., and Bonev, A. (1984). Etiological and clinical Study of chlamydial post-gonococcal urethritis in men. *Dermatol. Venerol XXIII* 4, 24–30.

[848] Richmond, S. J., Hilton, A. L., and Clarke, S. K. R. (1972). Chlamydial infection: Role of Chlamydia subgroup A in non-gonococcal and post-gonococcal urethritis. *Br. J. Vener. Dis.* 48, 437–444.

[849] Oriel, J. D., Reeve, P., Thomas, B. J., and Nicol, C. S. (1975). Infection with Chlamydia group A in men with urethritis due to *Neisseria gonorrhoeae*. *J. Inf. Dis.* 131, 376–382.

[850] Holmes, K. K., Handsfield, H. H., Wang, S.-P., Wentworth, B. B., Turck, M., Anderson, J. B., et al. (1975). Etiology of non-gonococcal urethritis. *N. Engl. J. Med.* 292, 1199–1206.

[851] Terho, P. (1978). *Chlamydia trachomatis* in gonococcal and post-gonococcal urethritis. *Br. J. Vener. Dis.* 54, 326–329.

[852] Perroud, H., and Miedzybrodzka, K. (1978). Chlamydial infection of the urethra in men. *Br. J. Vener. Dis.* 54, 45–49.
[853] Belli, L., Castro, J. M., and Casco, R. (1985). "Ciprofloxacin in the treatment of non-complicated gonococcal urethritis," in *Proceedings of the 14th Congress of Chemotherapy*, Kyoto, 1687–1688.
[854] Mavrov, I. I., and Cletnoy, A. G. (1986). "Treatment of complicated urogenital chlamydiosis," in *Chlamydial Infections,* ed. A. A. Shatkin (Moscow: Medicine), 82–84.
[855] Bagdasarov, A. B. (1982). *Clinical and Laboratory Diagnosis of Chlamydial Urethritis in Men*. Ph.D. thesis, Moscow.
[856] Krauss, H., Schiefer, H. G., Weidner, W., et al. (1983). Significance of *C. trachomatis* in "abacterial" prostatitis. *Zbl. Bak. Mikrobiol. Hyg. Orig. Abt. A.* 254, 545–551.
[857] Nilsson, S., Nilsson, B., Johannisson, G., and Lycke, E. (1984). Acute epididymitis and prostatitis caused by *Chlamydia trachomatis. Ann. Chir. Gynaecol.* 73, 42–44.
[858] Suominen, J., Grönross, M., Terho, P., and Wichmann, L. (1983). Chronic prostatitis. *Chlamydia trachomatis* and infertility. *Int. J. Androl.* 6, 405–413.
[859] Bruce, A. W. (1989). Prostatitis associated with *Chlamydia trachomatis* in 6 patients. *J. Urol.* 142, 1006–1007.
[860] Weidner, W., Diemer, Th., Huwe, P., Rainer, H., and Ludwig, M. (2002). The role of *Chlamydia trachomatis* in prostatitis. *Int. J. Antimicrob. Agents* 19, 466–470.
[861] Wagenlehner, F. M. E., Naber, K. G., and Weidner, W. (2006). Chlamydial infections and prostatitis in men. *BJWI* 97, 687–690.
[862] Mardh, P.-A., Ripa, K. T, Collin, S., Treharne, D., and Darougar, S. (1978). *Chlamydia trachomatis* in non- acute prostatitis. *Br. J. Vener. Dis.* 54, 330–334.
[863] Pandi, R. C. (2016). "Sexually transmitted diseases (Chapter 18), pp. 469–999," in *Comprehensive Approach to Infections in Dermatology,* eds A. Singal and G. Grover (New Delhi: JP Medical Ltd), 500.
[864] Popov, G., Martinov, S., and Dimitrov. K. D. (1985). "Complex etiological diagnostics of the *Chlamydia trachomatis* uro-genital infections," in *Proceedings of the 6th Comgress of Microbiology*, Varna, (Abs).
[865] Kristiancen, J. K., and Scheibel, J. (1984). Etiology of acute epididymitis presenting in a venereal disease clinic. *STD* 11, 32–34.

[866] Grant, J. B., Brooman, P. J., Chowdhury, S. D., Sequeira, P., and Blacklock. (1985). The clinical presentation of *Chlamydia trachomatis* in a urological practice. *Br. J. Urol.* 57, 218–221.
[867] Ostaszewska, I., Zdrodowska-Stefanow, B., Darewicz, J., Badyla, J., Pucilo, K., Bulhak, V., et al. (2000). Role of *Chlamydia trachomatis* in epididymitis. Part II: Clinical diagnostics. *Med. Sci. Monit.* 6, 1119–1121.
[868] Workowski, K. A., and Berman, S. (2010). "CDC: Epididymitis," in *Sexually Transmitted Diseases Treatment Guidelines, MMWR Recomm.* Rep 59 (RR-12), 67–69; 44–49.
[869] CDC. (2015). *Sexually Transmitted Diseases Treatment Guidelines. Epididymitis. Division of STD Prevention.* Atlanta: CDC.
[870] Bruncham, R. C., Paavonen, J., Stevens, C. E., Kiviat, N., Kuo, C. C., Critchlow, C. W., et al. (1984). Mucopurulent cervicitis-the ignored counpart in women of urethritis in man. *N. Engl. J. Med.* 311, 1–6.
[871] Paavonen, J., Vesterinen, E., Meyer, B., and Saksela, E. (1982). Colposcopic and histological findings in cervical chlamydial infections. *Obstet. Gynecol.* 59, 712–715.
[872] Schachter, J., Hanna, E. C., Hill, E. C., Massad, S., Sheppard, C. W., Conte, J. E., et al. (1975). Are chlamydial infections the most prevalent venereal diseases? *JAMA* 231, 1252–1253.
[873] Hobson, D., Johnson, F. W. A., Rees, E., and Taiti, T. A. (1974). Simplified method foe diagnosis of genital and ocular infections with Chlamydia. *Lancet* 2, 555–556.
[874] Hare, M., Taylor-Robinson, D., and Dates, J. K. (1981). *C. trachomatis* as a cause of follicular cervicitis. *Br. J. Obstet. Gynecol* 88, 174–180.
[875] Anonymous. (2016). *Chlamydia. Medical portal "Medguide book,"* 2014–2016.
[876] Rees, E., Tait, I. A., Hobson, D., and Johnson, F. W. A. (1977). Chlamydia in relation to cervical infection and pelvic inflammatory disease, pp. 67–76," in *NGU and Related Infections,* eds K. K. Holmes and D. Hobson (Washington, DC: Am. Soc. Microbiology).
[877] McCormack, W. M., Alpert, S., McComb, E., Nichols, R. L., Semine, D. Z., et al. (1979). Fifteen-months follow-up study of women infected with *C. trachomatis. N. Engl. J. Med.* 300, 123–125.
[878] Anonymous (2015). *CDC STD Guidelines. Chlamydil infections. Chlamydial infections in Adolescents and Adults.* Atlanta: CDC, 6A.

[879] Hill, L. H., Ruparela, H., and Embil, L. A. (1983). Nonspecific vaginitis and other genital infections in three clinical populations. *Sex. Transm. Dis.* 10, 114–118.

[880] Harrison, H. R., Costin, M., Meder, J. B., et al. (1985). Cervical *Chlamydia trachomatis* infection in university women: Relationship to history, contraception, ectopy, and cervicitis. *Am. J. Obstet. Gynecol.* 153, 244–251.

[881] Hanna, N. F., Taylor-Robinson, D., Kalodiki-Karamanoli, M., Hakrls, J. K. W., and McFadgen, I. R. (1985). The relation between vaginal Ph and the microbiological status in vaginitis. *J. Obstet. Gynecol.* 92, 1267–1271.

[882] Osborne, N. G., Grubin, L., and Pratson, L. (1982). Vaginitis to sexually active women: relationship to nine sexually transmitted organisms. *Am. J. Obstet. Gynecol.* 142, 962–967.

[883] Paavonen, J. (1979). *Chlamydia trachomatis*-induced urethritis in female partners of men with nongonococcal urethritis. *Sex. Transm. Dis.* 6, 69–71.

[884] Stamm, W. E., Koutsky, L. A., Benedetti, J. K., Jourdan, J. L., Brunham, R. C., and Holmes, K. K. (1984). *Chlamydia trachomatis* urethral infections in men: prevalence, risk factors, and clinical manifestations. *Ann. Int. Med.* 100, 47–51.

[885] Dunlop, E. M. C., Vaughan-Jackson, Dorougar, S., and Jones, B. R. (1972). Incidence of "non-specific urethritis. *Br. J. Vener. Dis.* 48, 425–428.

[886] Skolnik, A., Lynn, A., and Woodwarde, J. (eds). (2013). *Sexually Transmitted Diseases: A Practical Guide for Primary Care.* Dordrecht: Springer Science and Business Media, 185.

[887] Davis, J. A., Rees, E., Hobson, D., and Karayinnis, P. (1978). Isolation of *Chlamydia trachomatis* from Bartolin's ducks. *Br. J. Vener. Dis.* 54, 409–413.

[888] Bleker, O. P., Smalbraak, D. J. C., and Schutte, M. F. (1990). Bartolin's abscess: the role of *Chlamydia trachomatis*. *Gentrourin. Med.* 66, 24–25.

[889] Saul, H. M., and Grossman, M. P. (1988). The role of *Chlamydia trachomatis* in Bartolin's abscess. *Am. J. Obstet. Gynecol.* 158, 576–577.

[890] Garutii, A., Tangerini, A., Rossi, R., and Cireli, C. (1994). Bartolin's abscess and *Chlamydia trachomatis*. *Case Rep. Clin. Exp. Obstet. Gynecol.* 21, 103–104.

[891] Tanaka, K., Mikato, H., Ninomiya, M., Tamaya, T., Izumi, K., Ito, K., et al. (2005). Microbiology of Bartolin's gland abscess in Japan. *J. Clin. Microbiol.* 43, 4258–4261.

[892] Rivlin, M. E. (2016). Endometritis. *Medscape Medical News*. Available at: https://www.medscape.com/

[893] Paavonen, J., Kuviat, N., Brunham, C., Stevens, C. E., Kuo, C. C., Stamm, W., et al. (1985). Prevalence and manifestation of endometritis among women with cervicitis. *Am. J. Obstet. Gynecol.* 152, 280–286.

[894] Paavonen, J., Aine, R., Teisala, K., Heinonen, P. K., Punnonen, R., Lentinen, M., et al. (1985). Chlamydia endometritis. *J. Clin. Pathol.* 38, 726–732.

[895] Punnonen, R. (1985). *Chlamydia trachomatis* endometritis in infertile women. *Arch. Gynecol.* 237 (Suppl), 314.

[896] Moller, B. R., Weström, L., Ahrons, S., Ripa, K. T., Svensson, L., von Mecklenburg, C., et al. (1979). *Chlamydia trachomatis* of the fallopian tube. Histological findings in two patients. *Br. J. Vener. Dis.* 55, 422–428.

[897] Ingerslev, H. J., Moller, B. R., and Mardh, P.-A (1985). *Chlamydia trachomatis* in acute and chronic endometritis. *Scand. J. Infect. Dis. (Suppl.)* 32, 59–63.

[898] Faro, S. (2003). *Sexually Transmitted Diseases in Women*. Philadelphia: Lippicott W. Wilkins, 275.

[899] Buttner, H. H., and Barter, G. (1976). Histological and diagnostic criteria and incidence of endometritis in abrasion material. *Zentrl. Att fur Gynacologic*. 98, 1515–1517.

[900] Krettek, J. E., Arkin, S. I., Chaisilwattana, P., Monif, G. R. (1993). *Chlamydia trachomatis* in patients who used oral contraceptives and intermenstrual spotting. *Obstet. Gynecol.* 81, 728–731.

[901] Jones, R. B., Mammel, J. B., Shepard, M. K., and Fisher, R. R. (1986). Recovery of *Chlamydia trachomatis* from the endometrium of women at risk for chlamydial infection. *Am. J. Obstet. Gynecol.* 155, 35–39.

[902] Mardh, P.-A., Moller, B. R., Ingerslev, H. J., Nüssler, E., Weström, L., and Wølner-Hanssen, P. (1981). Endometritis, caused by *Chlamydia trachomatis*. *Br. J. Vener. Dis.* 57, 191–195.

[903] Sweet, R. L. (1982). Chlamydial salpingitis and infertility. *Fertil. Steril.* 38, 530–533.

[904] Mardh, P.-A., Paavonen, J., and Puolakkainen, M. (2012). *Chlamydia*. Dordrecht: Springer Science and Business Media, 388.

[905] Weström, L., Iosif, S., Svensson, L., and Mardh, P.-A. (1979). Infertility after acute salpingitis: results of treatment with different antibiotics. *Curr. Ther. Res. Suppl.* 26:712.

[906] Cianci, A., Tempera, G., Roccasalva, L., Fumeri, P. M., Nicoleti, G., and Palumbo, G. (1994). Laparoscopy diagnosis of chlamydial salpingitis using a tubal cytobrush. *Fertil. Steril.* 61, 181–184.

[907] Wolner-Hansen, P., Svensson, L., Mardh, P.-A., Weström, L. (1985). Laparoscopic findings and contraceptive use in women with signs and symptoms suggestive of acute salpingitis. *Obstet. Gynecol.* 66, 233–238.

[908] Bowie, W. R., and Jones, H. (1981). Acute pelvic inflammatory disease in out- patients: association with *Chlamydia trachomatis* and *Neisseria gonorrhoeae*. *Ann. Intern. Med.* 95, 685–688.

[909] Mardh, P.-A., and Svensson, L. (1982). Chlamydial salpingitis. *Scand. J. Infect. Dis. Suppl.*, 32, 64–72.

[910] Svensson, L., Weström, L., and Mardh, P.-A. (1981). Acute salpingitis with *Chlamydia trachomatis* isolated from fallopian tubes-clinical, cultural and serological findings. *Sex. Transm. Dis.* 8:51.

[911] Mitchell, C., and Prabhu, M. (2013). Pelvic inflammatory disease, a current concept in pathogenesis, diagnosis and treatment. *Infect Dis. Clin. N. Am.* 27, 793–809.

[912] Campion, E. W., Brunham, R. C., Sami, L. S., and Paavonen, J. (2015). Pelvic inflammatory disease. *N. Engl. J. Med.* 372, 2039–2048.

[913] Paavonen, J., Valtonen, V. V., Kasper, D. L., Malkamäki, M., Mäkelä, P. H. (1981). Serological evidence for the role of Bacteroides fragilis and Enterobacteriaceae in the pathogenesis of acute pelvic inflammatory disease. *Lancet* 1, 293–295.

[914] Forslin, L., Falk, V., and Danielsson, D. (1978). Changes in the incidence of acute gonococcal and non- gonococcal salpingitis. *Br. J. Vener. Dis.* 54:247.

[915] Anonymous. (2017). *Pelvic Inflammatory Disease (PID). CDC Fact Sheet*. Atlanta: CDC.

[916] Puolakkainen, M., Vesterinen, E., Purola, E., Saikku, U., and Paavonen, J. (1986). Persistence of chlamydial antibodies after pelvic inflammatory disease. *J. Clin. Microbiol.* 23, 924–928.

[917] Guderian, A. M., and Trobough, G. E. (1986). Residues of pelvic inflammatory disease in intrauterine device users: a result of the

intrauterine device or *Chlamydia trachomatis* infection? *Am. J. Obstet. Gynecol.* 154, 497–503.

[918] Mardh, P.-A., and Wolner-Hansen, P. (1986). Three novel manifestations of *Chlamydia trachomatis* infection-endometritis, perihepatitis and meningoencephalitis. *Infection* 10 (Suppl 1), 57–60.

[919] Anonymous. (2017). *Mayo Clinic Staff. Pelvic inflammatory disease (PID). MFMER.*

[920] Lee, M. (2017). Pelvic inflammatory disease. *J. Am. Acad. Phys. Assist.* 30:47.

[921] Peipert, J. F., Barbieri, R. L., and Falk, S. J. (eds). (2017). Long-term complications of pelvic inflammatory disease. [Accessed: January 20, 2017]. Available at: http: //www.uptodate.com/home

[922] Shepherd, S. M. (2017). Pelvic inflammatory disease clinical presentation. *Medscape Medical News*. Available at: https://emedicine.med scape.com/article/256448-clinical

[923] Anonymous. (2017). *Fitz Hugh Curtis Syndrome*. Danbury, CT: National Organization for Rare Disorders (NORD).

[924] Piscaglia, F., Vidili, G., Ugolini, G., Ramini, R., Montroni, I., De Iaco, P., et al. (2005). Fitz Hugh Curtis syndrome mimicking acute cholecystitis: vaue of new ultrasound findings in the differential diagnosis. *Ultraschall. Med.* 26, 227–230.

[925] Peter, N. G., Clark, L. R., and Jaeger, R. (2005). Fitz Hugh Curtis syndrome: a diagnosis to consider in women with right upper quadrant pain. *Cleveland Clin. J. Med.* 71, 233–239.

[926] Fitz Hugh, T. (1934). Acute gonococcal peritonitis of the right upper quadrant in women. *JAWA* 102, 2094–2095.

[927] Curtis, A. H. (1930). A cause of adhesion in the right upper quadrant. *JAWA* 94, 1221–1222.

[928] Woo, S., Kim, J. Il, Cheung, D. Y., Cho, S. H., Park, S.-H., et al. (2008). Clinical outcome of Fitz Hugh Curtis syndrome mimicking acute biliary disease. *World J. Gastroenterol.* 14, 6975–6980.

[929] Wolner-Hansen, P. (1982). *Proc. Scand. Symp. on C. trachomatis in genital and related infections*. 32, 77–82,

[930] Darougar, S., Forsey, T., Wood, J. J., Bolton, J. P., and Allan, A. (1981). Chlamydia and the Curtis- Fitz Hugh syndrome. *Br. J. Ven. Dis.* 57, 391–394.

[931] Puolakkainen, M., Saikku, P., Leinonen, M., Nurminen, M., Väänänen, P., Makelä, P. (1985). Comparative sensitivity of different

sertological tests for detecting chlamydial antibodies in perihepatitis. *J. Clin. Pathol (London)* 38, 929–932.

[932] Paavonen, J., Saikku, P., Von Knorring, J. K., Aho, K., and Wang, S. P. (1981). Association of infection with *C. trachomatis* with Fitz Hugh Curtis syndrome. *J. Inf. Dis*. 114:176.

[933] Huang, H. H., Tsai, C. M., and Tyan, Y. S. (2011). Unusual cause should be kept in mind of abdominal pain in female patient. Fitz-Hugh -Curtis syndrome. *Gastroenterology* 140, e7–e8.

[934] Antonie, F., Billlion, C., and Vic, C. (2013). *Chlamydia trachomatis* Fitz- Hugh -Curtis syndrome in a female adolescent. *Arch. Pediatr.* 20, 289–291.

[935] Le Moigne, F., Lamboley, J. L., and Vitry, T. (2009). Usefulness of contrst-enhanced CT scan for diagnosis of Fitz- Hugh -Curtis syndrome. Gastroenterol. *Clin. Biol*. 33, 1176–1178.

[936] Wang, P. Y., Zhang, L., Wang, X., Lin, X.-J., Chen, l., and Wang, B. (2015). Fitz-Hugh-Curtis syndrome: clinical diagnostic value of dynamic enhanced MSCT. *J. Phys. Ther. Sci.* 27, 1641–1644.

[937] Theofanakis, C. P., and Kyriakidis, A. V. (2011). Fitz- Hugh -Curtis syndrome. *Gynecol. Survey* 8, 129–134.

[938] Pankuch, G., Appelbaum, P. C., Lorenz, R. P., Botti, J. J., Schachter, J., and Naeye, R. L. (1984). Placental microbiology and histology and the pathogenesis of chorioamnionitis. *Obstet. Gynecol*. 64, 802–806.

[939] Rours, G. I. J. G., De Krijger, R. R., Ott, A., Willenise, H. F. M., De Groot, R., Zimmerman, J. I., et al. (2011). *Chlamydia trachomatis* and placental inflammation in early preterm delivery. *Eur. J. Epidemiol.* 26, 421–428.

[940] Baud, D., and Greub, G. (2011). Intracellular bacteria and adverse pregnancy outcomes. *Clin. Microbiol. Infect.* 17, 1312–1322.

[941] Kovacs, L., Nagy, E., Berbik, I., Meszaros, G., Deak, J., and Nyari, T. (1998). The frequency and the role of *Chlamydia trachomatis* infection in premature labor. *Int. J. Obstet. Gynecol*. 62, 42–54.

[942] Gencay, M., Koskiniemi, M., Fellman, V., Ammala, P., Vaheri, A., and Puolakkainen, M. (2001). *Chlamydia trachomatis* infection in mothers with preterm delivery and their newborn infants. *APMIS* 109, 636–640.

[943] Stears, S. (2009). Best evidence topic report. BETY. Treating Chlamydia in pregnancy. *Emerg. Med. J.* 26, 120–122.

[944] Martinov, S., Popov, G., and Dimitrov, K. (1985). Etiological and clinical studies of the chlamydial infection in Reiter's Syndrome. *Dermatol. Venerol. XXIV* 3, 5–14.

[945] Martinov, S. P., Popov, G. V., and Dimitrov, K. (1985). Chlamydial infection in the Oculo-Uretro-Synovial Reiter's syndrome. *Compt. Trend. Acad. Bulg. Sci*. 38, 787–790.

[946] Ford, D. K. (1977). "Reiter's syndrome," in *NGU and Related Infections,* eds D. Hobson and K. K. Holmes (Washington, DC: Am. Soc. Microbiol.), 64.

[947] Ostler, H. B., Dawson, C. R., Schachter, J., and Engleman, E. P. (1971). Reiter's syndrome. *Am. J. Ophthalmol*. 71, 986–991.

[948] Stoilov, R., Andreev, T., Martinov, S., Nankova, V., and Trankova, V. (1990). Unlocking bacterial infections in reactive arthritis. *Inter. Dis.* 29, 74–78.

[949] Anonymous. (2009). Urostim. *Health.*

[950] Usunova, Y., Simeonov, P., Mitov, I., Markova, B., and Martinov, S. (1998). Application of the preparation urostim in the andrological practice. *Urology* 4, 56–59.

[951] Brinke, A. (2001). Persistent airflow limitation in asthma is associated with serologic evidence for *Chlamydia pneumoniae* infection. *J. Allergy Clin. Immunol*. 107, 449–454.

[952] Hahn, D., Dodge, R., and Golubjatnikov, R. (1991). Association of *Chlamydia pneumoniae* infection with wheezing, asthmatic bronchitis, and adult-onset asthma. *JAMA* 266, 225–230.

[953] Hahn, D. (1998). Evidence for *Chlamydia pneumoniae* infection. in steroid-dependent asthma. *Ann. Asthma Allergy Immunol*. 80, 45–49.

[954] Hammerschlag, M. R. (1992). Persistent infection with *Chlamydia pneumoniae* following acute respoiratory illness. *Clin. Infect Dis*. 14, 178–182.

[955] Emre, U. (1994). The association of *Chlamydia pneumoniae* infection and reactive airway diseases in children. *Pediatr. Adolesc. Med*. 148, 727–792.

[956] Hahn, D. L., and Golubjatnikov, R. (1994). Asthma and chlamydial infection: a case series. *J. Fam. Pract*. 38, 589–595.

[957] Myashita, N. (1998). *Chlamydia pneumoniae* and exacerbations of asthma in adults. *Ann. Asthma Allergy Immunol*. 80, 405–409.

[958] Thom, D. H. (1994). Respiratory infection with *Chlamydia pneumoniae* in middle-aged and older adult out-patients. *Eur. J. Clin. Microbiol. Infect Dis*. 13, 785–792.

[959] Mitscherlich, E. (1954). Der Virusabort des Schafes in Detschland. *DTW* 61, 42–44.

[960] Kennedy, P. C., Olander, H. J., and Howardth, J. A. (1960). Pathology of epizootic bovine abortion. *Cornell Vet.* 50, 417–421.

[961] Storz, J., and McKercher, D. G. (1962). Etiological studies on epizootic bovine abortion. *Zbl. Veterinaermed.* 9, 411–427, 520–541.

[962] Pavlov, N., and Vesselinova, A. (1965). Morphology of the natural infection in lambs with the virus of lamb abortion. *Zbl. Veterinaermed.* 12, 517–526.

[963] Novilla, M. N. (1965). *Placental Pathology of Experimental Enzootic Abortion in Ewes.* Ph.D. thesis, Colorado State University, Fort Collins.

[964] Lincoln, S. (1968). Pathology of experimental epizootic bovine abortion. Ph.D. thesis, Colorado State University, Fort Collins.

[965] Schoilev, C., and Martinov, S. (1993). Pathomorphological changes caused by chlamydial abortion in sheep. *Mac. Vet. Rev.* 22, 17–24.

[966] Martinov, S., and Schoilev, C. (2008). Etiological and pathomorphological investigations upon chlamydial abortion in sheep. Traditions and Modern Times in the Veterinary Medicine, Vet. Med Faculty. Forestry University, Sofia, 214–223.

[967] Maley, S. W., Livingstone, M., Rodger, S. M., Longbottom, D., and Buxton. D. (2009). Identification of *Chlamydophila abortus* and the development of lesions in placental tissues of experimentally infected sheep. *Vet Microbiol.* 135, 122–127.

[968] Sammin, D., Markey, B., Bassett, H., and Buxton. D. (2009). The ovine placenta and placentitis-A review. *Vet. Microbiol.* 135, 90–97.

[969] Martinov, S., Popov, G., and Panova, M. (1981). Chlamydial polyarthritis in calves. *Vet. Sci.* 18, 83–91.

[970] Storz, J., and Spears, P. (1980). Chlamydienbedingte Polyarthritis bei Kälbern und Schafen: Pathogenese und Erregeigen-schhaften. Wiss. Z. der Humboldt-Universitat zu Berlin. *Math.-Nat. R. XXIX* 1, 53–56.

[971] Pavlov, N. (1967). Histopathology of neorickettsial pneumonias in sheep. *Zbl. Veterinaermed.* 14, 343–355.

[972] Blood, D. C., and Henderson, J. A. (1974). *Veterinary Medicine*, 4th Edn. *London Bail-litre*. London: Tindal and Cassel.

[973] Littlejohn, J. (1976). "Chlamydial diseases in sheep and cattle, other than abortions," in: *Infectious Diseases in the Twilight Zone. Course for veterinarians*, University of Sydney, Sydney, 18–110.

[974] Thoma, R., Guscetti, I., Schiller, I., Schmeer, N., Corboz, L., and Pospischil, A. (1997). Chlamydiae in porcine abortion. *Vet. Pathol.* 34, 467–469.

[975] Käser, T., Pastertnak, J. A., Delgado-Ortega, M., Hamonic, G., Lai, K., Erickson, J., et al. (2017). *Chlamydia suis* and Chlamydia trachomatic induce multifunctional CD 4T cells in pigs. *Vaccine* 35, 91–100.

[976] Filipovich, A., Ghasemian, E., Inic-Kanala, Lukic, I., Stain, E., et al. (2017). The effect of infectious dose on humoral and cellular immune responses in *Chlamydophila caviae* primary ocular infection. *PLoS ONE* 12:e0180551.

[977] Martinov, S. P., Popov, G. V., and Shoylev, H. I. (1985). "Etiological and pathomorphological comparisons in chlamydial miscarriage in sheep," in *Proceedings of the Sixth Congress of Microbiology I.*, Varna, 592–596.

[978] Novilla, M. N., and Jensen, R. (1970). Placental pathology of experimental enzootic abortion in ewes. *Am. J. Vet. Res.* 111:1983.

[979] Storz, J. (1967). Psittacosis agent as a cause of polyarthritis in cattle and sheep. *Vet. Med. Rev.* 2/3, 125–139.

[980] Cultip, R. C. (1974). Ultrastructure of the synovial membrane of lambs affected with chlamydial polyarthritis. *Am. J. Vet. Res.* 35, 171–176.

[981] Shupe, J. L., and Storz, J. (1964). Pathologic study of psittacosis lymphogranuloma polyarthritis in lambs. *Am. J. Vet. Res.* 25, 943–951.

[982] Norton, W. L., and Storz, J. (1967). Observation on the polyarthritis of sheep produced by an agent of the psittacosis - lymphogranuloma venereum - trachoma group. *Arthritis Rheum.* 10, 1–12.

[983] Hunt, H., Orbell, G. M. B., Buckle, K. N., Ha, H. J., Lawrence, K. E., et al. (2016). First report and histological features of *Chlamydia pecorum* encephalitis in calves in New Zealand. *New Zealand Vet. J.* 64, 364–368.

[984] Neykov, P. (1988). *Pathomorphological Studies on Chlamydial, Rickettsial and Toxoplasmic Abortions in Sheep.* Ph.D. dissertation, CRVMI, Sofia.

[985] Punke, G., Stellmacher, H., and Schulz. (1980). Klinish-epizootiologische und pathomorphologische Untersuchungen zur Chlamydien infection in Schafanlagen. Wiss. Z. der Humboldt-Universitat zu Berlin. *Math.-Nat. R. XXIX* 1, 81–84.

[986] Philip, J. I. H., Omar, A. R., Popovici, V. Lamond, P. H., and Darbyshire, D. H. (1964). Pathogenesis and pathology in calves of infection by bedsonia alone and by bedsonia and reovirus together. *J. Comp. Pathol.* 78, 89–99.

[987] Caswell, J. L., and Williams, K. J. (2016). "Chapter 5. Respiratory system. Major causes of nasal and sinus diseases in domestic animals. Chlamydial diseases," in *Jubb, Kenedy & Palmer's Pathology of Domestic Animals, Vol. 2, Sixth Edn*, ed. M. G. Maxie (London: Elsevier Health Sciences).

[988] Jelocnic, M., Forshaw, D., Cotter, J., Roberts, D., Timms, P., and Polkighorne, A. (2014). Molecular and pathological insights into *Chlamydia pecorum*-associated sporadic encephalomyelitis (SBE) in Western Australia. *BMC Vet. Res.* 10:121.

[989] Maxie, M. G., and Youssef (2007). *Jubb, Kenedy & Palmer's Pathology of Domestic Animals, Fifth Edition.* London: Elsevier Health Sciences, 167–168.

[990] Piercy, D. W., Griffits, P. C., and Teale, C. J. (1999). Encephalitis related to *Chlamydia psittaci* infection in a 14-week-old calf. *Vet. Rec.* 144, 126–128.

[991] Aiello, S. E., and Moses, M. A. (2014). *Chlamydial polyarthritis-serositis in large animals. The Merck Veterinary*. Manual 2012 (Accessed: August 2014).

[992] Kessell, A. E., Finnie, J. W., and Windsor, P. A. (2011). Neurological diseases in ruminant livestock in Australia. III. Bacterial and protozoal infections. *Austr. Vet. J.* 89, 289–296.

[993] Parkinson, T. J., Vermund, J. I., and Malmo (2010). "Sporadic bovine encephalomyelitis, pp. 832–833," in *Diseases of Cattle in Ausralasia, A Comprehensive Textbook,* First Edn, eds T. J. Parkinson, J. I. Vermund, Malmo (Wellington, NZ: Vet. Learn).

[994] Biberstein, E. L., Nishbet, D. I., and Thompson (1967). Experimental pneumonia in sheep. *J. Comp. Pathol.* 77, 181–192.

[995] Reggiarlo, C., Fuhrmann, T. J., Meerding, G. L., and Bicknell, E. J. (1989). Diagnostic features of Chlamydia infection in dairy calves. *J. Vet. Diagn. Invest.* 1, 305–308.

[996] Altera, K. P., and Storz, J. (1967). "Enteritis and alimentary tract responses of cattle to psittacosis (chlamydial) infection," in *Proceedings of the 18th World Vet. Congress*, Paris, 2, 537.

[997] Eugster, A. K. (1970). *Pathogenetic Studies on Intestinal Chlamydial Infections in Cattle*. Ph.D. thesis, Colorado State University, Fort Collins.

[998] Edison, M. (2002). Psittacosis (Avian Chlamydiosis). *J. Am. Vet. Med. Assoc.* 221, 1710–1712.

[999] Eugster, A. K., Joyce, B. K., and Storz, J. (1979). Immunofluorescent studies on the pathogenesis of intestinal chlamydial infections in calves. *Infect. Immun.* 2, 351–359.

[1000] Pavlak, M., Vlahovic, K., Gregoric, J., Zupancic, Z., Jercic, J., and Bozikov, J. (2000). An epidemiological study of *Chlamydia* sp. In feral pigeons (Columbia livia domestica). *Z. Jadgdwiss.* 46, 84–95.

[1001] Surdan, C., and Sorodoc. (1960). Research on the virus of enzootic pneumonia in swine. *Stud. Cerc. Inframicrobiol.* 11, 149–165.

[1002] Okada, N., Murakami, S., Niwa, R., Hara, Y., Murashima, T., Ito, N., et al. (1992). Occurrance of Chlamydial abortion in Swine. *J. Jap. Vet. Med. Assoc.* 45, 655–659.

[1003] Van Wetere, A. J. (2016). *Overview of Avian Chlamydiosis (Psittacosis, Ornithosis, Prrot fever)*. MSD Veterinary Manual.

[1004] Schmidt, R. E., Reavill, D. R., and Phalen, D. N. (2003). *Pathology of Pet and Aviary Birds*. Ames, IA: Blackwell Publishers.

[1005] Vanrompay, D., Hakinezdat, T., Van de Valle, M., Beckman, V., Drogenbrock, V., Verminen, K., et al. (2007). *Chlamydophila psittaci* transmission from pet birds to humans. *Emerg. Infect. Dis.* 13, 1108–1110.

[1006] Martinov, S. P. (1990). "Other chlamydial infections in sheep, pp 131–132," in *Viral Infections in Intensive Livestock Farming,* eds H. Haralambiev and M. Dilovski (Sofia: Zemizdat).

[1007] Brand, C. J. (1989). Chlamydial infections in free-living birds. *J. Am. Vet. Med. Assoc.* 195, 1531–1535.

[1008] Vanrompay, D., Ducatelli, R., and Haeserbronck, F. (1995). *Chlamydia psittaci* infections: a review with emphasis on avian chlamydiosis. *Vet Microbiol.* 45, 93–119.

[1009] Andersen, A. A., Grimes, J. F., and Wyrick, R. B. (1997). Chlamydiosis (Psittacosis, Ornithosis), pp. 333–349," in *Diseases of Poultry*, 10th Edn, ed. B. W. Calnec (Ames, IA: Iowa State University Press).

[1010] Vanrompay, D., Cox, E., Vandenbussche, F., Volckaert, G., and Goddeeris, B. (1999). Protection of turkeys against *Chlamydia psittaci* challenge by gene-based DNA immunization. *Vaccine* 17, 2628–2635.

[1011] Vanrompay, D., Cox, E., Volckaert, G., and Goddeeris, B. (1999). Turkeys are protected from infection with *Chlamydia psittaci* by plasmid DNA vaccination against the major outer membrane protein. *Clin. Exp. Immunol.* 118, 49–55.

[1012] Andersen, A. A., and Vanrompay, D. (2000). Avian Chlamydiosis. *Rev. Sci. Tech.* 19, 396–404.

[1013] Wachendorfer, G., Valder, W.-A, and Lüthgen, W. (1980). Experiment for the vaccination against chlamydial abortion in sheep. Wiss. Z. der Humboldt-Universitat zu Berlin. *Math.-Nat. R. XXIX* 1, 105–111.

[1014] Foggie, A. (1973). Preparation of vaccines against enzootic abortion of ewes. A review of the research work at the Moredun institute. *Vet. Bull.* 43, 587–590.

[1015] McEwen, A. D., Stamp, J. T., and Littlejohn, A. I. (1951). Foggie, A., II. Immunization and infection experiments. *Vet. Rec.* 63, 197–201.

[1016] Littlejohn, A. I., Foggie, A., and McEwen, A. D. (1952). Enzootic abortion in ewes. Field trials of vaccine. *Vet. Rec.* 65, 858–862.

[1017] McEwen, A. D., and Foggie, A. (1954). Enzootic abortion in ewes. Comparative studies of different vaccines. *Vet. Rec.* 66, 393–397.

[1018] McEwen, A. D., Dow, J. B., and Anderson, R. D. (1955). Enzootic abortion in ewes: an adjuvant vaccine prepared from eggs. *Vet. Rec.* 67, 393–394.

[1019] McEwen, A. D., and Foggie, A. (1956). Prolonged immunity following the injection of adjuvant vaccine. *Vet. Rec.* 68, 686–689.

[1020] Kauker, E., and Minners, P. (1956). Virusabort des Schafes und Versuche zur Bekämpfung mittels einer Formol-Vakzine. *Berlin München Tieraerztl. Wschr.* 69, 265–267.

[1021] Seffner, W. (1960). Erfahrungen zum Virusabort der Schafe unter besonderer Berücksichtigungder Diagnose mittels Komplementbindungsreaktion. *Berlin München Tieraerztl. Wschr.* 73, 284–287.

[1022] Frank, F. W., Scriver, L. H., Thomas, L., Waldhalm, D. G, and Meinershagen, W. A. (1968). Artificially induced immunity to enzootic abortion in ewes. *Am. J. Vet. Res.* 29, 1441–1447.

[1023] Popovici, V. (1964). Experimentalle über den Immunisierungswert von Vakzinen gegen Virusabort der Schafe. *Probl. Epiz. (Bucuresti)* 12, 15–30.

[1024] Hulet, C. V., Frank, F. W., Ercanlerack, S. K., Kuttler, A. K., and Meinershagen, W. A. (1965). Beobachtungen über das Ablammungsverhalten von Schafen, die gegen den Enzootischer Schafabort vakziniert sind. *Am. J. Vet. Res.* 26, 1464–1466.

[1025] Philip, R. N., Frank, F. W., Lackman, Price, D. A., Casper, E. A., and Meinershagen, W. A. (1968). Lambing performance of Idaho sheep vaccinated against naturally occurring enzootic abortion in ewes. *Am. J. Vet. Res.* 29, 1153–1159.

[1026] Sarateanu, D., Surdan, C., Sorodoc, G., and Fuhner-Anagnoste, B. (1961). Pararickettsial abortion in ewes: the experimental disease and vaccinations. *Stud. Cercet. Inframicrobiol.* 12, 441–450.

[1027] Sarateanu, D. (1963). Active Immunisierung gegen den Enzootischer Schafabort, Immunologische Untersuchungen unter verschiedenen epidemiologischen Bedingungen. *Rev. Sci. med. (Bucuresti)* 8, 167–171.

[1028] Shukal, R. (1969). *Beitrag zum Studium der antigenen und immunologischen Eigenschaften von Rakeia.* Ph.D. thesis, Frac. Sci., Paris, 169.

[1029] Sorodoc, G. (1979). *Rev. Roum. Med. Virologie* 30, 131–134.

[1030] Schoop, G., Wachendorfer, G., Krüger-Hansen-Schoop, U., and Berger, J. (1968). Erfahrungen mit einer Lebendvakzine zur Bekämfung des Miyagawanellen abortes des Schafe. *Zbl. Veterinaermed.* 15, 209–223.

[1031] Nevjestic, A., and Forsek, Z. (1969). Active Immunisierung gegen den Enzootischer Schafabort. I. Attenuierung des Erregers. *Vet. Glasn.* 23, 423–427.

[1032] Nevjestic, A., Rukavina, L., and Forsek, Z. (1970). Active Immunisierung bei der Prophylaxe des Enzootischen Schafabortes. II. Gebrauch und Effektivität einer lebenden attenuierten Vakzine. *Vet. Glasn.* 24, 503–508.

[1033] Mitscherlich, E. (1963). Vergleichende Prüfung von Impfstoffen gegen den Virusabort der Schafe. Im Mäusevakzinationsversuch. *Berlin München Tieraerztl. Wschr.* 76, 411–415.

[1034] Mitscherlich, E. (1965). Die Bekämpfung des Virusabortes der Schafe. *Berlin München Tieraerztl. Wschr.* 78, 81–88.

[1035] Mitscherlich, E. (1966). Die Bekämpfung des Virusabortes der Schafe. *Zbl. Vet. Med.* 13B, 180–184.

[1036] Yilmaz, S. (1966). *Der Virusabortes des Schafes und seine Bekämpfung.* Göttingen: Tierazztl. Institut.

[1037] Yilmaz, S. (1968). Enzootischer Schafabort und seine Kontrolle. *Etlik Vet. Bact. Enstit. Derg.* 3, 31–83.

[1038] Yilmaz, S., and Mitscherlich, E. (1973). Erfahrungen bei der Bekämpfung des Enzootischer Abortes der Schafe mit einer Lebendvakzine aus dem abgeschwächten Chlamydien ovis-Stamm "P". *Berlin München Tieraerztl. Wschr.* 86, 361–366.

[1039] Meinershagen, W. A., Waldhalm, D. G, Frank, F. W., Thomas, L., and Philip, R. H. (1971). Effektivitat einer Kombinierten Vakzinpräparation zur experimentellen Immunisierung von Schafen gegen dieSchafvibriose und den Epizootischen Schafabort. *Am. J. Vet. Res.* 32, 51–57.

[1040] Valder, W.-A., and Wachendorfer, G. (1975). Untersuchungen zur Wirksamkeit der Vakzination gegen den Chlamydienabort des Schafes. *DTW* 82, 221–225.

[1041] Polidorou, K. (1981). The control of enzootic abortion in sheep and goats in Cyprus. *Br. Vet. J.* 137, 411–415.

[1042] Linklater, K. A., and Dyson, D. A. (1979). Field studies of enzootic abortion in ewes. In south east Scotland. *Vet. Rec.* 105, 387–389.

[1043] Wheelhouse, N., Aitchison, K., Laroucau, K., Thomson, J., and Longbottom. D. (2010). Evidence of *Chlamydophila abortus* vaccine strain 1B as a possible cause of ovine enzootic abortion. *Vaccine* 28, 5657–5663.

[1044] Nietfeld, J. C. (2001). Chlamydial infections in small ruminants. Update on Small Ruminant Medicine, 17:2.

[1045] Popov, G., and Martinov, S. (1983). A method for the production of a concentrated and purified vaccine against the chlamydial abortion in sheep and its application in guinea pigs. *Vet. Sci. XX* 1, 9–16.

[1046] Martinov, S., and Popov, G. (1985). The use of a concentrated and purified vaccine against chlamydial abortion in sheep. *Vet. Sci. XXII* 5, 25–31.

[1047] Popov, G. V., and Martinov, S. P. (1985). Obtaining and testing a concentrated and purified vaccine against chlamydial abortion in sheep. *Compt. Rend. Acad. Bulg. Sci.* 38, 429–432.

[1048] Popov, G. V., and Martinov. S. P (1985). *A Concentrated Purified and Absorbed Vaccine against Chlamydial Abortion in Sheep.* Edinburgh: Int. Meeting, Moredun Research Instiute.

[1049] Martinov, S. P., and Popov, G. V. (1991). "A vaccine PM-3 against chlamydial abortion in sheep," in *Proceedings of the XXIV World Veterinary Congress*, Rio de Janeiro (abs), 100.

[1050] Martinov, S. P., and Popov, G. V. (1992). "Preparation and testing a vaccine against chlamydial abortion in sheep," in *Proceedings of the 5th Int. Congress for Infect. Diseases, Nairobi* (abs), 71.
[1051] Martinov, S. P., and Popov, G. V. (1992). "Vaccines against chlamydial abortion in sheep," in *Proceedings of the 12th Inter. Symposium of the W.A.V.M.I*, Davis, CA, 395.
[1052] Martinov, S., and Popov, G. (1993). A vaccine against chlamydial abortion in sheep. *Maced. Vet. Rev.* 22, 25–33.
[1053] Snyder, J. C., Bell, S. D., Murray, E. S., Thygeson, P., and Hadad, N. (1962). Attempt to immunize a volunteer (with formalin-inactivated virus) against experimental trachoma induced by Saudi Arabian strain 2. *Proc. N. Y. Acad. Sci.* 98, 368–375.
[1054] Grayston, J. T., Wooridge, R. L., and Wang, S. P. (1962). Trachoma vaccine studies in Taiwan. *Ann. N. Y. Acad. Sci.* 98, 352–366.
[1055] Collier, L. H., and Blyth, W. A. (1966). Immunogenicity of experimental trachoma vaccines in baboons. I. Experimental methods and preliminary tests with vaccines prepared in chick embryos and in HeLa cells. *J. Hyg. (Camb.)* 64, 513–528.
[1056] Grayston, J. T. (1967). "Immunization against trachoma," in *First Int. Conf. on Vaccines against Viral and Rickettsial Diseases of Man*, WHO, 546–559.
[1057] Wooridge, R. L, Gheng, K. H., Ghang, I. H., Yang, C. Y., Hsu, T. C., and Grayston, J. T. (1967). Failure of trachoma treatment with ophthalmic antibiotics and systemic sulfonamides used alone or in combination with trachoma vaccine. *Am. J. Ophthalmol.* 63, 1577–1583.
[1058] Jones, G. E., Jones, K. A., Machell, J., Brebner, J., Anderson, I. E., and How, S. (1995). Efficacy trials with tissue-culture grown, inactivated vaccines against chlamydial abortion in sheep. *Vaccine* 13, 715–723.
[1059] Longbottom, D., and Livingstone, M. (2006). Vaccination against chlamydial infections in man and animals. *Vet. J.* 171, 263–275.
[1060] Rodolakis, A. (1986). "Use of a temperature-sensitive vaccine in experimental and natural infections," in *Chlamydial Diseases of Ruminants*, ed. I. D. Aitken (Luxemburg: Commission of the European Communitis), 71–97.
[1061] Babiuk, L. A., Pontarollo, R., Babiuk, S., Loehr, B., Van Drunen, S., and Van den Hurk, L. (2003). Induction of immune responses by DNA vaccines in large animals. *Vaccine* 21, 641–658.

[1062] McKercher, D. G., Wada, E. M., Robinson, E. A., and Howard, J. A. (1966). Epizootiologic and immunologic studies of epizootic bovine abortion. *Cornell Vet*. 56, 433–450.

[1063] McKercher, D. G., Robinson, E. A., Wada, E. M., Sato, J. K., and Franti, C. E. (1969). Vaccination of cattle against epizootic bovine abortion. *Cornell Vet*. 59, 211–226.

[1064] McKercher, D. G., Crenshaw, G. L., Theis, J. H., Wada, E. M., and Mauris, C. M. (1973). Experimentally induced Immunity to Chlamydial Abortion of Cattle. *J. Inf. Dis*. 128, 231–234.

[1065] De Graves, F. J., Stemke-Hale, K., Huang, J., Johnston, S. A., Sykes, K. F., Schlapp, T., et al. (2002). "Vaccine identified by in vivo genomic screening enhances fertility in cattle during environmental challenge with Chlamydia," in *Chlamydial Infections, ed.* Schachter, J., in *Proceedings of the 10th Int. Symposium on Numan Chlamydial Infections*, Antalia, 265–268.

[1066] Kaltenboeck, B., Hehnen, R. H., and Vaglenov, A. (2005). Bovine *Chlamydophila* spp. infection: Do we understand the impact of fertility? *Vet. Res. Comm*. 29 (Suppl 1), 1–15.

[1067] Wills, J. M., Gruffydd-Jones, T. J., Richmond, S. I., Gaskell, R. M., and Bourne, F. J. (1987). Vaccination of Feline *Chlamydia psittaci* infection. *Infect. Immun*. 55, 2653–2657.

[1068] McKercher, D. G. (1952). Feline pneumonitis. *Am. J. Vet. Res*. 13, 557–561.

[1069] Mitzel, J. R., and Starting, A. (1977). Vaccionation against feline pneumonitis. *Am. J. Vet. Res*. 38, 1361–1363.

[1070] Shewen, P. E., Povey, R. C., and Wikson, M. R. (1980). A comparison on the efficacy of a live and four inactivated vaccine praparations for the protection of cats against experimental challenge with *Chlamydia psittaci*. *Can. J. Comp. Med*. 44, 244–251.

[1071] Cello, R. M. (1971). Microbiological and immunological aspect of feline pneumonitis. *J. Am. Vet. Med. Assov*. 158, 932–938.

[1072] Anonymous. (2014). *Chlamydia Vaccine for C ats. Vetstrret, 33–36.* Available at: http://vetstreet.com/care/chlamydia vaccine for cats.

[1073] Pierson, L. A. (2016). *Vaccines for Cats. We Need to Stop Overvaccinating Cat (Overview)*. Available at: CatInfo.org.

[1074] Anonymous. (2014). *Feline Chlamydophila*. MSD Animal Health.

[1075] Tasker, S. (2016). *Chlamydia felis*. European Advisory Board of cat diseases (ABCD).

[1076] Howie, S. E., Horner, P. J., Horne, E. W., and Entrican, G. (2011). Immunity and vaccines again sexually transmitted *Chlamydia trachomatis* infection. *Curr. Opin. Infect. Dis.* 24, 56–61.

[1077] Bulir, D. C., Liang, S., Lee, A., Chong, C., Simms, E., Stone, C., et al. (2016). Immunization with chlamydial type III secretion antigens reduces vaginal shedding and prevent fallopian tube pathology following live *C. muridarum* challenge. *Vaccine* 34, 3979–3985.

[1078] Hafner, L. M., and Timms, P. (2015). *Development of a Vaccine for Chlamydia Trachomatis: Challenges and Current Progress.* Auckland: Dove Medical Press, 5, 45–58.

[1079] Andrew, D. W., Hafner, L. M., Beegley, K. W., and Timms, P. (2011). Partial protection against chlamydial reproductive tract infection by a recombinant major outer membrane protein CpG/cholera toxin intranasal vaccine in the guinea pigs *Chlamydia caviae* model. *J. Reprod. Immunol.* 91, 9–16.

[1080] Wali, S., Gupta, R., Veselenak, R. L., Li, Y., Yu, J.-J., Murthy, A. K., et al. (2014). Use of a Guinea pig specific transcriptome array for evaluation of protective immunity against genital chlamydial infection following intranasal vaccination in Guinea pigs. *PLoS One*, 9:e114261.

[1081] Neuendorf, E., Gajer, P., Bowlin, A. K., Marques, P. X., Ma, B., Yang, H., et al. (2015). Chlamydia caviae infection alters abudance but not composition of the guinea pig vaginal microbiota. *Pathog. Dis.* 73:4.

[1082] Cooper, C. (2015). Chlamydia: US researchers claim to have developed vaccine against the disease. Available at: http: //www.indepen dent.co.uk; charley cooper8.

[1083] Newman, L., Rowley, J., Hoom, S. V., Wjesooriya, N. S., Unemu, M., Low, N., et al. (2015). Global estimates in the prevalence and incidence of four curable sexually transmitted infections in 2012 based on systematic review and global reporting. *PLoS One* 10:e0143304.

[1084] Devereaux, T. N., Polkinghorne, A. P., Meijer, A., and Timms, P. (2003). Molecular evidence for novel chlamydial infections in the coala (*Phascolarctos cinereus*). *Syst. Appl. Microbiol.* 26, 245–253.

[1085] Vogelnest, L., and Woods, R. (eds). (2008). *Medicine of Australian Mammals.* Clayton: CSIRO Publishing, 770.

[1086] Cockram, F. A., and Jackson, A. R. B. (1981). Keratoconjunctivitis of the koala Phascolarchos cinereus, caused by *Chlamydia psittaci. J. Wildlife Dis.* 17, 497–504.

[1087] Brown, A. S., Girjes, A. A., Lavin, M. F., Timms, P., and Woolcock, J. B. (1987). Chlamydia disease in koalas. *Austr. Vet. J.* 64, 345–350.

[1088] Canfield, P. (1989). A survey of urinary tract disease in New South Wales koalas. *Austr. Vet. J.* 66, 103–106.
[1089] Canfield, P., and Spencer, A. J. (1993). Renal complications of cystitis in koalas. *Austr. Vet. J.* 70, 310–311.
[1090] Obendorf, D. L. (1981). Pathology of the female reproductive tract in the koala (*Phascolarctos cinereus*) (Gold fuss), from Victoria, Australia. *J. Wildlife Dis.* 17, 587–592.
[1091] Jacobson, E., Origgi, F., and Heard, D. J., Detrisac, C. (2002). An outbreak of chlamydiosis in emerald tree boas, *Corallus caninus. Assoc. Reptilian Amphibian Veterinar.* 47–48.
[1092] Jacobson, E., Gaskin, J. M., and Mansell, J. (1989). Chlamydial infection in Puff Adders (*Bitis arientas*). *J. Zoo Wildlife Med.* 20, *Infect. Dis. Issue*, 364–369.
[1093] Soldati, G., Lu, Z. H., Vaughan, L., Polkinghorne, A. P., Zimmerman, D., Huder, J. B., et al. (2004). Detection of mycobacteria and chlamydiae in granulomatous inflammation of reptiles: a retrospective study. *Vet. Pathol.* 41, 388–397.
[1094] Bodetti, T. J., Jacobson, E., Wan, C., et al. (2002). Molecular evidence to support the expansion of the hostrange of *Chlamydia pneumoniae* to include reptiles, as well humans, horses, koalas and amphibians. Syst. Appl. Microbiol. 25 (1), 146–152
[1095] Taylor-Brown, A., Ruegg, S., Polkinghorne, A., and Borel, N. (2015). Characterization of *Chlamydia pneumoniae* and other novel chlamydial infection in captive snakes. *Vet. Microbiol.* 178, 88–93.
[1096] Bradley, J., Bachmann, N. L., Jelocnic, M., Myers, G. S., and Polkinghorne, A. (2016). Australian human and parrot *Chlamydia psittaci* strains cluster within the highly virulent 6BC clade of this important zoonotic pathogen. *Sci Rep.* 6:30019.
[1097] Taylor-Brown, A., and Polkinghorne, A. (2017). New and emerging chlamydial infections of creatures great and small. *New Microbes New Infect.* 18, 28–33.
[1098] Vorimore, F., Hsia, R. C., Huot-Creasy, H., Bastian, S., Deruyter, L., Passet, A., et al. (2013). Isolation of a new *Chlamydia* species from the feral sacred ibis (*Threskiornis aethiopicus*): *Chlamydia ibidis. PLoS One* 8:e74823.
[1099] Taylor-Brown, A., Bachmann, N. J., Borel, N., and Polkinghorne, A. (2016). Culture - independent genomic characterization of of *Candidatus* Chlamydia sanzinia, a novel uncultured bacterium infecting snakes. *BMC Genomics* 17:710.

[1100] Pillonel, T., Bertelli, C., Salamin, N., and Greub, G. (2015). Taxogenomics of the order Chlamydiales. *Int. J. Syst. Evol. Microbiol.* 65, 1381–1393.

[1101] Burnard, D., and Polkinghorne, A. (2016). Chlamydial infection in wildlife-conservation threats and/or reservoirs of "spil-over" infections. *Vet. Microbiol.* 196, 78–84.

[1102] Taylor, S. K., Green, D. E, Wright, K. M., and Whitaker, B. R. (2001). "Bacterial diseases, pp. 159–179," in *Amphibian Medicine and Captive Husbandry*, eds K. M. Wright and B. R. Whitaker (Malabar, FL: Krieger Publishing Company).

[1103] Blumer, C., Zimmerman, D. R., Wellenmann, L., Vaughan, L., and Pospischil, A. (2007). Chlamydiae in free-ranging and captive frogs in Switzerland. *Vet Pathol.* 44, 144–150.

[1104] Whitaker, B. R. (2016). *Infectious Diseases of Amphibians*. MSD Vet. Manual.

[1105] Densmore, C. L., and D. E. Green. (2016). Diseases of Amphibians. *ILAR eJournal* 48:20. Available at: www.ilarjournal.com

[1106] Mitchell, C. M., Hutton, S., Myers, G. S. A., Bruncham, R., and Timms, P. (2010). *Chlamydia pneumoniae* is genetically diverse in animals and appears to have crossed the host barier to humans on (at least) two occasdions. *PLoS Pathog*. 6:e1000903.

[1107] Kutlin, A., Roblin, P. M., Kumar, S., Kohlhoff, S., Bodetti, T., et al. (2007). Molecular characterization of *Chlamydophila pneumoniae* isolates from Western barred bandicoots. *J. Med. Microbiol.* 56, 407–417.

[1108] Krauss, H., Schmeer, N., and Wittenbrink, M. M. (1988). Significance of *Chlamydia psittaci* infection in animals in the F.R.G., p. 65/, in *Proceedings of the Eur. Soc. Chlamydia Res.*, Bologna.

[1109] Jelocnic, M., Bradley, Heller, J., Raidal, S., Anderson, S., Galea, F., et al. (2017). Multilocus sequence typing identifies an avian-like *Chlamydia psittaci* strain, involved in equine placentitis and associated with subsequent humans psittacosis. *Emerg. Microbes. Infect.* 6:e7.

[1110] Elwell, C., Mirrashich, K., and Engel, J. (2016). Chlamydial cell biology and pathogenesis. *Nat. Rev. Microbiol.* 14, 385–400.

[1111] Grant, J. B., Brooman, P. J., Chowdhury, S. D., Sequeira, P., Blacklock, N. J. (1985). The clinical presentation of *Chlamydia trachomatis* in a urological practice. Br. J. Urol. 57, 218–221.

Index

D

E

About the Author

Professor Svetoslav Petrov Martinov, DVM, PhD, D.Sci.
Director of the Central Research Veterinary Medical Institute – Sofia, Bulgaria (1990–2001). Longtime Head of the National Reference Laboratory of Chlamydia and Rickettsia at the institute. Chairman of the Committee on Innovative Development and Implementation of Scientific and Technical Achievements. He was Head of the Department of especially dangerous infections and zoonoses, Head of the Department of virology and viral diseases at the National Diagnostic and Research Veterinary Institute – Sofia and Professor of environmental microbiology at New Bulgarian University – Sofia. He works in the fields of microbiology, virology, infectious diseases in animals and zoonoses. Publications: more than 300 in these directions. Manages or participates in numerous national and international research projects. Over the years, the activities of Professor Martinov also includes his work as first Vice-President of the Union of Veterinarians in Bulgaria, founder and first chairman of the Veterinary Association of the countries from Balkan and Black Sea region, member of the Supreme Veterinary Council at the Ministry of Agriculture in Bulgaria and Vice-President, Bulgarian Society of Medical Geography. He is a member of several national and international scientific societies. He has received national and international awards. Biographical data about Dr. Martinov are included in several editions of the American Biographical Institute, USA and the International Biographical Centre, Cambridge, England.